Lecture Notes in Computer Science   4654

Commenced Publication in 1973
Founding and Former Series Editors:
Gerhard Goos, Juris Hartmanis, and Jan van Leeuwen

Il-Yeol Song   Johann Eder
Tho Manh Nguyen (Eds.)

Data Warehousing and Knowledge Discovery

9th International Conference, DaWaK 2007
Regensburg Germany, September 3-7, 2007
Proceedings

 Springer

Volume Editors

Il-Yeol Song
Drexel University
College of Information Science and Technology
Philadelphia, PA 19104, USA
E-mail: song@drexel.edu

Johann Eder
University of Vienna
Faculty of Computer Science
Dr.-Karl-Lueger-Ring 1, 1010 Wien, Austria
E-mail: johann.eder@univie.ac.at

Tho Manh Nguyen
Vienna University of Technology
Institute of Software Technology and Interactive Systems
Favoriten Strasse 9-11/188, 1040 Wien, Austria
E-mail: tho@ifs.tuwien.ac.at

Library of Congress Control Number: 2007933373

CR Subject Classification (1998): H.2, H.3, H.4, C.2, H.2.8, H.5, I.2, J.1

LNCS Sublibrary: SL 3 – Information Systems and Application, incl. Internet/Web and HCI

ISSN	0302-9743
ISBN-10	3-540-74552-1 Springer Berlin Heidelberg New York
ISBN-13	978-3-540-74552-5 Springer Berlin Heidelberg New York

Springer is a part of Springer Science+Business Media

springer.com

© Springer-Verlag Berlin Heidelberg 2007
Printed in Germany

Typesetting: Camera-ready by author, data conversion by Scientific Publishing Services, Chennai, India
Printed on acid-free paper   SPIN: 12114931   06/3180   5 4 3 2 1 0

Preface

Data Warehousing and Knowledge Discovery have been widely accepted as key technologies for enterprises and organizations to improve their abilities in data analysis, decision support, and the automatic extraction of knowledge from data. With the exponentially growing amount of information to be included in the decision-making process, the data to be processed become more and more complex in both structure and semantics. Consequently, the process of retrieval and knowledge discovery from this huge amount of heterogeneous complex data constitutes the reality check for research in the area.

During the past few years, the International Conference on Data Warehousing and Knowledge Discovery (DaWaK) has become one of the most important international scientific events bringing together researchers, developers and practitioners. The DaWaK conferences served as a prominent forum for discussing latest research issues and experiences in developing and deploying data warehousing and knowledge discovery systems, applications, and solutions. This year's conference, the Ninth International Conference on Data Warehousing and Knowledge Discovery (DaWaK 2007), built on this tradition of facilitating the cross-disciplinary exchange of ideas, experience and potential research directions. DaWaK 2007 sought to disseminate innovative principles, methods, algorithms and solutions to challenging problems faced in the development of data warehousing, knowledge discovery and data mining applications.

The DaWaK 2007 submissions covered broad research areas on both theoretical and practical aspects of data warehousing and knowledge discovery. In the areas of data warehousing, the topics covered by submitted papers included advanced techniques in OLAP and multidimensional modeling, real-time OLAP, innovation of ETL processes and data integration problems, materialized view optimization, data warehouses and data mining applications integration, multidimensional analysis of text documents, and data warehousing for real-world applications such as medical applications, spatial applications, and tourist data warehouses. In the areas of data mining and knowledge discovery, the topics covered by submitted papers included data mining taxonomy, Web information discovery, stream data analysis and mining, ontology-based data and text mining techniques, constraint-based mining, and traditional data mining topics such as mining frequent item sets, clustering, association, classification, ranking, and applications of data mining technologies to real-world problems. It was especially notable to see that some papers covered emerging real-world applications such as bioinformatics, geophysics, and terrorist networks, as well as integration of multiple technologies such as conceptual modeling of knowledge discovery process and results, integration of the Semantic Web into data warehousing and OLAP technologies, OLAP mining, and imprecise or fuzzy OLAP. All these papers show that data warehousing and knowledge discovery technologies are becoming mature, making impact on real-world applications.

From 202 submitted abstracts, we received 150 papers from 38 countries. The Program Committee finally selected 44 papers, making an acceptance rate of 29.33 % of submitted papers.

We would like to express our gratitude to all the Program Committee members and the external reviewers who reviewed the papers very profoundly and in a timely manner. Due to the high number of submissions and the high quality of the submitted papers, the reviewing, voting and discussion process was an extraordinary challenging task. We would also like to thank to all the authors who submitted their papers to DaWaK 2007, as their contributions formed the basis of this year's excellent technical program.

Many thanks go to Gabriela Wagner for providing a great deal of assistance in administering the DaWaK management issues as well as to Amin Andjomshoaa for the conference management software.

September 2007

Il Yeol Song
Johann Eder
Tho Manh Nguyen

Organization

Conference Program Chairs

Il-Yeol Song, *Drexel University, USA*
Johann Eder, *University of Vienna, Austria*
Tho Manh Nguyen, *Vienna University of Technology, Austria*

Program Committee Members

Alberto Abello, *Universitat Politecnica de Catalunya, Spain*
Paulo Azevedo, *Universidade do Minho, Portugal*
Jose Luis Balcazar, *University Polytechnic of Catalunya, Spain*
Elena Baralis, *Politecnico di Torino, Italy*
Ladjel Bellatreche, *ENSMA - Poitiers University, France*
Bettina Berendt, *Humboldt University Berlin, Germany*
Petr Berka, *University of Economics Prague, Czech Republic*
Jorge Bernardino, *Polytechnic Institute of Coimbra, Portugal*
Elisa Bertino, *Purdue University, USA*
Sourav S Bhowmick, *Nanyang Technological University, Singapore*
Francesco Bonchi, *ISTI - C.N.R., Italy*
Henrik Boström, *Stockholm University and Royal Institute of Technology, Sweden*
Jean-Francois Boulicaut, *INSA Lyon, France*
Mokrane Bouzeghoub, *University of Versailles, France*
Stephane Bressan (SoC), *National University of Singapore, Singapore*
Peter Brezeny, *University of Vienna, Austria*
Robert Bruckner, *Microsoft Coperation, USA*
Luca Cabibbo, *Università Roma Tre, Italy*
Tru Hoang Cao, *Ho Chi Minh City University of Technology, Vietnam*
F. Amilcar Cardoso, *University of Coimbra, Portugal*
Jesús Cerquides, *University of Barcelona, Spain*
Chan Chee-Yong, *National University of Singapore, Singapore*
Arbee L.P. Chen, *National Chengchi University, Taiwan*
Rada Chirkova, *NC State University, USA*
Sunil Choenni, *University of Twente and Dutch Ministry of Justice, Netherlands*
Ezeife Christie, *University of Windsor, Canada*
Frans Coenen, *The University of Liverpool, UK*
Graham Cormode, *Bell Labs, USA*
Bruno Crémilleux, *Université de Caen, France*
Alfredo Cuzzocrea, *University of Calabria, Italy*
Honghua Dai, *Deakin University, Australia*
Agnieszka Dardzinska, *Bialystok Technical University, Poland*
Karen C. Davis, *University of Cincinnati, USA*

Markus Schneider, *University of Florida, USA*
Michael Schrefl, *University Linz, Austria*
Giovanni Semeraro, *University of Bari, Italy*
Manuel Serrano, *University of Castilla - La Mancha, Spain*
Alkis Simitsis, *IBM Almaden Research Center, USA*
Dan Simovici, *University of Massachusetts at Boston, USA*
Andrzej Skowron, *Warsaw University Banacha 2, Poland*
Carlos Soares, *University of Porto, Portugal*
Il-Yeol Song, *Drexel University, USA*
Min Song, *New Jersey Institute of Technology, USA*
Jerzy Stefanowski, *Poznan University of Technology, Poland*
Olga Stepankova, *Czech Technical University, Czech Republic*
Reinhard Stolle, *BMW Car IT, Germany*
Jan Struyf, *Katholieke Universiteit Leuven, Belgium*
Ah-Hwee Tan, *Nanyang Technological University, Singapore*
David Taniar, *Monash University, Australia*
Evimaria Terzi, *University of Helsinki, Finland*
Dimitri Theodoratos, *New Jerseys Science and Technology University, USA*
A Min Tjoa, *Vienna University of Technology, Austria*
Riccardo Torlone, *Roma Tre University, Italy*
Juan Trujilo, *University of Alicante, Spain*
Jaideep Vaidya, *Rutgers University, USA*
Panos Vassiliadis, *University of Ioannina, Greece*
Roland Wagner, *Johannes Kepler University Linz, USA*
Wei Wang, *University of North Carolina, USA*
Marek Wojciechowski, *Poznan University of Technology, Poland*
Wolfram Wöß, *Johannes Kepler University Linz, Austria*
Illhoi Yoo, *University of Missouri, Columbia, USA*
Mohammed J. Zaki, *Rensselaer Polytechnic Institute, USA*
Carlo Zaniolo, *University of California, Los Angeles, USA*
Shichao Zhang, *Sydney University of Technology*
Xiuzhen Jenny Zhang, *RMIT University, Australia*
Djamel Abdelkader Zighed, *University of Lyon, France*

External Reviewers

Alípio M. Jorge	Marek Kubiak
Antonella Guzzo	Marko Robnik Šikonja
Arnaud Soulet	Massimo Ruffolo
Celine Vens	Matjaž Kukar
Chiara Renso	Mirco Nanni
Christian Thomsen	Nguyen Hua Phung
Daan Fierens	Oscar Romero
Dawid Weiss	Paolo Garza
Dimitrios Skoutas	Paulo Gomes
Domenico Ursino	Quan Thanh Tho

Ester Zumpano
Fajar Ardian
Francesco Bonchi
Francois Deliege
George Papastefanatos
Joris van Velsen
Juan Manuel Perez Martinez
Kamel Boukhalfa
Le Ngoc Minh
Luka Šajn
Marcos Domingues

Riccardo Dutto
Salvatore Iiritano
Salvatore Ruggieri
Show-Jane Yen
Silvia Chiusano
Tania Cerquitelli
Wugang Xu
Xiaoying Wu
Xiaoyun He
Zoran Bosnic

Table of Contents

Query Optimization

Data Warehousing and Data Mining

Clustering

Association Rules

Healthcare and Biomedical Applications

Classification

Partitioning

Privacy and Crytography

Miscellaneous Knowledge Discovery Techniques

A Hilbert Space Compression Architecture for Data Warehouse Environments

Todd Eavis and David Cueva

Concordia University, Montreal, Canada
eavis@cs.concordia.ca,
cueva@cs.concordia.ca

Abstract. Multi-dimensional data sets are very common in areas such as data warehousing and statistical databases. In these environments, core tables often grow to enormous sizes. In order to reduce storage requirements, and therefore to permit the retention of even larger data sets, compression methods are an attractive option. In this paper we discuss an efficient compression framework that is specifically designed for very large relational database implementations. The primary methods exploit a Hilbert space filling curve to dramatically reduce the storage footprint for the underlying tables. Tuples are individually compressed into page sized units so that only blocks relevant to the user's multi-dimensional query need be accessed. Compression is available not only for the relational tables themselves, but also for the associated r-tree indexes. Experimental results demonstrate compression rates of more than 90% for multi-dimensional data, and up to 98% for the indexes.

1 Introduction

Over the past ten to fifteen years, data warehousing (DW) has become increasingly important to organizations of all sizes. In particular, the representation of historical data across broad time frames allows decision makers to monitor evolutionary patterns and trends that would simply not be possible with operational databases alone. However, this accumulation of historical data comes at a price; namely, the enormous storage requirements of the tables that house process-specific measurements. These *fact* tables can in fact hold hundreds of millions of records or more, with each record often housing ten or more distinct *feature* attributes.

The growth of the core tables creates two related problems. First, given a finite amount of storage space, we are limited in both the number of fact tables that can be constructed, as well as the length of the historical period. Second, as fact tables grow in size, real time query costs may become unacceptable to end users. A partial solution to the latter constraint is the materialization of summary views that aggregate the fact table data at coarser levels of granularity. But even here, the space available for such views is limited by the size of the underlying fact tables. As a result, we can conclude that a reduction in fact table size not only improves space utilization but it allows for the construction of

I.Y. Song, J. Eder, and T.M. Nguyen (Eds.): DaWaK 2007, LNCS 4654, pp. 1–12, 2007.
© Springer-Verlag Berlin Heidelberg 2007

additional summary views that may dramatically improve real time performance on common user queries.

In this paper, we propose a DW-specific compression model that is based upon the notion of tuple differentials. Unlike previous work in which the compression of tables and indexes has been viewed as separate processes, we present a framework that provides an integrated model for the compression of both elements. Specifically, we employ a Hilbert space filling curve to order data points in a way that allows for the subsequent construction of pre-packed, Hilbert ordered r-tree indexes. The system is therefore able to provide superior performance with respect to multi-dimensional data warehouse queries, while at the same time dramatically reducing storage requirements. Furthermore, our methods have been extended to external memory and can support fact tables of virtually unlimited size. Experimental results demonstrate compression rates of 80%-95% on both skewed and uniform data and up to 98% compression on multi-dimensional r-tree indexes.

The paper is organized as follows. In Section 2 we discuss related work. Section 3 presents relevant supporting material while Section 4 describes the algorithms used for data and index compression, along with a number of performance considerations. Experimental results are presented in Section 5, with Section 6 offering concluding remarks.

2   Related Work

General compression algorithms have been well studied in the literature and include *Statistical* techniques, such as Huffman Coding [9] and Arithmetic Compression [17], as well as *dictionary* techniques like LZ77 [21]. Database-specific methods are presented by Ray and Harista, who combine column-specific frequency distributions with arithmetic coding [16]. Westmann at al. [20] propose a model that utilizes integer, float, and string methods, as well as simple dictionaries. DW-specific data characteristics are not exploited however.

Ng and Ravishankar propose a technique called Tuple Differential Coding (TDC), in which the records of the data set are pre-ordered so as to identify tuple *similarity* [14]. Tuple difference values are then calculated and compactly stored with Run-Length Encoding [4]. Unfortunately, integrated indexing is not provided, so that it is difficult to assess the query performance on TDC compressed data. A second differential technique, hereafter referred to as GRS, was subsequently presented by Goldstein et al [3]. GRS computes and stores column-wise rather than row-wise differentials. In contrast to TDC, the authors of GRS also discuss the use of indexes in conjunction with their compressed data sets. Both technique are particularly relevant in the DW setting as they specifically target multidimensional spaces.

With respect to multi-dimensional indexing, we note that this has historically been an active area of research [2], though few of the experimental methods have found their way into production systems. The r-tree itself was proposed by Guttman [6], while the concept of pre-packing the r-tree for improved storage and performance was first discussed by Roussopoulos and Leifker [18]. Packing strategies were reviewed by Leutenegger et al. [12] and Kamel and Faloutsos [11]. The former promoted a technique they called Sort Tile Recursion (STR), while the latter identified Hilbert ordering as the superior approach.

The concept of representing a multi-dimensional space as a single dimensional, non-differentiable curve began with Peano [15]. Jagadish provides an analysis of four such curves — snake scan, gray codes, z-order, and Hilbert — and identifies the Hilbert curve as providing the best overall clustering properties [10], a result duplicated in [1].

Finally, it is worth noting that *non-tabular* storage has also been investigated. For example, the Dwarf cube presented by Sismanis et. al [19] is a tree-based structure that represents the 2^d *group-bys* found in the d-dimensional data cube [5]. Compression is achieved via the reduction of path redundancies. However, the complexity of the Dwarf structure decreases the likelihood of mainstream DBMS integration. Moreover, the Dwarf tends to be most effective at compressing high dimensional views, which are least likely to be queried by end users. For both of these reasons, a more practical solution may be an efficiently compressed fact table coupled with a subset of small, heavily aggregated summary views.

3   Preliminary Material

As noted in Section 2, Ng and Ravishankar presented a technique called Tuple Differential Coding (TDC) that was specifically targeted at multi-dimensional data sets [14]. The algorithm works as follows. Given a relation $\mathcal{R}$, we may define a schema $< A_1, A_2, ..., A_d >$ as as sequence of attribute domains $A_i = \{1, 2, ..., |A_i|\}$ for $1 \leq i \leq d$. A d-dimensional tuple in this space exists within a *bounding box* of size $|A_1| \times |A_2| \times ... \times |A_d|$. In addition, the points in $\mathcal{R}$ may be associated with an *ordering rule*. This rule is defined by a mapping function that converts each tuple into a unique integer representing its ordinal position within the space. In the case of TDC, the mapping function utilizes a standard lexicographic ordering defined as $\varphi : R \rightarrow \mathcal{N}_R$ where $\mathcal{N}_R = \{0, 1, ...||\mathcal{R}|| - 1\}$ and $||\mathcal{R}|| = \prod_{i=1}^{d} A_i$: $\varphi(a_1, a_2, ...a_d) = \sum_{i=1}^{d} \left(a_i \prod_{j=i+1}^{d} |A_j| \right)$.

In effect, the TDC ordering algorithm converts a series of tuple attributes into a single integer that uniquely identifies the position of the associated tuple in the multi-dimensional space. Consequently, it is possible to convert the ordinal representation of each tuple into a differential value — that represents the distance between successive ordinals — with the full sequence relative to a single reference value stored in each disk block. Simple RLE techniques can then be used to minimize the number of bits actually required to store the difference values themselves.

While the lexicographic orderings utilized in TDC (and GRS) allow us to uniquely order the values of the point space prior to differential calculation, space filling curves represent a more sophisticated approach to this same general problem. Such curves were first described by Peano [15] and can be defined as non-differentiable curves of length s^d representing a *bijective* mapping on the curve C, with s equivalent to the *side width* of the enclosing space, and d referring to the number of spatial dimensions. Space filling curves have been investigated in the database context as they allow for the conversion of a multi-dimensional space to the linear representation of disk blocks. The Hilbert curve, in particular, has been extensively studied since it was first proposed by David Hilbert [8]. Simply put, the Hilbert curve traverses all s^d points of the d-dimensional grid, making unit steps and turning only at right angles. We say that a d-dimensional space has order k if it has a side length $s = 2^k$. We use the notation $\mathcal{H}_k^d$ to denote the k-th order approximation of a d-dimensional curve, for $k \geq 1$ and $d \geq 2$. Figure 1(a), for example, represents the $\mathcal{H}_3^2$ space. The *self-similar* property of the Hilbert curve should be obvious.

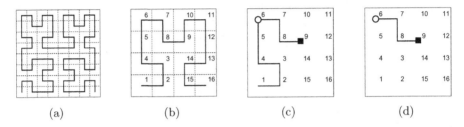

| | | | |
| (a) | (b) | (c) | (d) |

Fig. 1. (a)The $\mathcal{H}_3^2$ curve (b) The ordinal mapping (c) Basic differentials (d) Compressed form

4   Hilbert Space Differential Compression

In this section,we describe our new compression model. We note at the outset that our primary objectives are to provide (i) an intuitive tuple-oriented model that can be easily integrated into current RDBMS servers (ii) competitive compression facilities coupled with the most efficient indexing methods available for static environments.

4.1   Compressing the Point Data

As has been noted in previous research [10,13,1], the Hilbert curve has consistently been shown to provide superior point clustering properties, as well as very impressive r-tree packing characteristics. For this reason, we chose to investigate the use of the Hilbert curve within a multi-dimensional differential model. The differential mechanism was selected as it allows us to maintain tuple form, along with page level compression granularity.

To begin, we assume a d-dimensional space and a fact table $\mathcal{R}$ consisting of n tuples of $d + 1$ fields. The first d fields represent the feature attributes of $\mathcal{R}$, having cardinalities $\{|A_1|, |A_2|, ..., |A_d|\}$, while the final field provides the *measure* value. Note that our focus in this paper is upon compression of the d feature attributes, not the floating point measure (the same is true of previous research in the area).

Algorithm 1. Data compression

Input: An aggregated and arbitrarily sorted relation R, with n tuples consisting of k feature attributes and one measure. A temporary buffer P. A required block disk block size B.

Output: A fully compressed cuboid $R\prime$ with page level granularity.

1: **for all** n tuples of R **do**
2:    convert each k-tuple to its Hilbert ordinal value
3: re-sort the n indexes of R
4: **for** each tuple i, with $i \leq n$ **do**
5:    read next Hilbert ordinal h_i
6:    **while** estimated page size $P_{est} < B$ **do**
7:      **if** P is currently empty **then**
8:        record h_i as the uncompressed reference value
9:        set the page bit count $\rho = 0$
10:     **else**
11:        calculate differential δ as $h_{i+1} - h_i$
12:        compute bit count β required to represent δ
13:        **if** $\beta > \rho$ **then**
14:          $\rho = \beta$
15:    write page P to disk, appending the measures as required

---

Recall that the TDC algorithm converts a multi-dimensional tuple into a unique integer via a lexicographic ordering rule. This *invertible* mapping function is at the heart of differential encoding. The Hilbert curve, however, is completely incompatible with lexicographic ordering. That being said, the existence of a Hilbert space bijective mapping can be exploited in order to form a new invertible ordering rule. Specifically, the s^d-length Hilbert curve represents a unique, strictly increasing order on the s^d point positions, such that $C_{HIL} : \{1, ..., s^d\} \rightarrow \{1, ..., s\}^d$. We refer to each such position as a *Hilbert ordinal*. The result is a two-way mapping between d-dimensional tuples and their associated ordinals in Hilbert space defined as $\varphi(a_1, a_2, ...a_d) = C_{HIL}(a_1, a_2, ...a_d)$. Figure 1(b) provides a simple illustration of the 16 unique point positions in $\mathcal{H}_2^2$. Here, for example, we would have $\varphi(2, 2) = C_{HIL}(2, 2) = 3$.

Algorithm 1 presents the method for providing point compression based upon this Hilbert ordering rule. We assume the existence of a table R with arbitrary tuple order. We begin by first converting the multi-dimensional records into their Hilbert ordinal representaion and then sorting the result. Our immediate objective is to now fill page size units with compressed tuples, where each page is equivalent to some multiple of the physical block size. As each new page is created, we initialize the first record as the uncompressed *ordinal reference value* that represents the starting differential sequence. Next, we compute ordinal differentials $h_{i+1} - h_i$ until the page has been filled to capacity. We determine saturation by dynamically adjusting the *minimal bit count* δ required to represent the largest differential ρ in the running sequence. Once the capacity has been reached, the bit compacted indexes and their measures are written to a physical disk block.

To illustrate the concept, Figure 1(c) displays two 2-d points along the same simple $\mathcal{H}_2^2$ Hilbert curve. The initial point represents the reference ordinal stored in the block header, while the second point is the ordinal corresponding to the next tuple. Figure 1(c) provides the standard differential representation of the second point which, in this case, would be found nine steps form the origin. In Figure 1(d), we see the Bit Compacted Differential form for the second point. Storage required is now just $\rho = 9 - 6 = 3_{10}$, or 11_2 in binary form. This is 62 bits less that the default encoding for a two dimensional value (assuming 32-bit integers).

In addition to the tuple count and the uncompressed reference, the block's meta data includes the ρ value that indicates the minimum number of bits required to represent the highest tuple difference seen in the block. Each block has a unique ρ value, and therefore includes a varying number of compressed tuples. Note that differences are stored across byte or word boundaries, so as to maximize the compression rate. To de-compress, the query engine need only retrieve the reference tuple and then consecutively add each of the differences to obtain the original Hilbert ordinals. Subsequently, the ordinals are transformed back to multi-dimensional tuple form as required.

Finally, we note that in multi-dimensional environments, difference values can exceed the capacity of standard 32-bit or even 64-bit registers. Specifically, for a d-dimensional space, the maximum differential value is equivalent to $\prod_{i=1}^{d} |A_i|$ for dimension cardinalities $|A_1|, |A_2|, ..., |A_d|$. In a 10-dimensional space, for example, differentials can reach 100 bits, assuming uniform cardinalities of just 1000. Consequently, the Hilbert differential model has been extended to support integers of arbitrary length. To this end, the GNU Multiple Precision Arithmetic Library (GMP) has been integrated into our Hilbert transformation functions and seamlessly provides efficient index-to-tuple and tuple-to-index transformations for arbitrarily large Hilbert spaces.

4.2   Compression of the Indexes

One of the primary advantages of the use of the Hilbert curve is that it allows for the clean integration of data and index compression. Specifically, the same Hilbert order that provides the differential based mapping for point compression can also be used to support state-of-the-art pre-packing of multi-dimensional r-trees [11]. Moreover, the packed indexes can then be further processed in order to produce extremely high compression ratios. Recall that the GRS method also provides indexing but, in this case, relies on a very simple linearization of points that ultimately depends upon an unrealistic uniform distribution assumption. As we will see in Section 5, indexing efficiency suffers as a result.

Algorithm 2 illustrates the method for generating and compressing the index. As with data compression, the algorithm fills page size units, equivalent to a multiple of the physical block size. Each of the pages corresponds to a node n of the r-tree. Geometrically, a node constitutes a *hyper-rectangle* covering a delimited section of the point space. Therefore, compressed data blocks are

---

Algorithm 2. Index compression

Input: A cuboid $C\prime$ containing b blocks andcompressed with the Hilbert Differential
    method. Temporary buffers P and T. A required disk block size B.

Output: A fully compressed r-tree index I with one node per page.

1: **for all** b blocks of $R\prime$ **do**

2:    calculate V_{min} and V_{max} vertexes of b,
      $V = \{A_1, A_2, ..., A_k\}$

3:    write node $N_0^b = \{V_{min}, V_{max}\}$ to T uncompressed

4: **repeat**

5:    retrieve number of nodes nN from level L

6:    **for each** i, with $i \leq nN$ **do**

7:      read one node $N_L = \{V_{min}, V_{max}\}$ from level L

8:      **if** page P is empty **then**

9:        create node N_{L+1} in level $L + 1$

10:       initialize $N_{L+1} = N_L$

11:      **else**

12:        include vertexes of N_L in N_{L+1}

13:        recalculate pivot of N_{L+1}: $G = min_{L+1}$

14:        recalculate bit count β needed to represent V_{max} in N_{L+1}

15:        **if** estimated page size $P_{est} > B$ **then**

16:          rollback last inclusion

17:          write N_{L+1} to page P uncompressed

18:          **for all** N_L nodes of level L **do**

19:            **for all** k attribute of each vertex V_L of N_L **do**

20:              calculate differential ψ as $V_L(k) - G(k)$

21:              write ψ to P using Bit Compaction

22: **until** number of nodes $nN == 1$

---

naturally defined as the leaf nodes of the r-tree. The rest of the index structure
is built bottom-up from level-0 (i.e., the leaf level) to the root.

We start with a relation $R\prime$, which has already been compressed using our
Hilbert Differential method. For each of the leaf nodes, we record its enclosing
level-0 hyper-rectangle as a pair of d-dimensional vertexes. These two points
represent the attribute values corresponding to the block's local minima, which
we refer to as the *pivot*, and the maxima. Next, we create the nodes in level-1
by including each hyper rectangle from level-0. We initialize the pivot of the
node in level-1 with the one located in the current leaf. For each dimension, we
calculate the bits β necessary to represent the difference between the maxima
and minima. For each new hyper-rectangle, the pivot and bit strings are updated,
along with the point count, until saturation is achieved. Once the page is full, the
vertexes are bit compacted and written to a physical block that simply stores
the *per-dimension* differences with respect to the pivot value. This process is
then repeated until we reach a single root node at the tree's apex.

Index block meta data includes the following values: the number of child
nodes, the bit offset of the first child, and an array indicating the number of

bits ψ used to represent each dimension in the current block. As with the data compression, blocks may house a varying number of compressed vertexes. Note that the *pivot* value is not included in the block, but rather compressed in the node immediately above. Because searching the index always requires traversing its nodes top down, we can guarantee that the pivots required to uncompress nodes in level L have already been decompressed in level $L + 1$.

Note that the compressed indexing model supports query resolution for both range and point queries (if required). To retrieve data the query engine performs a standard top-down index traversal. Each time a node is visited, exactly one page P is read and decompressed. To decompress P we read each hyper-rectangle N formed by the two vertexes V_{min}, V_{max}, adding to every dimension attribute the value of its corresponding attribute in the current pivot vertex G_L. If the query lies inside the boundary formed by V_{min} and V_{max}, a new *task* is created. The task is included in a FIFO queue as $\{G_{L+1}, o\}$, where V_{min} becomes the pivot G_{L+1} when the task is to be resolved on the next tree level, and o is the offset of N in the disk file. When the query engine reaches a leaf node, a data page is loaded. At that point, tuple decompression is performed as per the description in Section 4.1.

4.3   Optimizations for Massive Data Sets

Given the size of contemporary data warehouses, practical compression methods must concern themselves not only with expected compression ratios, but with performance and scalability issues as well. As such, the framework supports a comprehensive buffering subsystem that allows for *streaming* compression of arbitrarily large data sets. This is an *end-to-end* model that consists of the following *pipelined* stages: generation of Hilbert indexes from multi-dimensional data, secondary memory sorting, Hilbert index alignment, Hilbert differential data compression, and r-tree index compression.

To minimize sorting costs during the initial compression phase, we note the following. Given an $O(n \lg n)$ comparison based sort, we can assume an average of $\lg n$ *Hilbert* comparisons for each of the n tuples. Since tuple conversions are identical in each case, signficant performance improvements can be realized by minimizing the number of Hilbert transformations. Consequently, our framework uses a pre-processing step in which the n tuples of R are converted into ordinal form *prior* to sorting. This single optimization typically reduces costs by a factor of 10 or more, depending on dimension count. To support larger data sets, we extend the compression pipeline with an external memory sorting module that employs a standard P-way merge on intermediate partial runs. P is maximized as $\lceil m/b \rceil$, where b is the available in-memory buffer size and m is the size of the disk resident file. With the combination of the streaming pipeline, ordinal pre-processing, and the P-way sort, there is currently no upper limit on the size of the files to be compressed other than the limitations of the file system itself.

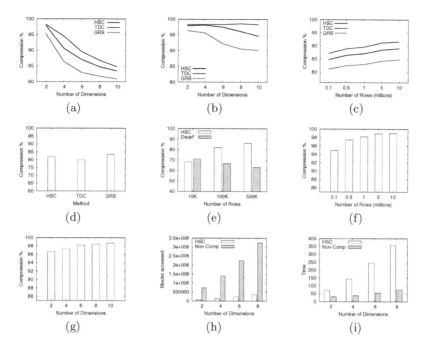

Fig. 2. (a) Typical skew (b) large skew (c) data set size (d) real data (e) Dwarf Cube (f) indexes, by rows (g) indexes, by dims (h) query blocks (i) query decompression

5   Experimental Results

In this section, we provide experimental results that assess the compression ratios on both data and indexes, as well as comparative results versus the original TDC algorithm, GRS, and the Dwarf cube. All tests are conducted a 3.2 GHz workstation, with 1 GB of main memory and a 160 GB disk drive operating at 7200 RPM. We utilize both real and synthetically generated data. For the query evaluations, we use the standard approach of timing queries in batch mode. In our case, an automated query generator constructs batches of 1000 range queries, in which high/low ranges are randomly generated for each of k attributes, randomly selected from the d-dimensional space, $k \subseteq d$. For all tests, compression is expressed as a percentage indicating the effective size reduction with respect to the original table, where the fields of each record are initially represented as 32 bit integers. High numbers are therefore better.

We begin by looking at the effect of compression on data sets of one million records, with cardinalities randomly chosen in the range 4–10000, as dimension count increases from two to ten. The methods evaluated include TDC, GRS, and our own Hilbert Space Compression methods, referred to as HSC. Figure 2(a) displays the results for a fairly typical zipfian skew of one, while Figure 2(b) illustrates the effect of more extreme clustering produced by a zipfian value of three. Several observations can be made. First all differential algorithms demonstrate compression ratios in excess of 80%, with TDC slightly above the Hilbert method. This difference is primarily attributable to the fact that the Hilbert axes must be rounded up to the nearest "power of two", thereby sacrificing a half bit on average. Second, the increased clustering produced by data skew leads to extremely high compression rates. TDC and HSC, in particular, hover between 95% and 98%. GRS trails the other two methods, largely due to its unrealistic uniformity assumption. Third, there is a general decline in the compression ratio as dimensions increase due to the point space becoming significantly larger.

In practice, record counts in data warehouse environments are extremely large. Figure 2(c), therefore illustrates the effect of increasing the data set size from 100,000 to 10 million. Here we see a marked increase in compression ratios as record count grows. Again, this is expected for differential methods as increased record count leads to elevated density that, in turn, reduces the distance between points. For TDC and HSC in particular, compression rates reach approximately 90% percent at just 10 million rows. We also performed a single HSC compression test on a data set of four dimensions and 100 million rows (skew = 1, mixed cardinalities). The rows occupied approximately 1.6 GB in uncompressed form, but were reduced to just 58 MB with the Hilbert methods. The final compression rate was 96.36%. Again, the very high compression rates are heavily influenced by the increased density of large sets. We note that the associated compressed index occupied just 553 KB.

Figure 2(d) provides another direct comparison, this time on a weather data set commonly used in DW evaluations [7]. The set consists of 1.1 million records, and nine dimensions with cardinality varying between 2 and 7037. In this case, the positions of TDC and GRS are reversed, a result reflective of the fact that the modest skew of the real data set closely models the uniformity assumption of GRS. The compression for HSC is almost completely consistent with Figure 2(a), as one would expect with the Hilbert curve.

In Figure 2(e), we see a direct compression comparison with HSC and the Dwarf cube showing the two techniques on data sets of varying record counts. We note that we use a smaller data set here due to a limited Dwarf implementation. Given the current parameters, we see a simple trend indicating improved performance for the Hilbert methods at higher record counts. This is likely due to the increased ability to exploit space density. We note that additional testing did show advantages for the Dwarf in very high dimensions, though again we stress that users seldom issue queries in this region of the space.

Figure 2(f) illustrates the compression rates for the index components for record counts varying from 100000 to 10 million (zipf = 1, mixed cardinalities). Here, we can see ratios from 94.90% up to 99.20% for the largest data sets. We note that comparable index compression numbers for GRS were never provided. Figure 2(g) examines the effect of varying the dimension count. As dimensions increase from two to 12, the compression ratio varies from 96.64% to 98.50%. In contrast to data compression, increased dimension count is not a problem for the indexing methods. Specifically, the compact pivot-based representation for the bounding boxes does not deteriorate as we increase the the number of axes of the enclosing hyper-rectangle.

Finally, we examine the impact of compression on query performance (using the data sets of test [a]). Figure 2(h) indicates that the average number of disk blocks retrieved during a query batch when using compressed Hilbert r-trees is about 10% of the number retrieved when using uncompressed data. Recall the GRS is the only other differential method that describes a possible indexing approach. In [3], the authors indicate that IO costs for their compressed methods are approximately 40%-50% of the costs of the uncompressed benchmark. The elevated I/O costs, relative to HSC, likely result from a linearization strategy that less effectively exploits the increased fanout of the compressed blocks. This is a crucial observation with respect to practical implementations.

Finally, in Figure 2(i), we look at the time taken during these same query runs to uncompress records and place them into the result set, versus the time to do so with completely uncompressed records. Here, we see the ratio increase from about a factor of two in low dimensions to about a factor of 4 in higher dimensions. We note, however, that we currently working with a simple Hilbert library and that further performance optimizations would almost certainly reduce the decompression costs. We expect that even modest improvements would eventually allow the decompression methods to keep pace with the raw IO.

6 Conclusions

In this paper, we have presented a broad compression framework for use in practical data warehouse settings. Building upon the extensively studied Hilbert space filling curve, the new methods identify multi-dimensional point clustering in order to minimize differential calculations. Impressive compression ratios are provided for a variety of parameter sets. Most importantly, the Hilbert ordering logic is incorporated into the index compression process so as to create a single coordinated framework. The indexes in particular are extremely small and are associated with an order of magnitude reduction in the number of blocks drawn from the disk. Various external memory extensions allow the framework to function in environments of almost unlimited size. Taken as a whole, we believe the current model represents the most effective and most comprehensive compression framework for today's vast multi-dimensional environments.

References

1. Faloutsos, C., Roseman, S.: Fractals for secondary key retrieval. In: ACM Symposium on Principles of Database Systems, pp. 247–252. ACM Press, New York (1989)
2. Gaede, V., Gunther, O.: Multidimensional access methods. ACM Computing Surveys 30(2), 170–231 (1998)
3. Goldstein, J., Ramakrishnan, R., Shaft, U.: Compressing relations and indexes. In: International Conference on Data Engineering (ICDE), pp. 370–379 (1998)
4. Golomb, S.W.: Run-length encodings. IEEE Transactions on Information Theory 12(3), 399–401 (1966)

5. Gray, J., Bosworth, A., Layman, A., Pirahesh, H.: Data cube: A relational aggregation operator generalizing group-by, cross-tab, and sub-totals. In: International Conference On Data Engineering (ICDE), pp. 152–159 (1996)
6. Guttman, A.: R-trees: A dynamic index structure for spatial searching, pp. 47–57 (1984)
7. Hahn, C., Warren, S., Loudon, J.: Edited synoptic cloud reports from ships and land stations over the globe, Available at http://cdiac.esd.ornl.gov/cdiac/ndps/ndpo26b.html
8. Hilbert, D.: Ueber die stetige abbildung einer line auf ein flchenstck. Mathematische Annalen 38(3), 459–460 (1891)
9. Huffman, D.: A method for the construction of minimum redundancy codes. Proceedings of the Institute of Radio Engineers (IRE) 40(9), 1098–1101 (1952)
10. Jagadish, H.: Linear clustering of objects with multiple attributes. In: ACM SIGMOD, 332–342 (1990)
11. Kamel, I., Faloutsos, C.: On packing r-trees. In: International Conference on Information and Knowledge Management (CIKM), pp. 490–499 (1993)
12. Leutenegger, S., Lopez, M., Eddington, J.: STR: A simple and efficient algorithm for r-tree packing. In: International Conference on Data Engineering (ICDE), pp. 497–506 (1997)
13. Moon, B., Jagadish, H., Faloutsos, C., Saltz, J.: Analysis of the clustering properties of the hilbert space-filling curve. Knowledge and Data Engineering 13(1), 124–141 (2001)
14. Ng, W., Ravishankar, C.V.: Block-oriented compression techniques for large statistical databases. IEEE Transactions on Knowledge and Data Engineering 9(2), 314–328 (1997)
15. Peano, G.: Sur une courbe, qui remplit toute une aire plane. Mathematische Annalen 36(1), 157–160 (1890)
16. Ray, G., Haritsa, J.R., Seshadri, S.: Database compression: A performance enhancement tool. In: International Conference on Management of Data (COMAD) (1995)
17. Rissanen, J.: Generalized kraft inequality and arithmetic coding. IBM Journal of Research and Development 20(3), 198–203 (1976)
18. Roussopoulos, N., Leifker, D.: Direct spatial search on pictorial databases using packed r-trees, pp. 17–31 (1985)
19. Sismanis, Y., Deligiannakis, A., Roussopoulos, N., Kotidis, Y.: Dwarf: shrinking the petacube. In: ACM SIGMOD, pp. 464–475 (2002)
20. Westmann, T., Kossmann, D., Helmer, S., Moerkotte, G.: The implementation and performance of compressed databases. SIGMOD Record 29(3), 55–67 (2000)
21. Ziv, J., Lempel, A.: A universal algorithm for sequential data compression. IEEE Transactions on Information Theory 23(3), 337–343 (1977)

Evolution of Data Warehouses' Optimization: A Workload Perspective

Cécile Favre, Fadila Bentayeb, and Omar Boussaid

University of Lyon (ERIC-Lyon 2) - Campus Porte des Alpes
5 av. Pierre Mendès-France, 69676 Bron Cedex, France
{cfavre|bentayeb}@eric.univ-lyon2.fr, omar.boussaid@univ-lyon2.fr
http://eric.univ-lyon2.fr

Abstract. Data warehouse (DW) evolution usually means evolution of its model. However, a decision support system is composed of the DW and of several other components, such as optimization structures like indices or materialized views. Thus, dealing with the DW evolution also implies dealing with the maintenance of these structures. However, propagating evolution to these structures thereby maintaining the coherence with the evolutions on the DW is not always enough. In some cases propagation is not sufficient and redeployment of optimization strategies may be required. Selection of optimization strategies is mainly based on workload, corresponding to user queries. In this paper, we propose to make the workload evolve in response to DW schema evolution. The objective is to avoid waiting for a new workload from the updated DW model. We propose to maintain existing queries coherent and create new queries to deal with probable future analysis needs.

Keywords: Data warehouse, Model, Evolution, Optimization strategy, Workload, Query rewriting.

1 Introduction

Data warehouses (DW) support decision making and analysis tasks by providing consolidated views. These consolidated views are built from the multi-dimensional model, based on heterogeneous data sources, of the DW. The multi-dimensional model being the most important component of a decisional architecture, the key point for the success of this type of architecture is the design of this model according to two factors: available data sources and analysis needs. As business environment evolves, several changes in the content and structure of the underlying data sources may occur. In addition to these changes, analysis needs may also evolve, requiring an adaptation of the existing multi-dimensional model. Thus the DW model has to be updated.

DWs containing large volume of data, answering queries efficiently required efficient access methods and query processing techniques. One issue is to use redundant structures such as views and indices. Indeed, among the techniques adopted in relational implementations of DWs to improve query performance,

I.Y. Song, J. Eder, and T.M. Nguyen (Eds.): DaWaK 2007, LNCS 4654, pp. 13–22, 2007.

view materialization and indexing are presumably the most effective ones [1]. One of the most important issues in DW physical design is to select an appropriate set of materialized views and indices, which minimizes total query response time, given a limited storage space.

A judicious choice in this selection must be cost-driven and influenced by the workload experienced by the system. Indeed, it is crucial to adapt the performance of the system according to its use [2]. In this perspective, the workload should correspond to a set of users' queries. The most recent approaches syntactically analyze the workload to enumerate relevant candidate (indices or views) [3]. For instance, in [4], the authors propose a framework for materialized views selection that exploits a data mining technique (clustering), in order to determine clusters of similar queries. They also propose a view merging algorithm that builds a set of candidate views, as well as a greedy process for selecting a set of views to materialize. This selection is based on cost models that evaluate the cost of accessing data using views and the cost of storing these views.

The workload is supposed to represent the users' demand on the DW. In the literature, most of the proposed approaches rely on the existence of a reference workload that represents the target for the optimization [5]. However, in [6], the authors argue that real workloads are much larger than those that can be handled by these techniques and thus view materialization and indexing success still depends on the experience of the designer. Thus, they propose an approach to build a clustered workload that is representative of the original one and that can be handled by views and indices selection algorithms.

In views selection area, various works concern their dynamical selection [7,8,9]. The dynamic aspect is an answer to workload evolution. However, workload evolution can also be considered from different points of views. In [9], the authors consider that the view selection must be performed at regular maintenance intervals and is based on an observed or expected change in query probabilities. In [7,8], the authors suppose that some queries are added to the initial workload. Thus these works assume that previous queries are always correct. However, we affirm that it is not always the case.

In this paper, we want to define a different type of workload evolution. Indeed, we have to consider that the DW model update induces impacts on the queries composing the workload. Our key idea consists in providing a solution to make the workload queries evolve. The objective is to maintain the queries coherent with the DW model evolution and to create new queries expressing forthcoming analysis needs if required. Since the processing time for answering queries is crucial, it is interesting to adopt a pro-active behaviour. That is we propose to make the workload evolve in response to the model evolution. In this paper, we present our preliminary work concerning this issue.

The remainder of this paper is organized as follows. First, we detail the problem of query evolution in Section 2. Then, we present the DW schema evolutions and their impacts on the workload queries in Section 3. In Section 4, we focus on our approach to adapt queries of an existing workload, as a support for optimization strategy evolution. Next, we present an example of our approach applied to a simplified case study in Section 5. We finally conclude and provide future research directions in Section 6.

2   The Query Evolution Problem

The problem of query evolution in DWs has been addressed in an indirect way. We evoked in the introduction the problem of materialized view maintenance. However, we have to consider the duality of views, which are both sets of tuples according to their definition of extension and queries according to their definition of intention. Indeed, a view corresponds to the result of a query. Thus the question of view evolution can be treated as the query evolution problem. This perspective is followed in [10], where the impact of data sources schema changes is examined to each of the clauses of the view query (structural view maintenance).

However, in data warehousing domain, the issue of query evolution has not yet been addressed as a problem in itself. This point is important because queries are not only used to define views; for instance, in a decisional architecture, queries are also used in reporting (predefined queries) or to test the performance of the system when they form a workload.

The problem of query maintenance has been evoked in the database field. Some authors think that model management is important for the entire environment of a database. Indeed, queries maintenance should also involve the surrounding applications and not only be restricted to the internals of a database [11]. Traditional database modeling techniques do not consider that a database supports a large variety of applications and provides tools such as reports, forms. A small change like the deletion of an attribute in this database might impact the full range of applications around the system: queries and data entry forms can be invalidated; application programs accessing this attribute might crash.

Thus, in [12], the authors first introduce and sketch a graph-based model that captures relations, views, constraints and queries. Then, in [13], the authors extend their previous work by formulating a set of rules that allow the identification of the impact of changes to database relations, attributes and constraints. They propose a semi-automated way to respond to these changes. The impact of the changes involves the software built around the database, mainly queries, stored procedures, triggers. This impact corresponds to an annotation on the graph, requiring a tedious work for the administrator.

In this paper, we focus on the query evolution aspect in data warehousing context. More particularly, we focus on the performance optimization objective through the evolution of the workload. As compared to previous presented work, we do not consider annotations for each considered model. We propose a general approach to make any workload evolve. These evolutions are made possible by using the specificity of multidimensional model that encapsulates semantic concepts such as dimension, dimension hierarchy. Indeed, these concepts induce roles that are recognizable in any model.

To achieve our workload evolution objective, we first define a typology of changes applied to the model (more precisely we focus on schema evolution) and their impacts on the workload. In this context, we tend to follow two goals: (1) maintaining the coherence of existing queries according to schema updates; (2) defining new queries when the schema evolution induces new analysis possibilities.

3   Data Warehouse Schema Evolution and Consequences

In this paper, we focus on the DW schema changes. We do not deal with data updating.

In a relational context, schema changes can occur on two levels: table or attribute. In the DW context, we cannot deal only with this concept. We have to introduce the DW semantic. Indeed, we have to consider that updates on a dimension table and on a fact table do not have the same consequences.

In Figure 1, we represent these changes and their impact on queries. We define what "concept" is modified, what kind of modification it is and the consequence on the workload. We consider three consequences: (1) updating queries; (2) deleting queries; (3) creating queries.

Role in the model	Type	Operation	Query Updating	Query Deleting	Query Creating
Dimension and level	Table	Creating			YES
Dimension and level	Table	Deleting	YES	YES	
Dimension and level	Table	Updating (Renaming)	YES		
Fact	Table	Updating (Renaming)	YES		
Measure	Attribute	Creating			
Measure	Attribute	Deletion	YES	YES	
Measure	Attribute	Updating (Renaming)	YES		
Dimension descriptor	Attribute	Creating			YES
Dimension descriptor	Attribute	Deleting	YES	YES	
Dimension descriptor	Attribute	Updating (Renaming)	YES		

Fig. 1. Schema evolution possibilities and consequences on the workload queries

First, we note that we do not deal with the creation or the deletion of a fact table. In this case, there is an impact on dimension tables also. We think that these changes make the DW model evolve in an important way, so that the previous queries expressing analysis needs are no longer coherent or we are not able to define future analysis needs due to a large number of possibilities.

Concerning the query updating, it consists in propagating the change on the syntax of the implied queries. More precisely, the syntax of the queries' clauses have to be rewritten when they use a concept that has changed. For the creation of a level, several possible queries can be created. However, we suppose in this paper that we only create one query that integrates the created level.

For the deleting operation, we can observe two consequences: propagation and deletion. This means that if rewriting (propagation) is not possible, the query is deleted.

Furthermore, when a measure is created, we propose no consequence on the workload. Indeed, these changes do not induce coherence problems on existing queries. Thus it implies neither query updating nor query deletion. Moreover, since the workload allows for performance study, we do not create additional queries. Indeed, choosing one or another measure does not have any impact on the query performance.

4   Workload Evolution: Our Approach

In this section, we detail various aspects of our proposal. First, we provide the general architecture of our approach. Then, we detail the algorithm for the workload evolution. Finally, we provide some explanations about the implementation.

4.1   General Architecture

Our proposal of workload evolution is placed in a global architecture (Figure 2). According to the use of a DW, an initial workload is defined, containing the users' queries. A selection of the optimization strategy can be carried out, on the basis of this workload. When the administrator makes the DW model evolve, the queries can become incoherent, i.e. they can no longer be executed on the updated DW.

Taking into account both the initial workload and the changes applied on the DW model, a workload evolution process is applied. This allows for the creation

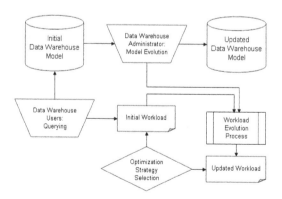

Fig. 2. Architecture for workload updating

of an updated workload that contains queries which can be executed on the updated DW. This workload contains not only queries that have been updated but also new queries if the DW changes induce new analysis possibilities. Note that this framework can be used for each DW model evolution. In the remainder of this paper we focus on schema (i.e. structure) evolution.

4.2   Algorithm

To achieve the workload evolution, we propose an algorithm which considers the DW schema evolutions and the initial workload as the input parameters (Algorithm 1).

First of all, concerning the rewriting process after the renaming of a concept, it consists in analyzing each clause of each query in the workload (SELECT, FROM, WHERE, GROUP BY) to find the old names and to replace them by the new ones.

Concerning the deletion of a concept, if a rewriting process fails, the corresponding query is deleted. The rewriting process can fail if the query no longer makes sense. For instance, if a dimension is deleted in the model and the existing query is an analysis according this dimension and no other dimension, we have to delete this one. Note that, as mentioned in the algorithm, we do not consider the deletion of measure even if it is the only one.

Concerning the creation of a concept (dimension table or attribute), we define a decisional query according to this new attribute or table. In such a query, we apply the operator AVG by default on a measure of the fact table. In the case of a table creation (resp. attribute), we can deal with the primary key of the new table (resp. this attribute) as a group by element.

4.3   Implementation

We implemented our approach according to a client/server architecture, with the Oracle 10g DBMS interfaced with PHP scripts. We suppose that a tool is used by the administrator to manage the DW model, and this tool provides the various changes that are carried out on the model. These changes are stored into two relational tables: one for changes on tables, and one for changes on attributes. These tables present some useful information such as the evolution type, the name of the object that has changed.

Furthermore, we exploit the DW meta-model which is stored in a relational way. For instance, it allows for determining what are the dimension hierarchies, the links between tables in order to write new queries. This meta-model contains the semantic induced by the DW model, particularly for the dimension hierarchies. Indeed, a dimension hierarchy corresponds to a set of semantical relations between values.

The queries of the workload are prepared to be transformed into a relational representation. More precisely, a table gathers the ID of the query, the SELECT

Algorithm 1. Workload Evolution Process

Require: workload W, set of DW schema evolutions E
Ensure: updated workload W'
1: Copying the queries of W into W'
2: **for all** evolution $e \in E$ **do**
3:     **if** e= "renaming dimension, level table" OR "renaming fact table" OR "renaming measure" OR "renaming dimension descriptor" **then**
4:         Rewriting the implied decisional queries according to the new names
5:     **end if**
6:     **if** e= "creating granularity level, dimension" OR "creating a dimension descriptor" **then**
7:         Searching for relational link from the fact table to the concerned level table
8:         Writing a decisional query according to the new level: (created attribute or table key in the SELECT clause)
9:         Adding the query to the workload W'
10:     **end if**
11:     **if** e= "deleting granularity level or dimension" **then**
12:         Searching for another level in the same dimension
13:         Rewriting the implied decisional queries according to the replaced level or deleting the queries if rewriting is not possible
14:     **end if**
15:     **if** e= "deleting a measure" **then**
16:         Searching for another measure in the same fact table
        {We suppose that there is another measure, on the contrary, it is a fact table deletion and not only a measure deletion }
17:         Rewriting the implied decisional queries according to the replaced measure
18:     **end if**
19:     **if** e= "deleting a dimension descriptor" **then**
20:         Searching for another dimension descriptor in the same level (table key by default)
        {Rewriting the implied decisional queries according to the replaced attribute}
21:     **end if**
22: **end for**
23: **return** workload W'

---

clause, the FROM clause, the WHERE clause, the GROUP-BY clause. The changes are carried out on this table, before computing the workload file.

5 Example

To illustrate our approach, we use a case study defined by Le Crédit Lyonnais french bank[1]. The DW model concerns data about the annual Net Banking Income (NBI). The NBI is the profit obtained from the management of customers account. It is a measure observed according to several dimensions: CUSTOMER, AGENCY and YEAR (Figure 3). The dimension AGENCY comprises a hierarchy with the level COMMERCIAL_DIRECTION. Each commercial direction corresponds to a group of agencies.

According to the use of the DW, a workload is defined from users' queries. Usually, in an OLAP environment, queries require the computation of aggregates over various dimension levels. Indeed, given a DW model, the analysis process allows to aggregate data by using (1) aggregation operators such as SUM or AVG; and (2) GROUP BY clauses. Here, we consider a simple example with a small workload comprising five queries that analyze the sum or the average of NBI according to various dimensions at various granularity levels (Figure 4).

---

[1] Collaboration with LCL-Le Crédit Lyonnais (Rhône-Alpes Auvergne).

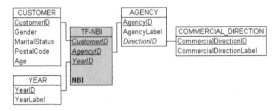

Fig. 3. Initial DW model for the NBI analysis

Q1: **SELECT** AgencyLabel, YearLabel, AVG (NBI) **FROM** AGENCY, YEAR, TF-NBI WHERE AGENCY.AgencyID=TF-NBI.AgencyID AND YEAR.YearID=TF-NBI.YearID **GROUP BY** AgencyLabel, YearLabel;

Q2: **SELECT** CommercialDirectionLabel, YearLabel, AVG (NBI) **FROM** AGENCY, COMMERCIAL_DIRECTION, TF-NBI WHERE AGENCY.AgencyID=TF-NBI.AgencyID AND COMMERCIAL_DIRECTION.CommercialDirectionID= AGENCY.DirectionID **GROUP BY** CommercialDirectionLabel, YearLabel;

Q3: **SELECT** MaritalStatus, YearLabel, SUM (NBI) **FROM** CUSTOMER, TF-NBI WHERE CUSTOMER.CustomerID=TF-NBI.CustomerID **GROUP BY** MaritalStatus, YearLabel;

Q4: **SELECT** CommercialDirectionLabel, YearLabel, AVG (NBI) **FROM** AGENCY, COMMERCIAL_DIRECTION, TF-NBI WHERE AGENCY.AgencyID=TF-NBI.AgencyID AND COMMERCIAL_DIRECTION.CommercialDirectionID=AGENCY.DirectionID AND YearLabel='2000' **GROUP BY** CommercialDirectionLabel, YearLabel;

Q5: **SELECT** AgencyLabel, SUM (NBI) **FROM** AGENCY, COMMERCIAL_DIRECTION, TF-NBI WHERE AGENCY.AgencyID=TF-NBI.AgencyID AND COMMERCIAL_DIRECTION.CommercialDirectionID=AGENCY.DirectionID AND YearLabel='2000' AND CommercialDirectionLabel='Lyon' **GROUP BY** AgencyLabel;

Fig. 4. Initial Workload for the NBI analysis

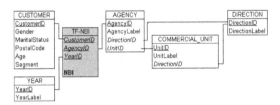

Fig. 5. Updated DW model for the NBI analysis

Due to evolutions of data sources and analysis needs, the following changes are applied on the DW model, resulting an updated model (Figure 5):

- adding the attribute Segment in the CUSTOMER dimension;
- renaming the COMMERCIAL_DIRECTION level to DIRECTION;
- renaming the CommercialDirectionID and the CommercialDirectionLabel attributes respectively to DirectionID and DirectionLabel;
- inserting the COMMERCIAL_UNIT level between AGENCY and DIRECTION levels.

After the schema evolution, our algorithm is applied, and an updated workload is provided (Figure 6). We can note that queries' syntax is updated to remain coherent with the DW model. Indeed, the schema's updates have been propagated to the required queries. Moreover, two queries have been added, to take into account potential analysis needs which concern the newly created COMMERCIAL_UNIT level and the newly created Segment attribute.

Q1: **SELECT** AgencyLabel, YearLabel, AVG (NBI) **FROM** AGENCY, YEAR, TF-NBI WHERE AGENCY.AgencyID=TF-NBI.AgencyID AND YEAR.YearID=TF-NBI.YearID **GROUP BY** AgencyLabel, YearLabel;

Q 2: **SELECT** DirectionLabel, YearLabel, AVG (NBI) **FROM** AGENCY, DIRECTION, TF-NBI WHERE AGENCY.AgencyID=TF-NBI.AgencyID AND DIRECTION.DirectionI D=AGENCY.DirectionID **GROUPBY** DirectionLabel, YearLabel;

Q3: **SELECT** MaritalStatus, YearLabel, SUM (NBI) **FROM** CUSTOMER, TF-NBI WHERE CUSTOMER.CustomerID=TF-NBI.CustomerID **GROUP BY** MaritalStatus, YearLabel;

Q4: **SELECT** DirectionLabel, YearLabel, SUM (NBI) **FROM** AGENCY, DIRECTION, TF-NBI WHERE AGENCY.AgencyID=TF-NBI.AgencyID AND DIRECTION.DirectionI D=AGENCY.DirectionID AND YearLabel='2000' **GROUP B Y** DirectionLabel, YearLabel;

Q5: **SELECT** AgencyLabel, AVG (NBI) **FROM** AGENCY, DIRECTION, TF-NBI WHERE AGENCY.AgencyID=TF-NBI.AgencyID AND DIRECTION.DirectionID=AGENCY.DirectionID AND YearLabel='2000' AND DirectionLabel='Lyon' **GROUP BY** AgencyLabel;

Q6: **SELECT** UnitLabel, AVG (NBI) **FROM** AGENCY, COMMERCIAL_UNIT, TF-NBI WHERE AGENCY.AgencyID=TF-NBI.AgencyID AND COMMERCIAL_UNIT.UnitID=AGENCY.UnitID **GROUP BY** UnitLabel;

Q7: **SELECT** Segment, AVG (NBI) **FROM** CUSTOMER, TF-NBI WHERE CUSTOMER.CustomerID=TF-NBI.CustomerID AND **GROUP BY** Segment;

Fig. 6. Updated Workload for the NBI analysis

6   Conclusion

In this paper, we presented a preliminary work to address the problem of query evolution in DWs. We focused particularly on the problem of workload evolution to support optimization strategy evolution. We proposed a global approach to achieve this task and an algorithm. We presented a simplified example based on a case study to illustrate our approach. The main advantage of our approach is to support the administrator by a pro-active method. Firstly it ensures that the existing queries remain coherent according to the schema evolution. Secondly it provides new queries to take into account possible analysis needs.

There are still many aspects to be explored. First of all, we want to develop a tool based on our proposal that helps the DW administrator in the evolution process. In this case, the workload evolution process can be extended to propose several consequences to be applied in a semi-automatic way and to create more complex queries by involving the administrator. For instance, for the deletion of a level which is between two others, the administrator can choose to rewrite the queries according to the higher level, or the lower level or both. Furthermore, in addition to the schema evolutions, we have to integrate the evolution of data in our approach. For instance, we can consider value evolution of a foreign key. This type of changes induces update on WHERE clauses of the queries. Moreover, we have to carry out a performance study in order to precisely determine in which context our approach is more interesting (instead of a manual evolution made by the administrator himself). We may study parameters such as the number of queries in the workload, the importance of changes in the DW. Finally, we want to integrate the evolution aspect in a benchmark. Indeed a benchmark allows for deciding of optimization strategies before the real use of the DW. And it can help to evaluate the impact of a change and to decide whether the change has to be applied or not. This issue requires to equip benchmarks with evolution operators in order to make the benchmark evolve.

Acknowledgments. The authors would like to thank Mohamed Bekhouche, Lorène Ducommun and Salim Guergouri for their contribution to this work.

References

1. Rizzi, S., Saltarelli, E.: View Materialization vs. Indexing: Balancing Space Constraint in Data Warehouse Design. In: Eder, J., Missikoff, M. (eds.) CAiSE 2003. LNCS, vol. 2681, pp. 502–519. Springer, Heidelberg (2003)
2. Gallo, J.: Operations and maintenance in a data warehouse environment. DM Review Magazine (2002), http://www.dmreview.com/article_sub. cfm?articleId=6118
3. Agrawal, S., Chaudhuri, S., Narasayya, V.R.: Automated selection of materialized views and indexes in sql databases. In: XXVIth International Conference on Very Large Data Bases (VLDB 00), Cairo, Egypt, pp. 496–505. Morgan Kaufmann, San Francisco (2000)
4. Aouiche, K., Jouve, P., Darmont, J.: Clustering-Based Materialized View Selection in Data Warehouses. In: Manolopoulos, Y., Pokorný, J., Sellis, T. (eds.) ADBIS 2006. LNCS, vol. 4152, pp. 81–95. Springer, Heidelberg (2006)
5. Theodoratos, D., Bouzeghoub, M.: A General Framework for the View Selection Problem for Data Warehouse Design and Evolution. In: IIIrd ACM International Workshop on Data Warehousing and OLAP (DOLAP 00), Washington, Columbia, USA, pp. 1–8. ACM Press, New York, NY, USA (2000)
6. Golfarelli, M., Saltarelli, E.: The Workload You Have, the Workload You Would Like. In: VIth ACM International Workshop on Data Warehousing and OLAP (DOLAP 03), New Orleans, Louisiana, USA, pp. 79–85. ACM Press, New York, NY, USA (2003)
7. Kotidis, Y., Roussopoulos, N.: DynaMat: A Dynamic View Management System for Data Warehouses. SIGMOD Rec. 28(2), 371–382 (1999)
8. Theodoratos, D., Sellis, T.: Incremental Design of a Data Warehouse. Journal of Intelligent Information Systems 15(1), 7–27 (2000)
9. Lawrence, M., Rau-Chaplin, A.: Dynamic view selection for olap. In: Tjoa, A.M., Trujillo, J. (eds.) DaWaK 2006. LNCS, vol. 4081, pp. 33–44. Springer, Heidelberg (2006)
10. Bellahsene, Z.: Schema Evolution in Data Warehouses. Knowledge and Information Systems 4(3), 283–304 (2002)
11. Vassiliadis, P., Papastefanatos, G., Vassiliou, Y., Sellis, T.: Management of the Evolution of Database-Centric Information Systems. In: Ist International Workshop on Database Preservation (PresDB 07), Edinburgh, Scotland, UK (2007)
12. Papastefanatos, G., Kyzirakos, K., Vassiliadis, P., Vassiliou, Y.: Hecataeus: A Framework for Representing SQL Constructs as Graphs. In: XXth International Workshop on Exploring Modeling Methods for Systems Analysis and Design (EMMSAD 05), in conjunction with the XVIIth International Conference on Advanced Information Systems Engineering (CAiSE 05), Oporto, Portugal (2005)
13. Papastefanatos, G., Vassiliadis, P., Vassiliou, Y.: Adaptive Query Formulation to Handle Database Evolution. In: Dubois, E., Pohl, K. (eds.) CAiSE 2006. LNCS, vol. 4001, Springer, Heidelberg (2006)

What-If Analysis for Data Warehouse Evolution

George Papastefanatos[1], Panos Vassiliadis[2], Alkis Simitsis[3], and Yannis Vassiliou[1]

[1] National Technical University of Athens, Dept. of Electr. and Comp. Eng., Athens, Hellas
{gpapas,yv}@dbnet.ece.ntua.gr
[2] University of Ioannina, Dept. of Computer Science, Ioannina, Hellas
pvassil@cs.uoi.gr
[3] IBM Almaden Research Center, San Jose, California, USA
asimits@us.ibm.com

Abstract. In this paper, we deal with the problem of performing what-if analysis for changes that occur in the schema/structure of the data warehouse sources. We abstract software modules, queries, reports and views as (sequences of) queries in SQL enriched with functions. Queries and relations are uniformly modeled as a graph that is annotated with policies for the management of evolution events. Given a change at an element of the graph, our method detects the parts of the graph that are affected by this change and indicates the way they are tuned to respond to it.

1 Introduction

Data warehouses are complicated software environments where data stemming from operational sources are extracted, transformed, cleansed and eventually loaded in fact or dimension tables in the data warehouse. Once this task has been successfully completed, further aggregations of the loaded data are also computed and stored in data marts, reports, spreadsheets, and other formats. The whole environment involves a very complicated architecture, where each module depends upon its data providers to fulfill its task. This strong flavor of inter-module dependency makes the problem of evolution very important in a data warehouse environment.

Figure 1 depicts a simplified version of an Extraction-Transformation-Loading (ETL) process. Data are extracted from sources and they are transferred to the Data Staging Area (DSA), where their contents and structure are modified; example transformations include joins, addition of new attributes produced via functions, and so on. Finally, the results are stored in the data warehouse (DW) either in fact or dimension tables and materialized views. During the lifecycle of the DW it is possible that several counterparts of the ETL process may be evolved. For instance, assume that an attribute is deleted from the underlying database S1 or it is added to the source relation S2. Such changes affect the entire workflow, possibly all the way to the warehouse (tables T1 and T2), along with any reports over the warehouse tables (abstracted as queries over view V3).

Research has extensively dealt with the problem of schema evolution, in object-oriented databases [1, 11, 15], ER diagrams [22], data warehouses [6, 16, 17, 18] and materialized views [2, 5, 7, 8]. However, to the best of our knowledge, there is no global framework for the management of evolution in the described setting.

I.Y. Song, J. Eder, and T.M. Nguyen (Eds.): DaWaK 2007, LNCS 4654, pp. 23–33, 2007.

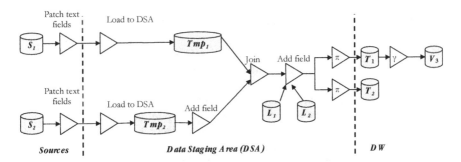

Fig. 1. A simple ETL workflow

In this paper, we provide a general mechanism for performing what-if analysis [19] for potential changes of data source configurations. We introduce a graph model that uniformly models relations, queries, views, ETL activities, and their significant properties (e.g., conditions). Apart from the simple task of capturing the semantics of a database system, the graph model allows us to predict the impact of a change over the system. We provide a framework for annotating the database graph with policies concerning the behavior of nodes in the presence of hypothetical changes. In addition, we provide a set of rules that dictate the proper actions, when additions, deletions or updates are performed to relations, attributes, and conditions. (All the above concepts are treated as first-class citizens in our model.) Assuming that a graph construct is annotated with a policy for a particular event (e.g., an activity node is tuned to deny deletions of its provider attributes), the proposed framework has the following features: (a) it performs the identification of the affected subgraph; and (b) if the policy is appropriate, it automates the readjustment of the graph to fit the new semantics imposed by the change. Finally, we experimentally assess our proposal.

Outline. Section 2 presents the graph model for databases. Section 3 proposes a framework of graph annotations and readjustment automation for database evolution. Section 4 presents the results of a case study for our framework. Section 5 discusses related work. Finally, Section 6 concludes and provides insights for future work.

2   Graph-Based Modeling of ETL Processes

In this section, we propose a graph modeling technique that uniformly covers relational tables, views, ETL activities, database constraints, and SQL queries as first class citizens. The proposed technique provides an overall picture not only for the actual source database schema but also for the ETL workflow, since queries that represent the functionality of the ETL activities are incorporated in the model.

The proposed modeling technique represents all the aforementioned database parts as a directed graph $G=(V,E)$. The nodes of the graph represent the entities of our model, where the edges represent the relationships among these entities. Preliminary versions of this model have been presented in our previous work [9,10].

The constructs that we consider are classified as *elementary*, including relations, conditions and queries/views and *composite*, including ETL activities and ETL processes. Composite elements are combinations of elementary ones.

Relations, R. Each relation $R(\Omega_1, \Omega_2, ..., \Omega_n)$ in the database schema can be either a table or a file (it can be considered as an external table). A relation is represented as a directed graph, which comprises: (a) a *relation node*, R, representing the relation schema; (b) n *attribute nodes*, $\Omega_i \in \Omega$, i=1..n, one for each of the attributes; and (c) n *schema relationships*, E_s, directing from the relation node towards the attribute nodes, indicating that the attribute belongs to the relation.

Conditions, C. Conditions refer both to *selection conditions* of queries and views, and *constraints* of the database schema. We consider three classes of atomic conditions that are composed through the appropriate usage of an operator op belonging to the set of classic binary operators, **Op** (e.g., <, >, =, ≤, ≥, !=, IN, EXISTS, ANY): (a) Ω op constant; (b) Ω op Ω'; and (c) Ω op Q. (Ω, Ω' are attributes of the underlying relations and Q is a query). A *condition node* is used for the representation of the condition. The node is tagged with the respective operator and it is connected to the *operand nodes* of the conjunct clause through the respective *operand relationships*, O. Composite conditions are easily constructed by tagging the condition node with a Boolean operator (e.g., AND or OR) and the respective edges, to the conditions composing the composite condition.

Queries, Q. The graph representation of a Select - Project - Join - Group By (SPJG) query involves a *query node* representing the query and *attribute nodes* corresponding to the schema of the query. Thus, the query graph is a directed graph connecting the query node with all its schema attributes, through *schema relationships*. In order to represent the relationship between the query graph and the underlying relations, the query is resolved into its essential parts: SELECT, FROM, WHERE, GROUP BY, HAVING, and ORDER BY, each of which is eventually mapped to a subgraph.

Select part. Each query is assumed to own a schema that comprises the attributes appearing in the SELECT clause, either with their original or alias names. In this context, the SELECT part of the query maps the respective attributes of the involved relations to the attributes of the query schema through *map-select relationships*, E_M, directing from the query attributes towards the relation attributes.

From part. The FROM clause of a query can be regarded as the relationship between the query and the relations involved in this query. Thus, the relations included in the FROM part are combined with the query node through *from relationships*, E_F, directing from the query node towards the relation nodes.

Where and Having parts. We assume the WHERE and/or HAVING clauses of a query in conjunctive normal form. Thus, we introduce two directed edges, namely *where relationships*, E_W, and *having relationships*, E_H, both starting from a query node towards an operator node corresponding to the conjunction of the highest level.

Group and Order By part. For the representation of aggregate queries, two special purpose nodes are employed: (a) a new node denoted as $GB \in \mathbf{GB}$, to capture the set of attributes acting as the aggregators; and (b) one node per aggregate function labeled with the name of the employed aggregate function; e.g., COUNT, SUM, MIN. For the aggregators, we use edges directing from the query node towards the GB node that are labeled <group-by>, indicating *group-by relationships*, E_G. The GB node is connected

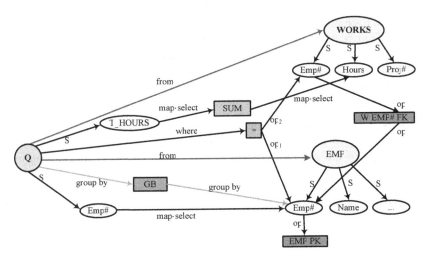

Fig. 2. Graph of an example aggregate query annotated with policies [10]

with each of the aggregators through an edge tagged also as <group-by>, directing from the GB node towards the respective attributes. These edges are additionally tagged according to the order of the aggregators; we use an identifier i to represent the i-th aggregator. Moreover, for every aggregated attribute in the query schema, there exists an edge directing from this attribute towards the aggregate function node as well as an edge from the function node towards the respective relation attribute. Both edges are labeled <map-select> and belong to $\mathbf{E_M}$, as these relationships indicate the mapping of the query attribute to the corresponding relation attribute through the aggregate function node. The representation of the ORDER BY clause of the query is performed similarly, whereas nested queries and functions used in queries are also incorporated in our model [21].

Views, V. Views are considered either as queries or relations (materialized views), thus, $\mathbf{V \subseteq R \cup Q}$.

Figure 2 depicts the proposed graph representation for an example aggregate query.

```
Q:  SELECT EMP.Emp# as Emp#, Sum(WORKS.Hours) as T_Hours
    FROM    EMP, WORKS
    WHERE EMP.Emp#=WORKS.Emp#
    GROUP BY EMP.Emp#
```

As far as modification queries are concerned, their behavior with respect to adaptation to changes in the database schema can be captured by SELECT queries. For lack of space, we simply mention that (a) INSERT statements can be dealt as simple SELECT queries and (b) DELETE and UPDATE statements can also be treated as SELECT queries, possibly comprising a WHERE clause.

ETL activities, A.. An ETL activity is modeled as a sequence of SQL views. An ETL activity necessarily comprises: (a) one (or more) *input view(s)*, populating the input of the activity with data coming from another activity or a relation; (b) an *output view*, over which the following activity will be defined; and (c) a *sequence of views* defined over the input and/or previous, internal activity views.

ETL summary, S. An ETL summary is a directed acyclic graph $G_s = (V_s, E_s)$ which acts as a zoomed-out variant of the full graph G [20]. V_s comprises of activities, relations and views that participate in an ETL process. E_s comprises the edges that connect the providers and consumers. Conversely, to the overall graph where edges denote dependency, edges in the ETL summary denote data provision. The graph of the ETL summary can be topologically sorted and therefore, execution priorities can be assigned to activities. Figure 1 depicts an ETL summary.

Components. A component is a sub-graph of the graph in one of the following patterns: (a) a relation with its attributes and all its constraints, (b) a query with all its attributes, functions and operands. Modules are disjoint and they are connected through edges concerning foreign keys, map-select, where, and so on.

3   Adapting ETL Workflows for Evolution of Sources

In this section, we formulate a set of rules that allow the identification of the impact of changes to an ETL workflow and propose an automated way to respond to these changes. Evolution changes may affect the software used in an ETL workflow (like queries, stored procedures, and triggers) in two ways: (a) syntactically, a change may evoke a compilation or execution failure during the execution of a piece of code; and (b) semantically, a change may have an effect on the semantics of the software used.

The proposed rules annotate the graph representing the ETL workflow with actions that should be taken when a change event occurs. The combination of events and annotations determines the policy to be followed for the handling of a potential change. The annotated graph is stored in a metadata repository and it is accessed from a what-if analysis module. This module notifies the designer or the administrator on the effect of a potential change and the extent to which the modification to the existing code can be fully automated, in order to adapt to the change. To alleviate the designer from the burden of manually annotating all graph constructs, a simple extension of SQL with clauses concerning the evolution of important constructs is proposed in the long version of this paper [21].

3.1   The General Framework for Schema Evolution

The main mechanism towards handling schema evolution is the annotation of the constructs of the graph (i.e., nodes and edges) with elements that facilitate what-if analysis. Each such construct is enriched with policies that allow the designer to specify the behavior of the annotated construct whenever events that alter the database graph occur. The combination of an event with a policy determined by the designer/administrator triggers the execution of the appropriate action that either blocks the event or reshapes the graph to adapt to the proposed change.

The space of potential events comprises the Cartesian product of two subspaces. The space of hypothetical actions (addition, deletion, and modification) over graph

Algorithm *Propagate changeS (PS)*	

Input: an ETL summary **S** over a graph $G_o = (V_o, E_o)$ and an event e
Output: a graph $G_n = (V_n, E_n)$
Variables: a set of events **E**, and an affected node A
Begin
    dps(**S**, **G₀**, **Gₙ**, {e}, A)
End
	``
	dps(S, Gn, Go, E, A) {
	  I = Ins_by_policy(affected(E))
	  D = Del_by_policy(affected(E))
	  Gn = Go - D ∪ I
	  E = E-{e}∪action(affected(E))
	  if consumer(A)≠nil
	    for each consumer(A)
	      dps(S,Gn,Go,E,consumer(A))
	}
	``

Fig. 3. Algorithm Propagate changeS (PS)

constructs sustaining evolution changes (relations, attributes, and conditions). For each of the above events, the administrator annotates graph constructs affected by the event with policies that dictate the way they will regulate the change.

Three kinds of policies are defined:

(a) *propagate* the change, meaning that the graph must be reshaped to adjust to the new semantics incurred by the event;

(b) *block* the change, meaning that we want to retain the old semantics of the graph and the hypothetical event must be blocked or, at least, constrained, through some rewriting that preserves the old semantics [8, 14]; and

(c) *prompt* the administrator to interactively decide what will eventually happen.

For the case of blocking, the specific method that can be used is orthogonal to our approach, which can be performed using any available method [8, 14].

Our framework prescribes the *reaction* of the parts of the system affected by a hypothetical schema change based on their annotation with policies. The mechanism that determines the reaction to a change is formally described by the algorithm *Propagate changeS (PS)* in Figure 3. Given an ETL summary S over a graph **G₀** and an event e, *PS* produces a new ETL summary **Gₙ**, which has absorbed the changes.

Example. Consider the simple example query SELECT * FROM EMP as part of an ETL activity. Assume that provider relation EMP is extended with a new attribute PHONE. There are two possibilities:

- The * notation signifies the request for any attribute present in the schema of relation EMP. In this case, the * shortcut can be treated as "return all the attributes that EMP has, independently of which these attributes are". Then, the query must also retrieve the new attribute PHONE.
- The * notation acts as a macro for the particular attributes that the relation EMP originally had. In this case, the addition to relation EMP should not be further propagated to the query.

A naïve solution to a modification of the sources; e.g., addition of an attribute, would be that an impact prediction system must trace all queries and views that are potentially affected and ask the designer to decide upon which of them must be modified to incorporate the extra attribute. We can do better by extending the current

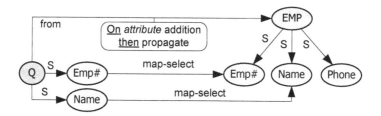

Fig. 4. Propagating addition of attribute PHONE

modeling. For each element affected by the addition, we annotate its respective graph construct (i.e., node, edges) with the policies mentioned before. According to the policy defined on each construct the respective action is taken to correct the query.

Therefore, for the example event of an attribute addition, the policies defined on the query and the actions taken according to each policy are:

- *Propagate attribute addition*. When an attribute is added to a relation appearing in the FROM clause of the query, this addition should be reflected to the SELECT clause of the query.
- *Block attribute addition*. The query is immune to the change: an addition to the relation is ignored. In our example, the second case is assumed, i.e., the SELECT * clause must be rewritten to SELECT A1,...,An without the newly added attribute.
- *Prompt*. In this case (default, for reasons of backwards compatibility), the designer or the administrator must handle the impact of the change manually; similarly to the way that currently happens in database systems.

The graph of the query SELECT * FROM EMP is shown in Figure 4. The annotation of the FROM edge as *propagating addition* indicates that the addition of PHONE node will be propagated to the query and the new attribute is included in the SELECT clause of the query. If a FROM edge is not tagged with this additional information, then a *default case* is assumed and the designer/administrator is prompted to decide.

4 Case Study

We have evaluated the effectiveness of our setting via the reverse engineering of real-world ETL processes, extracted from an application of the Greek public sector. We have monitored the changes that took place to the sources of the studied data warehouse. In total, we have studied a set of 7 ETL processes, which operate on the data warehouse. These processes extract information out of a set of 7 source tables, namely S_1 to S_7 and 3 lookup tables, namely L_1 to L_3, and load it to 9 tables, namely T_1 to T_9, stored in the data warehouse. The aforementioned scenarios comprise a total number of 53 activities.

Table 1 illustrates the changes that occurred on the schemata of the source and lookup tables, such as renaming source tables, renaming attributes of source tables, adding and deleting attributes from source tables, modifying the domain of attributes and lastly changing the primary key of lookup tables. After the application of these changes to the sources of the ETL process, each affected activity was properly readjusted (i.e., rewriting of queries belonging to activities) in order to adapt to the changes. For each event, we counted: (a) the number of activities affected both semantically and syntactically, (b) the number of activities, that have automatically been adjusted by our framework (*propagate* or *block* policies) as opposed to those (c) that required administrator's intervention (i.e., a *prompt* policy).

Table 1. Analytic results

Source Name	Event Type	Activities (53)			
		Affected	Autom. adjusted	Prompt	%
S_1	Rename	14	14	0	100%
	Rename Attributes	14	14	0	100%
	Add Attributes	34	31	3	91%
	Delete Attributes	19	18	1	95%
	Modify Attributes	18	18	0	100%
S_4	Rename	4	4	0	100%
	Rename Attributes	4	4	0	100%
	Add Attributes	19	15	4	79%
	Delete Attributes	14	12	2	86%
	Modify Attributes	6	6	0	100%
S_2	Rename	1	1	0	100%
	Rename Attributes	1	1	0	100%
	Add Attributes	4	4	0	100%
	Delete Attributes	4	3	1	75%
S_3	Rename	1	1	0	100%
	Rename Attributes	1	1	0	100%
S_5	Modify Attributes	3	3	0	100%
S_6	*NO_CHANGES*	0	0	0	-
S_7	Rename	1	1	0	100%
	Rename Attributes	1	1	0	100%
L_1	*NO_CHANGES*	0	0	0	-
L_2	Add Attributes	1	0	1	0%
L_3	Add Attributes	9	0	9	0%
	Change PK	9	0	9	0%

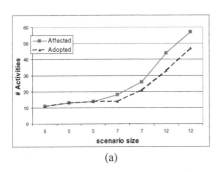

(a)

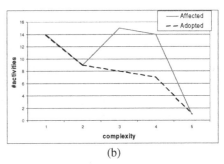

(b)

Fig. 5. Evaluation of our method

Figure 5a depicts the correlation between the average number of affected activities versus automatically adapted activities w.r.t. the total number of activities contained in ETL scenarios. For ETL processes comprising a small number of activities most affected activities are successfully adjusted to evolution changes. For longest ETL processes, the number of automatically adjusted activities increases proportionally to the number of affected activities.

Furthermore, Table 1 shows that our framework can successfully handle and propagate evolution events to most activities, by annotating the queries included in them with policies. Activities requiring administrator's intervention are mainly activities executing complex joins, e.g., with lookup tables, for which the administrator must decide upon the proper rewriting. Figure 5b presents the average amount of automatically adapted activities w.r.t. the complexity of activities. Complexity refers to the functionality of each activity; e.g., the type of transformation it performs or the types of queries it contains. In general, our framework handles efficiently simple activities. More complex activities, e.g., pivoting activities, are also adequately adjusted by our approach to evolution changes.

5   Related Work

Evolution. Related research work has studied in the past the problem of database schema evolution. A survey on schema versioning and evolution is presented in [13], whereas a categorization of the overall issues regarding evolution and change in data management is presented in [12]. The problem of view adaptation after redefinition is mainly investigated in [2, 5, 7], where changes in views definition are invoked by the user and rewriting is used to keep the view consistent with the data sources. In [6], the authors discuss versioning of star schemata, where histories of the schema are retained and queries are chronologically adjusted to ask the correct schema. The warehouse adaptation for SPJ views is studied in [2]. Also, the view synchronization problem considers that views become invalid after schema changes in the underlying base relations [8]. Our work in this paper builds mostly on the results of [8], by extending it to incorporate attribute additions and the treatment of conditions. The treatment of attribute deletions in [8] is quite elaborate; we confine to a restricted version to avoid overcomplicating both the size of requested metadata and the language extensions. Still, the [8] tags for deletions may be taken into consideration in our method. Finally, the algorithms for rewriting views when the schemas of their source data change (e.g., [2, 5]), are orthogonal to our approach. Thus, our approach can be extended in the presence of new results on such algorithms.

Model mappings. Model management provides a generic framework for managing model relationships, comprising three fundamental operators: match, diff, and merge [3,4]. Our proposal assigns semantics to the match operator for the case of model evolution, where the source and target models of the mapping are the original and resulting database graph, respectively, after evolution management has taken place. A similar framework for the management of evolution has been proposed [14]. Still, the model of [14] is more restrictive, in the sense that it is intended towards retaining the original semantics of the queries. Our work is a larger framework that allows the restructuring of the database graph (i.e., model) either towards keeping the original semantics or towards its readjustment to the new semantics.

6 Conclusions and Future Work

In this paper, we have discussed the problem of performing what-if analysis for changes that occur in the schema/structure of the data warehouse sources. We have modeled software modules, queries, reports and views as (sequences of) queries in SQL extended with functions. Queries and relations have uniformly been modeled as a graph that is annotated with policies for the management of evolution events. We have presented an algorithm that detects the parts of the graph that are affected by a given change and highlights the way they are tuned to respond to it. Finally, we have evaluated our approach over cases extracted from real world scenarios.

Future work may be directed towards many goals, with patterns of evolution sequences being the most prominent one.

References

1. Banerjee, J., et al.: Semantics and implementation of schema evolution in object-oriented databases. In: SIGMOD (1987)
2. Bellahsene, Z.: Schema evolution in data warehouses. Knowledge and Information Systems 4(2) (2002)
3. Bernstein, P., Levy, A., Pottinger, R.: A Vision for Management of Complex Models. SIGMOD Record 29(4) (2000)
4. Bernstein, P., Rahm, E.: Data Warehouse Scenarios for Model Management. In: Laender, A.H.F., Liddle, S.W., Storey, V.C. (eds.) ER 2000. LNCS, vol. 1920, Springer, Heidelberg (2000)
5. Gupta, A., Mumick, I.S., Rao, J., Ross, K.A.: Adapting materialized views after redefinitions: Techniques and a performance study. Information Systems (26) (2001)
6. Golfarelli, M., Lechtenbörger, J., Rizzi, S., Vossen, G.: Schema Versioning in Data Warehouses. In: ECDM 2004, pp. 415–428 (2004)
7. Mohania, M., Dong, D.: Algorithms for adapting materialized views in data warehouses. In: CODAS (1996)
8. Nica, A., Lee, A.J., Rundensteiner, E.A.: The CSV algorithm for view synchronization in evolvable large-scale information systems. In: Schek, H.-J., Saltor, F., Ramos, I., Alonso, G. (eds.) EDBT 1998. LNCS, vol. 1377, Springer, Heidelberg (1998)
9. Papastefanatos, G., Vassiliadis, P., Vassiliou, Y.: Adaptive Query Formulation to Handle Database Evolution. In: Dubois, E., Pohl, K. (eds.) CAiSE 2006. LNCS, vol. 4001, Springer, Heidelberg (2006)
10. Papastefanatos, G., Kyzirakos, K., Vassiliadis, P., Vassiliou, Y.: Hecataeus: A Framework for Representing SQL Constructs as Graphs. In: EMMSAD (2005)
11. Ra, Y.G., Rundensteiner, E.A.: A transparent object-oriented schema change approach using view evolution. In: ICDE (1995)
12. Roddick, J.F., et al.: Evolution and Change in Data Management - Issues and Directions. SIGMOD Record 29(1) (2000)
13. Roddick, J.F.: A survey of schema versioning Issues for database systems. Information Software Technology 37(7) (1995)

14. Velegrakis, Y., Miller, R.J., Popa, L.: Preserving mapping consistency under schema changes. VLDB J. 13(3) (2004)
15. Zicari, R.: A framework for schema update in an object-oriented database system. In: ICDE (1991)
16. Blaschka, M., Sapia, C., Höfling, G.: On Schema Evolution in Multidimensional Databases. In: Mohania, M.K., Tjoa, A.M. (eds.) DaWaK 1999. LNCS, vol. 1676, Springer, Heidelberg (1999)
17. Kaas, C., Pedersen, T.B., Rasmussen, B.: Schema Evolution for Stars and Snowflakes. In: ICEIS (2004)
18. Bouzeghoub, M., Kedad, Z.: A Logical Model for Data Warehouse Design and Evolution. In: Kambayashi, Y., Mohania, M.K., Tjoa, A.M. (eds.) DaWaK 2000. LNCS, vol. 1874, Springer, Heidelberg (2000)
19. Golfarelli, M., Rizzi, S., Proli, A.: Designing what-if analysis: towards a methodology. In: DOLAP (2006)
20. Simitsis, A., Vassiliadis, P., Terrovitis, M., Skiadopoulos, S.: Graph-Based Modeling of ETL Activities with Multi-level Transformations and Updates. In: Tjoa, A.M., Trujillo, J. (eds.) DaWaK 2005. LNCS, vol. 3589, Springer, Heidelberg (2005)
21. Papastefanatos, G., Vassiliadis, P., Simitsis, A., Vassiliou, Y.: What-if Analysis for Data Warehouse Evolution (Extended Version). Working Draft (April 2007), url: www. dbnet. ece.ntua.gr/g̃papas/Publications/ DataWarehouseEvolution-Extended.pdf
22. Liu, C.T., Chrysanthis, P.K., Chang, S.K.: Database schema evolution through the specification and maintenance of changes on entities and relationships. In: Loucopoulos, P. (ed.) ER 1994. LNCS, vol. 881, Springer, Heidelberg (1994)

An Extensible Metadata Framework for Data Quality Assessment of Composite Structures

José Farinha and Maria José Trigueiros

ISCTE/ADETTI, Department of Science and Information Technology,
Av. Forças Armadas,
1649-026 Lisbon, Portugal
{Jose.Farinha,Maria.Jose.Trigueiros}@iscte.pt

Abstract. Data quality is a critical issue both in operational databases and in data warehouse systems. Data quality assessment is a strong requirement regarding the ETL subsystem, since bad data may destroy data warehouse credibility. During the last two decades, research and development efforts in the data quality field have produced techniques for data profiling and cleaning, which focus on detecting and correcting bad values in data. Little efforts have been done considering data quality when it relates to the well-formedness of coarse grained data structures resulting from the assembly of linked data records. This paper proposes a metadata model that supports the structural validation of linked data records, from a data quality point of view. The metamodel is built on top of the CWM standard and it supports the specification of data structure quality rules in a high level of abstraction, as well as by means of very specific fine grained business rules.

Keywords: Data Quality, Metadata, Metamodel, CWM.

1 Introduction

Low data quality (DQ) has been recognized as a major problem associated with the information systems inside an organization, as it can induce low productivity, increase costs and reduce customer satisfaction. Specifically, considering the impact data has in the effectiveness of the decision process, low data quality may prevent an organization from devising a good business strategy, as important business facts may be obscured by incorrect, missing or badly structured data records. Therefore, regarding decision support systems, special emphasis must be put on data quality features. Specifically regarding data warehousing systems, significant requirements for data quality and cleaning are generally advisable and implemented along the Extract Transform and Load (ETL) chain, in order to ensure the quality of the data flowing from operational data sources to the data warehouse.

Whether the goals regarding data quality are for the highest performing data cleaning procedures, or for simple data profiling features intended to filter out data bellow an acceptable quality level, the ability to detect bad data is a fundamental requirement for the ETL layer.

The last decades' data quality R&D efforts have focused mainly on detecting incorrect values, null values and duplicate records [3]. On the contrary, very few

I.Y. Song, J. Eder, and T.M. Nguyen (Eds.): DaWaK 2007, LNCS 4654, pp. 34–44, 2007.

pieces of work have considered data quality as an issue related to the well-formedness of composite data structures, that result from the assemblage of multiple, interrelated data records.

Data quality has been characterized in literature as a concept that is founded on several criteria. Data *correctness, completeness, consistency* and *timeliness*, for instance, are all requirements for data quality. Some of these assessment dimensions may be negatively affected by incorrectly linked data records. For instance, in an academic data context, if a student data record is connected to an exam result of a course not belonging to that student's degree, the database includes *incorrect* information about that student. In another example, in a retailing context, if an invoice includes 8 invoice items and the corresponding order record is linked to 9 items, either the invoice is incomplete or a return took place and it was not registered in the database. In either case there are missing records, resulting in a penalty regarding the *completeness* criteria for data quality. This paper is focused on detecting data problems such as the ones mentioned.

In a previous paper [6] we focused on the problem of specifying metadata that could support the validation and cleaning of attribute values in individual data records. In the present paper we are mainly focused on detecting composite data structures containing badly or poorly connected data records. This paper aims at proposing a metadata model (metamodel) for this purpose. The presented metamodel is intended to be used in an ETL context and for this reason it was developed on top of CWM, a reference metadata standard for data warehousing [9].

Most of the data quality problems associated with semantically cohesive groups of connected data records could be treated as well as data integrity problems. Although some data quality problems are not typical data integrity problems, the ones we are concerned about in this paper are. Data integrity/quality is traditionally specified in one of three ways:

1st By means of structural relationships, as the ones provided by traditional modeling techniques, like Entity-Relationship and UML;

2nd By means of high level of abstraction structural constraints, such as UML's *Subset, Xor* and *Redefines* constraints [2];

3rd By means of detailed and specific business rules, typically expressed using some constraint language, like SQL/DDL or OCL (Object Constraint Language) [10].

Although those methods and tools are in place to support the development of information systems' capabilities for ensuring data integrity on data entry, too often data integrity is overlooked and becomes a data quality issue, which must be dealt with in a later phase of the data management process. Data integrity becoming a data quality issue is also common when it comes to non-relational data sources, as well as when it comes to multiple data source integration, both being commonplace for ETL.

In this paper we are mainly concerned with the 2nd and 3rd types of data integrity constraints, for the following reasons: (1) they tend to be more complex, (2) they are usually application bound instead of directly encoded in the data layer of the information system and (3) they are more prone to be overlooked by analysts and project managers. Therefore, the need for proper tools that ease their validation is clearly more urgent.

The 2nd type of constraint specification provides an adequate high level of abstraction view of the data structure and it is fairly supported by database design tools. However, such support mainly happens for documentation purposes, as the corresponding code generation facilities are usually absent from such tools. Regarding the 3rd type, business rules are flexible and commonly supported by database systems. However, their specification tends to be laborious, and difficult to read and interpret. Therefore, it seemed to us that the appropriate approach to data structure quality assessment would be one that integrates those two types of specification, in order to capitalize on the advantages of both. We propose a method supported on metadata that promotes the definition of new constraints of the 2nd type through generalization of constraints of the 3rd type.

The metamodel proposed in this paper allows for the definition and storage of DQ Rule Templates, which are high level of abstraction patterns of data integrity constraints that can be applied to business data models, under certain, predefined conditions. Rule templates are then mapped to generic, fine grained business rules, which are supported by means of expression metadata based on the OCL metamodel proposed in [12]. Once represented as expression metadata, data structure quality rules are then easily translated to SQL/DML.

The paper is structured as follows: section 2 presents some related work; section 3 presents the metamodel for rule template definition; section 4 illustrates the applicability of rule templates to some recurring data quality problems; and finally, section 5 concludes.

The case study examples presented in this paper are part of an academic data repository under a relational database whose structure is modeled in UML. It is assumed that every modeling construct in the UML business data model is already enforced by the database.

Some conventions used along this paper are as follows. Class names are always capitalized and printed in italic. When referring to objects of a specific class, the name of the class is used in small caps and italic; for instance, "a *structural feature*" denotes an object belonging to class *StructuralFeature* or any of its subclasses. Classes belonging to CWM are presented in diagrams with a shortcut arrow symbol on the bottom-left corner.

2   Related Work

There are plenty of tools and methods for DQ assessment and management, both from the industry and from the research fields. In the research field, the most notable are [5] and [11].

Specifically regarding cross-referencing DQ with metadata, scientific approaches are very sparse. We presented the foundational layer of our DQ metamodel in [6]. [7] and [8] are other identifiable pieces of work, although somehow lacking in thoroughness ([7]) or in formality ([8]).

Approaches to data quality assessment and management through business rule programming are very common, [4] for instance. Several approaches to data constraint pattern specification exist in research efforts associated to OCL, like [1] and [13].

3 Data Structure Quality Rules Defined by Template

The integrity and quality of data structures is highly prone to be business domain specific [4]. The proper way to specify quality conformance rules in such cases is using expressions that capture the semantics of the corresponding business rules. Such expressions may be expressed in a constraint language, like OCL, and may be stored as metadata. CWM's *Expressions* package standardizes a general data model for this purpose. However, we think higher levels of readability and computational efficiency may be achieved if a metamodel more specific to OCL is used. Therefore, we developed an extension to this package that closely follows the OCL metamodel proposed in [12].

Regardless of this considerable tendency to be business specific, there are patterns of quality rules that apply recurrently. Therefore, it seemed to us that an adequate approach to data quality rule specification should be one that, while supporting very specific rules, would also allow the setting up of high level specification concepts that could be applied in a more abstract way. We developed such an approach introducing the concept of rule templates.

3.1 Rule Template Declaration Metaclasses

A **rule template** is an abstract expression that establishes structure conformance rules without specifying the exact classes and structural features (attributes and associations) those rules actually apply. The process of applying such structure conformance rules to specific structural features is called the **instantiation** of the rule template, and this is the process that really produces an actual data quality rule.

Rule Template is the metaclass where rule templates are recorded (Fig. 1). Since rule templates always include an underlying defining expression, *Rule Template* is connected to *Boolean Expression*. *Rule Template* inherits from *ModelElement*, CWM's top most class, which provides some CWM's general features to rule templates, as for instance a *name* attribute and the ability to map to business nomenclature terms and concepts. The metamodel also accepts the definition of new rule templates as a refinement of others, through inheritance.

A rule template always possesses at least one parameter. Template parameters, which are stored in class *Template Parameter*, generically represent the profile of structural features the template may be applied to. Template parameters have a role similar to traditional function parameters in programming languages. However, differently from parameters in programming languages, template parameters within a single rule template are allowed to be structurally interrelated. Our metamodel allows for the establishment of structural relationships among parameters within a single rule template (or within multiple ones related by inheritance) in order to effectively profile the metadata that template(s) may be applied to.

There are two possible types of template parameters, represented by the subclasses *Structural Feature Template* and *Navigation Parameter*.

Structural Feature Template stores parameters that profile individual structural features. On template instantiation, such parameters should be replaced by a single

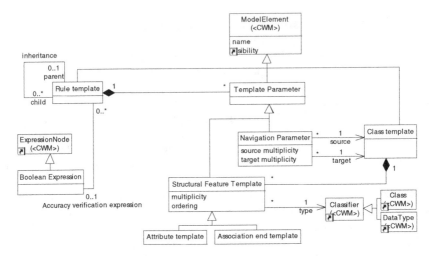

Fig. 1. Rule templates and parameters metaclasses

feature (attribute or association end), which must comply with certain, specified requirements. These may be type requirements and other structural requirements, whose specification is done by means of *Structural Feature Template* combined with *Class Template.*

Structural Feature Template's attributes are used to declare requirements for an actual feature if it is to be assigned to a template parameter on instantiation. If a *structural feature template* metaobject contains the *multiplicity* attribute filled in with value *1..** (for instance) it states a requirement on the multiplicity of an actual feature. Other attributes in that metaobject with assigned values serve as additional constraints on actual features. Contrarily, attributes containing null values mean there are no requirements regarding the aspect they represent. For instance, if *ordering* is null, actual features may be either ordered or not ordered.

Class Template allows to further profile template parameters, namely, it is intended to specify how actual structural features must aggregate into classes. As an example, if two *structural feature templates* are connected to a single *class template* this means the rule template may only be applied to two structural features belonging to the same class in the business data model.

Navigation Parameter stores the second type of template parameters. A navigation parameter may be replaced by any sequence of association ends representing a navigation chain (or join path) between objects of two specified classes. Every *navigation parameter* may be constrained regarding two aspects: the required profiles for the classes at the beginning and at the end of the actual association chain, which is done through *source* and *target* (Fig. 1); and the overall multiplicity profile for the association chain, done by assigning values to the *source multiplicity* and *target multiplicity* attributes.

As a full example, Fig. 2 shows the object diagram of a rule template that has three parameters, corresponding to two attributes and one navigation chain; these elements must comply with the following constraints in the business data model: both attributes

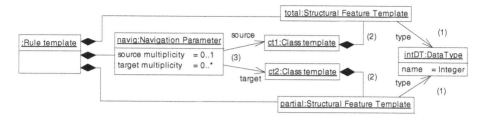

Fig. 2. Example of a rule template and associated parameters

must be of type *Integer* (1); those attributes must belong to different classes (2); those classes must be connected by a one-to-many association chain (3).

3.2 Extending the CWM's *Expressions* Package

Rule templates establish DQ constraints by means of **expression templates**, which are expressions that don't refer to ordinary features, only to template parameters. This leads to the need of extending the CWM's *Expressions* package (Fig. 3).

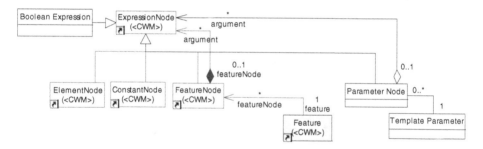

Fig. 3. Extending CWM's *Expression* package towards expression templates

CWM's *Expressions* package allows for the build of tree-like metadata structures for expressions [9]. The novelty in our metamodel is class ***Parameter Node***, which allows building expression node trees that include references to template parameters. A well-formedness constraint states that the expression nodes that belong to an expression template may not include CWM's *FeatureNode* objects. A second constraint states that it must at least include one *parameter node*.

3.3 Instantiating Rule Templates

Actual data quality rules are then specified by instantiating rule templates. The instantiation is performed by storing a new object in class *Rule Defined by Template* , connecting it to the desired *rule template* and then assigning actual structural features to template parameters, which is performed by inserting objects in classes ***Actual Struct Feature*** and ***Actual Navigation Chain*** (Fig. 4). An *actual struct feature* must always be connected to a *StructuralFeature* object, in order to identify the attribute or

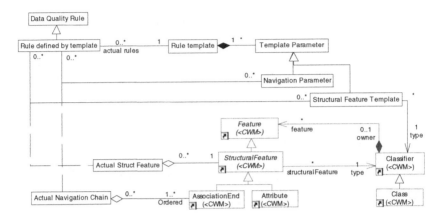

Fig. 4. Metamodel for rule template instantiation

association end that should be assigned to the corresponding parameter, while every *actual navigation chain* must identify a sequence of association ends.

Rule Defined by Template inherits from *Data Quality Rule*, the top most metaclass in our DQ extension to CWM. *Data Quality Rule* provides several DQ related metadata features. This includes, for instance, the ability to specify a weight representing the relevance of every DQ rule, which is important for DQ assessment of a data source, both globally and specifically to individual business processes; also includes the ability to record actual DQ problems found in business data, to log data cleaning transformations, etc. [6].

4   Examples of Rule Templates and Their Instantiation

This section shows some data structure quality rules produced through template instantiation.

Subset. The UML *Subset* constraint among associations ([2], p. 88) is frequently applicable in data models. Although advisable, its enforcement on data entry as well as on data integration is quite frequently absent. Therefore, its verification becomes a data quality issue for the sake of data correctness. Our first example is a specification of *Subset* as a rule template, defined as a constraint between two generic association chains.

Fig. 5 shows a business data model fragment regarding which the *Subset* constraint is advisable. In this database, it is expected that, for every object *e* ∈ *Exam*, the set of students that can be reached joining data through association class *Result* is constrained by the record set that can be reached through the chain *e.course.degree.students*. Specifically, it is expected that the former set of students is a subset of the latter, since a student can only have a result at an exam if he/she is enrolled in the corresponding degree.

This constraint can be specified by the following OCL expression:

```
context Exam inv:
  self.student.forAll (s | self.course.degree.students->includes (s))
```

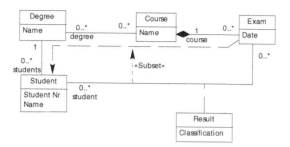

Fig. 5. Business example of a *Subset* rule

This OCL invariant might be expressed through the following rule template:

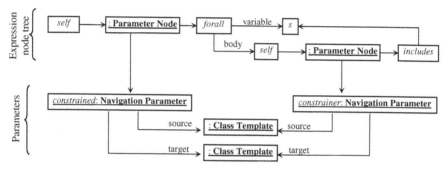

Fig. 6. Expression template and parameters for the *Subset* rule template. For clarity reasons, the expression tree is presented by its logical structure, and not by the real instance diagram.

In the expression node tree, *self* refers to an object belonging to the source class of the corresponding *navigation parameter*.

Such a rule template might then be instantiated as shown by the following object diagram:

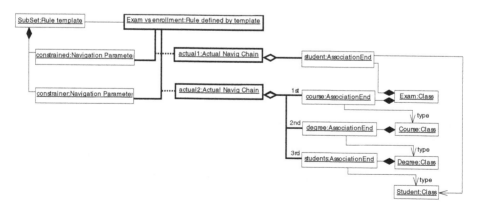

Fig. 7. Instantiation of the *Subset* rule template. The bold lines depict instantiation metadata.

Enclosure. Our second example is a rule template that states that a numeric value in a data object must be enclosed within a numeric range whose highest and lowest values are established by a second object. We decided to name this template *Enclosure*. Regarding the data model fragment in Fig. 8, the following business rule would apply: exam dates must be within the corresponding exam period.

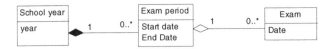

Fig. 8. Case study example of an *Enclosure* rule

This constraint can be specified by the following OCL expression:

```
context Exam inv:
  self.date >= self.examPeriod.startDate and
  self.date <= self.examPeriod.endDate
```

This OCL invariant may be generalized by the following rule template:

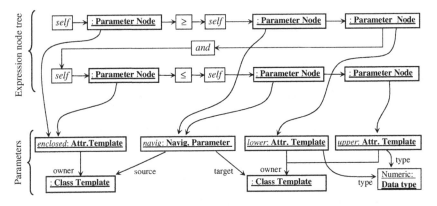

Fig. 9. Metadata instance diagram of the *Enclosure* rule template. For clarity reasons, the expression tree is presented by its logical structure and not by the real instance diagram.

This rule template might then be instantiated as shown by the following object diagram:

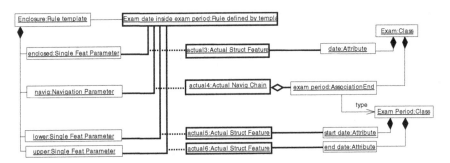

Fig. 10. Instantiation of the *Enclosure* rule template. Bold lines depict instantiation metadata.

5 Conclusions and Future Work

In this paper we presented a metadata model for storing data quality rules regarding structures of linked data records. The concept of rule template was presented as a mean to specify general data quality patterns that occur in a recurrent way. Some rule template application examples were presented in the paper, which were taken from a wider set of cases developed for the purpose of experimentally prove the effectiveness and expressiveness of rule templates.

Although rule templates are a concept with a larger applicability than the focus of this paper, the need for rule templates has been instilled by issues of data structure quality management, because of the need to balance requirements for supporting very specific rules with requirements for providing an highly abstract way of specification.

On going work aims at extending the metamodel for supporting the specification of duplicate detection and merging. Also, the application layer on top of the metamodel is under development, as experimental work that has been done so far was based on an API. Specifically, user interface paradigms for efficient manipulation of metadata are under special focus.

References

1. Ackermann, J., Turowski, V.: A Library of OCL Specification Patterns to Simplify Behavioral Specification of Software Components. In: Dubois, E., Pohl, K. (eds.) CAiSE 2006. LNCS, vol. 4001, pp. 255–269. Springer, Heidelberg (2006)
2. Booch, G., Rumbaugh, J., Jacobson, I.: The Unified Modeling Language User Guide. Addison-Wesley, Reading, MA (1999)
3. Dasu, T., Johnson, T.: Exploratory Data Mining and Data Cleaning. John Wiley & Sons, Chichester (2003)
4. Dasu, T., Vesonder, G., Wright, J.: Data quality through knowledge engineering. In: Proc. 9th ACM SIGKDD (KDD '03), Washington, D.C, pp. 705–710. ACM Press, New York (2003)
5. Galhardas, H., Florescu, D., Shasha, D., Simon, E.: AJAX: An Extensible Data Cleaning Tool. In: Proc. ACM SIGMOD Conf., Dallas, Texas, p. 590 (2000)
6. Gomes, P., Farinha, J., Trigueiros, M.J.: A Data Quality Metamodel Extension to CWM. In: Roddick, J.F., Hinze, A. (eds.) Proc. 4th Asia-Pacific Conference on Conceptual Modelling (APCCM2007), Ballarat, Australia. CRPIT, 67, pp. 17–26. ACS (2007)
7. Kimball, R., Caserta, J.: The Data Warehouse ETL Toolkit. Wiley Publishing, Inc, Chichester (2004)
8. Olson, J.: Data Quality: The Accuracy Dimension. Morgan Kaufman, San Francisco (2003)
9. OMG (ed.): Common Warehouse Metamodel (CWM), Version 1.1, Object Management Group, Inc., (2003) Internet: http://www.omg.org/technology/documents/formal/cwm.htm
10. OMG (ed.): Object Constraint Language Specification, Version 2.0, Object Management Group, Inc., (2006) Internet: http://www.omg.org/technology/documents/formal/ocl.htm
11. Raman, V., Hellerstein, J.: Potter's wheel: An interactive data cleaning system. In: Proc. 27th VLDB, Roma, Italy, pp. 381–390 (2001)

12. Richters, M., Gogolla, M.: A Metamodel for OCL. In: France, R.B., Rumpe, B. (eds.) UML 1999. LNCS, vol. 1723, pp. 156–171. Springer, Heidelberg (1999)
13. Wahler, M., Koehler, J., Bruckner, A.: Model-Driven Constraint Engineering. In: Proc. 6th OCL Workshop on OCL for (Meta-)Models (OCLApps 2006)/ MoDELS Conferences, Genova, Italy (2006)

Automating the Schema Matching Process for Heterogeneous Data Warehouses

Marko Banek[1], Boris Vrdoljak[1], A. Min Tjoa[2],
and Zoran Skočir[1]

[1] Faculty of Electrical Engineering and Computing, University of Zagreb,
Unska 3, HR-10000 Zagreb, Croatia
{marko.banek,boris.vrdoljak,zoran.skocir}@fer.hr
[2] Institute of Software Technology and Interactive Systems,
Vienna University of Technology, Favoritenstr. 9-11/188, A-1040 Wien, Austria
amin@ifs.tuwien.ac.at

Abstract. A federated data warehouse is a logical integration of data warehouses applicable when physical integration is impossible due to privacy policy or legal restrictions. In order to enable the translation of queries in a federated approach, schemas of the federated and the local warehouses must be matched. In this paper we present a procedure that enables the matching process for schema structures specific to the multidimensional model of data warehouses: facts, measures, dimensions, aggregation levels and dimensional attributes. Similarities between warehouse-specific structures are computed by using linguistic and structural comparison, where calculated values are used to create necessary mappings. We present restriction rules and recommendations for aggregation level matching, which builds the most complex part of the process. A software implementation of the entire process is provided in order to perform its verification, as well as to determine the proper selection metric for mapping different multidimensional structures.

1 Introduction

Increasing competitiveness in business and permanent demands for greater efficiency (either in business or government and non-profit organizations) enforce independent organizations to integrate their data warehouses. Data warehouse integration enables a broader knowledge base for decision-support systems, knowledge discovery and data mining than each of the independent warehouses could offer separately. Large corporations integrate their separately developed regional warehouses, newly merged companies integrate their warehouses to enable the business to be run centrally, while independent organizations join their warehouses leading to a significant benefit to all participants and/or their customers.

Copying data from the source heterogeneous warehouses to a new central location would be the easiest way to perform the integration. However, the privacy policy of the organizations, as well as the legal protection of sensitive data in the majority of cases prohibits the data to be copied outside the organizations where they are stored. Hence, a federated data warehouse is created, which implies that the integration must be performed from a logical point of view, using a common conceptual model, while

I.Y. Song, J. Eder, and T.M. Nguyen (Eds.): DaWaK 2007, LNCS 4654, pp. 45–54, 2007.

the heterogeneous source warehouses exist physically. The user of such a federated solution observes the whole federation as a single unit i.e. she must not notice that several heterogeneous parts of the warehouse actually exist. The logical existence of the federated warehouse does not have any impact on the users of local component warehouses [13]. In [3, 12], we developed the conceptual model of a federated data warehouse that unifies data warehouses of different health insurance organizations.

Queries on the federated data warehouse must be translated into sub-queries that correspond to the schemas of local warehouses. Since the multidimensional data model prevails in data warehouse design, it is necessary for the query translation to discover mappings between particular structures of the federated multidimensional conceptual model (facts, measures, dimensions, aggregation levels and dimensional attributes) and their matching counterparts in the multidimensional conceptual models of the local warehouses.

In this paper, the process of creating the mappings between the matching warehouse-specific components will be automated as much as possible in order to shorten the data warehouse integration process as a whole. The contribution of our work is to describe a match discovery procedure for data warehouse-specific structures, with particular emphasis to aggregation levels. We enhance the match discovery techniques for database schemas, enabling their application to data warehouse schemas. Software that automates the integration process is provided and the entire procedure evaluated on an example.

The paper is structured as follows. Section 2 gives an overview of the related work. Basic strategies for matching multidimensional structures are presented in Section 3. Similarity functions for multidimensional structures are explained in Section 4, while the mapping strategies are shown in Section 5. Section 6 evaluates the performance of the algorithm that matches data warehouse schemas automatically. Conclusions are drawn in Section 7.

2 Related Work

There exist two basic approaches to automated matching of database schemas, semi-structured data schemas and ontologies: schema-based and instance-based [10].

Schema-based matching considers only schema information, not instance data. The most prominent approaches are ARTEMIS-MOMIS [1], Cupid [7] and "similarity-flooding" [8, 9]. ARTEMIS/MOMIS and Cupid are based on linguistic matching. Using a word thesaurus, structure names are compared and their similarity is expressed as a probability function. Complex structures are decomposed into substructures where linguistic comparison can be performed and thereafter a formula is used to compute the similarity of complex structures. The "similarity-flooding" algorithm translates database tables or semi-structured sources into graphs whose structures are then compared. Names of the graph vertices are simply compared as strings and no semantic knowledge is required.

Instance-based approaches apply various learning and mining techniques to compare instance data, together with schema metadata (thus being able to outperform schema-based techniques). iMAP [5] is a comprehensive approach for semi-automatic discovery of semantic matches between database schemas, which uses neural networks and text searching methods.

An automated check of match compatibility between data warehouse dimensions is performed in [4], but match candidates must first be proposed manually by the integration designer. To the best of our knowledge, none of the existing frameworks for automated schema matching can be successfully applied to match data warehouses, due to the specific features of the multidimensional conceptual model of data warehouses.

3 Matching Strategy for Solving Heterogeneities Among Multidimensional Structures

Our algorithm aims at automating the matching of data warehouse schemas in cases when the access to data warehouse content is prohibited (e.g. medical data warehouses). Hence, we develop a matching algorithm that is based exclusively on analyzing data warehouse schemas.

3.1 Classification of Schema Heterogeneities

Heterogeneities in data warehouse schemas arise either when (1) different structures (relational, multidimensional etc.) are used to represent the same information, or (2) different specifications (i.e. different interpretations) of the same structure exist. A survey of heterogeneities that arise in "multidatabase systems", i.e. the integration of different relational databases is given in [6]. Attribute conflicts are due to different names and domains and can be divided into one-to-one (see Fig. 1: *city* in table *insurant* corresponding to *municipality* in table *patient* is an example of naming conflict while two *country* attributes with different data types is an example of domain conflict), and -to-many (*street_name* and *street_number* in table *insurant* corresponding to a single attribute *address*). Table conflicts occur when two tables in different databases describe the same information, but different names, data types or constraints are used to represent the information (there is a table conflict between *insurant* and *patient*).

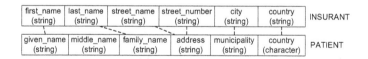

Fig. 1. To compatible tables with different attribute and table conflicts

Heterogeneities that are specific to data warehouses and the multidimensional conceptual model are analyzed in [2]. *Diverse aggregation hierarchies* can manifest either as *the lowest level conflict* or *inner level conflict*. In the first case, two semantically corresponding dimensions (e.g. time dimensions in DWH_1 and DWH_2 in Fig. 2) have different lowest (i.e. basic) grain level (*hour* and *day*) which means that the granularity of their facts is also different. In the second case there is a common aggregation level (not necessarily the lowest), but the aggregation at coarser grain levels takes different ways (levels *month* and *week* in DWH_1 and DWH_2).

The *dimensionality* conflict corresponds to a different number of dimensions associated to the same fact (the *hour* and *time* dimensions in DWH_1 corresponding to a single *time* dimension in DWH_2). Finally, *schema-instance conflicts* appear when some context of the fact in one data warehouse becomes the content (value) of dimensions in the other (*insurance* and *patient* are part of measure names in DWH_1, while being values of *cost_type* dimension in DWH_2).

Fig. 2. Data warehouses with heterogeneities specific to the multidimensional model

3.2 The Matching Algorithm

The matching process determines that two multidimensional structures S_1 and S_2, belonging to data warehouses DWH_1 and DWH_2, respectively, are mutually equivalent. The two structures, S_1 and S_2, may be either be facts, measures, dimensions, aggregation levels or dimensional attributes, but we state that both of them must belong to the same type of multidimensional structures. Mapping cardinalities can be one-to-one, one-to-many or many-to-many. -to-many mappings are more to be expected for attributes and measures than for aggregation levels, dimensions or facts.

The algorithm that automatically matches heterogeneous data warehouse schemas consists of two basic phases: (1) comparison of match target structures and (2) creation of mappings. During the first phase the equivalence of multidimensional structures is determined as the value of a probability function called similarity function. Similarity between multidimensional structures is calculated (using a heuristic algorithm given in Section 4) by comparing their names as well as their data types (for attributes and measures) or substructures (for aggregation levels, dimensions and facts). While the same basic idea is used by ARTEMIS/MOMIS and Cupid, these frameworks cannot solve the heterogeneities that specifically occur in the multidimensional model of data warehouses, especially the hierarchical organization of aggregation levels in dimensions. In the second phase, heuristic rules are applied to determine which structures should be mapped as equivalent, among a much larger number of possible matches. The result of the automated matching process must as much as possible commensurate with the solution that a data warehouse designer would produce manually.

Our matching algorithm recognizes four levels of matching: (1) facts, (2) dimensions and measures, constructing the facts, (3), aggregation levels, constructing the dimensions, and (4) dimensional attributes, constructing the aggregation levels.

Similarity calculation, as the first part of the matching algorithm, starts with atomic structures, dimensional attributes and measures, as their similarity is computed from the similarity of their names and data types. Similarity between aggregation levels is calculated next, taking into account their names and the already calculated similarity between attributes of which they consist. The process continues with dimension similarity, while similarity between facts is determined at the end.

On the other hand, we opine that mapping (the second phase of the automated matching process) must start with the most complex multidimensional structures: the facts. If we determine that two facts are compatible (i.e. that they match), we map their measures and dimensions in the next phase. The mapping candidates are only dimensions and measures of the two facts and no other measures and dimensions in the warehouse schema. Next, we proceed to aggregation levels and then, finally, to attributes.

3.3 Mapping Aggregation Levels

Mapping facts, measures, dimensions and dimensional attributes is similar to matching database structures. For instance, given two corresponding aggregation levels L_1 and L_2, an attribute A_{1i}, which belongs to L_1 can be mapped to any attribute A_{2j} belonging to L_2. Each attribute is mapped to its most similar counterpart according to the value of the similarity function.

Hierarchical structure imposes several inherent limits to aggregation level matching, as the existing partial order must be preserved. The limitations and additional heuristic recommendations are shown in the rest of this section.

Prohibition of mappings that violate the partial order in hierarchies. Let D_1 and D_2 be two mapped dimensions, each of them (for reasons of simplicity) consist of a single hierarchy. Let L_{1i} be an aggregation level in D_1 and L_{2j} an aggregation level in D_2. Let L_{1i} and L_{2j} be equivalent, matching levels and let their mapping be already registered, as shown in the left part of Fig. 3. No level in D_1 representing finer granularity than L_{1i} can be mapped to a level in D_2 representing coarser granularity than L_{2j}. Similarly, no level in D_1 representing coarser granularity than L_{1i} can be mapped to a level in D_2 representing finer granularity than L_{2j}. The mapping between L_{1i} and L_{2j} is represented by a solid line. Invalid mappings are showed as dashed lines. The coherence of the partial orders in dimensions has also been stated as a necessary condition for dimension compatibility in [4].

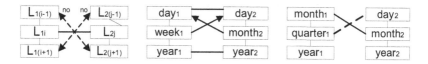

Fig. 3. Preserving the partial order in hierarchies

Mapping unmatched levels to the counterparts of their parent levels. This strategy is used to solve inner level conflicts. Let us suppose that the equivalent levels day_1 and day_2 have already been mapped (see the middle part of Fig. 3). The low value of the similarity function between $week_1$ and $month_2$ proves their incompatibility. Week records in D_1 and month records in D_2 are both aggregations obtained by summing the day records. Thus, $week_1$ can be mapped to the counterpart of its parent (i.e. the nearest finer) level, day_1, which is day_2. Similarly, $month_2$ can be mapped to day_1. When a query on D_1 concerning $week_1$ is translated in order to correspond to D_2, records of the level day_2 need to be summed.

Prohibition of mapping unmatched levels to the counterparts of their descendants.
The rule is not inherent to the multidimensional model like both previous rules (it would
be possible to create valid mappings that are not in accordance with it), but is experi-
ence-based. Mapping L_{1i} to $L_{2(j+1)}$ or $L_{2(j+2)}$, as well as L_{2j} to $L_{1(i+1)}$ or $L_{1(i+2)}$ (see the left
part of Fig. 4) should be forbidden. An explanation can be obtained by watching the
matching process in the other direction. The goal of the mapping process is to map
every level to the coarsest possible counterpart that semantically corresponds to the
target (in order to preserve the richness of the hierarchy). Supposing that $L_{1(i+2)}$ is
equally similar to L_{2j} and $L_{2(j+2)}$ we prefer mapping it to $L_{2(j+2)}$. Back to the initial problem, mapping L_{2j} to $L_{1(i+2)}$, can be viewed in the opposite direction, as mapping $L_{1(i+2)}$ to
L_{2j} (the right part of Fig. 4). $L_{1(i+2)}$ is already mapped to $L_{2(j+2)}$, which is coarser than L_{2j},
so mapping it also to L_{2j} would be redundant and useless.

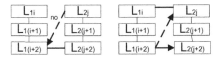

Fig. 4. Mapping unmatched levels to the counterparts of their descendants

4   Similarity Functions for Multidimensional Structures

As stated in Section 3, the process of schema matching starts with calculating similarities between various multidimensional structures and then applying a match selection algorithm based on those similarities.

The basic idea for similarity calculation is the fact that complex structures' similarity is computed from the similarities between their less complex substructures, by
means of some mathematical expression. Therefore, the similarity between any two
complex multidimensional structures of the same kind can recursively be translated
into a set of calculations at the atomic level: the level of dimensional attributes and
measures.

We introduce a new formula for calculating similarity between two attributes or
two measures. Their similarity is obtained from the similarity of their names (*nsim*)
and data type compatibility coefficient *tcoeff* raised to the power determined by a non-
negative real number *texp* (the latter enables us to calibrate the formula):

$$sim_{att}(A_1, A_2) = nsim(A_1, A_2) \cdot tcoeff(A_1, A_2)^{texp}, \qquad texp \in [0, \infty). \qquad (1)$$

Similarity of names is actually the semantic similarity of words that stem from the
name. Semantic similarity functions present the degree of relatedness of two input
words (word senses), which is calculated by using WordNet [14], a large thesaurus of
English language, hand-crafted by psycholinguists. WordNet organizes terms according to human, native speaker's perception, providing a list of synonyms, antonyms
and homonyms, as well as the subordination-superordination hierarchy and part-
whole relations.

We use the semantic similarity calculation method presented by Yang and Powers
[15] to calculate name similarity. Words (word senses) in WordNet are interpreted as
graph vertices, connected by edges representing subordination-superordination and

part-whole relations (each edge is given a weight according to the relation type). All possible paths between two target vertices are constructed and weights multiplied across paths. The highest weight product becomes the linguistic similarity between two target words.

We assume the notion of data type compatibility such that attributes sharing their data type are more related than those that do not. We reduce the data types to four basic ones: *numeric*, *string*, *datetime* and *boolean*. Rarely appearing data types (date, boolean) are more indicative than the frequent string and numeric types, either when their compatibility indicates a higher similarity or when their incompatibility suggest that attributes do not correspond. We empirically determine the following values of *tcoeff* for two compatible basic types: 1.0 for two *datetime* or *boolean* attributes, and 0.9 for two *numeric* or two *string* attributes. Different combinations of incompatible data types imply the following values of *tcoeff*: 0.8 for *numeric-string*, 0.7 for *datetime-boolean* and 0.75 for other combinations.

Similarity between complex multidimensional structures S_1 and S_2 is expressed as a weighted sum (i.e. linear combination) of their name similarity (*nsim*) and structural similarity (*ssim*), as stated in by Madhavan et al. [7]:

$$sim(S_1,S_2) = w_{name} \cdot nsim(S_1,S_2) + (1 - w_{name}) \cdot ssim(S_1,S_2), \qquad w_{name} \in [0, 1]. \qquad (2)$$

In [7] structure similarity is calculated recursively, with some initial values being adjusted in several steps. We use a different approach, adapting the formula for calculating semantic similarity among entity classes of different ontologies [11]. *Neighborhood similarity* between two entity classes not only takes into account their names, but also other classes surrounding them within a certain radius. We therefore make an analogy between aggregation levels (and dimensions and facts, respectively) and entity classes. Attributes (and aggregation levels and dimensions/measures, respectively) surrounding them correspond to neighborhood classes within radius one (each attribute is directly related to the target aggregation level).

5 Mapping Multidimensional Structures

Creation of mappings is the second main phase of the algorithm for automated warehouse schema matching. This process is determined by two basic factors: (1) constraints and (2) selection metrics. While there are no particular constraints to mapping facts, dimensions, measures and dimensional attributes, the aggregation level mapping is restricted by the rules given in Section 3.3, in order to preserve the partial order in hierarchies. A selection metric defines how the calculated similarity values are used to determine which mappings are "better" or "best" among all possible mappings.

The selection metric issue has long been studied in graph theory as the *problem of stable marriage in bipartite graphs*. Bipartite graphs consist of two disjoint parts such that no edge connects any two vertices in the same part (i.e. vertices in the same part are of the same sex, while an edge symbolizes a possible marriage and can thus exists only between vertices in different parts of the graph). Each edge is given a weighting coefficient that corresponds to the similarity between multidimensional structures represented by the vertices. The task of a selection metric (which is also called a filter), is to select an appropriate subset of edges as mappings and eliminate all others. In the original version of the stable marriage problem, only 1:1 mappings (i.e. the monogamous ones) are allowed.

In [9] six different filters were tested. We implemented three of them: *best sum*, *threshold* and *outer*. The best sum filter maps combinations with the highest possible total sum of edge similarities. For the graph given in the left part of Fig. 5, it would produce combinations (a_1,b_2) and (a_2,b_1) as their sum, 1.20, is greater than the sum of the other possible combination (1.04). The threshold and outer filter allow each of the partners to make the choice of their own. First, *relative similarities* are computed as fractions of the absolute similarities of the best match candidates for any given element:

$$sim_R(a,b_j) = sim(a,b_j)/\max_j sim(a,b_j).\qquad(3)$$

Vertices with the highest relative similarity choose first (if there are several such vertices, one of the two vertices that are adjacent to the edge with the highest absolute similarity will take the lead; in particular case a_1 or b_1). The threshold filter creates a mapping if relative similarities at both sides of the selected edge are greater than a threshold. Thus, only mapping (a_1,b_1) is created (the right part of Fig. 5). The outer filter checks the threshold only at the side that "proposed" a match, mapping (a_2,b_2) as the second pair.

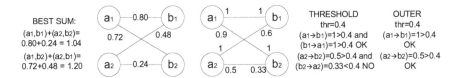

Fig 5. Mapping the same structures differently using the *best sum*, *threshold* and *outer* filter

Polygamous (i.e. 1:N and M:N) matches are allowed for multidimensional structures. The standard best sum, threshold and outer filter can be applied to map facts, dimensions, measures and attributes. Restriction rules to mapping aggregation levels create additional dynamic constraints. If we falsely map level $quarter_1$ to day_2 (see the right part of Fig. 3), we will not be able to produce the correct mapping between $month_1$ and $month_2$.

6 Evaluation of Algorithm Performance

The quality of the automated matching process is measured by its *accuracy*, as defined in [8, 9]. The measure estimates how much effort it costs the user to modify the automatically proposed match result $P=\{(x_1, y_1),...,(x_n, y_n)\}$ into the intended result $I=\{(a_1, b_1), ... ,(a_m, b_m)\}$, i.e. how many additions and deletions of map pairs have to be made. Let $c = |P \cap I|$ be the number of correct suggestions. The difference $(n-c)$ is the number of false positives to be removed from P, and $(m-c)$ the number of missing matches that need to be added. Assuming (for the reason of simplicity) that deletions and additions of match pairs require the same amount of effort, accuracy, i.e. the labor savings obtained by using our matching technique is defined as:

$$1 - [(n - c) + (m - c)] / m. \tag{4}$$

In a perfect match, $n = m = c$, resulting in accuracy 1. Negative accuracy (for $c/n < 0.5$ i.e. when more than half of the created matches are wrong) suggests that it would take the user more effort to correct the automatically proposed map than to perform the matching manually from scratch.

We evaluate the proposed algorithm by matching the schema of the federated data warehouse for health insurance organizations developed in [3, 12] to the local warehouse of one of the organizations. Both warehouses consist of mutually compatible three facts representing patient encounters, performed therapies and medicine prescriptions. In both cases facts share dimensions (many, but not all which are compatible).

The values of match accuracy for each of the three filters are given in Table 1. The outer and threshold filter perform best with relative similarity thresholds between 0.8 and 0.7. For the most critical part of the matching process, aggregation levels, the outer and best sum filter outperform threshold. Threshold outperforms other filters when mapping other multidimensional structures.

Table 1. Comparison of three mapping selection metrics (filters)

	OUTER				THRESHOLD				BEST SUM			
	n	m	c	acc	n	m	c	acc	n	m	c	acc
facts	3	3	3	**1.00**	3	3	3	**1.00**	3	3	3	**1.00**
measures	6	6	6	**1.00**	6	6	6	**1.00**	6	6	6	**1.00**
dimensions	8	7	7	0.86	7	7	7	**1.00**	7	7	7	**1.00**
agg. levels	24	24	22	**0.83**	22	24	20	0.75	24	24	22	**0.83**
dim. attrib.	122	57	51	-0.35	64	57	47	**0.52**	64	57	44	0.42

7 Conclusion

This paper presents an approach to automating the schema matching process for heterogeneous data warehouses in order to shorten the warehouse integration process. The approach is based on analyzing warehouse schemas and can be used when the access to *data content* is restricted. We defined a match discovery procedure capable of solving heterogeneities among data warehouse-specific structures: facts, measures, dimensions, aggregation levels and dimensional attributes. A particular relevance was given to aggregation level matching, as the partial order in dimension hierarchies must be preserved. A software implementation of the approach has been developed and different selection filters tested on a case study example. The best sum and outer filter performed better than threshold when mapping aggregation levels, while the latter was the best solution for other multidimensional structures.

References

1. Bergamaschi, S., Castano, S., Vincini, M.: Semantic Integration of Semistructured and Structured Data Sources. SIGMOD Record 28, 54–59 (1999)
2. Berger, S., Schrefl, M.: Analysing Multi-dimensional Data accross Autonomous Data Warehouses. In: Tjoa, A.M., Trujillo, J. (eds.) DaWaK 2006. LNCS, vol. 4081, pp. 120–133. Springer, Heidelberg (2006)
3. Banek, M., Tjoa, A.M., Stolba, N.: Integrating Different Grain Levels in a Medical Data Warehouse Federation. In: Tjoa, A.M., Trujillo, J. (eds.) DaWaK 2006. LNCS, vol. 4081, pp. 185–194. Springer, Heidelberg (2006)
4. Cabibbo, L., Torlone, R.: Integrating Heterogeneous Multidimensional Databases. In: Proc. Int. Conf. Scientific and Stat. Database Management '05, pp. 205–214, IEEE Comp. Soc. (2005)
5. Dhamankar, R., Lee, Y., Doan, A.-H., Halevy, A.Y., Domingos, P.: iMAP: Discovering Complex Mappings between Database Schemas. In: Proc. SIGMOD Conf. 2004, pp. 383–394. ACM Press, New York (2004)
6. Kim, W., Seo, J.: Classifying Semantic and Data Heterogeneity in Multidatabase Systems. IEEE Computer 24/12, 12–18 (1991)
7. Madhavan, J., Bernstein, P.A., Rahm, E.: Generic Schema Matching with Cupid. In: Proc. Int. Conf. on Very Large Data Bases '01, pp. 49–58. Morgan Kaufmann, San Francisco (2001)
8. Melnik, S., Garcia-Molina, H., Rahm, E.: Similarity Flooding: A Versatile Graph Matching Algorithm and Its Application to Schema Matching. In: Proc. Int. Conf. on Data Engineering 2002, pp. 117–128. IEEE Computer Society, Los Alamitos (2002)
9. Melnik, S., Garcia-Molina, H., Rahm, E.: Similarity Flooding: A Versatile Graph Matching Algorithm. Technical Report (2001), http://dbpubs.stanford.edu/pub/2001-25
10. Rahm, E., Bernstein, P.A.: A survey of approaches to automatic schema matching. VLDB J. 10, 334–350 (2001)
11. Rodríguez, M.A., Egenhofer, M.J.: Determining Semantic Similarity among Entity Classes from Different Ontologies. IEEE Trans. Knowl. Data Eng. 15, 442–456 (2003)
12. Stolba, N., Banek, M., Tjoa, A.M.: The Security Issue of Federated Data Warehouses in the Area of Evidence-Based Medicine. In: Proc. Conf. Availability, Reliability and Security '06, pp. 329–339. IEEE Computer Society, Los Alamitos (2006)
13. Sheth, A.P., Larson, J.A.: Federated Database Systems for Managing Distributed, Heterogeneous, and Autonomous Databases. ACM Computing Surveys 22, 183–236 (1990)
14. Princeton University Cognitive Science Laboratory: WordNet, a lexical database for English Language (Last access: March 25, 2007) http://wordnet.princeton.edu
15. Yang, D., Powers, D.M.W.: Measuring Semantic Similarity in the Taxonomy of WordNet. In: CRPIT 38, pp. 315–322, Australian Computer Society (2005)

A Dynamic View Materialization Scheme for Sequences of Query and Update Statements

Wugang Xu[1], Dimitri Theodoratos[1], Calisto Zuzarte[2],
Xiaoying Wu[1], and Vincent Oria[1,*]

[1] New Jersey Institute of Technology
wx2@njit.edu, dth@cs.njit.edu, xw43@njit.edu, oria@njit.edu
[2] IBM Canada Ltd.
calisto@ca.ibm.com

Abstract. In a data warehouse design context, a set of views is selected for materialization in order to improve the overall performance of a given workload. Typically, the workload is a set of queries and updates. In many applications, the workload statements come in a fixed order. This scenario provides additional opportunities for optimization. Further, it modifies the view selection problem to one where views are materialized dynamically during the workload statement execution and dropped later to free space and prevent unnecessary maintenance overhead. We address the problem of dynamically selecting and dropping views when the input is a sequence of statements in order to minimize their overall execution cost under a space constraint. We model the problem as a shortest path problem in directed acyclic graphs. We then provide a heuristic algorithm that combines the process of finding the candidate set of views and the process of deciding when to create and drop materialized views during the execution of the statements in the workload. Our experimental results show that our approach performs better than previous static and dynamic approaches.

1 Introduction

Data warehousing applications materialize views to improve the performance of workloads of queries and updates. The queries are rewritten and answered using the materialized views [10]. A central issue in this context is the selection of views to materialize in order to optimize a cost function while satisfying a number of constraints [14,2]. The cost function usually reflects the execution cost of the workload statements that is, the cost of evaluating the workload queries using possibly the materialized views and the cost of applying the workload updates to the affected base relations and materialized views. The constraints usually express a restriction on the space available for view materialization [11], or a restriction on the maintenance cost of the materialized views [9], or both [13]. Usually the views are materialized before the execution of the first statement and remain materialized until the last statement is executed. This is the static view selection problem which has been studied extensively during the last decade [11,15,9]. Currently most of the commercial DBMSs (e.g. IBM DB2, MS SQL Server,

---

* Research partially supported by a grant from the Army Research Laboratory.

I.Y. Song, J. Eder, and T.M. Nguyen (Eds.): DaWaK 2007, LNCS 4654, pp. 55–65, 2007.
© Springer-Verlag Berlin Heidelberg 2007

Oracle) provide tools that recommend a set of views to materialize for a given workload of statements based on a static view materialization scheme [18,1,6].

If a materialized view can be created and dropped later during the execution of the workload, we face a dynamic version of the view selection problem. A dynamic view selection problem is more complex than its static counterpart. However, it is also more flexible and can bring more benefit since a materialized view can be dropped to free useful space and to prevent maintenance overhead. When there is no space constraint and there are only queries in the workload, the two view materialization schemes are the same: any view that can bring benefit to a query in the workload is materialized before the execution of the queries and never dropped.

Although in most view selection problems the workload is considered to be unknown or a set of statements without order, there are many applications where the workload forms a sequence of statements. This means that the statements in the workload are executed in a specific order. For example, in a typical data warehousing application, some routine queries are given during the day time of every weekday for daily reports; some analytical queries are given during the weekend for weekly reports; during the night the data warehouse is updated in response to update statements collected during the day. This is, for instance, a case where the workload is a sequence of queries and updates. Such a scenario is shown in Figure 1. The information on the order of the statements in the workload is important in selecting materialized views.

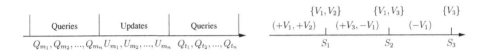

Fig. 1. A workload as a sequence   **Fig. 2.** Output of the problem

The solution to the view selection problem when the input is a sequence of statements can be described using a sequence of create view and drop view commands before the execution of every statement. An alternative representation determines the views that are materialized during the execution of every statement. Figure 2 shows these two representations. $+V_i$ $(-V_i)$ denotes the materialization (de-materialization) of view V_i.

In this paper we address the dynamic view selection problem when the workload is a sequence of query and update statements. This problem is more complex than the static one because we not only need to decide about which views to materialize, but also when to materialize and drop them with respect to the workload statements. Our main contributions are as follows:

1. We exploit the problem as a shortest path problem in a directed acyclic graph (DAG) [4]. Unlike that approach, our approach generate the DAG in a dynamic way. There- fore, it is autonomous in the sense that it does not rely on external modules for constructing the DAG.
2. In order to construct the nodes in the DAG, we extract "maximal" common subex- pressions of queries and/or views. We also produce rewritings of the queries using

these common subexpressions thus avoiding the application of expensive processes that match queries to views.
3. We suggest a heuristic approach that controls the generation of nodes for the DAG based on nodes generated in previous steps.
4. We have implemented our approach and conducted an extensive experimental evaluation. Our results show that our approach performs better than previous static and dynamic approaches.

The next section reviews related work. In Section 3, we formally defined the problem. In Section 4, we present its modeling as a shortest path problem and provide an optimal solution. In Section 5, we show our heuristic approach for the case where no update statements are presented in the workload and in Section 6, we discuss how it can be extended to include also update statements. We present our experiment results in Section 7 and conclude in Section 8.

2   Related Work

In order to solve a view selection problem, one has to determine a search space of candidate views from which a solution view set is selected [16]. The most useful candidate views are the common subexpressions on queries since they can be used to answer more than one query. Common subexpression for pure group-by queries can be determined in a straightforward way [11]. In [8] this class of queries is extended to comprise, in addition, selection and join operations and nesting. In [5] the authors elaborate on how to common subexpressions of select-project-join queries without self-join can be found. This results are extended in [16] to consider also self-joins. Currently, most major commercial DBMSs provide utilities for constructing common subexpressions for queries [3,17]. More importantly, they provide utilities to estimate the cost of evaluating queries using materialized views. The What-If utility in Microsoft SQL Server [1] and the EXPLAIN utility in IBM DB2 [19] are two such examples.

One version of the dynamic view selection problem has been addressed in the past in [7] and [12]. Kotidis et al. [12] show that using a dynamic view management scheme, the solution outperforms the optimal solution of the static scheme. Both approaches focus on decomposing and storing the results of previous queries in order to increase the opportunity of answering subsequent queries partially or completely using these stored results. However, both approaches are different to ours since they assume that the workload of statements is unknown. Therefore, they focus on predicting what views to store in and what views to delete from the cache.

Agrawal et al. [4] consider a problem similar to ours. They model the problem as a shortest path problem for a DAG. However, their approach assumes that the candidate view set from which views are selected for materialization is given. This assumption is problematic in practice because there are too many views to consider. In contrast, our approach assumes that the input to the problem is a sequence of queries and update statements from which it constructs candidate views by identifying common subexpressions among the statements of the sequence.

3   Problem Specification

We assume a sequence $S = (S_1, S_2, ..., S_n)$ of query and update statements is provided as input. The statements in the sequence are to be executed in the order they appear in the sequence. If no views are materialized, the total execution cost of S includes (a) the cost of evaluating all the queries in S over the base relations, and (b) the cost of updating the base relations in response to the updates in S. Let's now assume that we create and materialize a view just before the execution of statement S_i and drop it before the execution of statement S_j. Then, the total execution cost includes (a) the cost of evaluating the queries in $(S_1, ..., S_{i-1})$ and in $(S_j, ..., S_n)$ over the base relations, (b) the cost of materializing view V (i.e., the cost of computing and storing V), (c) the cost of evaluating all queries in $(S_i, ..., S_{j-1})$ *using the materialized view V* (V is used only if it provides some benefit), (d) the cost of updating view V in response to the changes of the base relations resulting by the update statements in $(S_i, ..., S_{j-1})$, and (e) the cost of updating the base relations in response to the updates in S. We ignore here for simplicity the cost of dropping the materialized view V. The total execution cost of S when multiple materialized views are created and dropped on different positions of sequence S is defined in a similar way. Since the cost of updating the base relations in response to updates in S is fixed, we consider it an overhead cost, and we ignore it in the following.

The problem we are addressing in this paper can be now formulated as follows. Given a sequence of n queries and updates $(S_1, S_2, ..., S_n)$ and a space constraint B, find a sequence $(O_1, O_2, ..., O_n)$ of sets of "create view" and "drop view" statements such that (a) the total execution cost of S is minimized and, (b) the space used to materialize views during the execution of S does not exceed B. Each create view statement contains also the definition of the view to be created. The set O_i of create view and drop view statements is executed before the execution of the statement S_i

The output of this dynamic view selection problem can also be described by a sequence of sets of views $C = (C_1, C_2, ..., C_n)$. Each set C_i contains the views that have been created and not dropped before the evaluation of the statement S_i. We call C a solution to the problem. Further, if the views to be materialized can only be selected from a set of candidate views $\mathcal{V}$, we call C a solution to problem for the view set $\mathcal{V}$.

4   Modeling the Dynamic View Selection Problem

In this section, we show how the dynamic view selection problem for a sequence of query and update statements can be modeled as a shortest path problem on a directed acyclic graph. We follow the approach introduced by [4]. That approach assumes that the set of candidate views (i.e. the pool from which we can choose view to materialize) is given. Our approach is different. As we show later in this section, the candidate views are dynamically constructed from the workload statement by considering common subexpressions among them.

We show below how [4] models the problem assuming that a candidate set of views $\mathcal{V}$ is given and contains only one view V. Then, there are only two options for each statement S_i in the workload: either the view V is materialized or not. Those two options are represented as $C_i^0 = \{\}$ and $C_i^1 = \{V\}$ respectively in a solution where C_i is the set of views that are materialized before the execution of S_i. We can now construct a directed acyclic graph (DAG) as follows:

1. For each statement S_i in the workload, create two nodes N_i^0 and N_i^1 which represent the execution of statement S_i without and with the materialized view V respectively. Create also two virtual nodes N_0 and N_{n+1} which represent the state before and after the execution of the workload respectively. Label the node N_i^0 by the empty set $\{\}$ and the node N_i^1 by the set $\{V\}$.
2. Add an edge from each node of S_i to each node of S_{i+1} to represent the change in the set of materialized views. If both the source node and the target node are labeled by the same set, then label the edge by a empty sequence "()". If the source node is labeled by $\{\}$ and the target node is labeled by $\{V\}$, then label the edge by a sequence $(+V)$ to represent the operation "create materialized view V". If the source node is labeled by $\{V\}$ and the target node is labeled by $\{\}$, then label the edge by $(-V)$ to represent the operation "drop materialized view V".
3. Compute the cost of each edge from a node of S_i to a node of S_{i+1} as the sum of: (a) the cost of materializing V, if a $(+V)$ labels the edge, (b) the cost of dropping V, if $(-V)$ labels the edge, and (c) the cost of executing the statement S_{i+1} using the set of views that label the target node of the edge.

Figure 3 shows the DAG constructed the way described above. Each path from the node N_0 to the node N_{n+1} represents a possible execution option for the workload. The shortest path among them represents the optimal solution for the dynamic view selection problem. The labels of the edges in the path denote the solution represented as a sequence of "create view" and "drop view operations". The labels of nodes in the path denote the solution represented as a sequence of sets of materialized views.

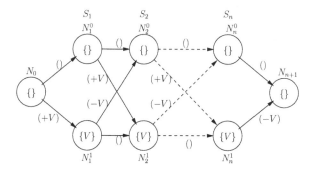

Fig. 3. Directed acyclic graph for candidate view set $\{V\}$

For a candidate set $\mathcal{V}$ of m views, we can construct a DAG in a similar way. Instead of having two nodes for each statement in the workload, we create 2^m nodes, one for each subset of $\mathcal{V}$. Again, the shortest path represents the optimal solution for the dynamic view selection problem.

The shortest path problem for a DAG can be solved by well known algorithms in linear time on the number of edges of the DAG. Therefore the complexity of the process is $O(n \cdot 2^{2m})$ where m is the number of candidate views and n is the number of statements in the workload. To compute the cost of each edge, an optimizer can be used to assess the cost of executing the target statement using the corresponding set of materialized views. This is too expensive even for a small number of candidate views. In practice, the set of candidate views is expected to be of substantial size. The dynamic view selection problem is shown to be NP-hard [4]. Therefore, a heuristic approach must be employed to efficiently solve this problem.

In practice, we cannot assume that the set of candidate views is given. It has to be constructed. Tools that suggest candidate views for a static view selection problem [4] are not appropriate for determining candidate views for a dynamic view selection problem. Our approach constructs candidate views which are common subexpressions of queries in the workload and of views. More specifically, we consider subexpressions of two queries which represent the maximum commonalities of these two queries as these are defined in [16]. An additional advantage of this approach is that the rewritings of the queries using the common subexpressions are computed along with the common subexpressions. These rewritings can be fed to the query optimizer to compute the cost of executing the queries using the materialization of the common subexpressions. If a view V is defined as a common subexpression of a set of queries $\mathcal{Q}$, we call each query $Q \in \mathcal{Q}$ a parent of the V. Every $Q \in \mathcal{Q}$ can be answered using V. We consider only rewritings of a query Q using its common subexpression with other queries and views. A major advantage of this approach is that we do not have to check whether a query matches a view (i.e., check if there is a rewriting of the query using the view) which is in general an expensive process. In the following, we ignore a rewriting of a query using a common subexpression if the cost of evaluating this rewriting is not less than the cost of evaluating the query over base relations.

To solve the dynamic view selection problem, we can generate different views that are common subexpressions of subsets of the queries in the workload. Finding the solution for this problem using the shortest path algorithm is expensive because of the large number of candidate views and the large number of nodes in the DAG. In the next section, we introduce a heuristic approach that combines the process of generating candidate views with the process of selecting views for a solution of the view selection problem.

5 A Heuristic Approach

For the heuristic approach, we start by considering that there are only queries in the workload. In the next section, we discuss how our method can be extended to the case where there are also update statements in the workload.

Our heuristic approach uses two solution merging functions $Merge1$ and $Merge2$. Each function takes as input two solutions to the dynamic view selection problem, each one for a specific set of candidate views, and outputs a new solution.

Consider two solutions $l_1 = (C_1^1, C_2^1, ..., C_n^1)$ and $l_2 = (C_1^2, C_2^2, ..., C_n^2)$, for the candidate view sets $\mathcal{V}_1$ and $\mathcal{V}_2$ respectively. Function $Merge1$ is analogous to function $UnionPair$ [4]. It first creates a DAG as follows: for each statement S_i in the workload, it creates two nodes, one labeled by C_i^1, the other labeled by C_i^2. If the new view set

---

Function 1. $Merge2$

Input: solution l_1, solution l_2, space constraint B
Output: solution l

1: Create a list of solutions $\mathcal{L}$ which initially contains l_1 and l_2
2: **for** each view V_1 in l_1 and each view V_2 in l_2 **do**
3:    Find the set of common subexpressions $\mathcal{V}_{12}$ of V_1 and V_2
4:    **for** each view $V \in \mathcal{V}_{12}$ **do**
5:       Find the solution for the candidate view set $\{V\}$ and add it into $\mathcal{L}$
6:    **end for**
7: **end for**
8: Find the solution l from $\mathcal{L}$ with the lowest execution cost and remove it from $\mathcal{L}$
9: **for all** solutions in $\mathcal{L}$ **do**
10:    Find the solution l' with the lowest execution cost and remove it from $\mathcal{L}$
11:    $l = Merge1(l, l')$
12: **end for**
13: Return the solution l

$C_i = C_i^1 \cup C_i^2$ satisfies the space constraint B, it also creates another node labeled by C_i. In addition it creates two virtual nodes representing the starting and ending states. Finally it creates edges as explained earlier from nodes of query S_i to nodes of query S_{i+1}. Once the DAG is constructed, it returns the solution corresponding to the shortest path in the DAG.

Function $Merge1$ does not add new views to the set of candidate views. Instead, for each statement, it might add one more alternative which is the union of two view sets. In contrast, function $Merge2$ shown in the previous page introduces new views into the set of candidate views.

Our heuristic approach for the dynamic view selection problem is implemented by Algorithm 2 shown above. Algorithm 2 first generates an initial set of candidate views which are common subexpressions of each pair of queries. For each view, it finds a solution. Then, it uses Function $Merge2$ to introduce new views into the set of candidate views.

Algorithm 2. A heuristic algorithm for the dynamic view selection problem

---

Input: list of queries $\mathcal{S}$, space constraint B
Output: solution l

1: Create a set of solutions $\mathcal{L}$ which is initially empty
2: **for** each pair of queries S_i and $S_j \in \mathcal{S}$ **do**
3:    Find the set of common subexpressions $\mathcal{V}_{ij}$ of S_i and S_j
4:    Find the solution for each view set $\mathcal{V} = \{V\}$ where $V \in \mathcal{V}_{ij}$ and add it into $\mathcal{L}$
5: **end for**
6: Find the solution l from $\mathcal{L}$ with the lowest execution cost and remove it
7: **for all** solutions in $\mathcal{L}$ **do**
8:    Find the solution l' lowest execution cost and remove it from $\mathcal{L}$
9:    $l = Merge2(l, l')$
10: **end for**
11: Return the solution l

---

6   Considering Update Statements in the Sequence

In general, an update statement that involves updating more than one base relations can be modeled by a sequence of update statements each of which updates one base relation. For the updates we follow an incremental view maintenance strategy.

Let us assume that a view V contains an occurrence of the base relation R, and an update statement U in the workload updates R. Then, if V is materialized when U is executed, V has to be updated. This incurs a maintenance cost. An optimizer can assess the cost for maintaining U. Roughly speaking, we define the unaffected part of a view V with respect to an update U to be the set of subexpressions of V resulting by removing R from V. If an expression of the unaffected part of a view is materialized or if it can be rewritten using another materialized view, the maintenance cost of V can be greatly reduced. For example, the unaffected part of a view $V = R \bowtie \sigma_{A>10}(S) \bowtie \sigma_{C=0}(T)$ with respect to R is the view $V' = \sigma_{A>10}(S) \bowtie \sigma_{C=0}(T)$. View V can be maintained in response to changes in R either incrementally or through re-computation much more efficiently if V' is materialized.

If there are update statements in the sequence, we still start the heuristic approach with a set of candidate views which are common subexpressions of pairs of queries in the workload sequence. Then, for each view and each update statement, we add to the candidate set of views the unaffected parts of the view with respect to the updates. If two update statements update the same base relation, only one is used to generate the unaffected parts of a view. Finally we apply the heuristic algorithm using the new set of views as a candidate view set.

7   Experimental Evaluation

We implemented the heuristic algorithm for the dynamic view selection problem for a sequence of query and update statements. In order to examine the effectiveness of our algorithm, we also implemented a static view selection algorithm similar to the one presented in [18]. Further, we implemented the greedy heuristic algorithm $GREEDY - SEQ$ presented in [4]. We provide as input to $GREEDY - SEQ$ a set of candidate views recommended by the static view selection algorithm. We consider select-project-join queries. We compute "maximal" common subexpressions of two queries using the concept of closest common derivator as is defined in [16]. In order to deal with update statements, we take into account the unaffected parts of the views with respect to the update statements. We use a cost model that assesses the cost of a query (or an update statement) when this is rewritten completely or partially using one or more materialized views.

We measure the performance of each approach as a percentage using the percentage of the total execution cost of a workload using the set of materialized view suggest by the approach to the execution cost of the same workload without using any view. For each experiment, we consider three schemes, $Static$ (the static view selection approach), $Dynamic$ (the view selection approach presented in this paper), $GREEDY$-SEQ (the algorithm in [4] fed with the result of Static).

First, we study the effect of the space constraint on the three algorithms. We consider two workloads W_1 and W_2 each of which consists of 25 queries (no updates). The

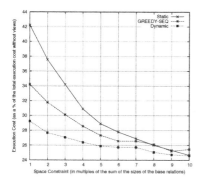

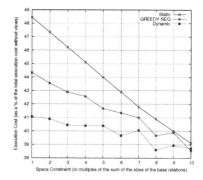

Fig. 4. Performance vs. space constraint, workload W_1

Fig. 5. Performance vs. space constraint, workload W_2

queries in W_1 have more overlapping than that of W_2. The space constraint varies from 1 to 10 times the total size of the base relations. The results are shown in Figures 4 and 5 for W_1 and W_2 respectively. When the space constraint is restrictive, the dynamic view selection schemes have better performance than the static one. This superiority is the result of the capacity of these approaches for creating and dropping materialized views dynamically. As expected, when the space constraint relaxes, all view selection schemes generate similar results. Among the two dynamic view selection approaches, $Dynamic$ performs better than the $GREEDY - SEQ$ algorithm. This shows that a statically selected view set is not appropriate for a dynamic view selection scheme. Our approach does not suffer from this shortcoming since its set of candidate views is constructed dynamically.

Then, we consider the effect of update statements on the three approaches. We consider two series of workloads WS_1 and WS_2, each workload contains the same 25 queries. However we vary the number of update statements in each workload from 1 to 10. Each update statement updates 30% tuples of base relation chosen randomly. In

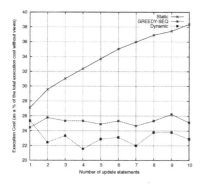

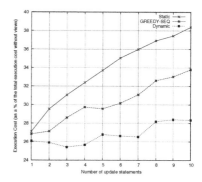

Fig. 6. Performance vs. number of update statements (workload Ws_1)

Fig. 7. Performance vs. space constraint (workload Ws_{1p})

the workloads of WS_1, all the update statements follow all the queries. In the work-loads of WS_2, the update statements are interleaved randomly with the queries. The space constraint is fixed to 10 times the total size of the base relations. The results are shown in Figures 6 and 7 respectively. In all cases, when the number of update state-ments in the workload increases, the dynamic view selection approaches perform better compared to the static one. This is expected since the dynamic algorithms can drop ma-terialized views before the evaluation of update statements and save the maintenance time of these views. The static view selection scheme does not depend on the order of query and update statements in the workload. Thus, for both workload series, the static view selection scheme performs the same. The dynamic view selection scheme depends on the order of query and update statements in the workload. When the up-date statements follow the queries, the dynamic view selection schemes perform better. The reason is that materialized views that are needed for queries are dropped after the queries are executed and therefore do not contribute to the maintenance cost. In any case, the $Dynamic$ outperforms the $GREEDY - SEQ$.

8   Conclusion

We addressed the problem of dynamically creating and dropping materialized views when the workload is a sequence of query and update statements. We modeled it as a shortest path problem in DAGs where the nodes of the DAG are dynamically con-structed by exploiting common subexpressions among the query and update statements in the workload. We designed a heuristic algorithm that combines the process of finding the candidate set of views and the process of deciding when to create and drop materi-alized views during the execution of the statements in the workload. An experimental evaluation of our approach showed that it performs better than previous static and dy-namic ones. We are currently working towards studying alternative algorithms and we are also addressing a similar problem where the input query and update statements form a partial order.

References

1. Agrawal, S., Chaudhuri, S., Kollár, L., Marathe, A., Narasayya, V., Syamala, M.: Database Tuning Advisor for Microsoft SQL Server 2005. In: Proc. of 30th Int. Conf. on VLDB (2004)
2. Yu, S., Atluri, V., Adam, N.R.: Selective View Materialization in a Spatial Data Warehouse. In: Tjoa, A.M., Trujillo, J. (eds.) DaWaK 2005. LNCS, vol. 3589, Springer, Heidelberg (2005)
3. Agrawal, S., Chaudhuri, S., Narasayya, V.R.: Automated Selection of Materialized Views and Indexes in SQL Databases. In: Proc. of 26th VLDB (2000)
4. Agrawal, S., Chu, E., Narasayya, V.R.: Automatic physical design tuning: workload as a sequence. In: Proc. ACM SIGMOD Int. Conf. on Management of Data (2006)
5. Chen, F.-C.F., Dunham, M.H.: Common Subexpression Processing in Multiple-Query Pro-cessing. IEEE Trans. Knowl. Data Eng. 10(3) (1998)
6. Dageville, B., Das, D., Dias, K., Yagoub, K., Zaït, M., Ziauddin, M.: Automatic SQL Tuning in Oracle 10g. In: Proc. of VLDB (2004)

7. Deshpande, P., Ramasamy, K., Shukla, A., Naughton, J.F.: Caching Multidimensional Queries Using Chunks. In: Proc. ACM SIGMOD (1998)
8. Golfarelli, M., Rizzi, S.: View materialization for nested GPSJ queries. In: Proc. Int. Workshop on Design and Management of Data Warehouses (2000)
9. Gupta, H., Mumick, I.S.: Selection of Views to Materialize Under a Maintenance Cost Constraint. In: Proc. 7th Int. Conf. on Database Theory (1999)
10. Halevy, A.Y.: Answering Queries Using Views: A survey. The International Journal on Very Large Data Bases 10(4), 270–294 (2001)
11. Harinarayan, V., Rajaraman, A., Ullman, J.D.: Implementing Data Cubes Efficiently. In: Proc. ACM SIGMOD Int. Conf. on Management of Data (1996)
12. Kotidis, Y., Roussopoulos, N.: DynaMat: A Dynamic View Management System for Data Warehouses. In: Proc. ACM SIGMOD Int. Conf. on Management of Data (1999)
13. Mistry, H., Roy, P., Sudarshan, S., Ramamritham, K.: Materialized View Selection and Maintenance Using Multi-Query Optimization. In: Proc. ACM SIGMOD (2001)
14. Theodoratos, D., Ligoudistianos, S., Sellis, T.K.: View selection for designing the global data warehouse. Data Knowl. Eng. 39(3) (2001)
15. Theodoratos, D., Sellis, T.K.: Data Warehouse Configuration. In: Proc. 23rd Int. Conf. on Very Large Data Bases (1997)
16. Theodoratos, D., Xu, W.: Constructing Search Spaces for Materialized View Selection. In: Proc. ACM 7th Int. Workshop on Data Warehousing and OLAP (2004)
17. Zaharioudakis, M., Cochrane, R., Lapis, G., Pirahesh, H., Urata, M.: Answering Complex SQL Queries Using Automatic Summary Tables. In: Proc. ACM SIGMOD, ACM Press, New York (2000)
18. Zilio, D., Rao, J., Lightstone, S., Lohman, G., Storm, A., Garcia-Arellano, C., Fadden, S.: DB2 Design Advisor: Integrated Automatic Physical Database Design. In: Proc. VLDB (2004)
19. Zilio, D., Zuzarte, C., Lightstone, S., Ma, W., Lohman, G., Cochrane, R., Pirahesh, H., Colby, L., Gryz, J., Alton, E., Liang, D., Valentin, G.: Recommending Materialized Views and Indexes with IBM DB2 Design Advisor. In: Proc. Int. Conf. on Autonomic Computing (2004)

Spatio-temporal Aggregations in Trajectory Data Warehouses

S. Orlando[1], R. Orsini[1], A. Raffaetà[1],
A. Roncato[1], and C. Silvestri[2]

[1] Dipartimento di Informatica - Università Ca' Foscari Venezia
[2] Dipartimento di Informatica e Comunicazione - Università di Milano

Abstract. In this paper we investigate some issues related to the design of a simple Data Warehouse (DW), storing several aggregate measures about trajectories of moving objects. First we discuss the loading phase of our DW which has to deal with overwhelming streams of trajectory observations, possibly produced at different rates, and arriving in an unpredictable and unbounded way. Then, we focus on the measure *presence*, the most complex measure stored in our DW. Such a measure returns the number of trajectories that lie in a spatial region during a given temporal interval. We devise a novel way to compute an approximate, but very accurate, presence aggregate function, which algebraically combines a bounded amount of measures stored in the base cells of the data cube.

1 Introduction

Modern location-aware devices and applications deliver huge quantities of spatio-temporal data concerning moving objects, which must be either quickly processed for real-time applications, like traffic control management, or carefully mined for complex, knowledge discovering tasks. Even if such data usually originate as timed, located observations of well identified objects, they must often be stored in aggregate form, without identification of the corresponding moving objects, either for privacy reasons, or simply for the sheer amount of data which should be kept on-line to perform analytical operations. Such an aggregation is usually a complex task, and prone to the introduction of errors which are amplified by subsequent aggregation operations.

For these reasons, we propose an approach to the problem which is based on classical Data Warehouse (DW) concepts, so that we can adopt a well established and studied data model, as well as the efficient tools and systems already developed for such a model. Our DW is aimed at defining, as basic elements of interest, not the observations of the moving objects, but rather their *trajectories*, so that we can study properties such as *average speed, travelled distance, maximum acceleration, presence of distinct trajectories*. We assume the granularity of the fact table given by a regular three-dimensional grid on the spatial and temporal dimensions, where the *facts* are the set of trajectories which intersect each cell of the grid, and the *measures* are properties related to that set.

I.Y. Song, J. Eder, and T.M. Nguyen (Eds.): DaWaK 2007, LNCS 4654, pp. 66–77, 2007.

One of the main issues to face is the efficient population of the DW from streams of trajectory observations, arriving in an unpredictable and unbounded way, with different rates. The challenge is to exploit a limited amount of buffer memory to store a few incoming past observations, in order to correctly reconstruct the various trajectories, and compute the needed measures for the base cells, reducing as much as possible the approximations.

The model of our DW and the corresponding loading issues have been introduced in [1]. In this paper we discuss in detail the loading and computation of a complex aggregate measure, the *presence*, which can be defined as the number of distinct trajectories lying in a cell. Such a measure poses non trivial computational problems. On one hand, in the loading phase, only a bounded amount of memory can be used for analysing the input streams. This suffices for trajectory reconstruction, but in some cases we may still count an object, with multiple observations in the same cell, more than once. Hence the measure presence computed for base cells is in general an approximation of the exact value. A second approximation is introduced in the roll-up phase. In fact, since the roll-up function for the measure presence is, as *distinct count*, a *holistic* function, it cannot be computed using a finite number of auxiliary measures starting from sub-aggregates. Our proposal is based on an approximate, although very accurate, presence aggregate function, which *algebraically* combines a bounded amount of other sub-aggregate measures, stored in the base cells of the grid.

The paper will provide a thorough analysis of the above mentioned technique, focusing, in particular, on the errors introduced by the approximations. From our tests, our method turned out to yield an error which is sensibly smaller than the one of a recently proposed algorithm [10], based on *sketches*, which are a well known probabilistic counting method in database applications [4].

The rest of the paper is organised as follows. In § 2 we discuss the issues related to the representation of trajectories. In § 3 we recall our trajectory DW model and in § 4 we discuss the loading phase of the measure presence and the aggregate functions supporting the roll-up operation. Then in § 5 the approximation errors of the loading and roll-up phases are studied, both analytically and with the help of suitable tests. Finally, § 6 draws some conclusion.

2   Trajectory Representation

In real-world applications the movements of a spatio-temporal object, i.e. its *trajectory*, is often given by means of a finite set of *observations*. This is a finite subset of points, called *sampling*, taken from the actual continuous trajectory. Fig. 1(a) shows a trajectory of a moving object in a 2D space and a possible sampling, where each point is annotated with the corresponding time-stamp. It is reasonable to expect that observations are taken at irregular rates for each object, and that there is no temporal alignment between the observations of different objects.

Formally, let $\mathcal{TS}$ be a stream of *samplings of 2D trajectories* $\mathcal{TS} = \{T_i\}_{i \in \{1,\dots,n\}}$. Each T_i is the sampling for an object trajectory: $T_i = (ID_i,\ L_i)$, where ID_i is the identifier associated with the object and $L_i = \{L_i^1, L_i^2, \ldots, L_i^{M_i}\}$ is a sequence of *observations*. Each observation $L_i^j = (x_i^j, y_i^j, t_i^j)$ represents the presence of an object at location (x_i^j, y_i^j) and time t_i^j. The observations are temporally ordered, i.e., $t_i^j < t_i^{j+1}$.

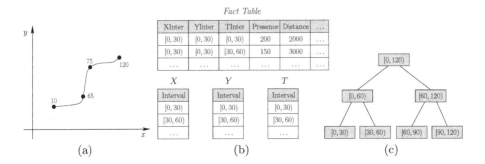

Fig. 1. (a) A 2D trajectory with a sampling, (b) a DW example, and (c) a concept (set-grouping) hierarchy

In many situations, e.g., when one is interested in computing the cumulative number of trajectories in a given area, an (approximate) reconstruction of each trajectory from its sampling is needed. Among the several possible solutions, in this paper we will use *linear local interpolation*, i.e., objects are assumed to move straight between two observed points with constant speed. The linear (local) interpolation seems to be a quite standard approach to the problem (see, for example, [9]), and yields a good trade-off between flexibility and simplicity.

Starting from the stream of samplings, $\mathcal{TS}$, we want to investigate if DW technologies [3,7], and thus the concept of multidimensional and multilevel data model, can be used to store and compute specific aggregate *measures* regarding trajectories.

3   Facts, Dimensions, Measures and Aggregate Functions

We define a DW model by means of a star schema, as simple and generic as possible. The *facts* of our DW are the set of trajectories which intersect each cell of the grid. Some typical properties we want to describe, i.e., the *measures*, are the number of trajectories, their average, maximum/minimum speed, the covered distance. The dimensions of analysis consist of the spatial dimensions X, Y ranging over spatial intervals, and the temporal dimension T ranging over temporal intervals. We assume a regular three-dimensional grid carried out by discretizing the corresponding values of the dimensions and then we associate with them a set-grouping hierarchy. A partial order can thus be defined among groups of values, as illustrated in Fig. 1(c) for a temporal or a spatial dimension.

Note that in the star model of Fig. 1(b), the basic information we represent concerns the set of trajectories intersecting the spatio-temporal cell having for each X, Y, or T dimensions as minimum granularity 30 units, measured in the corresponding unit of measure.

It is interesting to investigate whether some pre-computation must be carried out on the trajectory observations in order to feed the DW. Some measures require little pre-computation, and can be updated in the DW as soon as single observations of the various trajectories arrive, others need all the observations of a trajectory to be received, before updating the DW. In [1] we classified the measures according to an increasing amount of pre-calculation effort. For instance, in order to compute the average, maximum/minimum speed and the covered distance by the trajectories intersecting a cell we need two consecutive observations since we have to infer the trajectory route through interpolation. On the other hand, the measure *number of observations* falling in a cell can be updated directly using each single observation.

We can build the spatial data cube [6] as the lattice of cuboids, where the lowest one (*base cuboid*) references all the dimensions at the primitive abstraction level, while the others are obtained by summarising on different subsets of the dimensions, and at different abstraction levels along the concept hierarchy. In order to denote a component of the base cuboid we will use the term *base cell*, while we will simply use *cell* for a component of a generic cuboid.

In order to summarise the information contained in the base cells, Gray et al. [6] categorise the aggregate functions into three classes based on the space complexity needed for computing a super-aggregate starting from a set of sub-aggregates already provided, e.g., the sub-aggregates associated with the base cells of the DW. The classes are the following:

1. *Distributive.* The super-aggregates can be computed from the sub-aggregates.
2. *Algebraic.* The super-aggregates can be computed from the sub-aggregates together with a *finite* set of auxiliary measures.
3. *Holistic.* The super-aggregates cannot be computed from sub-aggregates, not even using any finite number of auxiliary measures.

For example, the super-aggregates for the *distance* and the *maximum/minimum speed* are simple to compute because the corresponding aggregate functions are *distributive.* In fact once the base cells have been loaded with the exact measure, for the distance we can accumulate such measures by using the function *sum* whereas for maximum/minimum speed we can apply the function *max/min*. The super-aggregate for the *average speed* is algebraic: we need the pair of auxiliary measures $\langle distance, time \rangle$ where *distance* is the distance covered by trajectories in the cell and *time* is the total time spent by trajectories in the cell. For a cell C arising as the union of adjacent cells, the cumulative function performs a component-wise addition, thus producing a pair $\langle distance_f, time_f \rangle$. Then the average speed in C is given by $distance_f / time_f$.

4   The Measure Presence

In this paper we focus on the measure *presence* which returns the number of *distinct* trajectories in a cell. A complication in dealing with such a measure is due to the *distinct count* problem: if an object is observed in the query region for several time-stamps during the query interval, we have to avoid counting it multiple times in the result. This has an impact on both the loading of the base cells and on the definition of aggregate functions able to support the roll-up operation.

4.1   Aggregate Functions

According to the classification presented in § 3, the aggregate function to compute the presence is *holistic*, i.e., it needs the base data to compute the result in all levels of dimensions. Such a kind of function represents a big issue for DW technology, and, in particular, in our context, where the amount of data is huge and unbounded. A common solution consists in computing holistic functions in an *approximate* way.

We propose two alternative and *non-holistic* aggregate functions that *approximate* the exact value of the Presence. These functions only need a small and constant memory size to maintain the information to be associated with each base cell of our DW, from which we can start computing a super-aggregate.

The first aggregate function is distributive, i.e., the super-aggregate can be computed from the sub-aggregate, and it is called $Presence_{Distributive}$. We assume that the only measure associated with each base cell is the exact (or approximate) count of all the *distinct* trajectories intersecting the cell. Therefore, the super-aggregate corresponding to a roll-up operation is simply obtained by summing up all the measures associated with the cells. This is a common approach (exploited, e.g., in [8]) to aggregate spatio-temporal data. However, our experiments will show that this aggregate function may produce a very inexact approximation of the effective *presence*, because the same trajectory might be counted multiple times. This is due to the fact that in the base cell we do not have enough information to perform a *distinct count* when rolling-up.

The second aggregate function is algebraic, i.e., the super-aggregate can be computed from the sub-aggregate together with a *finite* set of auxiliary measures, and it is called $Presence_{Algebraic}$. In this case each base cell stores a tuple of measures. Besides the exact (or approximate) count of all the *distinct* trajectories intersecting the cell, the tuple includes other measures which are used when we compute the super-aggregate. These are helpful to correct the errors, caused by the duplicates, introduced by the function $Presence_{Distributive}$.

More formally, let $C_{x,y,t}$ be a base cell of our cuboid, where x, y, and t identify intervals of the form $[l, u)$, in which the spatial and temporal dimensions are partitioned. The tuple associated with the cell consists of $C_{x,y,t}.presence$, $C_{x,y,t}.crossX$, $C_{x,y,t}.crossY$, and $C_{x,y,t}.crossT$.

- $C_{x,y,t}.presence$ is the number of *distinct* trajectories intersecting the cell.
- $C_{x,y,t}.crossX$ is the number of *distinct* trajectories crossing the *spatial* border between $C_{x,y,t}$ and $C_{x+1,y,t}$.

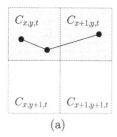
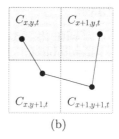

(a)   (b)

Fig. 2. A trajectory (a) that is correctly counted, and (b) that entails duplicates during the roll-up

- $C_{x,y,t}.crossY$ is the number of *distinct* trajectories crossing the *spatial* border between $C_{x,y,t}$ and $C_{x,y+1,t}$.
- $C_{x,y,t}.crossT$ is the number of *distinct* trajectories crossing the *temporal* border between $C_{x,y,t}$ and $C_{x,y,t+1}$.

Let $C_{x',y',t'}$ be a cell consisting of the union of two adjacent cells with respect to a given dimension, namely $C_{x',y',t'} = C_{x,y,t} \cup C_{x+1,y,t}$. In order to compute the super-aggregate corresponding to $C_{x',y',t'}$, we proceed as follows:

$$Presence_{Algebraic}(C_{x,y,t} \cup C_{x+1,y,t}) = C_{x',y',t'}.presence =$$
$$= C_{x,y,t}.presence + C_{x+1,y,t}.presence - C_{x,y,t}.crossX \tag{1}$$

The other measures associated with $C_{x',y',t'}$ can be computed in this way:

$$C_{x',y',t'}.crossX = C_{x+1,y,t}.crossX$$
$$C_{x',y',t'}.crossY = C_{x,y,t}.crossY + C_{x+1,y,t}.crossY$$
$$C_{x',y',t'}.crossT = C_{x,y,t}.crossT + C_{x+1,y,t}.crossT$$

Equation (1) can be thought of as an application of the well known Inclusion/Exclusion principle: $|A \cup B| = |A| + |B| - |A \cap B|$ for all sets A, B. Suppose that the elements included in the sets A and B are just the distinct trajectories intersecting the cells $C_{x,y,t}$ and $C_{x+1,y,t}$, respectively. Hence, their cardinalities $|A|$ and $|B|$ exactly correspond to $C_{x,y,t}.presence$ and $C_{x+1,y,t}.presence$. Then $C_{x,y,t}.crossX$ is intended to approximate $|A \cap B|$, but, notice that, unfortunately, in some cases $C_{x,y,t}.crossX$ is not equal to $|A \cap B|$, and this may introduce errors in the values returned by $Presence_{Algebraic}$. Figure 2(a) shows a trajectory that will be correctly counted, since it crosses the border between the two cells to be rolled-up. Conversely, Fig. 2(b) shows a very agile and fast trajectory, which will be counted twice during the roll-up, since it is not accounted in $C_{x,y,t}.crossX$, even if it should appear in $|A \cap B|$. In fact, the trajectory intersects both $C_{x,y,t}$ and $C_{x+1,y,t}$, but does not cross the border between the two cells.

4.2   Loading Phase

In this section we deal with the problem of feeding the DW, i.e., the base cells of our base cuboid, with suitable sub-aggregate measures, from which the aggregate functions compute the super-aggregates.

We recall that trajectory observations arrive in streams at different rates, in an unpredictable and unbounded way. In order to limit the amount of *buffer memory* needed, it is essential to store information only about *active*, i.e., not ended trajectories. In our simple model of trajectory sampling, since we do not have an end-mark associated with the last observation of a given trajectory, the system module that is responsible for feeding data decides to consider a trajectory as *ended* when for a long time interval no further observation for the object has been received.

In the following we present two options to compute the sub-aggregates stored in each base cell $C_{x,y,t}$, namely $C_{x,y,t}.presence$, $C_{x,y,t}.crossX$, $C_{x,y,t}.crossY$, and $C_{x,y,t}.crossT$, which are used by the aggregate functions $Presence_{Distributive}$ and $Presence_{Algebraic}$.

1. **Single observations.** Considering a single observation at a time, without buffering any past observations, we can only update/increment the measure $C_{x,y,t}.presence$. As a consequence, we cannot compute $Presence_{Algebraic}$, since $C_{x,y,t}.crossX$, $C_{x,y,t}.crossY$ and $C_{x,y,t}.crossT$ are not available.

2. **Pairs of observations.** We consider a pair of observations consisting of the currently received observation L_i^j of trajectory T_i, with the previously buffered L_i^{j-1}. Using this pair of points, we can *linearly interpolate* the trajectory, thus registering further presences for the cells only traversed by T_i.

   Moreover, if we store in the buffer not only L_i^{j-1}, but also the last base cell $C_{x,y,t}$ that was modified on the basis of trajectory T_i, we can reduce the number of duplicates by simply avoiding updating the measure when L_i^j falls into the same $C_{x,y,t}$. It is worth noting that since this method only exploits a very small buffer, it is not able to remove all the possible duplicates from the stored presence measures. Consider, for example, three observations of the same trajectory, all occurring within the same base time interval. If the first and the third points fall into the same cell $C_{x,y,t}$, but the second one falls outside (e.g., into $C_{x,y+1,t}$), this method will store a duplicate count in $C_{x,y,t}.presence$ when the third observation is encountered.

   Finally, by exploiting linear interpolation, we can also identify the cross points of each base cell, and accordingly update the various sub-aggregates $C_{x,y,t}.crossX$, $C_{x,y,t}.crossY$, and $C_{x,y,t}.crossT$.

4.3   Accuracy of the Approximate Aggregate Function

Let us give an intuitive idea of the errors introduced by the aggregate function $Presence_{Algebraic}$, starting from the measures loaded in the base cells by the method *Pairs of observations*. Notice that the overall error can be obtained as the sum of the errors computed for each single trajectory in isolation.

First of all we consider the simplest case: a trajectory is a line segment l. In this case, no roll-up errors are introduced in the computation of presence. In fact, let $C_{x',y',t'}$ be the union of two adjacent cells, $C_{x,y,t}$ and $C_{x+1,y,t}$, i.e., $C_{x',y',t'} = C_{x,y,t} \bigcup C_{x+1,y,t}$, and l intersect $C_{x',y',t'}$. Then by definition:

$$Presence_{Algebraic}(C_{x,y,t} \bigcup C_{x+1,y,t}) =$$
$$C_{x,y,t}.presence + C_{x+1,y,t}.presence - C_{x,y,t}.crossX$$

If l intersects the X border, then correctly $Presence_{Algebraic}(C_{x',y',t'}) = 1+1-1$. Otherwise, $C_{x,y,t}.crossX = 0$, but l intersects only one of the two cells. Hence $Presence_{Algebraic}(C_{x',y',t'}) = 1 + 0 - 0$ or $Presence_{Algebraic}(C_{x',y',t'}) = 0 + 1 - 0$.

This result can be extended to a trajectory which is composed of a set of line segments whose slopes are in the same octant of the three-dimensional coordinate system. Let us call *uni-octant sequence* a maximal sequence of line segments whose slopes are in the same octant. Clearly, every trajectory can be uniquely decomposed in uni-octant sequences and, in the worst case, the error introduced by the aggregate function $Presence_{Algebraic}$ will be the number of uni-octant sequences composing a trajectory.

5   Evaluating Approximate Spatio-temporal Aggregates

In this section we evaluate our method to approximate the measure *presence* and we are also interested in comparing it with a *distributive* approximation recently proposed [10]. The method is based on *sketches*, a traditional technique based on probabilistic counting, and in particular on the FM algorithm devised by Flajolet and Martin [4].

5.1   FM Sketches

FM is a simple, bitmap-based algorithm to estimate the number of distinct items. In particular, each entry in the sketch is a bitmap of length $r = \log UB$, where UB is an upper bound on the number of distinct items. FM requires a hash function h which takes as input an object ID i (in our case a trajectory identifier), and outputs a pseudo-random integer $h(i)$ with a geometric distribution, that is, $Prob[h(i) = v] = 2^{-v}$ for $v \geq 1$. Indeed, h is obtained by combining a uniformly distributed hash function h', and a function ρ that selects the least significant 1-bit in the binary representation of $h'(i)$. The hash function is used to update the r-bit sketch, initially set to 0. For every object i (e.g., for every trajectory observation in our stream), FM thus sets the $h(i)$-th bit. In the most naive and simple formulation, after processing all objects, FM finds the position of the leftmost bit of the sketch that is still equal to 0. If k is this position, then it can be shown that the overall object count n can be approximated with 1.29×2^k.

Unfortunately, this estimation may entail large errors in the count approximation. The workaround proposed in [4] is the adoption of m sketches, which are all updated by exploiting a different and independent hash function. Let

k_1, k_2, ..., k_m be the positions of the leftmost 0-bit in the m sketches. The new estimate of n is 1.29×2^{k_a}, where k_a is the mean of the various k_i values, i.e. $k_a = (1/m) \sum_{i=1}^{m} k_i$. This method reduces the standard error of the estimation, even if it increases the expected processing cost for object insertion. The final method that reduces the expected insertion cost back to O(1) is Probabilistic Counting with Stochastic Averaging (PCSA). The trick is to randomly select, for each object to insert, one of the m sketches. Each sketch thus becomes responsible for approximately n/m (distinct) objects.

An important feature of FM sketches is that they can be merged in a *distributive* way. Suppose that each sketch is updated on the basis of a different set of objects occurrences, and that each object can be observed in more than one of these sets. We can merge a pair of sketches together, in order to get a summary of the number of distinct items seen over the union of both sets of items, by simply taking the *bitwise-OR* of the corresponding bitmaps. We can find a simple application of these FM sketches in our trajectory DW. First, we can exploit a small logarithmic space for each base cuboid, used to store a sketch that approximates the distinct count of trajectories seen in the corresponding spatio-temporal region. The distributive nature of the sketches can then be used to derive the counts at upper levels of granularities, by simply OR-ing the sketches at the lowest granularities.

5.2   Experimental Setup

Datasets. In our experiments we have used several different datasets. Most of them are synthetic ones, generated by the traffic simulator described in [2], but also some real-world sets of trajectories, presented in [5], were used.

Due to space limitations, we present only the results produced for one of the synthetic datasets and for one real dataset. The synthetic dataset was generated using a map of the San Juan road network (US Census code 06077), and contains the trajectories of 10000 distinct objects monitored for 50 time units, with an average trajectory length of 170k space units. The real-world dataset contains the trajectories of 145 school buses. The average number of sample points per trajectory is 455, and, differently from the case of synthetic data, not all points are equally distant in time. Since the setting does not suggest a specific granularity, we define as a standard base granularity for each dimension the average difference of coordinate values between two consecutive points in a trajectory. We will specify granularities as multiples of this base granularity g. For the synthetic dataset $g = (1, 2000, 2000)$ and for the school buses one $g = (2, 100, 100)$, where the granularities of the three dimensions are in the order: t, x, y.

Accuracy assessment. Vitter et al. [11] proposed several error measures. We selected the one based on absolute error, which appears to be the most suited to represent the error on data that will be summed together. Instead of the $\infty-norm$, however, we preferred to use the $1-norm$ as a normalisation term, since it is more restrictive and less sensible to outliers.

The formula we used to compute the normalised absolute error for all the OLAP queries q in a set Q is thus the following:

$$Error = \frac{\left\|\widetilde{M} - M\right\|_1}{\|M\|_1} = \frac{\sum_{q \in Q} |\widetilde{M}_q - M_q|}{\sum_{q \in Q} M_q} \tag{2}$$

where $\widetilde{M}_q$ is the approximate measure computed for query q, while M_q is its exact value. The various queries q may either refer to base cells, or to cells at a coarser level of granularity. In particular, we used this error measure for evaluating the *loading errors* at the level of base cells, and the *roll-up errors* at the level of coarser cells. In all these cases, we always applied Equation (2) to evaluate the errors entailed by a set Q of queries at uniform granularity.

5.3   Experimental Evaluation

First we evaluate the errors in the loading phase of our DW. During this phase, the reconstruction of the trajectories takes place. Figures 3.(a) and 3.(b) show the accuracy of the loaded (sub-aggregate) presence, for different granularities of the base cells. In particular, the two figures illustrate the normalized absolute error as a function of the size of base cells for the two datasets.

Let us focus on the curves labelled *single observations*. In this case we simply fed the DW with each single incoming trajectory observation, without introducing any additional interpolated point. For small granularities, errors are mainly due to the fact that some cells are traversed by trajectories, but no observation falls into them. On the other hand, for very coarse granularities, the main source of errors is the duplicate observations of the same trajectory within the same cell. There exists an intermediate granularity (around 2 *g* for these particular datasets) which represents the best trade-off between the two types of errors, thus producing a small error.

Then consider the curves labelled as *observation pairs*. In this case the presence (sub-aggregate) measure of the base cells of the DW is loaded by exploiting a buffer, which stores information about active trajectories. For each of these trajectories, the buffer only stores the last received observation. This information is used to reduce the errors with respect to the previous feeding method, by adding interpolated points, and also avoiding to record duplicates in the base cells. The resulting normalized absolute error is close to 0 for small granularities, and it remains considerably small for larger values. For very large granularities, a buffer that only stores information about the last previously encountered observation does not suffice to avoid recording duplicates in the base cells.

The other curves in Figures 3(a) and 3(b) represent the error introduced in the base cell by approximating the presence (sub-aggregate) measure by using sketches. It is worth noting that also to load the sketch-base cells, we added further interpolated observations, by using the buffer during the loading phase. Although the error slightly decreases when a larger number of sketches, m, is used for each cell, it remains considerably larger than the curve labelled as *observation pairs*, except for coarse granularities and a large number of sketches ($m = 40$).

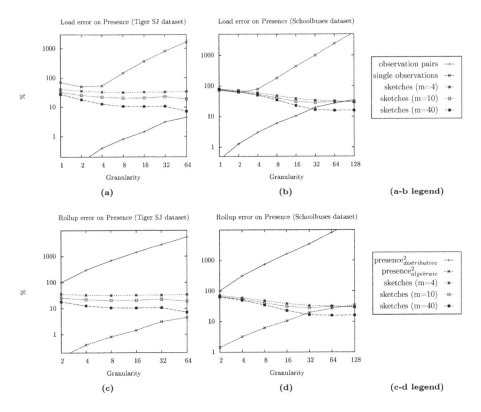

Fig. 3. Errors on different datasets during the load and roll-up phases as a function of the granularity. Granularity is expressed as a multiple of the minimal granularity g for load and of the base granularity g for roll-up (note logarithmic scale for both axes).

The second set of tests, whose results are reported in Figures 3(c) and 3(d), regards the errors introduced by rolling-up our cube, starting from the base cells at granularity g. In all the tests, the error is averaged by submitting many queries at a given granularity $g' > g$, $g' = n \cdot g$. Each query computes the presence measure associated with a distinct coarse-grain cell, obtained by aggregating different adjacent base cells. As an example of query, eight adjacent base cells are aggregated to make one cell of granularity $2 \cdot g$.

We evaluated $Presence_{Distributive}$ and $Presence_{Algebraic}$, assuming that the base cells have been loaded by using pairs of consecutive observations of each trajectory. As shown by the corresponding curves, the *distributive* aggregate function (*sum*) quickly reaches very large errors as the roll-up granularity increases. This is due to the duplicated counting of the same trajectories, when they are already counted by multiple sub-aggregates. Conversely, we obtained very accurate results with our *algebraic* method, with an error always smaller than 10%, for most granularities (except for very coarse ones and 40 sketches).

Finally, the *distributive* function (*or*) used to aggregate sketches produce much larger errors than our method, even because such large errors are already present in the sketches stored in the base cells. However, a nice property of sketches is that the cumulative errors do not increase, and, are actually often reduced when rolling-up, but sketches also exploit much more memory per cell. In case of $m = 40$, the memory usage of the sketch-based method is one order of magnitude larger than our algorithm. Thus it is remarkable that our method achieves a better accuracy in most cases, in spite of this unfavourable bias.

6   Conclusions

In this paper we discussed some relevant issues arising in the design of a DW devised to store measures regarding trajectories. In particular, we focused on the research problems related to storing and aggregating the holistic measure presence. We contributed a novel way to compute such a measure in an approximate, but nevertheless very accurate way. We also compared our approximate function with the sketch based method proposed in [10], showing, with the help of various experiments, that in our case the error is in general much smaller.

Acknowledgements. We thank the researchers involved in the EU GeoPKDD Project for many fruitful discussions and suggestions.

References

1. Braz, F., Orlando, S., Orsini, R., Raffaetà, A., Roncato, A., Silvestri, C.: Approximate Aggregations in Trajectory Data Warehouses. In: STDM Workshop, pp. 536–545 (2007)
2. Brinkhoff, T.: A Framework for Generating Network-Based Moving Objects. GeoInformatica 6(2), 153–180 (2002)
3. Chaudhuri, S., Dayal, U.: An overview of data warehousing and OLAP technology. SIGMOD Record, 30, 65–74 (1997)
4. Flajolet, P., Martin, G.: Probabilistic counting algorithms for data base applications. Journal of Computer and System Sciences 31(2), 182–209 (1985)
5. Frentzos, E., Gratsias, K., Pelekis, N., Theodoridis, Y.: Nearest Neighbor Search on Moving Object Trajectories. In: Bauzer Medeiros, C., Egenhofer, M.J., Bertino, E. (eds.) SSTD 2005. LNCS, vol. 3633, pp. 328–345. Springer, Heidelberg (2005)
6. Gray, J., Chaudhuri, S., Bosworth, A., Layman, A., Reichart, D., Venkatrao, M., Pellow, F., Pirahesh, H.: Data cube: A relational aggregation operator generalizing group-by, cross-tab and sub-totals. DMKD 1(1), 29–54 (1997)
7. Han, J., Chen, Y., Dong, G., Pei, J., Wah, B.W., Wang, J., Cai, Y.D.: Stream Cube: An Architecture for Multi-Dimensional Analysis of Data Streams. Distributed and Parallel Databases 18(2), 173–197 (2005)
8. Papadias, D., Tao, Y., Kalnis, P., Zhang, J.: Indexing Spatio-Temporal Data Warehouses. In: ICDE'02, pp. 166–175 (2002)
9. Pfoser, D., Jensen, C.S., Theodoridis, Y.: Novel Approaches in Query Processing for Moving Object Trajectories. In: VLDB'00, pp. 395–406 (2000)
10. Tao, Y., Kollios, G., Considine, J., Li, F., Papadias, D.: Spatio-Temporal Aggregation Using Sketches. In: ICDE'04, pp. 214–225 (2004)
11. Vitter, J.S., Wang, M., Iyer, B.: Data Cube Approximation and Histograms via Wavelets. In: CIKM'98, pp. 96–104. ACM Press, New York (1998)

Computing Join Aggregates over Private Tables

Rong She[1], Ke Wang[1], Ada Waichee Fu[2], and Yabo Xu[1]

[1] School of Computing Science, Simon Fraser University, Canada
{rshe,wangk,yxu}@cs.sfu.ca
[2] Dept. of Computer Science & Engineering, Chinese University of Hong Kong
adafu@cse.cuhk.edu.hk

Abstract. We propose a privacy-preserving protocol for computing aggregation queries over the join of private tables. In this problem, several parties wish to share aggregated information over the join of their tables, but want to conceal the details that generate such information. The join operation presents a challenge to privacy preservation because it requires matching individual records from private tables. We solve this problem by a novel sketching protocol that securely computes some randomized summary information over private tables. It ensures that during the query computation process, no party will learn other parties' private data, including the individual records and data distributions. Previous works have not provided this level of privacy for such queries.

1 Introduction

In many scientific and business applications, collaborations among various autonomous data holders are necessary in order to obtain global statistics or discover trends. When private sources are involved in such practices, their privacy concerns must be addressed. For example, a hospital keeps a table H of patients' medical histories, and a research institute has a table R of patients' DNA samples. Both tables contain a common attribute "Patient_Name" (not necessarily a key attribute if a patient has several diseases or DNA samples). To establish the relationships between diseases and DNA anomalies, the research institute wants the answer to the following query:

SELECT	H.Disease, R.DNA_Characteristics, COUNT(*)
FROM	H, R
WHERE	H.Patient_Name = R.Patient_Name
GROUP BY	H.Disease, R.DNA_Characteristics

This query returns the number of occurrences for each combination of disease and DNA characteristics, providing helpful insights into their relationships. For example, from $<d,c,70>$ and $<c,100>$, where d is a disease and c is a DNA characteristic, it can be learnt that if a patient has the DNA characteristic c, he/she has the disease d with a 70% chance. While there is need to answer such queries, due to privacy restrictions such as the HIPAA policies (http://www.hhs.gov/ocr/hipaa/), neither party is willing to disclose

I.Y. Song, J. Eder, and T.M. Nguyen (Eds.): DaWaK 2007, LNCS 4654, pp. 78–88, 2007.

local patient-specific information (such as patient names) to the other party. This is an example of *private join aggregate queries* that we want to consider in this paper.

Private Join Aggregate Queries. In general, a *join aggregate query* has the form:

SELECT	group-by-list, *agg(exp)*
FROM	$T_1,...,T_n$
WHERE	$J_1 = J_1'$ and ... and $J_m = J_m'$
GROUP BY	group-by-list

where each T_i is a table; J_i and J_i' are join attributes from different tables; group-by-list is a list of group-by attributes possibly from different tables; *exp* is an arithmetic expression over aggregation attributes; *agg* is an aggregation function. In this paper, we consider the aggregation functions COUNT and SUM. *Conceptually*, the tables in FROM are joined according to the predicates in WHERE, and the joined records are then grouped on the group-by-list. The query result contains one row for each group with *agg(exp)* being computed over the group. If the optional GROUP BY is missing, there is only one group and one row in the result. We can assume that a table contains only join attributes, aggregation attribute and group-by attributes; all other attributes are invisible to the query and thus are safe to be ignored for our purpose.

In a *private join aggregate query*, each T_i is a private table owned by a different party. These parties wish to compute and share the query result as specified in SELECT, but not any other information, including join attribute values and their distributions. This level of privacy was not provided by previous solutions, which will be discussed in more details in Section 2.

We assume the honest-but-curious behavior [12]: The parties follow the protocol properly with the exception that they may keep track of all intermediate computations and received messages, and try to induce additional information. Our focus is on privacy protection in the computation process, not against the answers to queries. Our assumption is that, when the participating parties wish to share the query result, information inferred from such results is a fair game [2]. Protection against query results has been studied in statistical databases [1] and is beyond the scope of this paper.

Our Contributions. We present a novel solution to private join aggregate queries based on the sketching technique previously studied for join size estimation and data stream aggregations [4][7]. None of these works involves privacy issues. The basic idea of sketching is that each table maintains a summary structure, called *atomic sketch*, which is later combined to estimate the query result. A striking property of atomic sketches is that they are computed locally without knowing any data in other tables, which makes them promising for privacy protection. However, we will show that a straightforward way of combining atomic sketches would allow a party to learn the distribution of join values owned by other parties. We will analyze the source of such privacy leakage and determine the types of information that need to be concealed. We then propose a private sketching protocol where each party holds a "random share" of the same atomic sketch so that collectively they represent the atomic sketch, but individually they are useless. We show how the query can be estimated directly from such random shares in a way such that no party learns private information (both individual values and their distributions) from other parties.

Due to space limit, we will mainly discuss our protocol for computing the join aggregate COUNT(*) with two parties. It should be noted that this protocol is extendable to multiple parties and on more general aggregates.

2   Related Work

In general secure computations, the trusted third party model [14] allows all parties to send data to a "trusted" third party who does the computation. Such a third party has to be completely trusted and is difficult or impossible to find. In the secure multi-party model [21], given two parties with inputs x and y, the goal is to compute a function $f(x,y)$ such that the two parties learn only $f(x,y)$ and nothing else. In theory, any multi-party computation can be solved by simulating a combinatorial circuit. However, its communication cost is impractical for data intensive problems.

In [8], privacy-preserving cooperative statistical analysis was studied for vertically or horizontally partitioned data. With vertically partitioned data, n records $\{(x_1,y_1),\ldots, (x_n,y_n)\}$ are distributed at two parties such that Alice holds $\{x_1,\ldots, x_n\}$ and Bob holds $\{y_1,\ldots,y_n\}$. In this case, the join relationship is one-to-one and is implicit by the sequential ordering of records. A similar data partition based on a common key identifier is assumed in [9][19]. In real world, it is odd that the data owned by two mutually un-trusted parties are about the exactly same set of entities. In [3], horizontally partitioned data is considered where each party possesses some records from the same underlying table. In this paper, we consider general join relationships specified by the SQL statement, which can be foreign-key based join or many-to-many join.

Another recent work [10] also discussed aggregations such as SUM queries over several private databases. However, it assumes all parties contain the same pair of attributes: a "key" and a "value" field. The SUM query is to aggregate the values with the same key from all parties. In contrast, we deal with tables with different attributes and general join relationships, where the aggregation is defined over the joined table. The closest work to ours is [2] that studied the private join size problem. It proposed a scheme for encrypting join values but required exchanging the frequency of encrypted values. As noted in [2], if the frequency of some join values is unique, the mapping of the encryption can be discovered by matching the frequency before and after the encryption. So the privacy of join values is compromised. We do not have such problem.

Another related problem is the restriction-based inference control in OLAP queries [20][22]. The goal of inference control is to prevent values of sensitive data from being inferred through answers to OLAP queries. These works mainly dealt with the privacy breaches that arise from the answers to multiple queries; they do not consider the privacy breaches during query processing. The inference control problem has been studied largely in statistical databases, for example, see [1].

3 Preliminaries

First, we review some basic techniques that are the building blocks of our solution.

Sketching. Sketching is a randomized algorithm that estimates the join aggregate result with a random variable, called *sketch*. The sketch is obtained by multiplying some *atomic sketches*, which are computed at each table. The expected value of the sketch is shown to be equal to the aggregate result with bounded variance [4][7].

As an example, consider the query SUM(A) over 3 tables T_1, T_2 and T_3, with join conditions $T_1.J_1=T_2.J_2$ and $T_2.J_3=T_3.J_4$, where A is an aggregation attribute in T_1. The table containing A (in this case, T_1) will be called the *aggregation table*. Each pair of the join attributes (J_1, J_2) or (J_3, J_4) is called a *join pair*. J_2 and J_3 may be the same attribute in T_2, but conceptually they belong to different join pairs and are treated separately. Let D_i denote the domain of J_i. Each join pair shares the same domain. For simplicity, we assume $D_i=\{1,...,|D_i|\}$. For any table T_i, let JS_i be the set of join attributes in T_i. Suppose JS_i contains m join attributes, then a *value instance V* on JS_i is a set that contains one value for each of the m attributes, i.e. $V=\{x_1,...,x_m\}$ where x_i is a value of a distinct join attribute in JS_i. Let $T_i(V)$ be the set of records in T_i having the value instance V on JS_i. For an aggregation table T_i, we define $S_i(V)$ to be the sum of aggregation attribute values over all records in $T_i(V)$; for any non-aggregation table T_i, we define $F_i(V)$ to be the number of records in $T_i(V)$. Thus, T_1 will have $S_1(V)$ defined; T_2 and T_3 will have $F_2(V)$ and $F_3(V)$ defined. The sketch is constructed as follows:

The ε family: For each join pair (J_i, J_j), select a family of 4-wise independent binary random variables $\{\varepsilon_k, k=1,...,|D_i|\}$, with each $\varepsilon_k \in \{1,-1\}$. That is, each join value k is associated with a variable ε_k whose value is randomly selected from $\{1,-1\}$ and any 4 tuple of such ε variables is jointly independent. The set of values for all ε_k variables is called a ε family. In this example, there are two independent ε families, one for each join pair. In table T_i, with a join value instance V, for each join value x in V (x is a value of some join attribute J), there is one ε_x variable from J's ε family. Let $E_i(V)= \Pi_{x \in V} \varepsilon_x$.

Atomic sketches: There is one atomic sketch for each table. For the aggregation table T_1, its atomic sketch $X_1=\Sigma_V[S_1(V)\times E_1(V)]$, i.e. the sum of $S_1(V)\times E_1(V)$ over all distinct V in T_1, called *S-atomic sketch* (S for summary); the atomic sketch for T_2 and T_3 is $X_2= \Sigma_V[F_2(V)\times E_2(V)]$ and $X_3=\Sigma_V[F_3(V)\times E_3(V)]$, called *F-atomic sketches* (F for frequency). *The sketch:* The sketch is defined as $\Pi_i(X_i)$, the multiplication of atomic sketches over all tables. The expected value of the sketch can be shown to be equal to SUM(A) with bounded variance. We refer interested readers to [4][7] for details.

Because sketch is a random variable, the above computation must be repeated many times to get a good average. [5] suggests a procedure of boosting where the number of trials is $\alpha\times\beta$. For every α trials, the average of their sketches is computed, resulting in β averages. The final estimator is the median of these β averages. Note the ε families are chosen independently in each trial. We will refer to this process as $\alpha\beta$-boosting. The time complexity of sketching with $\alpha\beta$-boosting is $O(\alpha\times\beta\times\Sigma_i|T_i|)$, where $|T_i|$ denotes the number of records in table T_i. Experiments from previous works and our experiences show it is usually accurate (error rate < 5%) with moderate size of α (~50) and β (~5).

Private Shared Scalar Product Protocol. The *private scalar product* protocol was first discussed in [8]. Given two d-dimensional vectors $\bar{U} = <U_1, ..., U_d>$ and $\bar{V} = <V_1, ..., V_d>$ owned by two honest-but-curious parties, this protocol computes $\bar{U} \times \bar{V}$ such that the two parties obtain no additional knowledge other than $\bar{U} \times \bar{V}$.

In some applications, a scalar product $\bar{U} \times \bar{V}$ is needed as a part of the computation, but the value of $\bar{U} \times \bar{V}$ needs to be concealed. Such problems can be addressed by the *private shared scalar product* (SSP) protocol [11]. Each party obtains a random share of $\bar{U} \times \bar{V}$, denoted as R_1 and R_2, such that $R_1 + R_2 = \bar{U} \times \bar{V}$. R_1 and R_2 are complementary to each other with their sum being $\bar{U} \times \bar{V}$, but the sum is unknown to both parties. The range of R_1 and R_2 can be the real domain, thus it is impossible to guess $\bar{U} \times \bar{V}$ from any single share. Efficient two-party SSP protocols are available with linear complexity [6][8][9]. The multi-party SSP protocol was studied in [11]. In the rest of this paper, we will use $SSP(\bar{V}_1, ..., \bar{V}_k)$ to denote the SSP protocol on input vectors $\bar{V}_1, ..., \bar{V}_k$.

4   Private Sketching Protocol

We illustrate our protocol on the basic join aggregate COUNT(*) over two private tables, i.e. the join size of two tables. Note that our protocol can be extended to other queries. We will first analyze the privacy breaches in the standard sketching process, from which we derive the requirements on the types of information that must be concealed. We then show how to conceal such information using our protocol.

Assume that Alice holds table T_1 and Bob holds T_2 with a common join attribute J. Let D_1 be the set of join values in T_1, D_2 be the set of join values in T_2. Thus, J's active domain $D = D_1 \cup D_2$. First, a ε family for J is selected. Then both parties use this same ε family to compute their atomic sketches. For illustration purposes, assume D has two values v_1 and v_2 ($|D|=2$); there are two variables $\{\varepsilon_1, \varepsilon_2\}$ in the ε family. Let $F_i(v)$ denote the number of records with join value v in table T_i. $F_1(v)$ belongs to Alice and should be concealed from Bob; $F_2(v)$ belongs to Bob and should be concealed from Alice.

The two parties compute their atomic sketches X_i as follows:

Alice (T_1): $X_1 = F_1(v_1) \times \varepsilon_1 + F_1(v_2) \times \varepsilon_2,$ (1)

Bob (T_2): $X_2 = F_2(v_1) \times \varepsilon_1 + F_2(v_2) \times \varepsilon_2.$ (2)

So far, the computation of X_i is done locally, using the shared ε family and locally owned $F_i(v_j)$ values without any privacy problem. Because the ε family is just some random value, knowing it will not lead to any private information about the other party.

Next, the sketch $X_1 \times X_2$ needs to be computed. Suppose Alice sends her atomic sketch X_1 to Bob. Now, Bob knows X_1, X_2 and the ε family. In Equation (1) and (2), with only $F_1(v_1)$ and $F_1(v_2)$ being unknown, Bob can infer some knowledge about $F_1(v_j)$. For example, knowing $\varepsilon_1=1$ and $\varepsilon_2=-1$, if X_1 is positive, Bob knows that v_1 is more frequent than v_2 by a margin of X_1 in T_1. The problem may be less obvious when there are more values in D, however, it still leaks some hints on $F_1(v_j)$. Furthermore, in the $\alpha\beta$-boosting process, the above computation is repeated $\alpha \times \beta$ times and there is one pair of Equation (1) and (2) for *each* of the $\alpha \times \beta$ trials. With the ε family being independently chosen in each trial, each pair of equations provides a new constraint on the unknown $F_1(v_j)$ values. If the number of trials is equal to or greater than |D|, Bob will have a sufficient number of Equation (1) to solve all $F_1(v_j)$ values. Therefore, even for a large domain D, the privacy breach is severe if Bob knows both atomic sketches.

On the other hand, even if Bob only knows his own X_2, given the result of $X_1 \times X_2$, he can easily get X_1. Now, if the individual sketch $X_1 \times X_2$ in each trial is also concealed from Bob, because both parties agree to share the final result which is an average of $X_1 \times X_2$, by comparing the final result with his own X_2's, Bob may still infer some approximate knowledge on X_1. The situation is symmetric with Alice. Therefore, to prevent any inference on other party's $F_i(v_j)$, all atomic sketches should be unknown to all parties. This implies that the ε families should also be concealed from all parties.

Now suppose all atomic sketches X_i and ε families are concealed. If the sketch $X_1 \times X_2$ is known to Bob, Bob may still learn X_1 in some extreme cases. For example, knowing $X_1 \times X_2 = 0$, and $F_2(v_1) = 10$ and $F_2(v_2) = 5$, since $X_2 \neq 0$ for any value of $ε_1$ and $ε_2$, Bob can infer that $X_1 = 0$. Additionally, from Equation (1), $X_1 = 0$ holds only if $F_1(v_1) = F_1(v_2)$ (because $ε_1, ε_2 \in \{1, -1\}$). Consequently, Bob learns that the two join values are equally frequent in T_1. To prevent this, the individual sketch in each trial should also be concealed from all parties. Therefore, the only non-local information that a party is allowed to know is the final query result which will be shared at the end. Because the final result is something that has been averaged over many independent trials, disclosing one average will not let any party infer the individual sketches or underlying atomic sketches. Note that with the current problem definition where parties agree to share the final result, we cannot do better than this.

Since the ε families must be concealed from Alice and Bob, we need a semi-trusted third party [15], called Tim, to generate the ε families. To fulfill its job, Tim must also be an honest-but-curious party who does not collude with Alice or Bob. In real world, finding such a third party is much easier than finding a trusted third party. The protocol must ensure that Tim does not learn private information about Alice or Bob or the final query result, i.e. Tim knows nothing about atomic sketches, sketches or $F_i(v_j)$ values. The only thing Tim knows is the ε families which are just some random variables. On the other hand, Alice owns $F_1(v_j)$ and Bob owns $F_2(v_j)$, both should know nothing about the other party's $F_i(v_j)$, the ε families, atomic sketches or individual sketches.

Information Concealing. Let Y denote an average of sketches over α trials in αβ-boosting. There will be β number of such Y's in total. A protocol satisfying the following requirements is called *IC-conforming*: Alice learns only Y's and local $F_1(v_i)$; Bob learns only Y's and local $F_2(v_i)$; Tim learns only the ε families; atomic sketches X_i and individual sketches $X_1 \times X_2$ are concealed from all parties.

Theorem 1. A IC-conforming protocol conceals $F_i(v)$ from all non-owning parties throughout the computation process.   ∎

Proof: First, Tim knows only the ε families and nothing about $F_i(v)$. Consider Alice and Bob. From IC-conformity, the only non-local knowledge gained by Alice or Bob is the value of Y, which is an average of sketches over α trials. With $α \geq 2$, such an average provides no clue on any individual sketch because each sketch is computed with an independent and random ε family. Even if there is a non-zero chance that Tim chooses the same ε family in all α trials, therefore Y is equal to each individual sketch, Alice or Bob will have no way of knowing it because the ε families are unknown to them.

Alice or Bob knows that Y is an approximation of the query result $F_1(v_1) \times F_2(v_1) + ... + F_1(v_k) \times F_2(v_k)$. However, this approximation alone does not allow any party to solve the other party's $F_i(v)$ because there are many solutions for the unknown $F_i(v)$. It does not help to use different averages Y in $\alpha\beta$-boosting because they are instances of a random variable and do not act as independent constraints.   ∎

4.1 IC-Conforming Protocol for Two-Party COUNT(*) Query

Assume Bob is the querying party who issues the query. The overall process of our protocol is shown below. The $\alpha \times \beta$ trials are divided into β groups, each containing α trials. Each trial has the ε-phase and the S-phase. The ε-phase generates the ε family and the S-phase computes atomic sketches. For each group, the α-phase computes the sketch average over α trials. Finally the β-phase finds the median of the β averages.

```
1. for i=1 to β do
2.       for j=1 to α do
3.             ε-phase;
4.             S-phase;
5.       α-phase;
6. β-phase;
```

ε-**phase.** In this phase Tim generates the ε family. Let D_1 be the set of join values in T_1 (Alice's table); D_2 be the set of join values in T_2 (Bob's table). $D = D_1 \cup D_2$. To generate the ε family, Tim needs $|D|$, $|D_1|$, $|D_2|$ and the correspondence between ε variables and join values. Alice and Bob can hash their join values by a cryptographic hash function H [18]. H is (1) *pre-image resistant*: given a hash value H(v), it is computationally infeasible to find v; (2) *collision resistant*: it is computationally infeasible to find two different inputs v_1 and v_2 with $H(v_1) = H(v_2)$. Industrial-strength cryptographic hash functions with these properties are available [16]. The ε-phase is as follows.

1. Alice and Bob agree on some cryptographic hash function H and locally compute the hashed sets of their join values $S_1 = \{H(v) | v \in D_1\}$ and $S_2 = \{H(v) | v \in D_2\}$ using H.
2. Alice sends S_1 and Bob sends S_2 to Tim.
3. Tim computes $S = S_1 \cup S_2$.
4. Tim assigns a unique ε variable to each value in S, generating a ε family $\vec{E}_1$ for S_1 and a separate ε family $\vec{E}_2$ for S_2.

Security analysis. Alice and Bob do not receive information from any party. With the cryptographic hash function H, Tim is not able to learn original join values from the hashed sets. Since Tim does not know H, it is impossible for Tim to infer whether a join value exists in T_1 or T_2 by enumerating all possible values. What Tim does learn is the domain size $|D_1|$, $|D_2|$, $|D_1 \cup D_2|$ and $|D_1 \cap D_2|$. But they will not help Tim to infer atomic sketches or sketches. Therefore, this phase is IC-conforming.

S-phase. This phase computes the atomic sketches X_1 for T_1 and X_2 for T_2. Let $\vec{F}_i$ be the vector of $F_i(v)$ values where $v \in D_i$, arranged in the same order as in $\vec{E}_i$. X_i is actually the scalar product $\vec{F}_i \times \vec{E}_i$ where $\vec{F}_i$ is owned by T_i and $\vec{E}_i$ is owned by Tim. To conceal $\vec{E}_i$, $\vec{F}_i$ and X_i, the three parties can use SSP protocol to compute X_i.

1. Alice and Tim compute SSP($\vec{E}_1, \vec{F}_1$), where Alice obtains RA and Tim obtains TA, with RA+TA=X_1.
2. Bob and Tim compute SSP($\vec{E}_2, \vec{F}_2$), where Bob obtains RB and Tim obtains TB, with RB+TB=X_2.

Security analysis. The SSP protocol ensures that $\vec{E}_i$, $\vec{F}_i$ and X_i are concealed. Tim obtains two non-complementary random shares of different atomic sketches, which are not useful to infer any atomic sketch. Thus, this phase is IC-conforming.

α-phase. This phase computes the average of sketches for every α trials. The sketch in the jth trial is $X_{1j} \times X_{2j}$, where X_{1j} and X_{2j} are atomic sketches for T_1 and T_2. However, at the end of S-phase, no party knows X_{1j} or X_{2j}; rather, Tim has TA_j and TB_j, Alice has RA_j and Bob has RB_j, such that $X_{1j}=TA_j+RA_j$ and $X_{2j}=TB_j+RB_j$. After α trials, let $\overline{RA}$ be the vector $<RA_1,...,RA_\alpha>$ and let $\overline{RB}, \overline{TA}, \overline{TB}$ be defined analogously. Alice owns $\overline{RA}$, Bob owns $\overline{RB}$, Tim owns $\overline{TA}$ and $\overline{TB}$. The sketch average Y over the α trials is:

$$Y = \frac{\sum_{j=1}^{\alpha}(X_{1j} \times X_{2j})}{\alpha} = \frac{\sum_{j=1}^{\alpha}[(RA_j + TA_j) \times (RB_j + TB_j)]}{\alpha}$$

$$= \frac{\sum_{j=1}^{\alpha}(RA_j \times RB_j + TA_j \times RB_j + RA_j \times TB_j + TA_j \times TB_j)}{\alpha} = \frac{\overrightarrow{RA} \times \overrightarrow{RB} + \overrightarrow{TA} \times \overrightarrow{RB} + \overrightarrow{RA} \times \overrightarrow{TB} + \overrightarrow{TA} \times \overrightarrow{TB}}{\alpha}$$

The numerator is the sum of several scalar products. To compute these scalar products, if we allow the input vectors to be exchanged among parties, a party obtaining both complementary random shares immediately learns the atomic sketch, thereby violating the IC-conformity. Therefore we use the SSP protocol again as follows.

1. Alice and Bob compute SSP($\overline{RA}, \overline{RB}$).
2. Tim and Bob compute SSP($\overline{TA}, \overline{RB}$).
3. Tim and Alice compute SSP($\overline{TB}, \overline{RA}$).
4. Tim computes $\overline{TA} \times \overline{TB}$ (no SSP is needed).
5. Tim sums up all his random shares and $\overline{TA} \times \overline{TB}$, sends the sum to Alice.
6. Alice adds all her random shares to the sum from Tim, forwards it to Bob.
7. Bob adds all his random shares to the sum from Alice, divides it by α. In the end, Bob has the average Y over the α trials.

Security analysis. The SSP protocols ensure that $\overline{RA}, \overline{RB}, \overline{TA}, \overline{TB}$ are concealed from a non-owning party; therefore, no party learns atomic sketches. After SSP computations, a party may obtain several non-complementary random shares. For example, Alice obtains one random share of $\overline{RA} \times \overline{RB}$ and one random share of $\overline{TB} \times \overline{RA}$, which will not help her learn anything. A party may receive a partial sum during sum forwarding. However, each partial sum always contains two or more non-complementary random shares. It is impossible for the receiver to deduce individual contributing random shares from such a sum. Therefore, this phase is IC-conforming.

β-phase. Repeating the α-phase β times would yield the averages $Y_1, ..., Y_\beta$ at Bob. In the β-phase, Bob finds the median of them, which is the final query estimator.

Security analysis. This phase is done entirely by Bob alone and there is no information exchange at all. Thus the level of privacy at all parties is unchanged.

Cost Analysis. Let $|T_i|$ be the number of records in T_i. Let C_H denote the computation cost of one hash operation. Let $C_{SP}(d)$ denote the computation cost and $S(d)$ denote the communication cost for executing the SSP protocol on d-dimensional vectors.

The running time of our protocol is as follows. (1) *ε-phase*: hashing and generating ε families takes $O(C_H \times |D| + \Sigma_i |T_i| + \alpha \times \beta \times |D|)$ time. Note that hashing is done only once for all trials, but the ε family is generated independently in each trial. (2) *S-phase*: computing atomic sketches takes $O(\alpha \times \beta \times \Sigma_i C_{SP}(|D|))$. (3) *α-phase*: computing the β averages takes $O(\beta \times C_{SP}(\alpha))$. (4) *β-phase*: finding the median of β averages takes $O(\beta)$ time.

The communication cost is $2|D|$ for generating ε families, $2\alpha\beta \times S(|D|)$ for computing atomic sketches, $3\beta \times S(\alpha)$ for computing averages.

4.2 Protocol Extensions

Our protocol can be extended to n-party queries. Using only one third party Tim, each party securely computes their atomic sketches with Tim and the sketch averages are computed over n parties using the n-party SSP protocol. Our protocol also works with SUM(A) queries. The only difference is that the table with attribute A will compute S-atomic sketches instead of F-atomic sketches. Such change only affects local computations in that table. We can even handle more general forms of aggregations like SUM(A×B×C) or SUM(A+B+C). Group-by operators can also be handled, where each group can be considered a partition of original tables and there is a sketch for each partition. In addition, our protocol is extendable to perform roll-up/drill-down operations [13], by rolling-up/drilling-down on local random shares. For details on our protocol extensions, please refer to a full version of our paper [17].

5 Experiments

We implemented the two-party protocol on three PCs in a LAN to simulate Alice (T_1), Bob (T_2) and Tim. All PCs have Pentium IV 2.4GHz CPU, 512M RAM and Windows XP. The cryptographic hash function was implemented using QuickHash library 3.0 (http://www.slavasoft.com/quickhash/). We use the SSP protocol in [9]. Tests were done on synthetic datasets with various table sizes and join characteristics. $|D|$ varies from 100 to 10000, $|T_1|$ from 10000 to 1 million, join values follow zipf distribution. T_2 was generated such that for every join value in T_1, B number (1~10) of records are generated in T_2 with the same join value. Thus, $|T_2|=B \times |D|$ and the join size $|T_1 \infty T_2|=B \times |T_1|$. In our experiments, with α ranging from 50 to 300 and β from 5 to 20, the error rate in all runs is no more than 6%. For large α and β, the error is usually less than 2%. This shows that the approximation provided by sketching is sufficient for most data mining applications where the focus is on trends and patterns, instead of exact counts. As analyzed in Section 4, the protocol is very efficient and finishes within seconds in all runs. Please see [17] for more details.

6 Conclusions

We proposed a privacy-preserving protocol for computing join aggregate queries over private tables. The capabilities of computing such queries are essential for collaborative data analysis that involves multiple private sources. By a novel transformation of the sketching technique, we achieve a level of protection not provided by the previous encryption method. The key idea is locally maintaining random shares of atomic sketches that provide no clue on the data owned by other parties.

References

1. Adam, N.R., Wortman, J.C.: Security-control methods for statistical databases. ACM Computing Surveys 21(4), 515–556 (1989)
2. Agrawal, R., Evfimievski, A., Srikant, R.: Information sharing across private databases. In: SIGMOD (2003)
3. Agrawal, R., Srikant, R., Thomas, D.: Privacy preserving OLAP. In: SIGMOD (2005)
4. Alon, N., Gibbons, P.B., Matias, Y., Szegedy, M.: Tracking join and self-join sizes in limited storage. In: PODS (1999)
5. Alon, N., Matias, Y., Szegedy, M.: The space complexity of approximating the frequency moments. In: STOC (1996)
6. Clifton, C., Kantarcioglu, M., Vaidya, J., Lin, X., Zhu, M.Y.: Tools for privacy preserving distributed data mining. In: SIGKDD Explorations (2002)
7. Dobra, A., Garofalakis, M., Gehrke, J., Rastogi, R.: Processing complex aggregate queries over data streams. In: SIGMOD (2002)
8. Du, W., Atallah, M.J.: Privacy-preserving cooperative statistical analysis. In: Computer Security Applications Conference (2001)
9. Du, W., Zhan, Z.: Building decision tree classifier on private data. In: Workshop on Privacy, Security, and Data Mining. In: ICDM (2002)
10. Emekci, F., Agrawal, D., Abbadi, A.E., Gulbeden, A.: Privacy preserving query processing using third parties. In: ICDE (2006)
11. Goethals, B., Laur, S., Lipmaa, H., Mielikainen, T.: On private scalar product computation for privacy-preserving data mining. In: International Conference in Information Security and Cryptology (2004)
12. Goldreich, O.: Secure multi-party computation. Working draft, Version 1.3 (2001)
13. Gray, J., Bosworth, A., Layman, A., Pirahesh, H.: Data cube: a relational aggregation operator generalizing group-by, cross-tab, and sub-totals, ICDE (1996)
14. Jefferies, N., Mitchell, C., Walker, M.: A proposed architecture for trusted third party services. In: Cryptography Policy and Algorithms Conference (1995)
15. Kantarcioglu, M., Vaidya, J.: An architecture for privacy-preserving mining of client information. In: Workshop on Privacy, Security and Data Mining, ICDM (2002)
16. National Institute of Standards and Technology (NIST), Secure hash standard, Federal Information Processing Standards Publication (FIPS). vol. 180(2) (2002)
17. She, R., Wang, K., Fu, A.W., Xu, Y.: Computing join aggregates over private tables. Technical report TR 2007-12, School of Computing Science, Simon Fraser University (2007), http://www.cs.sfu.ca/research/publications/techreports/
18. Stinson, D.R.: Cryptography: theory and practice, 3rd edn. Chapman & Hall/CRC (2006)

19. Vaidya, J.S., Clifton, C.: Privacy preserving association rule mining in vertically partitioned data. In: SIGKDD, pp. 639–644 (2002)
20. Wang, L., Jajodia, S., Wijesekera, D.: Securing OLAP data cubes against privacy breaches. In: IEEE Symposium on Security and Privacy (2004)
21. Yao, A.C.: How to generate and exchange secrets. In: 27th IEEE Symposium FOCS, IEEE Computer Society Press, Los Alamitos (1986)
22. Zhang, N., Zhao, W., Chen, J.: Cardinality-based inference control in OLAP systems: an information theoretical Approach. In: DOLAP (2004)

An Annotation Management System for Multidimensional Databases

Guillaume Cabanac[1], Max Chevalier[1,2], Franck Ravat[1], and Olivier Teste[1]

[1] IRIT (UMR 5505), SIG, 118 route de Narbonne, F-31062 Toulouse cedex 9, France
[2] LGC (ÉA 2043), 129A avenue de Rangueil, BP 67701, F-31077 Toulouse cedex 4, France
{cabanac,chevalier,ravat,teste}@irit.fr

Abstract. This paper deals with an annotation-based decisional system. The decisional system we present is based on multidimensional databases, which are composed of facts and dimensions. The expertise of decision-makers is modelled, shared and stored through annotations. These annotations allow decision-makers to make an active reading and to collaborate with other decision-makers about a common analysis project.

Keywords: Multidimensional Database, Annotation, Expertise Memory.

1 Introduction and Motivations

Multidimensional data analysis consists in manipulations through aggregations of data drawn from various transactional databases. This approach is often based on *multidimensional databases* (MDB). MDB schemas are composed of facts (subjects of analysis) and dimensions (axes of analysis) [11]. Decision-making consists in analysing these multidimensional data. Nevertheless, due to its numeric nature it is difficult to interpret business data. Decision-makers must analyse and interpret data. This expertise, which can be considered as an *immaterial capital*, is very valuable but it is not exploited in traditional multidimensional systems.

As paper annotations convey information between readers [10], we argue that *annotations* may support this immaterial capital on MDB. We consider an annotation as a high value-added component of MDB from the users' point of view. Such components can be used for a *personal use* to remind any information concerning studied data as well as for a *collective use* to share information that makes complex analyses easier. This collective use of annotations would be the base to build an expertise memory that stores previous decisions and commentaries.

This paper addresses the problem of integrating the annotation concept into MDB management systems. Annotations are designed with the objective of assisting decision-makers and making their expertise persistent and reusable.

Related works and discussion. To the best of our knowledge, the integration of annotations in MDB has not been studied yet. The closest works are related to annotation integration in Relational DataBase Management Systems (RDBMS). First, in the DBNotes system [13, 1, 2, 5] zero or several annotations are associated to a relation element. Annotations are transparently propagated along as data is being transformed

I.Y. Song, J. Eder, and T.M. Nguyen (Eds.): DaWaK 2007, LNCS 4654, pp. 89–98, 2007.

(through SQL queries). This annotation system traces the origin and the flow of data. Second, the authors in [6] and [7] specify an annotation oriented data model for the manipulation and the querying of both data and annotations. This model is based on the concept of block to annotate both a single value and a set of values. A prototype, called MONDRIAN, supports this annotation model. Third, similar to the previous systems, the works described in [3] and [4] consist in annotating relational data. DBNotes and MONDRIAN use relational data to express annotations whereas this last work models annotations using eXtensible Markup Language (XML). The model allows users to cross-reference related annotations.

As conceptual structures of a MDB are semantically richer, the outlined works cannot be directly applied to our context.

- Contrary to RDBMS where a unique data structure is used to both store and display data, in our MDB context, the storage structures are more complex and a specific display is required.
- In RDBMS framework, annotations are straightforwardly attached to tuples or cell values [1]. Due to the MDB structures, the annotations must be attached to more complex data; $e.g.$, dimension attributes are organised according to hierarchies and displayed decisional data are often calculated from aggregations.

To annotate MDB we specify a specific model having the following properties:

- An annotation is characterised by a type, an author, and a creation date.
- Annotations can be traced according to an ancestor link. These cross-references are stored to facilitate collaborative work through discussion threads.
- Each annotation is associated to an anchor, which is based on a path expression associating the annotations to the MDB components (structure or value). Thanks to this anchor, annotations can be associated to different data granularities.
- To facilitate user interactions, annotations are defined and displayed through a conceptual view of the MDB where the annotations are transparently propagated and stored into R-OLAP structures.

Paper outline. Section 2 extends the conceptual multidimensional model defined in [11] for integrating annotations. Section 3 describes an R-OLAP system to manage annotations into MDB.

2 An Annotation-Featured Multidimensional Model

2.1 Multidimensional Concepts

The conceptual model we define represents data as a constellation [9] regrouping several subjects of analysis (facts), which are studied according to several axes of analysis (dimensions).

Definition. A *constellation* is defined as (N^C, F^C, D^C, Star^C, **Annotate**^{C}) where N^C is the constellation name, F^C is a set of facts, D^C is a set of dimensions, Star^C: $F^C \rightarrow 2^{D^C}$ associates each fact to its linked dimensions, Annotate^C is a set of *global annotations* of the constellation elements (see section 3).

A dimension reflects information according to which subjects will be analysed. A dimension is composed of parameters organised through one or several hierarchies.

Definition. A *dimension* D^i is defined by $(N^{Di}, A^{Di}, H^{Di}, Ext^{Di})$ where N^{Di} is the dimension name, $A^{Di} = \{a_1,..., a_q\}$ is a set of dimension attributes, $H^{Di} = \{H^{Di}_1,..., H^{Di}_w\}$ is a set of hierarchies, $Ext^{Di} = \{i^{Di}_1,..., i^{Di}_Y\}$ is a set of dimension instances.

Dimension attributes are modelled according to one or several hierarchies.

Definition. A *hierarchy* H^{Di}_j is defined by $(N^{HDi}_j, P^{HDi}_j, WA^{HDi}_j)$ where N^{HDi}_j is the hierarchy name, $P^{HDi}_j = <Id, p_1,... ,p_s ,All>$ is an ordered set of dimension attributes, called parameters, $\forall k \in [1..s], p_k \in A^{Di}$. Each parameter specifies a granularity level of the analysis. The $WA^{HDi}_j : P^{HDi}_j \rightarrow 2^{ADi}$ function associates each parameter to a set of *weak attributes* for adding semantic information to the parameter.

A fact regroups indicators called measures that have to be analysed.

Definition. A *fact* F^i is defined by $(N^{Fi}, M^{Fi}, Ext^{Fi}, IStar^{Fi})$ where N^{Fi} is the fact name, $M^{Fi} = \{f_1(m_1),..., f_p(m_p)\}$ is a set of measures $m_1,..., m_p$ associated to aggregation functions $f_1,..., f_p$, $Ext^{Fi} = \{i^{Fi}_1,..., i^{Fi}_x\}$ is a set of fact instances. $IStar^{Fi}$: $Ext^{Fi} \rightarrow Ext^{Star(Fi)}$ associates each fact instance to its linked dimension instances.

Example. Fig. 1 shows a multidimensional schema that supports analyses about amount orders and ordered product quantities. It is composed of one fact, named *Order*, and three dimensions, named *Time*, *Customer*, and *Product*.

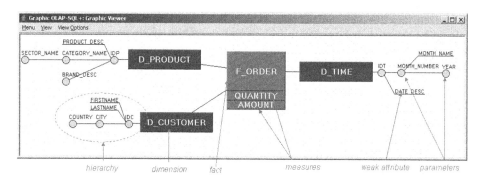

Fig. 1. Example of constellation schema

The visualisation of constellations consists in displaying one fact according to several dimensions into a multidimensional table (MT). A MT is more complex than relations because it is organised according to a non-clear separation between structural aspects and data contents [8].

Definition. A multidimensional table T is defined as $(S^T, L^T, C^T, R^T, \mathbf{Annotate^T})$

- $S^T = (F, \{f_1(m_1), ..., f_p(m_p)\})$ is the subject of analysis, which is represented by the fact and its displayed measures $f_1(m_1),...,f_p(m_p)$,
- $L^T = (DL, HL, PL)$ is the horizontal analysis axis where $PL=<p^{HL}_{max}, ..., p^{HL}_{min}>$ are displayed parameters of $DL \in Star^C(F)$ and $HL \in H^{DL}$ is the *current hierarchy*,
- $C^T = (DC, HC, PC>$ is the vertical analysis axis where $PC=<p^{HC}_{max}, ..., p^{HC}_{min}>$, $HC \in H^{DC}$ and $DC \in Star^C(F)$, HC is the *current hierarchy* of DC,
- $R^T = pred_1 \wedge ... \wedge pred_s$ is a normalised conjunction of predicates.
- $Annotate^T$ is a set of *local annotations* of the MT elements (see section 2.3).

Example. Fig. 2 depicts an MT example that displays amount orders according to the temporal axis and the customer axis. Note that a MT represents an excerpt of data recorded in a constellation. $T_1 = (S_1, L_1, C_1, R_1, \varnothing)$ with S_1=(F_ORDER, {SUM(Amount)}), L_1=(D_TIME, HTPS, <All, YEAR, MONTH_NUMBER>), C_1=(D_CUSTOMER, HGEO, <All, COUNTRY, CITY>) and R_1=true.

			D_CUSTOMER IH_GEO		
			COUNTRY	France		United Kingdom
	F_ORDER SUM (AMOUNT)		CITY	Paris	Toulouse	London
		YEAR	MONTH_NU...			
D_TIME IHTPS			4	(10697)	(2693)	
		2007	5	(2100)	(3868)	
			6	(3868)	(4293)	

Fig. 2. Example of MT

2.2 An Integrated Annotation Model

In order to annotate a MDB, we provide a specific annotation model that is directly integrated into the multidimensional model. The proposed annotation model is collaboration-oriented. It provides functionalities that allow users to share information and to discuss directly in context during the data analysis. Therefore, an annotation can be defined at two specific levels of the MDB.

– A *global* annotation concerns a conceptual component of a constellation (fact, measure, dimension, hierarchy, parameter, weak attribute). Such annotations are displayed in any MT including the globally annotated element.
– A *local* annotation concerns a specific component of a given MT. These annotations are only displayed in a specific context corresponding to a MT.

As for paper, an MDB annotation is twofold; it consists in:

– *subjective* information that corresponds to its content (e.g. a text typed by decision-makers) and at least one type to understand easier without having to read its content. We define some specific types: a commentary, a question, an answer to an existing annotation, a conclusion…
– *objective* data (also called meta-data) that corresponds to its unique identifier, its creation date, its creator identifier, a link to the parent annotation (when answering to another annotation) and an anchor to annotated data.

The system automatically generates the set of objective data whereas the annotation creator formulates the set of subjective information.

During an analysis, decision makers visualize synthesized data through MT. These MT can be manipulated by the use of commands defined by the related algebra [11]. Thus annotations should follow the MT changes. That is why anchors cannot be specified with a coordinate-based system. That is why we define a unique anchoring structure. This later relies on a path-like notation that can be generalized to any MT. An anchor attaches both local and global annotations.

In next definitions, ^ corresponds to an empty path. Let us consider *CONS* as a constellation, *MT* as a multidimensional table, *Fact* as a fact, *measure* as a measure, *f(m)* as an aggregation function applied to a measure *m*, *val* as a specific measure value, *Dim* as a dimension, *Hier* as a hierarchy, *param* as a dimension attribute (parameter or weak attribute) and *valueP* as a value of a dimension attribute.

Definition. An anchor is defined as (S, D1, D2) where

- S = ^ | {CONS | MT}[.Fact[[/measure | /f(m)][=val]?]*]? designates a path to any fact or measure used in a constellation or in a MT,
- D_1 = ^ | Dim[.Hier[/param[=valueP]?]*]? designates a path concerning the a first dimension (row or column of the MT),
- D_2 = ^ | Dim[.Hier[/param[=valueP]?]*]? designates a path concerning the second dimension (row or column of the MT).

If the two dimensions D_1 and D_2 are given, the system is able to identify a specific cell in the MT. Thanks to this anchoring structure and to the different combinations of values that it allows, annotations can be easily stored in the MDB be retrieved and displayed in a specific MT for instance.

Example. In Fig. 3, two users annotate the constellation C1 and a multidimensional table MT1. The first user, noted U1, annotates MT1 with A7, A8, A9, and A10. The annotation A9 is a question. This is the root of a discussion thread. The second user, noted U2, has annotated C1 with the annotations from A1 to A6. He also answers to A9 through the annotation A11. The anchor for each annotation is:

- *A1:* (^, D_CUSTOMER, ^) or (^, ^, D_CUSTOMER). This anchor supposes that the annotation concerns the D_CUSTOMER dimension. The annotation is displayed every time D_CUSTOMER is used. To associate this annotation to this dimension when it is only linked to the fact F_ORDER the anchor should be transformed in (C1.F_ORDER, D_CUSTOMER, ^).
- *A2:* (^,D_CUSTOMER.HGEO, ^) or (^, ^, D_CUSTOMER.HGEO).
- *A3:* (^, D_CUSTOMER.HGEO/YEAR, ^) or (^, ^, D_CUSTOMER.HGEO/YEAR).
- *A4:* (^,D_CUSTOMER.HGEO/YEAR/DATE_DESC, ^).
- *A5:* (C1.F_ORDER, ^, ^).
- *A6:* (C1.F_ORDER/AMOUNT, ^, ^).
- *A7:* (MT1, D_TIME.HTPS/YEAR='2007', ^).
- *A8:* (MT1, D_CUSTOMER.HGEO/COUNTRY='France'/CITY='Toulouse', ^).
- *A9:*   (MT1.F_ORDER/SUM(AMOUNT),   D_TIME.HTPS/YEAR   =   '2007'/ MONTH_NUMBER = '6', D_CUSTOMER.HGEO/COUNTRY = 'France'/CITY = 'Toulouse'). To annotate one value corresponding to a specific measure of this cell, the annotation should be transformed in: (MT1.F_ORDER/ AMOUNT.SUM = '4293',   D_TIME.HTPS/YEAR   =   '2007'/MONTH_NUMBER   =   '6', D_CUSTOMER.HGEO/ COUNTRY = 'France'/CITY = 'Toulouse')
- *A10:* (MT1, D_TIME.HTPS/YEAR='2007', ^).
- *A11:* The path of A11 is identical to A9, only its content is different.

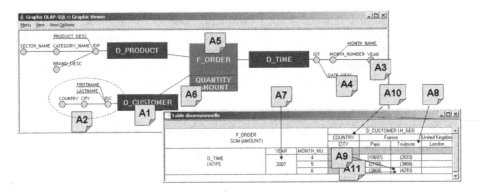

Fig. 3. Example of annotations on a MT as well as on the MDB schema

3 R-OLAP Implementation

As mentioned in Fig. 4, the architecture of our annotation management system is composed of three main modules.

- The display interfaces (GUI) allow decision-makers (1) to annotate the constellation schema and the MT via global and local annotations, (2) to display analyses through annotated MT.
- The query engine translates user interactions into SQL queries. Correctness of query expressions is validated through meta-data. These SQL queries are sent to the databases; results are sent back to the GUI.
- The R-OLAP data warehouse is an RDBMS storing multidimensional data, meta-data and annotations.

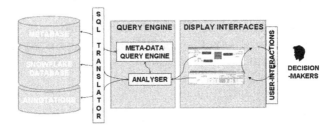

Fig. 4. Threefold annotation management system for MDB s architecture

3.1 Metabase

Constellations are implemented in an R-OLAP context. To store the multidimensional structures, we have defined meta-tables that describe the constellation (META_FACT, META_DIMENSION, META_HIERARCHY…). For example, Fig. 5 describes the constellation structure illustrated in Fig. 1.

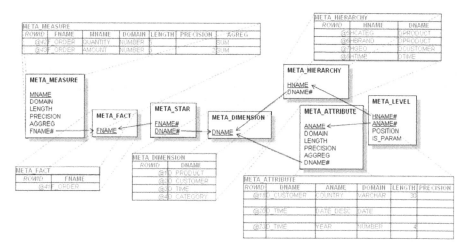

Fig. 5. Metabase for storing a constellation

3.2 Snowflake Database

An important challenge for storing annotations is the implementation of anchors. To associate each annotation to a unique row in the R-OLAP database, we adopt a snow-flake data schema [9]. It consists in normalising dimensions according to hierarchies to eliminate redundancy; the annotation anchors designate a unique data.

Example. Fig. 6 shows the R-OLAP implementation of the constellation illustrated in Fig. 1 according to a snowflake modelling.

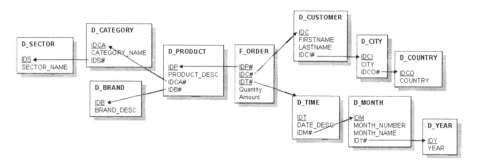

Fig. 6. R-OLAP snowflake schema

Note that these tables of the snowflake schema must be completed with pre-aggregated tables for improving query performances. Moreover, as argued in [1], adding and propagating annotations in RDBMS must drop down performances. Their experimentation results show that for larger databases (500MB and 1GB), the queries integrating annotations took only about 18% more time to execute than their corresponding SQL queries.

3.3 Annotation Storage

We have to provide a mechanism for storing global and local annotations into the same structure. The main problem consists in implementing the formal anchoring technique meanwhile providing a homogeneous way to manage the annotations such that some annotations may be anchored to detailed data, aggregated data or meta-data.

Our solution consists in storing annotations into a single table whose schema is composed of the following columns:

- PK is the annotation identifier,
- NTABLE is the table or a meta-table where the annotated data is stored,
- ROW is an internal row identifier used in the database system related to the annotated data of the NTABLE,
- COL stores the attribute name of annotated data. If the annotation is anchored to the multidimensional structure, it is anchored to a row in a meta-table (COL is null) whereas if the annotation is associated to a value, COL is valued.
- DESC stores the annotation content.
- LOCAL represents the annotation scope. When the annotation is local to a MT, then this attribute is valued.
- TYPE describes the annotation type (comment, question, answer…).
- DATE stores the creation date of the annotation.
- PARENT represents a relationship between annotations. The attribute is used to store a discussion thread (for example, an answer following a question).
- AUTHOR is the author of the annotation.

Example. The following table (Fig. 7) stores annotations defined in section 2.2.

ANNOTATION									
PK	NTABLE	ROW	COL	DESC	LOCAL	TYPE	DATE	PARENT	AUTHOR
1	META_DIMENSION	@2		A1		Comment	02/04/2007		U2
2	META_HIERARCHY	@7		A2		Comment	02/04/2007		U2
3	META_ATTRIBUTE	@23		A3		Comment	02/04/2007		U2
4	META_ATTRIBUTE	@20		A4		Comment	02/04/2007		U2
5	META_FACT	@41		A5		Comment	02/04/2007		U2
6	META_MEASURE	@43		A6		Comment	02/04/2007		U2
7	D_YEAR	@100	YEAR	A7		Comment	02/04/2007		U1
8	D_CITY	@101	CITY	A8	V1	Question	02/04/2007		U1
9	MV1	@200	AMOUNT	A9	V1	Comment	02/04/2007		U1
10	META_ATTRIBUTE	@18		A10	V1	Comment	02/04/2007		U1
11	MV1	@200	AMOUNT	A11	V1	Answer	03/04/2007	9	U2

Fig. 7. Storage of the annotations

These annotations are anchored to three levels.

- Annotations A1 to A6 as well as A10 are associated to the **meta-data tables**; e.g. A1, stored in the 1st row and conceptually noted (^, D_CUSTOMER, ^), is anchored to the row identified by @2 into META_DIMENSION table.
- The global annotation A7 and the local annotation A8 are anchored to **detailed values** (of parameters). The attributes named ROW and COL are used to locate these annotated data. In Fig. 7 we assume that D_YEAR and D_CITY contain respectively the rows [@100, y1, 2007] and [@101, ci1, Toulouse, co1].

- The annotations A9 and A11 are anchored to **aggregated values** of the measure AMOUNT. In order to define the anchor, aggregated data must be materialised. The MT is calculated from the following SQL query, noted V1:

```
SELECT year, month_number, country, city, SUM(amount) AS amount
FROM   F_ORDER  or, D_CUSTOMER  cu, D_CITY  ci, D_COUNTRY  co, D_TIME
ti, D_MONTH mo, D_YEAR ye
WHERE or.idc=cu.idc AND cu.idci=ci.idci AND ci.idco=co.idco
    AND or.idt=ti.idt AND ti.idm=mo.idm    AND mo.idy=ye.idy
GROUP BY year, month_number, country, city;
```

To store these annotations we define the materialized view of V1, noted MV1 that stores only annotated aggregated values as illustrated in Fig. 8.

MV1					
ROWID	year	month_number	country	city	amount
@200	2007	6	France	Toulouse	4293

Fig. 8. Storage of annotated aggregated values

4   Concluding Remarks and Future Works

This paper described an implementation of an annotation management system for MDB. Every piece of multidimensional data can be associated with zero or more superimposed information called annotations. We conceive annotations with the objective of storing decision-makers remarks about multidimensional data that would otherwise have not been kept in a traditional database. We provide annotations for a personal use to remind any information concerning analysed data as well as for a collective use to materialise and to share decision-makers' expertise to facilitate collaborative analyses and decisions.

Our solution allows decision-makers to annotate multidimensional data at various levels of granularities – fact, dimension, hierarchy, attributes, detailed or aggregated values. Annotations assist users in understanding MDB structures and decisional analysis expressed through MT. Global annotations are expressed on the MDB schema components and they are transparently propagated into the MT where local annotations are added according to the analysis context.

We investigated how the global and local annotations can be stored into a homogeneous structure. We develop a relational meta-database describing constellation components; these metadata are associated to global annotations. We also described an R-OLAP environment where multidimensional data are stored into snowflake relations. The normalised dimensions allow the system to annotate detailed multidimensional data. In this normalised framework, we are interested in determining which aggregated information to materialise annotated values. The implementation solution we describe provides straightforward, uniform and efficient storage structures of decisional annotations over multidimensional data.

In our current system, global annotations are propagated into local analysis contexts. In addition, it would also be interesting to prospect how to detect similarities between analyses in order to propagate annotations from the local analysis context to similar analysis contexts. We also investigate opportunities for integrating annotations into the lattice of materialised views to improve query computation in our current approach. Future works will revisit materialised view selection algorithms for determining relevant materialised views according to annotations. A new challenge raised compared to RDBMS context is the annotation propagation along aggregated data. The aggregation of decisional data implies aggregations of associated textual annotations with specific aggregation functions [12].

References

1. Bhagwat, D., Chiticariu, L., Tan, W-C., Vijayvargiya, G.: An Annotation Management System for Relational Databases. In: 30th International Conference on Very Large Databases, VLDB'04, Toronto, Canada (2004)
2. Bhagwat, D., Chiticariu, L., Tan, W.-C., Vijayvargiya, G.: An Annotation Management System for Relational Databases. VLDB Journal 14(4), 373–396 (2005)
3. Bhatnagar, N., Juliano, B.A., Renner, R.S.: Data Annotation Models. In: 3rd Int. Conference on Intelligent Computing and Information Systems, ICICIS'07, Cairo, Egypt (2007)
4. Bhatnagar, N., Juliano, B.A., Renner, R.S.: Data Annotation Models and Annotation Query Language. In: 4th Int Conference on Knowledge Mining, ICKM'07, Bankok, Thaïland (2007)
5. Chiticariu, L., Tan, W.-C., Vijayvargiya, G.: DBNotes: A Post-It System for Relational Databases based on Provenance. In: ACM SIGMOD International Conference on Management of Data, SIGMOD'05, pp. 942–944 (2005)
6. Cong, G., Fan, W., Geerts, F.: Annotation Propagation Revisited for Key Preserving Views. In: 15th Conf. on Information and Knowledge Management, CIKM'06, Arlington, USA, pp. 632–641 (2006)
7. Geerts, F., Kementsietsidis, A., Milano, D.: MONDRIAN: Annotating and querying databases through colors and blocks. In: 22th Int. Conf. on Data Engineering, p. 82 (2006)
8. Gyssens, M., Lakshmanan, L.V.S.: A Foundation for Multi-Dimensional Databases. In: 23rd Int Conference on Very Large Data Bases, VLDB'97, Athens, Greece, pp. 106–115 (1997)
9. Kimball, R.: The Data Warehouse Toolkit: Practical Techniques for Building Dimensional Data Warehouses. John Wiley & Sons, USA (1996)
10. Marshall, C.C.: Toward an Ecology of Hypertext Annotation. In: 9th ACM conference on Hypertext and Hypermedia, HYPERTEXT'98, pp. 40–49 (1998)
11. Ravat, F., Teste, O., Tournier, R., Zurfluh, G.: Algebraic and graphic languages for OLAP manipulations. Int. Journal of Data Warehousing and Mining (2007)
12. Ravat, F., Teste, O., Tournier, R.: OLAP Aggregation Function for Textual Data Warehouse. In: Int. Conf. on Enterprise Information Systems, Funchal, Portugal (2007)
13. Tan, W.-C.: Containment of Relational Queries with Annotation Propagation. In: Lausen, G., Suciu, D. (eds.) DBPL 2003. LNCS, vol. 2921, pp. 37–53. Springer, Heidelberg (2004)

On the Need of a Reference Algebra for OLAP

Oscar Romero and Alberto Abelló

Universitat Politècnica de Catalunya
{oromero,aabello}@lsi.upc.edu

Abstract. Although multidimensionality has been widely accepted as the best solution for conceptual modelling, there is not such agreement about the set of operators to handle multidimensional data. This paper presents a comparison of the existing multidimensional algebras trying to find a common backbone, as well as it discusses the necessity of a reference multidimensional algebra and the current state of the art.

1 Introduction

OLAP tools are conceived to exploit the data warehouse (DW) for analysis tasks based on *multidimensionality*, the main feature of these tools. The multidimensional conceptual view of data is distinguished by the *fact/dimension* dichotomy and it is characterized by representing data as if placed in an n-dimensional space, allowing us to easily analyze data in terms of facts and dimensions showing the different points of view where a subject can be analyzed from.

Lots of efforts have been devoted to multidimensional modelling, and up to now, several models have been introduced in the literature (see [1], [2]). However, we can not yet benefit from a standard multidimensional model, and a common framework in which to translate and compare the research efforts in the area is missing. As discussed in [3] and [4], experiences in the field of databases have proved that a common framework to work with is crucial for the evolution of the area: (1) conceptual modelling is vital for the design, evolution and optimization of a DW, whereas (2) a multidimensional algebra is crucial for a satisfactory navigation and analysis (i.e. querying) of data contained in the DW. Specifically, a reference set of operators would help to develop design methodologies oriented to improve querying, better and accurate indexing techniques as well as to facilitate query optimization; issues even more critical than in an operational database, due to the huge amount of data stored in a DW. However, although multidimensionality (i.e. to model in terms of facts and dimensions) has been widely accepted as the best solution to DW modelling, there is no such agreement about the set of operators to handle multidimensional data. To our knowledge, does a comparison of algebras not even exist in the literature.

Thus, section 3 compares existing multidimensional algebras trying to find a common backbone, whereas section 4 discusses why the relational algebra (used by ROLAP tools) does not *directly* fit to multidimensionality. Due to the lack of a reference model, section 2 presents the multidimensional framework used in this paper. Section 5 discusses about the necessity of a reference multidimensional algebra as well as the current state of the art, and section 6 concludes this paper.

I.Y. Song, J. Eder, and T.M. Nguyen (Eds.): DaWaK 2007, LNCS 4654, pp. 99–110, 2007.
© Springer-Verlag Berlin Heidelberg 2007

2   Our Framework

Due to the lack of a standard multidimensional model, and hence, the lack of a common notation, we need a reference framework in which to translate and compare the multidimensional algebras presented in the literature. Otherwise, a comparison among all those different algebras would be rather difficult. In this section we introduce a multidimensional data structure and a set of operators (introduced in detail in [5]) used in this paper to concisely and univocally define the multidimensional concepts, as well as to provide a common notation. From here on, these concepts will be **bold faced** for the sake of comprehension.

First, we introduce our framework data structure, where a **Dimension** contains a hierarchy of **Levels** representing different granularities (or levels of detail) to study data, and a **Level** contains **Descriptors**. We differentiate between identifier **Descriptors** (univocally identifying each instance of a **Level**) and non-identifier. On the other hand, a **Fact** contains **Cells** which contain **Measures**. One **Cell** represents those individual **cells** of the same granularity that show data regarding the same **Fact** (i.e. a **Cell** is a "Class" and **cells** are its instances). We denote by a **Base** a minimal set of **Levels** identifying univocally a **Cell**, that would result in "primary key" in ROLAP tools. A set of **cells** placed in the multidimensional space with regard to the **Base** is called a **Cube**. One **Fact** and several **Dimensions** to analyze it give rise to a **Star**.

Next, we present our reference multidimensional operations set that was proved to be closed, complete and minimal (see [5] for further information):

- **Selection:** By means of a logic clause C over a **Descriptor**, this operation allows to choose the subset of points of interest out of the whole n-dimensional space.
- **Roll-up:** It groups **cells** in the **Cube** based on an aggregation hierarchy. This operation modifies the granularity of data by means of a many-to-one relationship which relates instances of two **Levels** in the same **Dimension**, corresponding to a part-whole relationship. As argued in [6] about **Drill-down** (i.e. the inverse of **Roll-up**), it can only be applied if we previously performed a **Roll-up** and did not lose the correspondences between **cells**.
- **ChangeBase:** This operation reallocates exactly the same instances of a **Cube** into a new n-dimensional space with exactly the same number of points, by means of a one-to-one relationship. Actually, it allows two different kinds of changes in the space **Base**: we can just rearrange the multidimensional space by reordering the **Levels** or, if more than one set of **Dimensions** identifying the **cells** (i.e. alternative **Bases**) exist, by replacing the current **Base** by one of the alternatives.
- **Drill-across:** This operation changes the subject of analysis of the **Cube** by means of a one-to-one relationship. The n-dimensional space remains exactly the same, only the **cells** placed on it change.
- **Projection:** It selects a subset of **Measures** from those shown in the **Cube**.
- **Set Operations:** These operations allow to operate two **Cubes** if both are defined over the same n-dimensional space. We consider **Union**, **Difference** and **Intersection** as the most relevant ones.

Table 1. Summary of the comparison between multidimensional algebras

Algebra	Operator	Selection	Projection	Roll-up Drill-down	changeBase	Drill-across	Union Difference Intersection	Remarks
[7]	"Add Dimension"				$\checkmark_p$			
	"Transfer"				$\sim$			
	"Cube Aggr."			$\checkmark$				
	"Rc-join"	$\checkmark$						
	"Union"						$\checkmark$	
[8]	"Push"					$\checkmark_p$		Semantic Rels.
	"Pull"		$\mathcal{D}$		$\checkmark_p$			Semantic Rels.
	"Destroy Dimension"		$\mathcal{D}$		$\checkmark_p$			
	"Restriction"	$\checkmark$						
	"Join"					$\checkmark$		
	"Merge"			$\checkmark$				
[9]	"Selection"	$\checkmark$						
	"Projection"		$\checkmark$					
	"Cartesian Product"					$\sim$		
	"Union/Diff./Inters."						$\checkmark$	
	"Fold/Unfold"				$\checkmark_p$			
	"Classification"			$\mathcal{D}$				
	"Summarization"			$\mathcal{D}$				
[10]	"Restriction"	$\checkmark$						
	"Metric Projection"		$\checkmark$					
	"Aggregation"			$\checkmark$				
	"Cartesian Product"					$\sim$		
	"Join"					$\checkmark$		
	"Union/Diff."						$\checkmark$	
	"Extract"				$\checkmark_p$			Semantic Rels.
	"Force"					$\checkmark_p$		Semantic Rels.
[11]	"Slicing"	$\checkmark$						
	"Roll-up/Drill-down"			$\checkmark$				
	"Split/Merge"			$\sim$				
	"Implicit/Explicit Aggr."			$\checkmark_p$				
	"Cell Operators"							Derived Measures
[12]	"Cartesian Product"					$\sim$		
	"Natural Join"					$\checkmark$		
	"Roll-up"			$\mathcal{D}$				
	"Aggregation"			$\mathcal{D}$				
	"Level Description"				$\checkmark_p$			Semantic Rels.
	"Scalar Function App."							Derived Measures
	"Selection"	$\checkmark$						
	"Simple Projection"		$\checkmark$		$\checkmark_p$			
	"Abstraction"		$\checkmark_+$		$\checkmark_{p+}$			
[13]	"Restrict"	$\checkmark$						
	"Destroy"				$\checkmark_p$			
	"join"					$\checkmark$		
	"Join"			$\checkmark_+$		$\checkmark_+$		
	"Aggr"			$\checkmark$				
[14]	"Selection"	$\checkmark$						
	"Projection"		$\checkmark$					
	"Union/Diff."						$\checkmark$	
	"Identity-based Join"					$\sim$		
	"Aggregate Formation"			$\checkmark_p$				
	"Value-based Join"					$\checkmark$		
	"Duplicate Removal"							**Base** definition
	"SQL-like Aggr."				$\checkmark_p$			
	"Star-join"	$\checkmark_+$		$\checkmark_+$				
	"Roll-up/Drill-down"			$\checkmark$				
[15]	"Navigate"			$\checkmark$				
	"Selection"	$\checkmark$						
	"Split Measure"		$\checkmark$					
[16]	"Derived Measures"							Derived Measures
	"Join"					$\checkmark_p$		
	"Slice/Multislice"	$\checkmark$						
	"Union/Diff./Inters."						$\checkmark$	
[17]	"Selection Cube"	$\checkmark$						
	"Decoration"				$\checkmark_p$			
	"Fed. Gen. Projection"		$\checkmark_+$	$\checkmark_+$	$\checkmark_+$			

3   The Multidimensional Algebras

This section presents a thorough comparison among the multidimensional algebras presented in the literature. To the best of our knowledge, it is the first comparison of multidimensional algebras carried out. In [2], a survey describing the multidimensional algebras in the literature is presented. However, unlike us, it does not compare them.

Results presented along this section are summarized in table 1. There, rows, representing an algebraic operator, are grouped according to which algebra they belong to (also ordered chronologically), whereas columns represent multidimensional algebraic operators in our framework (notice **Roll-up** and **Drill-down** are considered together since one is the inverse of the other).

The notation used is the following: a ✓ cell means that those operations represent the same conceptual operator; a ~ stands for operations with similar purpose but different proceeding making them slightly different; a ✓$_p$ means that the operation partially performs the same data manipulation as the reference algebra operator despite the latter also embraces other functionalities, and a ✓$_+$ means that this operation is equal to combine the marked operators of our reference algebra, meaning it is not an atomic operator. Analogously, there are some reference operators that can be mapped to another algebra combining more than one of its operators. This case is showed in the table with a $\mathcal{D}$ (from *derived*). Keep in mind this last mark must be read vertically unlike the rest of marks. Finally, notice we have only considered those operations manipulating data and therefore, those aimed to manipulate the data structure are not included:

[7] introduces a multidimensional algebra as well as its translation to SQL. To do so, they introduce an ad hoc grouping algebra extending the relational one (i.e. with grouping and aggregation operators). This algebra was one of the first multidimensional algebras introduced, and authors main aim was to construct **Cubes** from local operational databases.

More precisely, it defines five multidimensional operators representing mappings between either **Cubes** or *relations* and **Cubes**. The "Add dimension" and "Transfer" operators are aimed to rearrange the multidimensional space similar to a **changeBase**: while "Add dimension" adds a new analysis **Dimension** to the current **Cube**, "Transfer" *transfers* a **Dimension** *attribute* (i.e. a **Descriptor**) from one **Dimension** to another via a "Cartesian Product". Since multidimensional concepts are directly derived from non-multidimensional relations, **Dimensions** may be rather vaguely defined, justifying the transfer operator; the "Cube Aggregation" operator performs grouping and aggregation over data, being equivalent to **Roll-up** and finally, the "Rc-join" operator, that allows us to join a *relational* table with a **Dimension** of the **Cube**, **Selects** those **Dimension** values also present in the table. This low level operator is tightly related to the multidimensional model presented, and it is introduced to relate non-multidimensional relations with relations modelling **Cubes**.

[8] presents an algebra composed by six operators rather relevant, since they inspired many following algebras. First, "Push" and "Pull" transform a **Measure** into a **Dimension** and viceversa, since in their model **Measures** and **Dimensions** are handled uniformly. In our framework they would be equivalent to define semantic relationships between the proper **Dimensions** and **Cells** and then, **Drill-across** and **changeBase** respectively; "Destroy Dimension" drops

a **Cube Dimension** rearranging the multidimensional space and hence, being equivalent to **changeBase**, whereas the "Restriction" operator is equivalent to **Selection**; "Merge" to **Roll-up** and "Join" to an unrestricted **Drill-across**. Consequently, the latter can even be performed without common **Dimensions** between two **Cubes**, giving rise to a "Cartesian Product". However, a "Cartesian Product" does not make any multidimensional sense if it is not restricted, since it would not preserve *disjointness* when aggregating data ([18]). Finally, notice we can **Project** data by means of "Pull"ing the **Measure** into a **Dimension** and performing a "Destroy Dimension" over it.

[9] presents an algebra based on the classical relational algebra operations. Therefore, it includes "Selection", "Projection", "Union" / "Intersection" / "Difference" and the "Cartesian Product"; all of them being equivalent to their analogous operators in our reference algebra, except for the latter which is mappable to an unrestricted **Drill-across** as discussed above. The "Fold" and "Unfold" operators add / remove a **Dimension**, like in a **changeBase**; whereas **Roll-up** is decomposed in two operators: "Classification of Tables" (i.e. grouping of data) and "Summarization of Tables" (aggregation of data). Hence, this algebra proposes to differentiate *grouping* (i.e. the conceptual change of **Levels** through a part-whole relationship or in other words, the result of mapping data into groups) from *aggregation* (i.e. aggregating data according to an aggregation function).

[10] and [19] present an algebra with eight operators based on [8]. Therefore, the "Restriction" operator is equivalent to **Selection**; the "Metric Projection" to **Projection**; the "Aggregation" to **Roll-up** and the "Union" / "Difference" operators to those with the same name in our reference algebra. Moreover, like in [8], **Measures** can be transformed into **Dimensions** and viceversa. Hence, the "Force" and "Extract" operators are equivalent to the "Push" and "Pull" ones. Finally, they rename the "Join" operator in [8] as "Cubic Product", and denote by "Join" an specific "Cubic Product" over two **Cubes** with common **Dimensions** (i.e. preserving disjointness if joined through their shared **Dimensions**) since, in general, a "Cartesian Product" does not make multidimensional sense.

[11] presents an algebra composed by five operators. "Roll-up" and "Drill-down" and the "Split" and "Merge" operators are equivalent to **Roll-up** and **Drill-down**. According to its model data structure that differentiates two analysis phases of data, these four operations are needed because "Roll-up" and "Drill-down" find and interesting context in a first phase, whereas "Split" and "Merge" modify the data granularity *dynamically* along the "dimensional attributes" (non-identifiers **Descriptors**) defined in the "classification hierarchies" nodes of the data structure. It also introduces two operators to aggregate data: the "Implicit" and the "Explicit" aggregation. The first one is *implicitly* used when navigating by means of "Roll-up"'s, whereas the second one can be *explicitly* stated by the end-user. Since they are equivalent, these operators are just differentiated because of the conceptual presentation followed in the paper. Finally, "Slicing" operator reduces the multidimensional space in the same sense as **Selection**, whereas the "Cell-oriented operator" derives new data preserving the same multidimensional space by means of "unary operators" (-, *abs* and *sign*) or "binary

operators" (*, +, -, /, *min* and *max*). "Binary operators" ask for two multidimensional objects aligned (i.e. over the same multidimensional space). In our framework it is obtained defining **Derived Measures** in design time.

[12], [20] and [21] present an algebra with nine operators where, similar to [9], **Roll-up** is decomposed into "Roll-up" (i.e. grouping) and "Aggregation". "Level description" is equivalent to **changeBase**: it changes a **Level** by another one related through a one-to-one relation to it. In our framework we should define a semantic relationship among **Levels** involved and perform a **changeBase**; "Simple projection" projects out selected **Measures** and reduces the multidimensional space by dropping **Dimensions**: it can just drop **Measures** (equivalent to **Projection**), **Dimensions** (to **changeBase**) or combine both. Finally, "Abstraction" is equivalent to the "Pull" operator in [8] and "Selection", "Cartesian Product" and "Natural Join" to those discussed along this section.

[13] presents an algebra based on Description Logics developed from [8]. Therefore, it also introduces "Restrict", "Destroy" (equivalent to "Destroy Dimension") and "Aggr" (equivalent to "Merge"). Furthermore, the "join" and "Join" operators can be considered an extension of the "Join" operator in [8]: both operators restrict it to make multidimensional sense and consequently, being equivalent to **Drill-across**; despite the second one also allows to group and aggregate data before showing it (i.e. being equivalent to **Drill-across** plus **Roll-up**).

[14] presents an algebra where "Selection", "Projection", "Union" / "Difference" and **Roll-up** and **Drill-down** are equivalent to those with the same name presented in our framework, whereas the "Value-based join" is equivalent to **Drill-across** and the "Identity-based join" to "Cartesian product". Moreover, it also differentiates the "Aggregate operation" from the "Roll-up" (i.e. grouping); the "Duplicate Removal" operator is aimed to remove **cells** characterized by the same combination of dimensional values. In our framework it can never happen because of the **Base** definition introduced. Finally, it presents a set of non-atomic operators; the "star-join" operator combines a **Selection** with a **Roll-up**, by the same aggregation function, over a set of **Dimensions**, and the "SQL-like aggregation" applies the "Aggregate operation" to a certain **Dimensions** and projects out the rest (that is, performs a **changeBase**).

[15] presents an algebra with three operators. "Navigation" allows us to **Roll-up**, and according to [22] it is performed by means of "Level-Climbing" (reducing the granularity of data), "Packing" (grouping data) and "Function Application" (aggregating by an aggregation function). Finally, "Split a Measure" is equivalent to **Projection** and "Selection" to **Selection**.

[17] presents an algebra over an XML and OLAP federation: "Selection Cube" allows us to **Select** data; "Decoration" adds new **Dimensions** to the **Cube** (i.e. mappable to a **changeBase**) and "Federation Generalized Projection" (FGP) **Roll-ups** the **Cube** and removes unspecified **Dimensions** (**changeBase**) and **Measures** (**Projection**). Notice that despite **Roll-up** is mandatory, FGP can combine it with a **Projection** or/and a **changeBase**.

An algebra with four operations is presented in [16]. "Slice" and "Multislice" **Select** a single or a range of dimensional values; "Union" / "Intersection" /

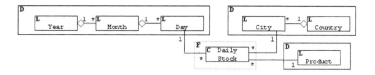

Fig. 1. Schema of a multidimensional Cube

"Difference" combine two aligned **Cubes**, whereas "Join" is rather close to **Drill-across** but in a more restrictive way, forcing both **Cubes** to share the same multidimensional space. "Derived Measures" derives new **Measures** from already existent. In our framework, as already said, derived **Measures** should be defined in design time. Finally, notice that **Roll-up** is not included in their set of operators, since it is considered in their model data structure.

Finally, to conclude our comparison, we would like to remark that some of these approaches have also presented an equivalent calculus besides the algebra introduced above (like [9] and [12]). Moreover, [23] presents a query language to define the expected workload for the DW. We have not included it in table 1 since it can not be smoothly compared to algebraic operators one per one. Anyway, analyzing it, we can deduce many of our reference operators are also supported by their model like **Selection**, **Projection**, **Roll-up**, **Union** and even a partial **Drill-across** as they allow to overlap *fact schemes*.

4   Multidimensional Algebra vs. Relational Algebra

In our study, we also need to place the relational algebra in our framework since, nowadays, ROLAP tools are the most widely spread approach to model multidimensionality and therefore, multidimensional queries are being translated to SQL and (eventually) to the relational algebra.

This section aims to justify the necessity of a semantic layer (the multidimensional algebra) on the top of an RDBMS (i.e. the relational algebra). Despite we believe ROLAP tools are a good choice to implement multidimensionality, we present, by means of a conceptual comparison between the multidimensional and the relational algebra operators, why the relational algebra (and therefore SQL) does not directly fit properly to multidimensionality. Furthermore, we emphasize on those restrictions and considerations needed to be made over the relational algebra with regard to multidimensionality.

In this comparison we consider the relational algebra presented in [24]. Thus, we consider "Selection" (σ), "Projection" (π), "Union" ($\cup$), "Difference" ($-$) and "Natural Join" ($\bowtie$) as the relational algebra operators. We talk about "Natural Join", or simply "Join", instead of "Cartesian Product" (the one presented in [24] and where "Join" can be derived from) since a "Cartesian Product" without restrictions is meaningless in the multidimensional model, as discussed in [18].

For the sake of comprehension, since we focus on a conceptual comparison, and to avoid messing results with considerations about the DW implementation, we can assume, without loss of generality, that each multidimensional **Cube** is implemented as a single relation (i.e. a denormalized relational table). So that, considering the **Cube** depicted in figure 1 we would get the following relation: {City, Day, Product, Daily Stock, Country, Month, Year}. The underlined fields denote the multidimensional **Base** and therefore, the relation "primary key". Along this section, we will refer to this kind of denormalized relation as the *multidimensional table*.

Table 2 summarizes the mapping between both set of algebraic operators. Notice we are considering the "group by" and "aggregation" as relational operators, and both will be justified consequently below. Since *multidimensional tables* contain (1) identifier fields (i.e. identifier **Descriptors**) identifying data, for instance: City, Day and Product in the above example, (2) numerical fields, Daily Stock, representing multidimensional data (i.e. **Measures**) and (3) descriptive fields, Country, Month and Year, (i.e. non-identifier **Descriptors**), we use the following notation in the table: $\checkmark_{Measures}$ if the multidimensional operator is equivalent to the relational one but it can be only applied over relation fields representing **Measures**, $\checkmark_{Descs}$ if the multidimensional operator must be applied over **Descriptors** fields and finally, $\checkmark_{Descs_{id}}$ if it can be only applied over identifier **Descriptors** fields. Consequently, a $\checkmark$ without restrictions means both operators are equivalent, without additional restrictions. If the translation of a multidimensional operator combines more than one relational operator, the subscript $+$ is added. Next, we clearly define the relational algebra *proper subset* mappable to the multidimensional algebra:

- The multidimensional **Selection** operator is equivalent to a restricted relational "Selection". It can only be applied over **Descriptors** and then, it is equivalent to restrict the relational "Selection" just over **Level** data. According to our notation, we express the multidimensional **Selection** in terms of the relational algebra as $\sigma_{Descriptors}$.

- Similarly, the multidimensional **Projection** operator is equivalent to the relational one restricted to **Measures**; that is, specific **Cell** data. In terms of the relational algebra we could express it as $\pi_{Measures}$.

- OLAP tools emphasize on flexible data grouping and efficient aggregation evaluation over groups, and it is the multidimensional **Roll-up** operator the one aimed to provide us with powerful grouping and aggregation of data. In order to support it, we need to extend the relational algebra to provide

Table 2. Comparison table between the relational and the multidimensional algebras

Reference Operator		"Selection"	"Projection"	"Join"	"Union"/"Diff."	"Group by"	"Aggregation"
Selection		$\checkmark_{Descs}$					
Projection			$\checkmark_{Measures}$				
Roll-up						$\checkmark_{Descs_{id}+}$	$\checkmark_{Measures+}$
Drill-across			$\checkmark_{Descs_{id}+}$	$\checkmark_{Descs_{id}+}$			
changeBase	Add Dim.			$\checkmark_{Descs_{id}}$			
	Remove Dim.		$\checkmark_{Descs_{id}}$				
	Alt. Base		$\checkmark_{Descs_{id}+}$	$\checkmark_{Descs_{id}+}$			
Union/Difference					$\checkmark$		

grouping and aggregation mechanisms. This topic has already been studied and previous works like [7], [25] and [26] have already presented extensions of the relational algebra to what is also called the *grouping algebra*. All of them introduce two new operators; one to group data and apply a simple addition, counting or maximization of a collection of domain values and the other one to compute the aggregation of a given attribute over a given *nested* relation. Following the [26] grouping algebra, we will refer to them as the "group by" and the "aggregation" operators. In terms of this grouping algebra, a **Roll-up** operator consists of a proper "group by" operation along with an "aggregation" of data.

– A consistent **Drill-across** typically consists of a "Join" between two *multi-dimensional tables* sharing the same multidimensional space. Notice that to "Join" both tables it must be performed over their common **Level** identifiers that must univocally identify each **cell** in the multidimensional space (the **Cube Base**). Moreover, once "joined", we must "project" out the columns in the *multidimensional table drill-acrossed* to, except for its **Measures**. Formally, Let $\mathcal{A}$ and $\mathcal{B}$ be the *multidimensional tables* implementing, respectively, the origin and the destination **Cells** involved. In the relational algebra it can be expressed as:

$$\pi_{Descriptors_A, Measures_A, Measures_B}(\mathcal{A} \bowtie \mathcal{B})$$

– As stated in section 2, **changeBase** allows us to rearrange our current multidimensional space either by changing to an alternative **Base** (adding / removing a **Dimension** or replacing **Dimensions**) or reordering the space (i.e. "pivoting" as presented in [3]).

When changing to an alternative **Base** we must assure it does not affect the functional dependency of data with regard to the **Cube Base**. Hence:

- To add a **Dimension** it must be done through its `All` **Level** (i.e. one instance representing the whole **Dimension**) or fixing just one value at any other **Level** by means of a **Selection**, to not lose **cells** (i.e. representing the whole **Dimension** as a unique instance as discussed in section 2). Therefore, in the relational algebra adding a **Dimension** is achieved through a "cartesian product" between the *multidimensional table* and the **Dimension** table (that would contain a unique instance). Specifically, if $\mathcal{C}$ is the initial *multidimensional table* and $\mathcal{D}$ the relation implementing the added **Dimension**, it can be expressed as:

$$\mathcal{C} \times \mathcal{D}, \quad where \; |\mathcal{D}| = 1$$

- To remove a **Dimension** it is just the opposite, and we need to get rid of the proper **Level** identifier projecting it out in the *multidimensional table*.

- To change the set of **Dimensions** identifying each **cell**, that is, choosing an alternative **Base** to display the data, we must perform a "join" between both **Bases** and project out the replaced **Levels Descriptors** in the *multidimensional table*. In this case, the "join" must be performed through the identifier **Descriptors** of **Levels** replaced and **Levels** introduced. Formally, let $\mathcal{A}$ be the *multidimensional table*, $\mathcal{B}$ the table showing the correspondence between both **Bases** and $d_1, ..., d_n$ the identifier **Descriptors** of those **Dimensions** introduced. In the relational algebra, it is equivalent to:

$$\pi_{Descriptors_{\mathcal{B}(d_1, ..., d_n)}, Measures_{\mathcal{A}}} (\mathcal{A} \bowtie \mathcal{B})$$

- Finally, pivoting just asks to reorder the **Levels** identifiers using the SQL "order by" operator, not mappable to the relational algebra. For that reason, it is not included in table 2.

– The multidimensional **Union** (**Difference**) unites (differences) two **Cubes** defined over the same multidimensional space. In terms of the relational algebra, it is equivalent to "Union" ("Difference") two *multidimensional tables*.

5   Discussion

By means of a comparison of the multidimensional algebras introduced in the literature, section 3 has been able to identify a *multidimensional backbone* shared by all the algebras. Firstly, **Selection**, **Roll-up** and **Drill-down** operators are considered in every algebra. It is quite reasonable since **Roll-up** is the main operator of multidimensionality and **Selection** is a basic one, allowing us to select a subset of multidimensional points of interest out of the whole n-dimensional space. **Projection**, **Drill-across** and **Set Operations** are included in most of the algebras. In fact, along the time, just two of the first algebras presented did not include **Projection** and **Drill-across**. We may include **Set Operations** in our algebra depending on the transformations that the model allows us to perform over data and indeed, it is a personal decision to make. However, we do believe that to unite, intersect or difference two **Cubes** is a kind of navigation desirable. Finally, **changeBase** is also partially considered in most of the algebras. Specifically, they agree on the necessity of modifying the n-multidimensional space by adding / removing **Dimensions**, and they include it as a first class citizen operator. Moreover, our framework provides additional alternatives to rearrange the multidimensional space (i.e. to change the multidimensional space **Base** and "pivoting"). In general, we can always rearrange the multidimensional space in any way, if we preserve the functional dependencies of the **cells** with regard to the **Levels** conforming the **Cube Base**; that is, if the replaced **Dimension**(s) and the new one(s) are related through a one-to-one relationship.

Finally, we would like to underline the need to work in terms of a multidimensional algebra. As shown in section 4, the multidimensional data manipulation should be performed by a restricted subset of the relational algebraic operators (used by ROLAP tools). Otherwise, the results of the operations performed either would not conform a **Cube** (since the whole relational algebra is not closed with regard to the multidimensional model) or would introduce aggregation problems ([18]). In other words, the multidimensional algebra represents the *proper subset* of the relational algebra applicable to multidimensionality.

Summing up, all the algebras surveyed are subsumed by our framework; that is also strictly subsumed by the relational algebra. Thus, we have been able to

(1) identify an implicit agreement about how multidimensional data should be handled and to (2) show that this common set of multidimensional operators can be expressed as a subset of the relational algebra.

6 Conclusions

The comparison of algebras presented in this paper has revealed many implicit agreements about how multidimensional data should be handled. We strongly believe that a reference set of operators such as the *multidimensional backbone* identified in our study could be used to develop design methodologies oriented to improve querying, better and accurate indexing techniques and to facilitate query optimization. That is, provide us with all the benefits of a reference framework. Moreover, we have shown that this common set of multidimensional operators can be expressed as a *proper subset* of the relational algebra; essential to give support to ROLAP tools.

Acknowledgments. This work has been partly supported by the Spanish Ministerio de Educación y Ciencia under project TIN 2005-05406.

References

1. Abelló, A., Samos, J., Saltor, F.: A Framework for the Classification and Description of Multidimensional Data Models. In: Mayr, H.C., Lazanský, J., Quirchmayr, G., Vogel, P. (eds.) DEXA 2001. LNCS, vol. 2113, pp. 668–677. Springer, Heidelberg (2001)
2. Vassiliadis, P., Sellis, T.: A Survey of Logical Models for OLAP Databases. SIGMOD Record 28(4), 64–69 (1999)
3. Franconi, E., Baader, F., Sattler, U., Vassiliadis, P.: Multidimensional Data Models and Aggregation. In: Jarke, M., Lenzerini, M., Vassilious, Y., Vassiliadis, P. (eds.) Fundamentals of Data Warehousing, Springer, Heidelberg (2000)
4. Rizzi, S., Abelló, A., Lechtenbörger, J., Trujillo, J.: Research in Data Warehouse Modeling and Design: Dead or Alive? In: Proc. of 9th Int. Workshop on Data Warehousing and OLAP (DOLAP 2006), pp. 3–10 (2006)
5. Abelló, A., Samos, J., Saltor, F.: **YAM**^{2} (Yet Another Multidimensional Model): An extension of UML. Information Systems 31(6), 541–567 (2006)
6. Hacid, M.-S., Sattler, U.: An Object-Centered Multi-dimensional Data Model with Hierarchically Structured Dimensions. In: Proc. of IEEE Knowledge and Data Engineering Exchange Workshop (KDEX 1997), IEEE Computer Society Press, Los Alamitos (1997)
7. Li, C., Wang, X.S.: A Data Model for Supporting On-Line Analytical Processing. In: Proc. of 5th Int. Conf. on Information and Knowledge Management (CIKM 1996), pp. 81–88. ACM Press, New York (1996)
8. Agrawal, R., Gupta, A., Sarawagi, S .: Modeling Multidimensional Databases. In: Proc. of the 13th Int. Conf. on Data Engineering (ICDE'97), pp. 232–243. IEEE Computer Society Press, Los Alamitos (1997)
9. Gyssens, M., Lakshmanan, L.: A Foundation for Multi-dimensional Databases. In: Proc. of 23rd Int. Conf. on Very Large Data Bases (VLDB 1997), pp. 106–115. Morgan Kaufmann, San Francisco (1997)

10. Thomas, H., Datta, A.: A Conceptual Model and Algebra for On-Line Analytical Processing in Data Warehouses. In: Proc. of the 7th Workshop on Information Technologies and Systems (WITS 1997), pp. 91–100 (1997)
11. Lehner, W.: Modelling Large Scale OLAP Scenarios. In: Schek, H.-J., Saltor, F., Ramos, I., Alonso, G. (eds.) EDBT 1998. LNCS, vol. 1377, pp. 153–167. Springer, Heidelberg (1998)
12. Cabibbo, L., Torlone, R.: From a Procedural to a Visual Query Language for OLAP. In: Proc. of the 10th Int. Conf. on Scientific and Statistical Database Management (SSDBM 1998), pp. 74–83. IEEE Computer Society Press, Los Alamitos (1998)
13. Hacid, M.-S., Sattler, U.: Modeling Multidimensional Database: A formal object-centered approach. In: Proc. of the 6th European Conference on Information Systems (ECIS 1998) (1998)
14. Pedersen, T.: Aspects of Data Modeling and Query Processing for Complex Multidimensional Data. PhD thesis, Faculty of Engineering and Science (2000)
15. Vassiliadis, P.: Data Warehouse Modeling and Quality Issues. PhD thesis, Dept. of Electrical and Computer Engineering (Nat. Tech. University of Athens) (2000)
16. Franconi, E., Kamble, A.: The GMD Data Model and Algebra for Multidimensional Information. In: Persson, A., Stirna, J. (eds.) CAiSE 2004. LNCS, vol. 3084, pp. 446–462. Springer, Heidelberg (2004)
17. Yin, X., Pedersen, T.B.: Evaluating XML-extended OLAP queries based on a physical algebra. In: Proc. of 7th Int. Workshop on Data Warehousing and OLAP (DOLAP 2004), ACM Press, New York (2004)
18. Romero, O., Abelló, A.: Improving Automatic SQL Translation for ROLAP Tools. Proc. of 9th Jornadas en Ingeniería del Software y Bases de Datos (JISBD 2005), 284(5), 123–130 (2005)
19. Thomas, H., Datta, A.: A Conceptual Model and Algebra for On-Line Analytical Procesing in Decision Suport Databases. Information Systems 12(1), 83–102 (2001)
20. Cabibbo, L., Torlone, R.: Querying Multidimensional Databases. In: Cluet, S., Hull, R. (eds.) Database Programming Languages. LNCS, vol. 1369, pp. 319–335. Springer, Heidelberg (1998)
21. Cabibbo, L., Torlone, R.: A Logical Approach to Multidimensional Databases. In: Schek, H.-J., Saltor, F., Ramos, I., Alonso, G. (eds.) EDBT 1998. LNCS, vol. 1377, pp. 183–197. Springer, Heidelberg (1998)
22. Vassiliadis, P.: Modeling Multidimensional Databases, Cubes and Cube operations. In: Proc. of the 10th Statistical and Scientific Database Management (SSDBM 1998), pp. 53–62. IEEE Computer Society Press, Los Alamitos (1998)
23. Golfarelli, M., Maio, D., Rizzi, S.: The Dimensional Fact Model: A Conceptual Model for Data Warehouses. Int. Journal of Cooperative Information Systems (IJCIS) 7(2-3), 215–247 (1998)
24. Codd, E.F.: Relational Completeness of Data Base Sublanguages. Database Systems, 65–98 (1972)
25. Klug, A.: Equivalence of Relational Algebra and Relational Calculus Query Languages Having Aggregate Functions. Journal of the Association for Computing Machinery 29(3), 699–717 (1982)
26. Larsen, K.S.: On Grouping in Relational Algebra. Int. Journal of Foundations of Computer Science 10(3), 301–311 (1999)

OLAP Technology for Business Process Intelligence: Challenges and Solutions

Svetlana Mansmann[1], Thomas Neumuth[2], and Marc H. Scholl[1]

[1] University of Konstanz, P.O.Box D188, 78457 Konstanz, Germany
{Svetlana.Mansmann,Marc.Scholl}@uni-konstanz.de
[2] University of Leipzig, Innovation Center Computer Assisted Surgery (ICCAS),
Philipp-Rosenthal-Str. 55, 04103 Leipzig, Germany
Thomas.Neumuth@medizin.uni-leipzig.de

Abstract. The emerging area of business process intelligence aims at enhancing the analysis power of business process management systems by employing data warehousing and mining technologies. However, the differences in the underlying assumptions and objectives of the business process model and the multidimensional data model aggravate a straightforward solution for a meaningful convergence of the two concepts.

This paper presents the results of an ongoing project on providing OLAP support to business process analysis in the innovative application domain of Surgical Process Modeling. We describe the deficiencies of the conventional OLAP technology with respect to business process modeling and formulate the requirements for an adequate multidimensional presentation of process descriptions. The modeling extensions proposed at the conceptual level are verified by implementing them in a relational OLAP system, accessible via a state-of-the-art visual analysis tool. We demonstrate the benefits of the proposed analysis framework by presenting relevant analysis tasks from the domain of medical engineering and showing the type of the decision support provided by our solution.

1 Introduction

Modern enterprises increasingly integrate and automate their business processes with the objective of improving their efficiency and quality, reducing costs and human errors. However, the prevailing business process management systems focus on the design support and simulation functionalities for detecting performance bottlenecks, with rather limited, if any, analysis capabilities to quantify performance against specific business metrics [1]. The emerging field of *Business Process Intelligence* (BPI), defined as the application of performance-driven management techniques from Business Intelligence (BI) to business processes, claims that the convergence of BI and business process modeling (BPM) technologies will create value beyond the sum of their parts [2]. However, the task of unifying the flow-oriented process specification and the snapshot-based multidimensional design for quantitative analysis is by far not trivial due to very different and even partially incompatible prerequisites and objectives of the underlying approaches.

I.Y. Song, J. Eder, and T.M. Nguyen (Eds.): DaWaK 2007, LNCS 4654, pp. 111–122, 2007.

This research was inspired by practical data warehousing issues encountered in the medical domain. An emerging field of surgical process analysis fosters intelligent acquisition of process descriptions from surgical interventions for the purpose of their clinical and technical analysis [3]. A medical engineering term *Surgical Workflows* describes the underlying methodological concept of this acquisition procedure. The process data is obtained manually and semi-automatically by monitoring and recording the course of a surgical intervention. Apparently, a well-defined formal recording scheme of the surgical process is required to support such data acquisition [4]. Surgical workflow use cases are manifold, ranging from the preoperative planning support by retrieving similar precedent cases to the postoperative workflow exploration, from discovering the optimization potential for instrument and device usage to verifying medical hypotheses, etc.

The prevailing process modeling standards, such as Business Process Modeling Notation (BPMN) [5] or the reference model of Workflow Management Coalition (WfMC) [6], tend to be too general to adequately address the domain-specific requirements. Multidimensional modeling seems a promising solution as it allows to view the data from different perspectives, define various business metrics and aggregate the data to the desired granularity.

To be admitted into an OLAP system, business process descriptions have to undergo the transformation imposed by the underlying multidimensional data model. This model categorizes the data into *facts* with associated numerical *measures* and descriptive *dimensions* characterizing the facts [7]. For instance, a surgical process can be modeled as a fact entry SURGERY characterized by the dimensions Location, Surgeon, Patient and Discipline. The values of a dimension are typically organized in a containment type hierarchy (e.g., location ↗ hospital ↗ city) to support multiple granularities.

1.1   Business Process Modeling

Business process models are employed to describe business activities in the real world. Basic BPM entities are objects, activities, and resources. Activities are the work units of a process that have an objective and change the state of the objects. Resources are consumed to perform activities. Relationships between the entities may be specified using control flow (consecutive, parallel, or alternative execution) and/or hierarchical decomposition. The differentiation between the notions *business process* and *workflow* consists in the levels of abstraction: business processes are mostly modeled in a high-level and informal way, whereas workflow specifications serve as a basis for the largely automated execution [8].

A process is "a co-ordinated (parallel and/or serial) set of process activity(s) that are connected in order to achieve a common goal. Such activities may consist of manual activity(s) and/or workflow activity(s)" [6]. Workflow refers to automation of business processes as a whole or in part, during which documents, information or tasks are passed from one participant to another for action, according to a set of procedural rules [6]. Different workflow specification methods coexist in practice, with *net-based*, or *graph-based* as the most popular ones. *Activity and state charts* are frequently used to specify a process as an oriented graph with nodes representing the activities and arcs defining the ordering in which these are performed. Other popular approaches are *logic-based*, or temporal, methods and *Event-Condition-Action* rules. [9]

Surgical Process Modeling, classified as a specific domain of BPM [3], adopts the concepts from BPM and Workflow Modeling (WfM). The WfM approach of decomposing a workflow into activities is useful for providing a task-oriented surgery perspective. However, since surgical work steps are predominantly manual and involve extensive organizational context, such as participants and their roles, patients and treated structures, instruments, devices and other resources, high-level BPM abstractions are used to model such domain-specific elements.

The remainder of the paper is structured as follows: Section 2 provides an overview of the related work, Section 3 describes our approach to gaining a multidimensional perspective of a process scheme, followed by the conceptual design and its relational implementation in Section 4. Section 5 demonstrates a prototype implementation of the model in a visual frontend tool and presents an exemplary analysis task. Concluding remarks are given in Section 6.

2   Related Work

In this section we describe the related contributions in the fields of BPI data warehousing and discuss their relationships to our approach.

Grigori et al. present a BPI tool suite based on a data warehouse approach [10]. The process data is modeled according to the star schema, with process, service, and node state changes as facts and the related definitions as well as temporal and behavioral characteristics as dimensions. While this approach focuses on the analysis of process execution and state evolution, we pursue the task-driven decomposition, with activities as facts and their characteristics, corresponding to the factual perspectives (e.g., behavior, organization, information etc.) proposed in [11] as well as domain ontologies as dimension hierarchies.

An approach to visual analysis of business process performance is given in [12]. The proposed visualization tool *VisImpact* is based on analyzing the schema and the instances of a process to identify business metrics, denoted *impact factors*. A metric along with its related process instances are mapped to a symmetric circular graph showing the relationships and the details of the process flows.

Pedersen et al. proposed an extended multidimensional data model for meeting the needs of non-standard application domains and used a medical application scenario as a motivating case study [13]. Jensen et al. formulated the guidelines for designing complex dimensions in the context of spatial data such as location-based services [14]. Our previous work [15] presents the limitations of current OLAP systems in handling complex dimensions and describes an extended formal model and its implementation in a visual analysis tool.

Interdisciplinary research in the field of surgical workflow modeling, analysis and visualization is carried out at the Innovation Center Computer Assisted Surgery (ICCAS) located in Leipzig, Germany [3,16,4].

3   Multidimensional Perspective of a Process

Surgeons, medical researchers and engineers are interested in obtaining a formal model of a surgical process that would lay a fundament for a systematic accumulation of the obtained process descriptions in a centralized data warehouse to enable its comprehensive analysis and exploration. The initial design phase aims at identifying a multidimensional perspective of the process scheme. Figure 1 shows the general structure of a surgical workflow in the UML class notation. The properties of a process are arranged according to vertical decomposition:

1. *Workflow level*: The upper part of the diagram contains the characteristics describing the surgery as a whole, such as Patient and Location, and thus building a *workflow granularity level*.
2. *Work step level*: Properties belonging to particular components (e.g., events, activities) within a workflow, such as Instrument used in Activity, refer to the *intra-workflow granularity level*.

Further, we account for two complimentary data acquisition practices, namely the *task-driven* and the *state-based* structuring. Activity describes a surgical task, or work step, as perceived by a human observer (e.g., "the irrigation of a vessel with a coagulator"). The state-based perspective uses the concepts of State and Event to describe the state evolution of involved subsystems and the events that trigger the transition. Subsystem is a heterogeneous class comprising participants and patients and their body parts, instruments and devices, etc. For instance, the gaze direction of the surgeon's eyes can be modeled as states, while the surgeon's directives to other participants may be captured as events.

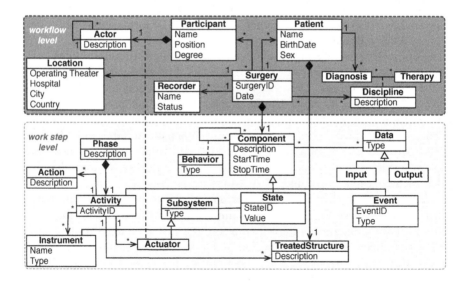

Fig. 1. Recording scheme of a surgical process model as a UML class diagram

Fig. 2. Examples of the multidimensional views of a surgical process

Vertical decomposition is followed by a horizontal one according to the *factual perspectives* [11], similar to identifying the dimensions of a data cube. For example, Instrument refers to the *operation* perspective, Input and Output build the *information* perspective, etc. Figure 2 presents examples of multidimensional views of a surgery as a whole (left) and at work step level (right), as well as a behavioral perspective of the relationships between components (e.g., input component 'Event X' caused output component 'State Y').

3.1 Modeling Challenges

In addition to classical OLAP constraints, such as summarizability of dimension hierarchies and disallowance of NULL values in fact entries, the context of BPM confronts the data warehouse designer with the following modeling challenges:

- *Many-to-many relationships between facts and dimensions* are very common. For instance, within a single work step, multiple surgical instruments may be used by multiple actuators.
- *Heterogeneity of fact entries.* Processes consist of heterogeneous components. Modeling Component as one fact type leads to loss of subclass properties, while mapping subclasses to separate fact types disables treating all components as the same class in part of their common properties.
- *Interchangeability of measure and dimension roles.* In conventional data warehousing, the measure attributes of a data cube, which form the focus of the analysis, are known at design time. However, "raw" business process data may contain no explicit quantitative characteristics and the measures of interest vary from one query to another. Therefore, it is crucial to enable runtime measure specification from virtually any attribute.
- *Interchangeability of fact and dimension roles.* Surgery has dimensional characteristics of its own (Location, Patient, etc.) and therefore, deserves to be treated as a fact type. However, with respect to single work steps, Surgery clearly plays the role of a dimension (e.g., Event rolls-up to Surgery).

3.2 Formalization

In relational OLAP systems, facts and dimensions are stored in relational tables and, therefore, each entry is a tuple that adheres to the schema of its relation. *Fact table schema* F is a set of attributes $A^F = \{A_i, i = 1, \dots, k\}$, and *dimension table schema* D is a set of attributes $A^D = \{A_j, j = 1, \dots, l\}$.

An *n-dimensional fact schema* $\mathcal{S}$ is composed of the fact schema $\mathcal{F}$ and a set of corresponding dimension schemata $\{\mathcal{D}_i, i = 1, \ldots, n\}$.

A *dimension schema* $\mathcal{D}$ is obtained by grouping the attributes of $\mathbf{A}^D$ into a set of hierarchical categories $\mathcal{C}^{\mathcal{D}} = \{\mathcal{C}_m, m = 1, \ldots, p\}$ and the partial order $\sqsubseteq_{\mathcal{D}}$ on those categories. A dimension has two distinguished categories $\top_{\mathcal{D}} \in \mathcal{C}^{\mathcal{D}}$ and $\bot_{\mathcal{D}} \in \mathcal{C}^{\mathcal{D}}$ as its respective top and bottom levels. $\top_{\mathcal{D}}$ serves as a root hierarchy element with the value 'all'. $\bot_{\mathcal{D}}$ represents the dimension's finest granularity, i.e., the one to which the fact entries are mapped. A *category schema* $\mathtt{C}$ is a subset of the dimension's attributes $\mathbf{A}^C \subseteq \mathbf{A}^D$ including a distinguished hierarchy attribute $\bar{\mathtt{A}}_C$ and property attributes, functionally dependent on the former: $\forall \mathtt{A}_j \in \mathbf{A}^C : \bar{\mathtt{A}}_C \rightarrow \mathtt{A}_j$.

A *fact schema* $\mathcal{F}$ is derived from $\mathbf{F}$ by assigning each attribute to either the set of dimensional attributes $Dim^{\mathcal{F}}$ or to the set of measures $Meas^{\mathcal{F}}$. In a standard measure definition, any measure attribute is functionally dependent on the set of the associated dimensional attributes: $\forall \mathtt{A}_i \in Meas^{\mathcal{F}} : Dim^{\mathcal{F}} \rightarrow \mathtt{A}_i$. We nullify this restriction to enable query-specific measure definition from any attribute. Instead, the measure is defined as a function $Measure(A_i, Aggr)$ where A_i is any actual or derived attribute from $\mathcal{S}$ and $Aggr$ is an aggregation function from the set of SQL functions $\{\mathtt{SUM}, \mathtt{COUNT}, \mathtt{AVG}, \mathtt{MIN}, \mathtt{MAX}\}$. By defining the measure as a function and not as a property, we enable symmetric treatment of fact and dimension roles as well as of measure and dimension attributes.

As the measure is specified at query time, the set of predefined measure attributes may be empty: $Meas^{\mathcal{F}} = \emptyset$, in which case the resulting fact schema is called *measure-free*. As long as no user-defined measures exist, the default measure $\mathtt{COUNT(*)}$, i.e. mere counting of the qualifying fact entries, is implied.

Optionally, $\mathcal{F}$ disposes of a distinguished *fact identifier attribute* $\bar{\mathtt{A}}_{\mathcal{F}}$, $\bar{\mathtt{A}}_{\mathcal{F}} \in \mathbf{A}^F$, which is a single-valued primary key of the respective fact table. For instance, Surgery fact entries are identified by SurgeryID. $\bar{\mathtt{A}}_{\mathcal{F}}$ owes its existence to the fact of data warehousing non-cumulative data: fact entries in a business process scenario are not some derived measurements, but the actual process data. Formally, fact identifiers may be assigned neither to $Dim^{\mathcal{F}}$ nor to $Meas^{\mathcal{F}}$.

4   Conceptual Design and Its Relational Mapping

The resulting structure of the surgical recording scheme in terms of facts ("boxed" nodes) and dimension hierarchies (upward directed graphs of "circular" category nodes) is presented in Figure 3 in the notation similar to the Dimensional Fact Model [17]. Solid arrows show the roll-up relationships while dashed arrows express the "is a" relationships between fact types, such as identity and generalization, explained later in this section. Shared categories are displayed without redundancy, thus helping to recognize valid aggregation and join paths in the entire scheme. The vertical ordering (bottom-up) of the facts and the category nodes corresponds to the descending order of their granularity.

Fact tables in our model have shared dimensional categories and, therefore, build a complex structure called *fact constellation*[18]. As expected, different

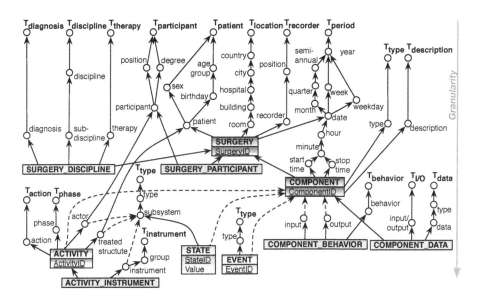

Fig. 3. The Dimensional Fact Model of a surgical workflow scheme

facts are potentially joinable to each other along their shared axes. In our scenario, apart from the shared dimensions, facts may be related to each other via fact identifiers. We define the following special cases of fact constellation:

- *Fact rollup*, or *fact hierarchy*, reflects the vertical decomposition, as is the case with COMPONENT that rolls-up to SURGERY.
- *Satellite fact* is a special case of fact rollup and is a result of "extracting" each many-to-many relationship between a fact and a dimension into a *subfact*, or *bridge table*. This solution has been inspired by a related concept of dimensional modeling, namely mapping non-strict hierarchies to *bridge tables* [19]. For example, SURGERY_DISCIPLINE handles multidisciplinarity of surgeries by storing each value in a separate tuple.
- *Fact generalization* results from unifying heterogeneous fact types into a common superclass, as may be observed with ACTIVITY, STATE, and EVENT, generalized as COMPONENT. Thereby, multiple classes can be treated as the same class in part of their common characteristics. A further advantage is the ability to uniformly model the relationships between the heterogeneous components in form of a satellite fact COMPONENT_BEHAVIOR. The concept of fact generalization has also been inspired by the related approach in the dimensional modeling [20,15].

In part of dimensional hierarchy modeling, the major challenges lie in identifying various hierarchy types and normalizing them for correct aggregation. In literature, numerous guidances for handling complex dimensions may be found [13,14,20,15], and, therefore, we skip the detailed description and proceed with presenting the implementation challenges.

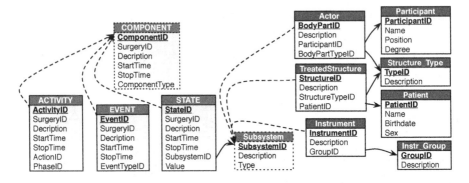

Fig. 4. Implementing generalization in facts (left) and dimensions (right)

4.1 Relational Implementation

Star schema and snowflake schema are the two options of the relational data warehouse design. Star schema places the entire dimension hierarchy into a single relation by pre-joining all aggregation levels, while snowflake schema decomposes complex dimensions into separate tables according to the relational normalization rules. Snowflake schema becomes the only option when dimensional hierarchies are prone to irregularities, such as heterogeneity, non-strictness, missing values, mixed granularity etc. We extend the classical snowflake schema to include additional table types, such as satellite fact and generalization tables. Figure 4 shows the examples of creating a superclass fact table COMPONENT and a superclass dimension Subsystem as materialized views, defined as a union of all subclass tables, projected to the subset of their common attributes.

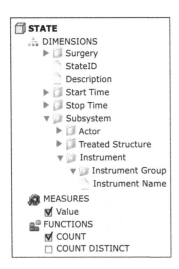

Fig. 5. Navigation hierarchy of the data cube STATE

Another benefit of the "schema-aware" snowflake design is the ability to automatically retrieve the metadata by tracing the outgoing foreign key relationships. Notice how the navigation structure of Subsystem in Figure 5 corresponds to its relational structure in Figure 4. Metadata, such as the description of the fact tables, their dimensions and valid aggregation paths, applicable aggregation functions, etc., provides the input for generating a navigation hierarchy for a visual frontend. Figure 5 demonstrates the intuition behind the visual exploration approach at the example of the navigation fragment for the data cube STATE. To facilitate the translation of the navigation events into database queries, we supplemented the snowflake schema by a star-schema view of each homogeneous subdimension.

5   Visual Analysis

Visual analysis and interactive exploration have grown to be the prevailing methods of modern data analysis at the end-user level. Therefore, the ultimate value of the proposed conceptual and relational model extensions is determined by the easiness of incorporating those extensions into visual OLAP tools. In this section we provide some insights into a prototypical implementation of an end-user interface for multidimensional business process analysis.

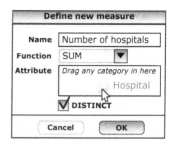

Fig. 6. Defining a measure

Visual query specification evolves in a predominantly "drill-down" fashion, i.e. gradually descending from coarsely grained overviews towards the desired level of detail. Queries are specified interactively via a navigation hierarchy, as the one depicted in Figure 5. Compulsory elements of any analytical query are 1) a measure specified as an aggregation function (e.g., sum, average, maximum etc.) and its input attribute, and 2) a set of dimension categories defining the granularity of the aggregation. In addition to the standard measures, i.e., those that have been pre-configured at the metadata level, in our application scenario it is imperative to enable runtime measure specification.

Our prototype provides a wizard for specifying a user-definde measure as a 3-step procedure, depicted in Figure 6:

1. Selecting the aggregation function from the function list;
2. Specifying the measure attribute via a "drag&drop" of a category from the navigation, as shown in Figure 6 where Hospital category is being dragged into the measure window;
3. Specifying whether the duplicates should be eliminated from the aggregation by activating the DISTINCT option. This option is crucial in the presence of many-to-many relationships between facts and dimensions.

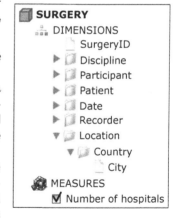

Fig. 7. Navigation fragment with a user-defined measure

Optionally, the newly defined measure may be supplied with a user-friendly name. Once defined, the measure is permanently added to the list of pre-configured measures. Let us consider an example of analyzing the number of hospitals,by choosing the category Hospital from Location as the measure attribute. Obviously, to support this measure, the data cube SURGERY needs to be aggregated to the Hospital level, Hospital category as well as the granularities below Hospital, i.e., Room and Building, are removed from the navigation path of their dimension Location because those granularities become invalid in the defined query context. The adapted navigation fragment of the affected data cube SURGERY is presented in Figure 7.

5.1 Experiments

In this section we present an application case of analyzing the use of instruments in the surgical intervention type discectomy. The intervention goal of a discectomy is the partial removal of the herniated intervertebral disc. A typical expert query in this scenario focuses on the occurrence of different instruments with the same surgical objective. During a discectomy parts of the vertebra are removed using different bone ablating surgical instruments: surgical punch, trephine, and/or surgical mallet/chisel. To query the frequency of the use of bone ablating instruments or the duration of the usage periods, one has to navigate to the entry 'bone ablating' in the category Instrument Group and apply filtering to reduce the selection to the instrument of interest mentioned above. Figure 8 contains a pivot-table with the result of the following two queries:

Query 1. *For each of the interventions of type discectomy, summarize the occurrences of using the specified bone ablating instruments by the surgeon (i.e., the number of surgery activities, in which the respective instrument is used).*

The measure of this query, i.e., the number of activities (COUNT(DISTINCT ActivityID)), is rolled-up by SurgeryID and Instrument with a selection condition along Instrument Group. The input data cube is obtained by joining the fact table ACTIVITY with its satellite fact ACTIVITY_INSTRUMENT.

Query 2. *For each of the interventions of type discectomy, calculate the average usage time for each of the specified instruments used by the surgeon (i.e., the average duration of the respective surgery activities).*

The duration of a work step equals the time elapsed between its start and end, therefore, the required measure is specified as AVG(StopTime-StartTime) and produces the values of data type time.

The above queries describe a real-world example from medical engineering. The aggregation results for the bone ablating instruments are used to predict the success of a new surgical instrument in this field [4], namely a milling system whose evolution speed is controlled by its spatial position in relation to the

		Measures							
		Occurrence				Average duration			
Dimensions		SurgeryID							
Instrument Group	Instrument	A	B	C	D	A	B	C	D
- bone ablating	mallet/chisel	0	3	0	1	00:00	00:23	00:34	00:50
	punch	9	22	10	9	02:38	00:35	00:46	01:27
	trephine	3	0	7	0	02:18	00:00	00:43	00:00
bone ablating Total		12	25	17	10	02:33	00:33	00:45	01:24

Fig. 8. Obtained query results on instrument usage statistics as a pivot table

patient's body [21]. This system is intended to replace the conventional instruments surgical punch, trephine and mallet/chisel. To predict the chance of its success, the pre-development analysis of the instrument usage patterns is crucial.

6 Conclusions

Motivated by the growing research interest to the evolving area of business process intelligence, we attempted to provide OLAP support to business process analysis. Conventional BPM tools are rather limited in the types of supported analytical tasks, whereas the data warehousing techniques appear more suitable when it comes to managing large amounts of data, defining various business metrics and running complex queries. As a challenging real-world application scenario, we chose an innovative and promising domain of Surgical Process Modeling, concerned with designing a recording scheme for acquiring process descriptions from surgical interventions for their subsequent analysis and exploration.

We demonstrated the deficiencies of the standard relational OLAP approach with respect to the requirements of our case study and proposed an extended data model that addresses such challenges as non-quantitative and heterogeneous facts, many-to-many relationships between facts and dimensions, runtime definition of measures, and interchangeability of fact and dimension roles. We also described a prototypical implementation of the proposed model extensions in a relational OLAP system, in which the data is stored according to the fact constellation schema and can be queried with standard SQL.

The feasibility of the proposed modeling solution is presented by demonstrating the results of its incorporation in an advanced visual analysis tool that supports flexible measure definition at runtime. The work is concluded by presenting a relevant analysis scenario from the domain of medical engineering and demonstrating what type of decision support in the field of process analysis is provided by the presented data warehouse framework.

Acknowledgement

We would like to thank Oliver Burgert from ICCAS at the University of Leipzig as well as Christos Trantakis and Juergen Meixensberger from the Neurosurgery Department at the University Hospital of Leipzig for the valuable domain expertise.

References

1. Dayal, U., Hsu, M., Ladin, R.: Business process coordination: State of the art, trends, and open issues. In: VLDB 2001: Proc. 27th International Conference on Very Large Data Bases, pp. 3–13 (2001)
2. Smith, M.: Business process intelligence. Intelligent Enterprise (December, 5 2002) Retrieved 22.10.2006 from http://www.intelligententerprise.com/021205/601feat2_1.jhtml
3. Neumuth, T., Strauß, G., Meixensberger, J., Lemke, H.U., Burgert, O.: Acquisition of process descriptions from surgical interventions. In: Bressan, S., Küng, J., Wagner, R. (eds.) DEXA 2006. LNCS, vol. 4080, pp. 602–611. Springer, Heidelberg (2006)

4. Neumuth, T., Trantakis, C., Eckhardt, F., Dengl, M., Meixensberger, J., Burgert, O.: Supporting the analysis of intervention courses with surgical process models on the example of fourteen microsurgical lumbar discectomies. In: CARS 2007: Proc. 21st International Conference on Computer Assisted Radiology and Surgery (to appear, 2007)
5. OMG (Object Management Group): BPMN (Business Process Modeling Notation) 1.0: OMG Final Adopted Specification (February 2006), Retrieved 15.03.2007 from `http://www.bpmn.org/`
6. WfMC (Workflow Management Coalition): WfMC Standards: The Workflow Reference Model, Version 1.1. (January 1995), Retrieved 15.03.2007 from `http://www.wfmc.org/standards/docs/tc003v11.pdf`
7. Pedersen, T.B., Jensen, C.S.: Multidimensional database technology. IEEE Computer 34(12), 40–46 (2001)
8. Muth, P., Wodtke, D., Weißenfels, J., Weikum, G., Kotz-Dittrich, A.: Enterprise-wide workflow management based on state and activity charts. In: Proc. NATO Advanced Study Institute on Workflow Management Systems and Interoperability, pp. 281–303 (1997)
9. Matousek, P.: Verification of Business Process Models. PhD thesis, Technical University of Ostrava (2003)
10. Grigori, D., Casati, F., Castellanos, M., Dayal, U., Sayal, M., Shan, M.C.: Business process intelligence. Computers in Industry 53(3), 321–343 (2004)
11. Jablonski, S., Bussler, C.: Workflow Management. Modeling Concepts, Architecture and Implementation. International Thomson Computer Press, London et al (1996)
12. Hao, M.C, Keim, D.A, Dayal, U., Schneidewind, J.: Business process impact visualization and anomaly detection. Information Visualization 5, 15–27 (2006)
13. Pedersen, T.B., Jensen, C.S., Dyreson, C.E.: A foundation for capturing and querying complex multidimensional data. Information Systems 26(5), 383–423 (2001)
14. Jensen, C.S., Kligys, A., Pedersen, T.B., Timko, I.: Multidimensional data modeling for location-based services. The VLDB Journal 13(1), 1–21 (2004)
15. Mansmann, S., Scholl, M.H.: Extending visual OLAP for handling irregular dimensional hierarchies. In: DaWaK 2006: Proc. 8th International Conference on Data Warehousing and Knowledge Discovery, pp. 95–105 (2006)
16. Burgert, O., Neumuth, T., Gessat, M., Jacobs, S., Lemke, H.U.: Deriving dicom surgical extensions from surgical workflows. In: SPIE MI 2007: Proc. of SPIE Medical Imaging 2007 - PACS and Imaging Informatics, CID 61450A (2007)
17. Golfarelli, M., Maio, D., Rizzi, S.: The dimensional fact model: A conceptual model for data warehouses. International Journal of Cooperative Information Systems 7(2-3), 215–247 (1998)
18. Chaudhuri, S., Dayal, U.: An overview of data warehousing and olap technology. SIGMOD Rec. 26(1), 65–74 (1997)
19. Kimball, R., Ross, M.: The Data Warehouse Toolkit: The Complete Guide to Dimensional Modeling. John Wiley & Sons, Inc., New York, NY, USA (2002)
20. Malinowski, E., Zimányi, E.: Hierarchies in a multidimensional model: From conceptual modeling to logical representation. Data & Knowledge Engineering 59(2), 348–377 (2006)
21. Jank, E., Rose, A., Huth, S., Trantakis, C., Korb, W., Strauss, G., Meixensberger, J., Krueger, J.: A new fluoroscopy based navigation system for milling procedures in spine surgery. In: CARS 2006: Proc. 20st International Conference on Computer Assisted Radiology and Surgery, pp. 196–198 (2006)

Built-In Indicators to Automatically Detect Interesting Cells in a Cube

Véronique Cariou[2], Jérôme Cubillé[1], Christian Derquenne[1], Sabine Goutier[1], Françoise Guisnel[1], and Henri Klajnmic[1]

[1] EDF Research and Development
1 avenue du Général de Gaulle, 92140 Clamart, France
firstname.lastname@edf.fr
[2] ENITIAA
Rue de la Géraudière, BP 82225, 44322 Nantes CEDEX 3, France
veronique.cariou@enitiaa-nantes.fr

Abstract. In large companies, On-Line Analytical Processing (OLAP) technologies are widely used by business analysts as a decision support tool. Nevertheless, while exploring the cube, analysts are rapidly confronted by analyzing a huge number of visible cells to identify the most interesting ones. Coupling OLAP technologies and mining methods may help them by the automation of this tedious task. In the scope of discovery-driven exploration, this paper presents two methods to detect and highlight interesting cells within a cube slice. The cell's degree of interest is based on the calculation of either test-value or Chi-Square contribution. Indicators are computed instantaneously according to the user-defined dimensions drill-down. Their display is done by a color-coding system. A proof of concept implementation on the ORACLE 10g system is described at the end of the paper.

Keywords: OLAP, Data cube, Olap Mining, Data Mining, Atypical Values, Chi-Square, Test-value, Oracle 10g.

1 Introduction

Most of large companies have identified the importance and the strategic value of the information contained in their Data Warehouses. But those Data Warehouses are of interest only if the knowledge they contain is correctly extracted, formatted, summarized, presented and shared by business analysts, in order to create added value. That is why EDF (Electricité de France), the main French energy supplier and electricity producer, has resort to On-Line Analytical Processing (OLAP) applications to analyze internal and external data. OLAP multidimensional view of data is easily understandable by business analysts who can become direct end-users of the corresponding commercial software. Nevertheless, as soon as multidimensional databases become very large, the number of cells of the cube dramatically increases. Analysts are rapidly confronted to a difficult and tedious task in order to identify the most interesting cells of a cube. Coupling Data Mining methods and OLAP applications allow to carry out such detection.

I.Y. Song, J. Eder, and T.M. Nguyen (Eds.): DaWaK 2007, LNCS 4654, pp. 123–134, 2007.

This work is in the scope of discovery-driven exploration of a data cube. In our approach, two analyses are considered. The first one consists in analyzing rapidly a slice, by highlighting the most significant cells. The second one facilitates the whole process of exploration of the data cube by giving the most relevant dimensions to drill-down. For example, in the marketing domain of the EDF business, a goal could be to automatically identify the criteria which influence customers' energy consumption or to detect unusual energy consumption for a specific market segment. In this paper, we focus on the first kind of analysis. Given a user-defined dimensions drill-down, it consists in identifying which cells are the most interesting. A cell is said to be interesting if its measure is very different from the other selected cells. Detection of interesting cells is performed thanks to built-in indicators computed instantaneously during exploration without any pre-computation.

The paper is organized as follows. In section 2, we outline relevant works in the context of discovery-driven exploration. In section 3, the two different operators are detailed. Finally, section 4 presents the implementation on the ORACLE 10g system.

2   Related Works

Since the last decade, there have been growing interests in coupling Data Mining methods and OLAP tools. Different approaches have been studied to perform Data Mining techniques on a data cube, to facilitate user-exploration and to enrich OLAP tools with statistical operators.

Historically, researches focused on the first aspect. In 1997, Han et al. [5] proposed a new tool, DB Miner, which aimed at mining multidimensional data at different levels of aggregation. Several functions were investigated such as association rules, classification or clustering. Nevertheless, most of the other contributions focused on the association rules algorithms [7], [13]. Closely related to this problem, Imielinski et al. [6] defined the cubegrade query in order to explain main trends in the data cube.

The second aspect concerns discovery-driven exploration of a data cube to facilitate user exploration [4], [9]. New operators were introduced to automatically detect relevant cells in a data cube (SELFEXP operator) or to identify the best path in exploring data. Operators are based on the notion of degree of exception of a cell. A cell measure is said to be an exception if it differs significantly from the value predicted with saturated log-linear model or analysis of variance one. Going on this approach, Sarawagi proposed several discovery-driven operators like DIFF, INFORM and RELAX [10], [11], [12].

More recently, some authors have studied how to enrich the data cube manipulation by statistical operators. The aim is to improve the data cube visualization thanks to model reorganization. This is achieved by values permutation within a dimension [2] or by dimension hierarchy building [3].

Our approach is similar to the SELFEXP operator. Nevertheless, we stress two main contributions compared to the previous ones. The degree of exception of a cell is not based on a saturated model. While considering a cross-product involving a great number of dimensions, we are often faced to sparse datacube. In such cases, saturated models lead to unstable solution. In our work, a straightforward approach is adopted since no hypothesis is made on the exploration path. Consequently, the degree of interest is computed without regard to interactions between dimensions. This guarantees the business analyst an easier and more robust interpretation of the most interesting cells. The other contribution lies in the way indicators are built. We take advantages of OLAP functionalities so that indicators are embedded within the OLAP engine. This approach differs from [9] who proposed a tool attachment to an OLAP data source. Implementation under Oracle 10g is discussed in part 4. The approach leads to an enriched OLAP tool without any need of pre-computation task.

3   Detection of Interesting Cells

Our framework is the static analysis of a cube slice (also called table). Given a current cube slice (or sub-cube in the case of restriction), this method is able to take into account, in an interactive way, the changes of criteria that the end-user could operate during his analysis.

3.1   Introduction

Our method is based on the computation of built-in indicators. Built-in indicators defined below are generic since they can be computed whenever any OLAP operations are performed (*roll-up*, *drill-down*, *restriction*, *slice*...). Within the data cube, dimensions are usually built with hierarchies that specify aggregate levels. Given dimensions hierarchies, built-in indicators can be evaluated either at a detailed dimension level or at an aggregated one. They are also insensitive to the *pivot* operator.

Two methods are developed, one for a count measure and the other for a quantitative measure. The first one is based on the computation of the cell contribution to the total Chi-Square of the cube slice. The second one is based on the calculation of the test-value [8].

The display of these methods is done by a color-coding system which proposes to the end-user several levels colors depending on the degree of interest of the cells.

3.2   Working Example

We illustrate our approach with an example, issued from EDF marketing context. In this example, we consider the following simplified customer detailed table CUSTOMER(#CUST,CONTRACT,DWELLING,HEATING,CONSUMPTION).

One tuple of Tab. 1 is related to one EDF customer and will be called a *statistical unit* in this article. On this detailed table, with added attributes (OCCUPATION, WATER_HEATING,   OLDNESS_OF_HOUSE,   DATE_OF_CONTRACT_SUBSCRIPTION,

Table 1. Example of customers detailed table

#CUST	CONTRACT	DWELLING	HEATING	CONSUMPTION
1	Tariff 1	House	Electrical	212
2	Tariff 2	Flat	Gas	153
3	Tariff 1	Flat	Electrical	225
4	Tariff 2	Flat	Gas	142
5	Tariff 3	House	Fuel	172
...	...	...	...	...

GEOGRAPHY), we build a datacube structured by 8 dimensions and the following measures: number of customers and mean of electrical consumption.

A hierarchy City→Province→Region is defined on the dimension GEOGRAPHY. In our snapshots, the category 'TOT' stands for the common word 'ANY' or 'ALL' which represents the total aggregated level of a dimension.

3.3   Notations

Let us note $\widetilde{X}$ one measure of the cube, structured by several dimensions (in our experiments, more than 10 dimensions may occur).

With the dimensions values selected on the page axes, we define a sub-cube (also called sub-population) E of the whole cube (also called population) P. If no selection is made on the page axes, $E = P$.

Once E is fixed, we call t the slice resulting from the crossing of the dimensions selected in lines and columns. A line or a column of the table where the value is TOT, is called a margin of the table (it corresponds to the ANY position). The margins of the table are not included in this slice t and are shaded when color-coding is applied. The slice t is defined by the crossing of k_t dimensions $D_t = \{d_1, \ldots, d_r, \ldots, d_{k_t}\}$ and the choice of a measure $\widetilde{X}$.

Example 1
On Fig. 1, E is the sub-population of the customers who are home owners (STOC=O) and who live in their main home (TYPRES=P). In this table with two dimensions, t is framed by a dotted line.

Let us note:

- n_i the size of the cell $i \in t$ as the number of statistical units in the cell i,
- n_E the size of the sub-population E, we notice that $\forall t, n_E = \sum_{i \in t} n_i$,
- M_{d_r} the number of distinct values on the dimension d_r at the current aggregated level of d_r hierarchy ('TOT' is excluded),
- $x_j^{(i)}$ the value of $\widetilde{X}$ for the statistical unit j in the cell i of the slice t.

The cell $i \in t$ is at the crossing of the values $m_i(d_1), \ldots, m_i(d_r), \ldots, m_i(d_{k_t})$ of the k_t dimensions.

Indicateur		STOC	TARPUI	CONTRAT	TYPRES
MOYENNE_CONSO	▼	0 ▼	TOT ▼	TOT ▼	P ▼

▼	TOT	MISS	E	A	G
▼ TOT	12 751.47	11 790.12	13 647.57	12 853.37	12 136.80
E	13 187.41	12 450.31	13 685.03	12 789.66	12 155.36
A	12 840.30	9 390.00	13 452.47	13 508.27	12 124.08
MISS	11 908.18	11 776.87	13 489.88	10 603.62	12 178.38

Fig. 1. Definition of sub-population and slice t

Example 2

For the cell i such as $i = d_1 \text{'}A\text{'} \times d_2\text{']}1975 - 1985]\text{'}$, $m_i(d_1) = \text{'}A\text{'}$ and $m_i(d_2) = \text{']}1975 - 1985]\text{'}$.

Let $n_i^{[d_r]}$ be the size of the cell i at the crossing of value $m_i(d_r)$ for dimension d_r and value 'TOT' for all other dimensions in $D_t \backslash \{d_r\}$. It corresponds to the number of statistical units with value $m_i(d_r)$ in sub-population E.
In addition, we will denote:

- the relative frequency of value $m_i(d_r)$ in the sub-population E as

$$f_i^{[d_r]} = \frac{n_i^{[d_r]}}{n_E},$$

- the relative frequency of cell i in the sub-population E as

$$f_i = \frac{n_i}{n_E},$$

- the product of the relative frequencies for all margins of a cell i as

$$\widetilde{f}_i = \prod_{r=1}^{k_t} f_i^{[d_r]},$$

- the average of the variable $\widetilde{X}$ for the cell i of the slice t as

$$\overline{X}_i = \frac{1}{n_i} \sum_{j=1}^{n_i} x_j^{(i)},$$

- the average of the variable $\widetilde{X}$ on all sub-population E as

$$\overline{X}_E = \frac{1}{n_E} \sum_{i \in t} \sum_{j=1}^{n_i} x_j^{(i)}.$$

Without any specific hypothesis, we can define the following variances:
- empirical variance on the sub-population E of study for variable $\overline{X}$ as

$$S_E^2 = \frac{1}{n_E - 1} [\sum_{i \in t} \sum_{j=1}^{n_i} [x_j^{(i)}]^2 - n_E \overline{X}_E^2],$$

- theoretical variance of $\tilde{X}_i$ (in the case where the dataset is not a sample of a population) as

$$\tilde{S}_i^2 = \frac{n_E - n_i}{n_E} \frac{S_E^2}{n_i}.$$

The last formula allows to calculate the theoretical variance of $\overline{X}_i$ in the case of a simple random sampling without replacement of the n_i individuals in cell i.

3.4   Case 1: Count Measure

The problem is to compute the level of interest of each cell of the current slice according to a count measure. Dealing with a count measure, $x_j^{(i)} = 1$ for all statistical units j in the cell i of the slice t.

Basic Idea. The method is based on the comparison between the observed value in the cell and the theoretical value computed under independence model hypothesis. If two dimensions d_1 and d_2 are independent, the frequency in the cells at each crossing of a value of d_1 and a value of d_2 only depends on the margins of the table.

Let us define N_i as the random variable associated to the theoretical number of statistical units in a cell i, whereas n_i is the observed number. The expected value of the frequency N_i in the cell i defined by the crossing of d_1 and d_2 can be written as follows:

$$\mathbb{E}(N_i) = \mathbb{E}(n_E P_i^{[d_1, d_2]}) = n_E \mathbb{E}(P_i^{[d_1, d_2]}) = n_E \mathbb{E}(P_i^{[d_1]} \cap P_i^{[d_2]}) = n_E f_i = n_i.$$

If d_1 and d_2 are independent, it could be rewritten:

$$\mathbb{E}(N_i) = n_E \mathbb{E}(P_i^{[d_1]}) \times \mathbb{E}(P_i^{[d_2]}) = n_E f_i^{[d_1]} \times f_i^{[d_2]} = n_E \frac{n_i^{[d_1]}}{n_E} \frac{n_i^{[d_2]}}{n_E} = \frac{n_i^{[d_1]} n_i^{[d_2]}}{n_E}.$$

Let us illustrate this idea with the following example.

Example 3

In this example (based on Tab. 2), the measure of the data cube is a count measure which represents the number of customers whose specific characteristics are organized by dimensions d_1 and d_2.

Let us focus on the pointed-out cell (d_1 'A' et d_2 'F'). If d_1 and d_2 were independent, instead of the observed value 19 062, we should have a value close to the product of the two margins values (underlined values) (43 195 and 47 628) divided by the size of the sub-population (i.e. here 172 075). So, in the case of independency between the two dimensions, that given cell should contain a value close to 11 956, instead of 19 062.

We generalize the independence concept to several dimensions $d_1, \ldots, d_{k_t}$ as

$$\mathbb{E}(N_i) = n_E \tilde{f}_i = n_E \prod_{r=1}^{k_t} f_i^{[d_r]}.$$

Table 2. Count measure with two dimensions

d_1 / d_2	A	B	C	D	Total
H	12 034	716	24 398	370	37 518
F	**19 062**	20 973	2 759	401	**43 195**
G	16 532	11 800	10 368	52 662	91 362
Total	**47 628**	33 489	37 525	53 433	172 075

Construction of the Indicator. Let us now define the theoretical framework of this formulation. Denote $C_{\chi^2}(i)$ the contribution of a cell i to the table chi-square. In the general case of k_t dimensions, this contribution can be written:

$$C_{\chi^2}(i) = \frac{(n_E f_i - n_E \widetilde{f}_i)^2}{n_E \widetilde{f}_i} = n_E \frac{(f_i - \widetilde{f}_i)^2}{\widetilde{f}_i}.$$

Based on this contribution, we define the level of interest of the cell i by the following built-in indicator as

$$IC(i) = \frac{C_{\chi^2}(i)}{\sum_{j \in t} C_{\chi^2}(j)}.$$

We notice that $\sum_{i \in t} IC(i) = 1$. IC will be the built-in indicator for a count measure in order to detect interesting cells.

General Interpretation of the Indicator IC. By definition $0 \leq IC(i) < 1$. If $IC(i) = 0$, the cell i corresponds to an homogeneous distribution of the statistical units with regard to the margins values. The more $IC(i)$ is close to 1, the farthest the distribution of the statistical units is from the situation of independence. It thus means that the cell i abnormally either attracts or rejects the statistical units of slice t. They are identified as interesting cells. Moreover, we can notice that

$$IC(i) = \frac{e^2(i)}{\sum_{j \in t} e^2(j)}$$

where $e(i)$ is the standardized residual of the log-linear model of independence [1] and can be written with our notations $e(i) = \sqrt{C_{\chi^2}(i)}$.

Highlighting Cells. *Research of interesting cells.* At this stage, the indicator $IC(i)$ is calculated for each cell. The problem to be solved now is to select the cells for which indicators values are the highest, in order to color them in a suitable way. To perform that, two ways are investigated in parallel: a statistical (based on a clustering method) and an analytical approach (based on the detection of gaps on the distribution curve of the $IC(i)$). In our experiments, the

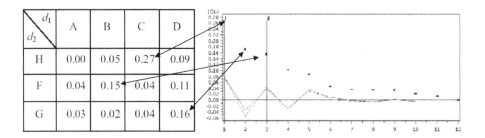

d_2 ╲ d_1	A	B	C	D
H	0.00	0.05	0.27	0.09
F	0.04	0.15	0.04	0.11
G	0.03	0.02	0.04	0.16

Fig. 2. Research of thresholds for $IC(i)$

results of both approaches are very close. With regard to time performance, the second approach gives better results while implemented in the OLAP management system. So we retained this latter. In our method, we actually have defined two thresholds (labeled 1 and 2 on Fig. 2) in order to distinguish two levels of interest for the cells. Cells having an indicator value greater than threshold 1 are colored in dark color; those having an indicator value between threshold 1 and threshold 2 are colored in light color. The other cells are not colored. The color will be specified in the following paragraph.

Example 4

In the working example, we have obtained 12 values for $IC(i)$ (as shown on left side of Fig. 2). We sort them decreasingly and plot them to obtain the distribution curve of IC values (as shown on right side of Fig. 2). This corresponds to the selection of two dimensions (as shown on Tab. 2). One cell of the slice will be colored in dark color while two cells in light color. The margin cells will be shaded to indicate that they do not contribute to analysis.

Definition of attraction and repulsion zones. For a business analyst, it may be relevant to know whether a highlighted cell corresponds to an attraction or a repulsion case. Each computed contribution can be signed:

- If $f_i - \widetilde{f}_i > 0$ then the cell is an attraction zone of the statistical units. It means that the cell contains more statistical units than expected under

Table 3. Colored cells

d_2 ╲ d_1	A	B	C	D	Total
H	12 034	716	**24 398**	370	37 518
F	19 062	**20 973**	2 759	401	43 195
G	16 532	11 800	10 368	**52 662**	91 362
Total	47 628	33 489	37 525	53 433	172 075

independence hypothesis. These cells are colored in green with two levels of intensity: light green and dark green.

- If $f_i - \widetilde{f}_i < 0$ then the cell is a repulsion zone of the statistical units. It means that the cell contains less statistical units than expected in the independence case. These cells are colored in red with two levels of intensity: light red and dark red.

Let us illustrate on the working example.

Example 5
In Tab. 3, the cell defined by d_1='C' and d_2='H' corresponds to $f_i = 0.14179$ and $\widetilde{f}_i = 0.04755$. This cell will be colored in dark green while two other ones will colored in light green (cells defined by d_1='B' and d_2='F' and by d_1='D' and d_2='G').

3.5 Case 2: Quantitative Measure

Basic Idea. Studying a quantitative measure $\widetilde{X}$ such as electric consumption, the test-value [8] is chosen to detect interesting cells. The general idea is to compare the real value of the measure $\widetilde{X}$ in a cell with its average on the whole sub-population for the given slice t. The test-value enables to order the difference between the estimated value (e.g. the mean value) and the cell aggregated one. It makes possible to compare the gap between the average on a sub-population (for example the average consumption of all the customers whatever their central heating energy is.) and the average for a combination of dimensions values (for example the average consumption of the customers with an electric heating system and hot water system). With regard to the measure $\widetilde{X}$, the more important this variation is, the more different the cell is. The test-value thus provides an indicator of exception.

Construction of the Indicator. The assumption for the use of the test-value is that the n_i statistical units of cell i are randomly sampled without replacement. The measure $\overline{X}_i$ represents the average of the measure in the cell with $\mathbb{E}(\overline{X}_i) = \overline{X}_E$ and $V(\overline{X}_i) = \widetilde{S}_i^2$.

According to the previous formulas, the test-value is defined as

$$vt(i) = \frac{\overline{X}_i - \overline{X}_E}{\widetilde{S}_i}.$$

Thus, if the test-value is high (corresponding to a low probability of equality between the studied averages), we can consider that the average in the cell significantly differs from the average of the population under study. We use this indicator to order the cells according to their level of interest.

Research of a Threshold. According to the test-value definition [8], the cells with an absolute test-value up to 1.96 are said to be relevant. We can indeed consider that all test-values are distributed among a normal distribution $\mathcal{N}(0,1)$ (1.96 is the 0.975 quantile). Nevertheless, while applying such a threshold, we are still confronted to a large number of interesting cells. At this stage, we have to find a method to identify the cells with the highest absolute test-value. We color a percentage of cells with the highest indicators value. This percentage is interactively chosen by the business analyst.

4   Implementation with Oracle

Some EDF OLAP databases are built with Oracle 10g. It enables to build a Web application with different reports and graphs on a specific part of the EDF activity such as marketing, human resources... The two presented built-in indicators (Chi-Square contribution and test-value) and thresholds detection (based on geometrical approach) have been implemented in a convenient way thanks to Oracle 10g OLAP DML language. For each cell of a given slice (explored interactively by the user), the appropriate indicator is calculated depending on the kind of the measure. Then, the cell is colored depending on the value of the indicator (red or green colored since they correspond to an abnormally low or high measure's value). On all reports, either a count measure or a quantitative measure is provided. When a user analyzes a cube slice, he or she may be helped using the extra "Detect and color" button. The exploration can be made even if a great number of dimensions has been drilled-down. All computations are done after the user has defined his data view. An example is given on Fig. 3 where the shaded margins help the understanding of the slice defined by the three dimensions. A facsimile of the Oracle schema is also shown in this

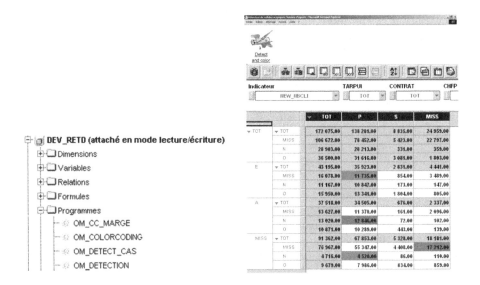

Fig. 3. Facsimile in Oracle 10g (Administration and web view)

figure. The program OM_CC_MARGE first detects the table's margins and shades those cells in order to facilitate the reading of the table. Then the program OM_DETECT_CAS is executed to choose the appropriate algorithm, whether the measure is a count or a quantitative one. The computation of all indicators (test-value or Chi-Square contribution) is done by the OM_DETECTION program. Finally, the program OM_COLORCODING is run to return the right color. When the business analyst has chosen the dimensions of interest, the display of the table with colored cells is instantaneous. Nevertheless, two factors can influence the processing time. As the system is based on a client-server architecture, computation is dependent on network performance. Furthermore, the number of chosen dimensions influences the system response time but it never exceeds 2-3 seconds.

5   Conclusion and Future Works

In this paper, we describe an original work to facilitate business analysts interactive multidimensional data exploration. We have devised a novel approach to detect and highlight interesting cells within a data cube slice. Our approach differs from previous work by the fact that the degree of exception of a cell is not based on a saturated model but on the computation of statistical built-in indicators embedded within the OLAP engine.

We consider two cases depending on the kind of measure. For a count measure, the indicator is based on the computation of the cell contribution to the total Chi-Square of the analyzed data cube slice. For a quantitative measure, we use test-value. In both cases, without any pre-computation, those indicators are computed on the fly according to the user-defined dimensions chosen to structure the slice. The display is done by a color-coding system. As a proof of concept, those algorithms have been implemented under Oracle 10g on a client-server architecture.

The work presented in this paper allows a static analysis of a multidimensional cube by highlighting the most relevant cells after the business analyst has performed the OLAP transformations he wished. We are extending the present work in order to facilitate the whole process of exploration of the data cube by developing a new method that proposes the most pertinent dimensions to drill-down. This future work will enable a discovery-driven dynamic exploration.

References

1. Agresti, A.: Categorical Data Analysis. Wiley, NY (1990)
2. Ben Messaoud, R., Boussaid, O., Rabaseda, S.: Efficient multidimensional data representations based on multiple correspondence analysis. In: Proceedings of the Twelfth ACM SIGKDD International Conference on Knowledge Discovery and Data Mining (KDD 2006), Philadelphia, Pennsylvania, USA, August 20–23, 2006, ACM Press, New York (2006)
3. Ben Messaoud, R., Rabaseda, S., Boussaid, O., Bentayeb, F.: OpAC: A New OLAP Operator Based on a Data Mining Method. In: Proceedings of the 6th International Baltic Conference on Databases and Information Systems (DB&IS04), June 2004, Riga, Latvia, (2004)

4. Chen, Q.: Mining exceptions and quantitative association rules in OLAP data cube. PhD Thesis of science, School of Computing Science, Simon Fraser University, British Columbia, Canada (1999)
5. Han, J.: OLAP Mining: An Integration of OLAP with Data Mining. In: Proceedings of the 1997 IFIP Conference on Data Semantics (DS-7), Leysin, Switzerland, October, 1997, pp. 1–11 (1997)
6. Imielinski, T., Khachiyan, L., Abdulghani, A.: Cubegrades: Generalizing association rules. Technical report, Dept. Computer Science, Rutgers University (August 2000)
7. Kamber, M., Han, J., Chiang, J.: Metarule-Guided Mining of Multi-Dimensional Association Rules Using Data Cubes. Knowledge Discovery and Data Mining 207–210 (1997)
8. Morineau, A.: Note sur la caractérisation statistique d'une classe et les valeurs-tests. Bulletin technique Centre Statistique Informatique Appliquées 2(1-2), 20–27 (1984)
9. Sarawagi, S., Agrawal, R., Megiddo, N.: Discovery-driven exploration of OLAP data cubes. Technical report, IBM Almaden Research Center, San Jose, USA (1998)
10. Sarawagi, S.: Explaining differences in multidimensional aggregates. In: Proceedings of the 25th International Conference On Very Large Databases (VLDB1999), September 7-10, 1999, Edinburgh, Scotland, UK (1999)
11. Sarawagi, S.: User-adaptative exploration of multidimensional data. In: Proceedings of the 26th International Conference On Very Large Databases (VLDB2000), September 10-14, 2000, Cairo, Egypt (2000)
12. Sathe, G., Sarawagi, S.: Intelligent Rollups in Multidimensional OLAP Data. In: Proceedings of the 27th International Conference On Very Large Databases (VLDB 2001), September 11-14, 2001, Roma, Italy (2001)
13. Zhu, H.: On-Line Analytical Mining of Association Rules. PhdThesis, Burnaby, British Columbia V5A 1S6, Canada (1998)

Emerging Cubes for Trends Analysis in OLAP Databases

Sébastien Nedjar, Alain Casali, Rosine Cicchetti, and Lotfi Lakhal

Laboratoire d'Informatique Fondamentale de Marseille (LIF),
CNRS UMR 6166, Université de la Méditerranée
Case 901, 163 Avenue de Luminy, 13288 Marseille Cedex 9, France
`lastname@lif.univ-mrs.fr`

Abstract. In various approaches, data cubes are pre-computed in order to efficiently answer OLAP queries. Such cubes are also successfully used for multidimensional analysis of data streams. In this paper, we address the issue of performing cube comparisons in order to exhibit trend reversals between two cubes. Mining such trend changes provides users with a novel and specially interesting knowledge. For capturing the latter, we introduce the concept of emerging cube. Moreover, we provide a condensed representation of emerging cubes which avoids to compute two underlying cubes. Finally, we study an algorithmic way to achieve our representation using cube maximals and cube transversals.

1 Introduction and Motivations

Computing data cubes for data warehouse or OLAP database management has been widely investigated. Such cubes provide summaries, according to various points of view or dimensions, of a given "population". However when various datasets are collected, for instance from different sources or at different periods along time, it is of great interest to perform comparisons between them. The decision maker can then be provided with significant differences between two populations or with meaningful evolutions of a population: apparition, disappearing or clear-cut change in the behavior or characteristics of the studied population. This kind of knowledge is also specially relevant in OLAP(Sarawagi et al., 1998; Imielinski et al., 2002) and data stream applications (Han et al., 2005; Han), where the reactions to changes must be performed in real time, or for dynamic classification in databases.

In this paper, inspired from the concept of emerging patterns in a classification context (Dong and Li, 2005), we introduce the concept of emerging cube which captures, in an aggregated way,the significant differences or trend reversals between two categorical database relations. Moreover, motivated by the huge amounts of results to be handled, we characterize a condensed representation of emerging cubes through borders using cube maximals ans cube transversals (Casali et al., 2003a). Finally we show that these borders can be computed by existing algorithms proved to be very efficient.

I.Y. Song, J. Eder, and T.M. Nguyen (Eds.): DaWaK 2007, LNCS 4654, pp. 135–144, 2007.
© Springer-Verlag Berlin Heidelberg 2007

The paper is organized as follows. In section 2, we summarize the framework in which our contribution fits: a suitable search space called the cube lattice. The following section is devoted to the emerging cube. Section 4 focuses on the computation of emerging cube borders. An outline and research perspectives are given in conclusion.

2   Background: Cube Lattice Framework

In this section, we recall the concepts of cube lattice (Casali et al., 2003b) and convex cube (Nedjar et al., 2006) which are used to formalize the new structure proposed in this paper.

Throughout the paper, we make the following assumptions and use the introduced notations. Let r be a relation over the schema $\mathcal{R}$. Attributes of $\mathcal{R}$ are divided in two sets (i) $\mathcal{D}$ the set of dimensions, also called categorical or nominal attributes, which correspond to analysis criteria and (ii) $\mathcal{M}$ the set of measures.

2.1   Search Space

The multidimensional space of the categorical database relation r groups all the valid combinations built up by considering the value sets of attributes in $\mathcal{D}$, which are enriched with the symbolic value ALL. The latter, introduced in (Gray et al., 1997) when defining the operator Cube-By, is a generalization of all the possible values for any dimension.

The multidimensional space of r is noted and defined as follows: $Space(r) = \{\times_{A \in \mathcal{D}}(Dim(A) \cup \text{ALL})\} \cup \{(\emptyset, \ldots, \emptyset)\}$ where $\times$ symbolizes the Cartesian product, $Dim(A)$ is the projection of r on the attribute A and tuple $(\emptyset, \ldots, \emptyset)$ stands for the combination of empty values. Any combination belonging to the multidimensional space is a tuple and represents a multidimensional pattern.

The multidimensional space of r is structured by the generalization/specialization order between tuples, denoted by $\preceq_g$. This order has been originally introduced by T. Mitchell (Mitchell, 1997) in the context of machine learning. In a datawarehouse context, this order has the same semantic as the operator ROLLUP/DRILLDOWN (Gray et al., 1997) and is used, in the quotient cube (Lakshmanan et al., 2002), to compare tuples (cells).

Let u, v be two tuples of the multidimensional space of r:

$$u \preceq_g v \Leftrightarrow \begin{cases} \forall A \in \mathcal{D} \text{ such that } u[A] \neq \text{ALL}, \\ \qquad\qquad\qquad u[A] = v[A] \\ \text{or } v = (\emptyset, \ldots, \emptyset) \end{cases}$$

If $u \preceq_g v$, we say that u is more general than v in $Space(r)$. In other words, u captures a similar information than v but at a rougher granularity level.

Example 1. Let us consider the relation DOCUMENT$_1$ (*cf* Table 1) yielding the quantities sold by Type, City and Publisher. In the multidimensional space of our relation example, we have: (Novel, ALL, ALL) $\preceq_g$ (Novel, Marseilles, Hachette),

Table 1. Relation example DOCUMENT$_1$

RowId	Type	City	Publisher	Qty
1	Novel	Marseilles	Hachette	2
2	Novel	Marseilles	Collins	2
3	Textbook	Paris	Collins	1
4	Essay	Paris	Collins	6
5	Textbook	Marseilles	Collins	1

i.e. the tuple (Novel, ALL, ALL) is more general than (Novel, Marseilles, Hachette) and (Novel, Marseilles, Hachette) is more specific than (Novel, ALL, ALL).

The two basic operators provided for tuple construction are: Sum (denoted by +) and Product (noted •). The Sum of two tuples yields the most specific tuple which generalizes the two operands. Let u and v be two tuples in $Space(r)$,

$$t = u + v \Leftrightarrow \forall A \in \mathcal{D}, t[A] = \begin{cases} u[A] \text{ if } u[A] = v[A] \\ \text{ALL otherwise.} \end{cases}$$

We say that t is the Sum of the tuples u and v.

Example 2. In our example, we have (Novel, Marseilles, Hachette) + (Novel, Marseilles, Collins)=(Novel, Marseilles, ALL). This means that the tuple (Novel, Marseilles, ALL) is built up from the tuples (Novel, Marseilles, Hachette) and (Novel, Marseilles, Collins).

The Product of two tuples yields the most general tuple which specializes the two operands. If it exists, for these two tuples, a dimension A having distinct and real world values (i.e. existing in the original relation), then the only tuple specializing them is the tuple $(\emptyset, \ldots, \emptyset)$ (apart from it, the tuple sets which can be used to retrieve them are disjoined). Let u and v be two tuples in $Space(r)$, then:

$$t = u \bullet v \Leftrightarrow \begin{cases} t = (\emptyset, \ldots, \emptyset) \text{ if } \exists A \in \mathcal{D} \text{ such that } u[A] \neq v[A] \neq \text{ALL}, \\ \text{otherwise } \forall A \in \mathcal{D} \begin{cases} t[A] = u[A] \text{ if } v[A] = \text{ALL} \\ t[A] = v[A] \text{ if } u[A] = \text{ALL}. \end{cases} \end{cases}$$

We say that t is the Product of the tuples u and v.

Example 3. In our example, we have (Novel, ALL, ALL) • (ALL, Marseilles, ALL) = (Novel, Marseilles, ALL). This means that (Novel, ALL, ALL) and (ALL, Marseilles, ALL) generalize (Novel, Marseilles, ALL) and this latter pattern participates to the construction of (Novel, ALL, ALL) and (ALL, Marseilles, ALL) (directly or not). The tuples (Novel, ALL, ALL) and (Textbook, ALL, ALL) have no common point apart from the tuple of empty values (*i.e.* the tuple $(\emptyset, \emptyset, \emptyset)$).

By providing the multidimensional space of r with the generalization order between tuples and using the above-defined operators Sum and Product, we define an algebraic structure which is called cube lattice. Such a structure provides a sound foundation for several multidimensional data mining issues.

Theorem 1. *Let r be a categorical database relation over $\mathcal{D} \cup \mathcal{M}$. The ordered set $CL(r) = \langle Space(r), \preceq_g \rangle$ is a complete, graded, atomistic and coatomistic lattice, called cube lattice in which Meet ($\bigwedge$) and Join ($\bigvee$) elements are given by:*

1. $\forall\, T \subseteq CL(r),\ \bigwedge T = +_{t \in T}\, t$
2. $\forall\, T \subseteq CL(r),\ \bigvee T = \bullet_{t \in T}\, t$

2.2   Convex Cubes

We recall the definitions of monotone/antimonotone constraints according to $\preceq_g$ and the convex cubes (Casali et al., 2007).

Definition 1 (Monotone/antimonotone constraints)

1. A constraint *Const* is monotone according to the generalization order if and only if: $\forall\, t, u \in CL(r) : [t \preceq_g u$ and $Const(t)] \Rightarrow Const(u)$.
2. A constraint *Const* is antimonotone according to the generalization order if and only if: $\forall\, t, u \in CL(r) : [t \preceq_g u$ and $Const(u)] \Rightarrow Const(t)$.

Theorem 2. *The cube lattice with monotone and/or antimonotone constraints is a convex space which is called convex cube, $ConvexCube(r) = \{t \in CL(r)$ such that $Const(t)\}$. Its upper set U and lower set L are: $L = \min_{\preceq_g}(ConvexCube(r))$ and $U = \max_{\preceq_g}(ConvexCube(r))$.*

The upper set U represents the most specific tuples satisfying the constraint conjunction and the lower set L the most general tuples respecting such a conjunction. Thus U and L result in condensed representations of the convex cube faced with a conjunction of monotone and/or antimonotone constraints.

3   Emerging Cubes

In this section, we introduce the concept of emerging cube. Such cubes capture trends which are not relevant for the users (because under a threshold) but which grow significant or on the contrary general trends which soften but not necessarily disappear. They are of particular interest for data stream analysis because they exhibit trend reversals (Han). For instance, in a web application where continuous flows of received data describe in a detailed way the user navigation, knowing the craze for (in contrast the disinterest in) such or such URL is specially important for the administrator in order to allow at best available ressources according to real and fluctuating needs.

In the remainder of the paper, we only consider the aggregative functions COUNT and SUM. Furthermore to preserve the antimonotone property of SUM, we assume that the measure values are strictly positive. Let us introduce a relative version of these functions.

Definition 2 (Relative Aggregate Functions). Let r be a relation, $t \in CL(r)$ a tuple, and $f \in \{\text{SUM}, \text{COUNT}\}$ an aggregative function. We call $f_{rel}(r, .)$ the relative aggregative function of f for the relation r. $f_{rel}(r, t)$ is the ratio between the value of f for the tuple t and the value of f for the whole relation r (in other words for the tuple $(\text{ALL}, \ldots, \text{ALL})$).

$$f_{rel}(t, r) = \frac{f(t, r)}{f((\text{ALL}, \ldots, \text{ALL}), r)}$$

For instance, the function $\text{COUNT}_{rel}(t, r)$ merely corresponds to $Freq(t, r)$ (the frequency of a multidimensional pattern t in the relation r). We use SUM_{rel} in all our examples because its semantics is intuitive and easy to catch.

Remark 1. According to our assumptions, f is additive and the measure values are strictly positive, we have $0 < f(t) < f((\text{ALL}, \ldots, \text{ALL}))$ and thus $0 < f_{rel}(t, r) < 1$.

Definition 3 (Emerging Tuple). A tuple $t \in CL(r_1 \cup r_2)$ is said emerging from r_1 to r_2 if and only if it satisfies the two following constraints C_1 and C_2:

$$\begin{cases} f_{rel}(t, r_1) < MinThreshold_1 \ (C_1) \\ f_{rel}(t, r_2) \geq MinThreshold_2 \ (C_2) \end{cases}$$

where $MinThreshold_1$ and $MinThreshold_2 \in]0, 1[$

Example 4. Let $MinThreshold_1 = 1/3$ be the threshold for the relation DOCUMENT$_1$ (*cf.* Table 1) and $MinThreshold_2 = 1/5$ the threshold for DOCUMENT$_2$ (*cf.* Table 2), the tuple $t_1 = (\text{Textbook, Paris, ALL})$ is emerging from DOCUMENT$_1$ to DOCUMENT$_2$ because $\text{SUM}_{rel}(t_1, r_1) = 1/12$ and $\text{SUM}_{rel}(t_1, r_2) = 3/15$. In contrast, the tuple $t_2 = (\text{Essay, Marseilles, ALL})$ is not emerging because $\text{SUM}_{rel}(t_2, r_2) = 1/15$.

Table 2. Relation example DOCUMENT$_2$

RowId	Type	City	Publisher	Qty
1	Textbook	Marseilles	Hachette	3
2	Textbook	Paris	Collins	3
3	Textbook	Marseilles	Collins	1
4	Novel	Marseilles	Hachette	3
5	Essay	Paris	Collins	2
6	Essay	Paris	Hachette	2
7	Essay	Marseilles	Collins	1

Definition 4 (Emerging Cube). We call emerging cube the set of all tuples of $CL(r_1 \cup r_2)$ emerging from r_1 to r_2. The emerging cube, noted $EmergingCube$

(r_1, r_2), is a convex cube with the hybrid constraint "t *is emerging from* r_1 *to* r_2". Thus it is defined by: $EmergingCube(r_1, r_2) = \{t \in CL(r_1 \cup r_2) \mid C_1(t) \wedge C_2(t)\}$, with $C_1(t) = f_{rel}(t, r_1) < MinThreshold_1$ and $C_2(t) = f_{rel}(t, r_2) \geq MinThreshold_2$.

$EmergingCube(r_1, r_2)$ is a convex cube with the conjunction of a monotone constraint (C_1) and an antimonotone one (C_2). Thus we can use its borders (*cf* Theorem 2) to assess whether a tuple t is emerging or not.

Definition 5 (Emergence Rate). Let r_1 and r_2 be two unicompatible relations, $t \in CL(r_1 \cup r_2)$ a tuple and f an additive function. The emergence rate of t from r_1 to r_2, noted $ER(t)$, is defined by:

$$ER(t) = \begin{cases} 0 \ if \ f_{rel}(t, r_1) = 0 \ and \ f_{rel}(t, r_2) = 0 \\ \infty \ if \ f_{rel}(t, r_1) = 0 \ and \ f_{rel}(t, r_2) \neq 0 \\ \dfrac{f_{rel}(t, r_2)}{f_{rel}(t, r_1)} \ otherwise. \end{cases}$$

Example 5. Table 3 presents the set of emerging tuples from DOCUMENT$_1$ to DOCUMENT$_2$ for the thresholds $MinThreshold_1 = 1/3$ and $MinThreshold_2 = 1/5$.

Table 3. Set of emerging tuples from DOCUMENT$_1$ to DOCUMENT$_2$

Produit	Ville	Publisher	ER
Textbook	ALL	ALL	2.8
Textbook	Paris	ALL	2.4
Textbook	ALL	Collins	1.6
Textbook	Paris	Collins	2.4
Textbook	ALL	Hachette	∞
Textbook	Marseilles	ALL	3.2
Textbook	Marseilles	Hachette	∞
ALL	ALL	Hachette	3.2
ALL	Marseilles	Hachette	2.4
Novel	ALL	Hachette	1.2
Novel	Marseilles	Hachette	1.2

We observe that when the emergence rate is greater than 1, it characterizes trends significant in r_2 and not so clear-cut in r_1. When the rate is lower than 1, on contrary it highlights immersing trends, relevant in r_1 and not in r_2.

Example 6. From the two relations DOCUMENT$_1$ and DOCUMENT$_2$, we compute $ER((\text{Textbook, Paris, ALL})) = 2.4$. Of course more the emergence rate is high, more the trend is distinctive. Therefore, the quoted tuple means a jump for the textbook sales in paris between DOCUMENT$_1$ and DOCUMENT$_2$.

Proposition 1. *Let* $MinRatio = \frac{MinThreshold_2}{MinThreshold_1}$, $\forall t \in EmergingCube(r_1, r_2)$, *we have* $ER(t) \geq MinRatio$.

The proposition consequence is that emerging tuples have an " *interesting* " emergence rate. For instance let us consider $MinThreshold_1 = 1/3$ and $Min Threshold_2 = 1/5$, then all the emerging tuples have an emergence rate greater than $5/3$.

The emerging cube being a convex cube, it can be represented by its borders without computing and storing the two underlying cubes. This capability is specially attractive because trend reversals, jumping or plunging down, can be very quickly isolated and at the lowest cost.

4   Mining Borders of Emerging Cubes

For computing borders of the cube $EmergingCube(r_1, r_2)$, we give another formulation of the constraints in order to take benefit of existing algorithms proved to be efficient: (*i*) Max-Miner (Bayardo, 1998) or GenMax (Gouda and Zaki, 2001) for the computation of cube maximals and (*ii*) Trans (Eiter and Gottlob, 1995), CTR (Casali et al., 2003a), MCTR (Casali, 2004) and (Gunopulos et al., 1997) for mining minimal cube transversals.

4.1   Cube Transversals

We present the concept of cube transversal, a particular case of hypergraph transversals (Berge, 1989; Eiter and Gottlob, 1995).

Definition 6 (Cube Transversal). Let T be a set of tuples ($T \subseteq CL(r)$) and $t \in CL(r)$ be a tuple. t is a cube transversal of T over $CL(r)$ iff $\forall t' \in T, t \not\preceq_g t'$. t is a minimal cube transversal iff t is a cube transversal and $\forall t' \in CL(r), t'$ is a cube transversal and $t' \preceq_g t \Rightarrow t = t'$. The set of minimal cube transversals of T are denoted by $cTr(T)$ and defined as follows:

- $cTr(T) = \min_{\preceq_g}(\{t \in CL(r) \mid \forall t' \in T, t \not\preceq_g t'\})$

Let $\mathbb{A}$ be an antichain of $CL(r)$ ($\mathbb{A}$ is a set of tuple such that all tuples of $\mathbb{A}$ are not comparable using $\preceq_g$). We can constrain the set of minimal cube transversals of r using $\mathbb{A}$ by enforcing each minimal cube transversal t to be more general than at least one tuple u of the antichain $\mathbb{A}$. The new related definitions are the following:

- $cTr(T, \mathbb{A}) = \{t \in cTr(T) \mid \exists u \in \mathbb{A} : t \preceq_g u\}$

Example 7. With our relation example, we have the following result: $cTr(\text{DOCU-MENT}_1) = \{$ (Novel, Paris, ALL), (Essay, ALL, Hachette), (Essay, Marseilles, ALL), (Textbook, ALL, Hachette), (ALL, Paris, Hachette) $\}$.

4.2   Characterizing Emerging Tuple Using Cube Maximals and Cube Transversals

We come down the constraint "*t is an emerging tuple*" in the search for the frequent cube maximals and minimal cube transversals. It is shown that (C_1) is

Table 4. Set M_1 of maximal frequent tuples of DOCUMENT$_1$

Type	City	Publisher
Novel	Marseilles	Hachette
Essay	Paris	Collins

Table 5. Set M_2 of maximal frequent tuples of DOCUMENT$_2$

Type	City	Publisher
Novel	Marseilles	Hachette
Textbook	Marseilles	Hachette
Textbook	Paris	Collins
Essay	Paris	ALL
Essay	ALL	Collins

a monotone constraint and (C_2) is an antimonotone constraint for the generalization order. The emerging tuples can be represented by the borders: U which encompasses the emerging maximal tuples and L which contains all the emerging minimal tuples.

$$\begin{cases} L = \min_{\preceq_g}(\{t \in CL(r_1 \cup r_2) \mid C_1(t) \wedge C_2(t)\}) \\ U = \max_{\preceq_g}(\{t \in CL(r_1 \cup r_2) \mid C_1(t) \wedge C_2(t)\}) \end{cases}$$

Proposition 2. *Let M_1 and M_2 be the maximal frequent tuples in the relations r_1 and r_2:*

$M_1 = \max_{\preceq_g}(\{t \in CL(r_1) \text{ such that } f_{rel}(t, r_1) \geq MinThreshold_1\})$
$M_2 = \max_{\preceq_g}(\{t \in CL(r_2) \text{ such that } f_{rel}(t, r_2) \geq MinThreshold_2\})$

We can characterize the borders U and L of the set of emerging tuples as follows:

1. $L = cTr(M_1, M_2)$
2. $U = \{t \in M_2 : \exists u \in L : u \preceq_g t\}$

Proof

1. $t \in L \Leftrightarrow t \in \min_{\preceq_g}(\{u \in CL(r_1 \cup r_2) : f_{rel}(u, r_1) \leq MinThreshold_1 \text{ and } f_{rel}(u, r_2) \geq MinThreshold_2\})$
   $\Leftrightarrow t \in \min_{\preceq_g}(\{u \in CL(r_1 \cup r_2) : f_{rel}(u, r_1) \leq MinThreshold_1\})$ and $\exists v \in M_2 : t \preceq_g v$
   $\Leftrightarrow t \in \min_{\preceq_g}(\{u \in CL(r_1 \cup r_2) : \nexists v \in M_1 : u \preceq_g v\})$ and $\exists v \in M_2 : t \preceq_g v$
   $\Leftrightarrow t \in cTr(M_1)$ and $\exists v \in M_2 : t \preceq_g v$
   $\Leftrightarrow t \in cTr(M_1, M_2)$
2. True because $EmergingCube(r_1, r_2)$ is a convex space; thus we can find $t \in U$ which does not generalize any $v \in L$. □

Remark 2. This new characterization using the minimal cube transversals can be applied, in a binary context, for computing emerging patterns (Dong and Li, 2005), as well as borders of patterns constrained by an hybrid conjunction (Raedt and Kramer, 2001) by using the classical concept of transversal

Table 6. Upper set of *EmergingCube* (DOCUMENT₁, DOCUMENT₂)

Type	City	Publisher
Textbook	Marseilles	Hachette
Novel	Marseilles	Hachette
Textbook	Paris	Collins

Table 7. Lower set of *EmergingCube* (DOCUMENT₁, DOCUMENT₂)

Type	City	Publisher
ALL	ALL	Hachette
Textbook	Marseilles	ALL

(Eiter and Gottlob, 1995). It is in the very same spirit of the theoritical connection between positive and negative borders introduce by *Mannila et al.* (Mannila and Toivonen, 1997) to study the complexity of frequent pattern mining.

Example 8. Let us consider the relations $r_1 = $ DOCUMENT₁ and $r_2 = $ DOCUMENT₂. The sets M_1 and M_2 are given in Tables 4 and 5.

The borders of $EmergingCube(r_1, r_2)$ with $MinThreshold_1 = 1/3$ and $MinThreshold_2 = 1/5$ are presented in Tables 6 and 7. Provided with these borders, we know that there is a jump in textbook sales in Marseilles because (Textbook, Marseilles, ALL) generalizes the tuple (Textbook, Marseilles, Hachette) which belongs to L. In contrast, nothing can be said about the tuple (Textbook, ALL, Collins) because it does not generalize any tuple of L.

5   Conclusion

In this paper, we have presented the concept of emerging cube which captures trend reversals between two categorical database relations along with a suitable and condensed representation through the borders U and L. We have shown that existing and efficient algorithms can be used for computing such borders. As a future work, we would like to answer the following questions. Are the borders U and L the smallest representation of emerging cubes? Does a more concise representation exist? The borders U and L do not make possible to retrieve the measures of any emerging tuple. Thus another research perspective is to investigate a new representation, the emerging closed cube, based on the cube closure (Casali et al., 2003a) and show how to extract an emerging cuboid from the emerging closed cube.

References

Bayardo, R.: Efficiently mining long patterns from databases. In: Proceedings of the International Conference on Management of Data, SIGMOD, pp. 85–93 (1998)

Berge, C.: Hypergraphs: combinatorics of finite sets, North-Holland, Amsterdam (1989)

Casali, A.: Mining borders of the difference of two datacubes. In: Kambayashi, Y., Mohania, M.K., Wöß, W. (eds.) DaWaK 2004. LNCS, vol. 3181, pp. 391–400. Springer, Heidelberg (2004)

Casali, A., Cicchetti, R., Lakhal, L.: Extracting semantics from datacubes using cube transversals and closures. In: Proceedings of the 9th ACM SIGKDD International Conference on Knowledge Discovery and Data Mining, KDD, pp. 69–78. ACM Press, New York (2003a)

Casali, A., Cicchetti, R., Lakhal, L.: Cube lattices: a framework for multidimensional data mining. In: Proceedings of the 3rd SIAM International Conference on Data Mining, SDM, pp. 304–308 (2003b)

Casali, A., Nedjar, S., Cicchetti, R., Lakhal, L.: Convex Cube: Towards a Unified Structure for Multidimensional Databases. In: Proceedings of the 18th international conference on database and expert systems applications, dexa (2007)

Dong, G., Li, J.: Mining border descriptions of emerging patterns from dataset pairs. Knowledge Information System 8(2), 178–202 (2005)

Eiter, T., Gottlob, G.: Identifying The Minimal Transversals of a Hypergraph and Related Problems. SIAM Journal on Computing 24(6), 1278–1304 (1995)

Gouda, K., Zaki, M.: Efficiently Mining Maximal Frequent Itemsets. In: Proceedings of the 1st IEEE International Conference on Data Mining, ICDM, pp. 3163–3170. IEEE Computer Society Press, Los Alamitos (2001)

Gray, J., Chaudhuri, S., Bosworth, A., Layman, A., Reichart, D., Venkatrao, M., Pellow, F., Pirahesh, H.: Data cube: A relational aggregation operator generalizing group-by, cross-tab, and sub-totals. Data Mining and Knowledge Discovery 1(1), 29–53 (1997)

Gunopulos, D., Mannila, H., Khardon, R., Toivonen, H.: Data mining, hypergraph transversals, and machine learning. In: Proceedings of the 16th Symposium on Principles of Database Systems, PODS, pp. 209–216 (1997)

Han, J.: Mining dynamics of data streams in multi-dimensional space, http://www-faculty.cs.uiuc.edu/ hanj/projs/streamine.htm

Han, J., Chen, Y., Dong, G., Pei, J., Wah, B.W., Wang, J., Cai, Y.D.: Stream cube: An architecture for multi-dimensional analysis of data streams. Distributed and Parallel Databases 18(2), 173–197 (2005)

Imielinski, T., Khachiyan, L., Abdulghani, A.: Cubegrades: Generalizing association rules. Data Mining and Knowledge Discovery, DMKD 6(3), 219–257 (2002)

Lakshmanan, L., Pei, J., Han, J.: Quotient cube: How to summarize the semantics of a data cube. In: Proceedings of the 28th International Conference on Very Large Databases, VLDB, pp. 778–789 (2002)

Mannila, H., Toivonen, H.: Levelwise Search and Borders of Theories in Knowledge Discovery. Data Mining and Knowledge Discovery 1(3), 241–258 (1997)

Mitchell, T.M.: Machine learning. MacGraw-Hill Series in Computer Science (1997)

Nedjar, S., Casali, A., Cicchetti, R., Lakhal, L.: Cocktail de cubes. In: Actes des 22ème journées Bases de Données Avancées, BDA (2006)

Raedt, L., Kramer, S.: The Levelwise Version Space Algorithm and its Application to Molecular Fragment Finding. In: Proceedings of the 17th International Joint Conference on Artificial Intelligence, IJCAI, pp. 853–862 (2001)

Sarawagi, S., Agrawal, R., Megiddo, N.: Discovery-driven exploration of olap data cubes. In: Schek, H.-J., Saltor, F., Ramos, I., Alonso, G. (eds.) EDBT 1998. LNCS, vol. 1377, pp. 168–182. Springer, Heidelberg (1998)

Domination Mining and Querying

Apostolos N. Papadopoulos, Apostolos Lyritsis, Alexandros Nanopoulos,
and Yannis Manolopoulos

Department of Informatics, Aristotle University,
Thessaloniki 54124, Greece
{papadopo,lyritsis,ananopou,manolopo}@csd.auth.gr

Abstract. Pareto dominance plays an important role in diverse application domains such as economics and e-commerce, and it is widely being used in multicriteria decision making. In these cases, objectives are
usually contradictory and therefore it is not straightforward to provide
a set of items that are the "best" according to the user's preferences.
Skyline queries have been extensively used to recommend the most dominant items. However, in some cases skyline items are either too few, or
too many, causing problems in selecting the prevailing ones. The number of skyline items depend heavily on both the data distribution, the
data population and the dimensionality of the data set. In this work, we
provide a dominance-based analysis and querying scheme that aims at
alleviating the skyline cardinality problem, trying to introduce ranking
on the items. The proposed scheme can be used either as a mining or as
a querying tool, helping the user in selecting the mostly preferred items.
Performance evaluation based on different distributions, populations and
dimensionalities show the effectiveness of the proposed scheme.

1   Introduction

Preference queries are frequently used in multicriteria decision making applications, where a number of (usually) contradictory criteria participate towards
selecting the most convenient answers to the user. Each item is represented as a
multidimensional point.

Assume that someone is interested in purchasing a PDA device. Unfortunately,
there are a lot of criteria that should be balanced before a wise choice is made.
Assume further that the customer focuses on two important parameters of a
PDA, the price, and the weight of the device. Therefore, the "best" PDAs are
the ones that are cheap and light-weighted. Unfortunately, these two criteria
are frequently contradictory and therefore, the number of candidates should be
carefully selected.

In this example, we have two attributes, and the user is interested in items
that have values in these attributes as minimum as possible. Depending on the
semantics of each attribute, in other cases the user may ask for maximization of
the attributes, or any combination (minimization in some attributes and maximization in the others). For example, if the user focuses on price and available
memory, the "best" PDAs are the ones that are as cheap as possible and have

I.Y. Song, J. Eder, and T.M. Nguyen (Eds.): DaWaK 2007, LNCS 4654, pp. 145–156, 2007.

the largest available memory. Without loss of generality, in the sequel we focus on minimizing the attributes of interest.

A fundamental preference query is the skyline query. The skyline of a set of points S comprises all points that are not dominated. A point p_i dominates another point p_j, if p_i is as good as p_j in all dimensions and it is better than p_j in at least one of the dimensions. Let d be the total number of attributes (dimensions) and $p_i.a_m$ denote the m-th dimension value of point p_i. Since we have assumed that "smaller is better", p_i is better than p_j in dimension a_m if $p_i.a_m < p_k.a_m$.

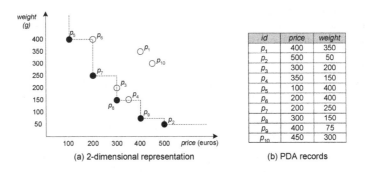

(a) 2-dimensional representation   (b) PDA records

Fig. 1. Skyline example

The PDA example is depicted in Figure 1. In Figure 1(a) each PDA is represented by a two-dimensional point (each dimension corresponds to an attribute). PDA records are shown in Figure 1(b). Points connected by the dashed line comprise the skyline of the set of points (PDAs). Any point falling on the right and top of the dashed line is dominated by at least on skyline point.

The black dots of Figure 1(a) represent the skyline. These points are not dominated by any other point. On the other hand, points p_6, p_1, p_3, p_4 are dominated by at least one other point. For example, PDA p_6 is dominated by p_5 because $p_5.weight = p_6.weight$, but $p_5.price < p_6.price$. Therefore, none of these points can be part of the skyline. The skyline is therefore consists of points p_5, p_7, p_8, p_9 and p_2.

The number of skyline points depends heavily on the dimensionality of the data set (number of attributes) and the data distribution. Therefore, according to these factors, in some cases the number of skyline points are too few, too many. Moreover, in some cases the skyline points may not be available. In our PDA example, it is expected that PDAs belonging to the skyline are too popular, and therefore purchase may not be possible due to lack of stock. In these cases, a ranking should be enforced to the multidimensional points, and interesting items should be presented to the user. This is exactly the topic of this paper. More specifically, we provide a meaningful way of ranking multidimensional points according to their domination power, towards presenting the items to the user in an incremental manner. The proposed techniques constitute a way to facilitate mining and querying for dominance relationships.

The rest of the work is organized as follows. Section 2 presents the related work in the area, and describes our contributions shortly. Our proposal is given in Section 3, whereas Section 4 gives some representative experimental results. Finally, Section 5 concludes the work and briefly describes future work in the area.

2   Related Work

Skyline queries have received considerable attention recently, due to their aid in selecting the most preferred items, especially when the selection criteria are contradictory. Although the problem has been attacked in the past by a number of researchers [2,8], only recently it has been tackled under a database point of view [3].

The current literature is rich in algorithms and organization schemes to facilitate skyline query processing. In [11] an efficient skyline query processing scheme has been proposed based on branch-and-bound, which utilizes the R-tree spatial access method [5,1]. This method shows significant improvement over previously proposed methods. This scheme however, assumes that the skyline is computed over the whole set of attributes, which in many cases may not be meaningful.

Recently, alternative solutions have been proposed towards helping the user even further in selecting the most promising items. The output of a skyline query may contain too few, or too many answers, posing difficulties in selecting the best items. Towards alleviating this problem, k-dominant skylines have been proposed in [4]. According to this method, the definition of dominance is relaxed, in order to enable some points to be dominated, reducing the cardinality of the skyline point-set.

Another approach has been followed by [6] where the proposed technique searches for *thick* skylines. The thick skyline is composed of some skyline points and some additional points which are close to skyline points but not necessarily contained in the skyline set. This way, only points in these dense areas are proposed to the user.

In the same lines, an algorithm is proposed in [10] for selecting skyline points according to their domination capabilities. More specifically, the algorithm selects a subset of the skyline points aiming at maximizing the number of dominated points. However, this method is NP-hard for high-dimensional spaces, and therefore approximation algorithms are required towards fast computation.

In [9] the authors study dominance relationships between different data sets (e.g., products and customers). The authors propose DADA cube, an organization scheme to support a number of significant query types towards analyzing dominance relationships. DADA cube has been designed in accordance to data cubes utilized in data warehouses.

Finally, [7] study algorithmic techniques to compute approximately dominating representatives (ADRs). ADRs can be computed by appropriately post-processing the result of a skyline query. The authors show that The authors show that the problem of minimizing the number of points returned can be solved in polynomial time in two dimensions, whereas it is NP-hard in higher dimensionalities. However, there is a polynomial-time logarithmic approximation.

Our research focuses on helping the user quantify the significance of items based on domination power. Therefore, the proposed scheme is mostly related to research contributions [6,10,9] because: (i) as in [6] items other than skyline items may be proposed, (ii) as in [10] our scheme returns items according their domination abilities, and (iii) as in [9] we investigate dominance relationships, However, our work differs significantly from the previous contributions. A transformation is applied which maps the items from the original d-dimensional space to a 2-dimensional one. Mining and ranking of items is performed to the new space (target space), avoiding the excess skyline cardinality that appears in high dimensions. Additionally, our proposed scheme handles cases where the skyline points are too few (e.g., in correlated data sets).

3   Domination Mining and Querying

Some skyline points are more significant than others. The quantification of the significance can be performed in a number of ways. For example, significance can be measured by means of the number of points dominated by a skyline point. A skyline point p_i is more significant than another skyline point p_j if the number of points dominated by p_i is larger than that dominated by p_j. This is the main concept of the scheme proposed in [10] for selecting the most representative skyline points.

However, in several cases we have to consider non-skyline points as well, which may also be of interest to the user. This becomes more clear by investigating Figure 1. It is evident that p_2 is a skyline point. However, although the weight of this PDA is very low, its price is high in relation to the other devices (in fact it has the largest price of all). It would be better, if PDAs p_3 and p_4 are also proposed to the user, since they better compromise the weight/price tradeoff. This example shows that it would be more appropriate to extend the concept of significance for all points in the data set, instead of considering skyline points only as in [10].

Another observation is that thick skylines that have been proposed in [6] do not solve our problem either, since a skyline point belonging to a dense area does not necessarily means that is more important than a skyline point lying in a sparse area. For example, consider a new PDA p with $p.price = 100$, and $p.weight = 50$. Evidently, although this point belongs to a sparse area, it is the most important skyline point since it has the smallest price and the smallest weight in comparison to all other PDAs. Moreover, we argue that sparse skyline points may be very important, since they are far from the competition.

3.1   Determining the Target Space

In the sequel we present three different transformation methods towards taking into account the *domination power* of each point. Points with large domination power should be proposed to the user before points with less domination power. According to our example, it is evident that the PDA represented by point p_8

has more domination power than that represented by p_5. In the sequel we will define domination power in a more formal way.

The first transformation (T_1), is the simplest to implement. Each multidimensional point p is transformed to a 2-dimensional point $T_1(p) = (da(p), 1 - DA(p)/a)$, where $da(p)$ is the area dominating p, whereas $DA(p)$ is the area dominated by p. Evidently, the values of $da(p)$ and $DA(p)$ are distribution independent, since they are computed based only on the location of p in the data space. This transformation may be used when only the domination *potential* of points is of interest.

$$T_1(p) = \left(\frac{da(p)}{a}, \ 1 - \frac{DA(p)}{a} \right) \tag{1}$$

The second transformation (T_2), tries to capture the fact that the significance of a point p is increased when, the number of points dominating p decreases, and the number of points dominated by p increases. Based on this observation, each point p is transformed to $T_2(p) = (dps(p), 1 - DSP(p)/n)$, where $dps(p)$ is the number of points dominating p, and $DPS(p)$ is the number of points dominated by p. In contrast to the previous transformation, this one is distribution dependent, since only domination information is being taken into account, whereas the absolute location of p in the data space is completely ignored. This transformation should be used when the domination relationships among points needs to be investigated.

$$T_2(p) = \left(\frac{dps(p)}{n}, \ 1 - \frac{DPS(p)}{n} \right) \tag{2}$$

The next transformation (T_3) tries to combine the previous two, by proposing a hybrid scheme. Each point p is transformed to $T_3(p) = (da(p)/a, DPS(p)/n)$, where again $da(p)$ is the area that dominates p and $DPS(p)$ is the set of points dominated by pp.

$$T_3(p) = \left(\frac{da(p)}{a}, \ 1 - \frac{DPS(p)}{n} \right) \tag{3}$$

Note that, all methods transform the d-dimensional points of P to 2-d, in order to reduce the cardinality of skyline sets in the target space, and provide a more convenient space for mining and querying dominance relationships.

Next we define two important measures, the *domination power* of a point and the *domination distance* between two points. The first one aims to quantify the significance of each point, whereas the second one quantifies the difference between two points in terms of their distance to the target space. Note that, since the target space is defined in the unit square, the measures assume values in the interval [0,2].

Definition 1 (*domination power*)
The domination power $P(p)$ of a point p is the cityblock distance of its image p' point to the origin of the target space. More formally:

$$P(p) = p'.x + p'.y \tag{4}$$

Definition 2 (*domination distance*)
The domination distance $D(p,q)$ between two points p and q, is the cityblock distance of their images p' and q' in the target space. More formally:

$$D(p,q) = |p'.x - q'.x| + |p'.y - q'.y| \tag{5}$$

3.2   Creating and Maintaining the Target Space

There are two important issues to be resolved regarding the generation of the target space: (i) target space generation and (ii) target space maintenance. Among the three different transformations described previously, the simplest to generate and maintain is the first one (T_1), since the image p' of a point p depends only on the location of the point itself in the original space. On the other hand, the most expensive transformation is the second (T_2), because the image of a point depends on the locations of the other points. Moreover, the insertion/deletion of a point may cause changes in the images of other points. In the sequel, we describe the generation of the target space according to T_2, since the process for T_1 and T_3 are similar and less demanding regarding computations.

Without loss of generality, we assume the existence of an R-tree access method [5,1] to organize the points in the original space. Other hierarchical access methods can be utilized equally well. The main issue regarding T_2 is that for every point p we need to calculate the values for $dps(p)$ and $DPS(p)$. A naive approach to follow is for each point to apply two range queries in the R-tree, one for the region $da(p)$ and one for the region $DA(p)$. Evidently, $da(p)$ contains all points that dominate p, whereas $DA(p)$ contains all points dominated by p. By using an arbitrary examination order, it is expected that significant I/O overhead will take place. Instead, we use an approach which better exploits locality of references. This approach is based on the Hilbert space filling curve. In the first phase the value $dps(p)$ is calculated for every point p. In the second phase the $DPS(p)$ value is calculated. Both phases respect the Hilbert order in visiting the points. Internal nodes are prioritized based on the Hilbert value of their MBR centroid.

A significant improvement can be applied to both algorithms towards reducing the number of I/O requests. This involves the utilization of batched processing. When a leaf node is reached, instead of executing multiple range queries for each point in the leaf, we execute only one in each phase. Our intuition regarding the efficiency of this variation is verified by the computational and I/O time required to build the target space as shown in the experimental results. The improved algorithm can be used both when all data are memory-resident or disk-based. The outline of the final algorithm is illustrated in Figure 2.

Since the suggested methodology is focused on analysis and mining, keeping the target space up-to-date upon every insertion and deletion is not our primary concern. However, if such a synchronization action is required this is easily achieved for T_1. More specifically, if a new point p is inserted in the original space, its image is directly calculated by using Equation 1. For T_2, two range queries must be executed in the R-tree, one for $da(p)$ and one for $DA(p)$ to produce the image of

---

Algorithm CreateTargetSpace (*Rtree*)

Rtree: the R-tree organizing the original space

minheap: the priority queue of tree entries prioritized by increasing Hilbert values

maxheap: the priority queue of tree entries prioritized by decreasing Hilbert values

---

1.   *treeNode* = root of the *Rtree*
2.   place entries of *treeNode* into *minHeap*
3.   **while** (*minHeap* not empty)
4.   *heapEntry* = top of *minHeap*
5.   **if** (*heapEntry* points to a leaf node)
6.   *leafNode* = the leaf pointed by *heapentry*
7.   execute range query using upper-right corner of leaf MBR
8.   update dominance information
9.   **else**
10.   *internalNode* = node pointed by *heapEntry*
11.   calculate Hilbert values and insert all entries of *internalNode* into *minHeap*
12.   **end if**
13. **end while**
14. *treeNode* = root of the *Rtree*
15. place entries of *treeNode* into *maxHeap*
16. **while** (*maxHeap* not empty)
17.   *heapEntry* = top of *maxHeap*
18.   **if** (*heapEntry* points to a leaf node)
19.   *leafNode* = the leaf pointed by *heapentry*
20.   execute range query using lower-left corner of leaf MBR
21.   update dominance information
22.   **else**
23.   *internalNode* = node pointed by *heapEntry*
24.   calculate Hilbert values and insert all entries of *internalNode* into *maxHeap*
25.   **end if**
26. **end while**

Fig. 2. Outline of CreateTargetSpace algorithm

p. Moreover, a complete synchronization requires additional updates because: (i) the total number of points has been changed, (ii) for each point p_x that dominated p the value $DPS(p_x)$ must be updated, and (iii) for each point p_x dominated by p the value $dps(p_x)$ must be updated. The case of T_3 is slightly more efficient, since only one range query is required, but several updates of images may be required. To avoid increased maintenance costs when T_2 or T_3 is being used, updates in the target space may be performed periodically. T_3 is the best compromise between maintenance efficiency and perception of domination information.

3.3   Utilizing the Target Space

Domination power and domination distance allow for a number of important querying and mining operations to be supported. The first important operation is the determination of the "best" points, by executing a skyline computation algorithm in the target space, such as the one proposed in [11]. If the user is not satisfied by the answers, the next layer of skylines can be computed, by ignoring the previously returned points. This process may continue according to the user's request. Note that, since the dimensionality of the target space is 2 and the distribution of the images show a high degree of correlation, the number of skyline points is not expected to be large. By supporting multiple skyline layers we alleviate the problem of the small skyline cardinality.

Another important operation is the top-k query, which given an integer k, returns the k best points, according to domination power. Since the target space is already computed, such a top-k query is immediately supported by applying a branch-and-bound search algorithm with a priority queue.

The third operation is related to nearest-neighbors in the target space. Given a point q, a k-nearest-neighbor (k-NN) query returns the k points closest to q', according to the domination distance measure. This way, one can determine which points "threaten" q with respect to domination distance. A similar query can be defined by given the query point q and a domination distance threshold e. In this case, we are interested in determining which point images p' are within distance e from q' (range query). Evidently, if two images are close in the target space, it is not necessary that the original points will be close to each other in the original space, and in fact this is one of the most important properties of the target space.

Apart from using the target space in a online fashion, data mining techniques can be applied to discover useful knowledge and support the online part in a number of ways. A useful operation is to detect the outliers in the target space. Points that are quite "isolated" from the competition are the candidates for being outliers. Regarding domination relationships, an outlier may be significant, and therefore it is important to give the chance to the user to check it. However, not all outliers need to be presented, but only the most significant ones with respect to domination power. It is up to the user to select all or the most significant ones for further inspection.

A data mining task related to outlier detection is clustering. In fact, many clustering algorithms support outlier detection (e.g., DBSCAN). The application of clustering in the target space will partition the images of points to a number of groups. Each group is expected to contain "similar" points regarding their domination characteristics. Additionally, clustering can be utilized towards partitioning the target space and then use the partitions to suggest items to users. This way, instead of proposing single points, a whole cluster of points is suggested to the user. Cluster selection may be applied by using the domination power of the cluster centroid. Therefore, the cluster whose centroid is closer to the origin of the target space is proposed first.

4   Experiments and Results

In the sequel, we present some representative results. All experiments have been conducted on a Pentium IV machine at 3GHz, with 1GB RAM running Windows

Table 1. Cost for target space generation (10,000 points in 2-d)

data set	**NAIVE** (IO/CPU)	**2P** (IO/CPU)	**2P-BATCH** (IO/CPU)
IND	1199297/117.73	973156/115.78	25719/4.13
ANTI	412939/32.80	137797/31.34	7873/1.53
CORR	1644380/165.68	1430990/162.70	33747/5.40
GAUSS	1253231/116.45	1030278/113.87	25849/4.00

XP Prof. We have used four synthetically generated data sets with different distributions: (i) anticorrelated (ANTI), (ii) indepednent (IND), (iii) gaussian (GAUSS) and (iv) correlated (CORR). The data sets have been generated by using the generation process reported in [3].

Initially, we present some representative results regarding the efficiency of target space generation, by comparing the naive method (algorithm NAIVE), with the two-phase Hilbert-based method (algorithm 2P) and the improved batch-enabled alternative (algorithm 2P-BATCH). Table 1 illustrates some representative performance comparison results for 2-dimensional data sets. CPU cost is measured in seconds, whereas the I/O cost corresponds to the number of disk accesses performed due to buffer misses. Figure 3 shows the scalability of the algorithms for different dimensionalities for the IND data set. An LRU-based buffer has been used which maintains the 25% of the total storage requirements. Each experiment has been executed on a data set containing 10,000 multidimensional points, whereas the page size of the R-tree has been set to 2 KBytes. It is evident, that 2P-BATCH has the best overall performance, both in terms of computational time and I/O cost.

In the sequel, we investigate how clustering algorithms can aid towards exploration of the target space. Figure 4 illustrates the distribution of the point images in the target space for all available transformation methods (T_1, T_2, and T_3), for the ANTI data set containing 1,000 points. We have chosen this set because the

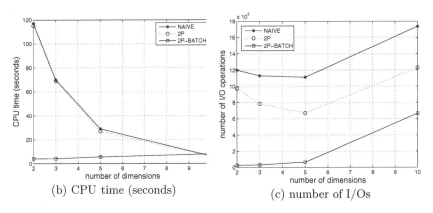

(b) CPU time (seconds)   (c) number of I/Os

Fig. 3. Comparison of methods for different dimensions (10,000 IND points)

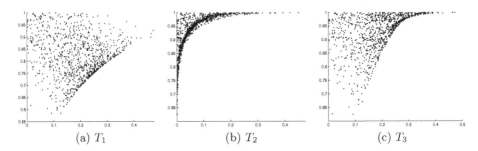

(a) T_1 (b) T_2 (c) T_3

Fig. 4. Target spaces for ANTI data set (1,000 points in 2 dimensions)

number of skyline points in the original space is quite large, and therefore there are difficulties in proposing some representative points to the user.

By inspecting Figure 4 we can see that the distribution of points in the target space is completely different than that in the original one. The transformation performed offers an opportunity to select the "best" points much more easily. This can be performed by other using the queries described in Section 3.3 or by performing a partitioning of the target space by means of clustering. To illustrate this issue, we have applied two fundamental clustering algorithms: (i) DBSCAN and (ii) K-means. By using DBSCAN we aim at discovering outliers whereas the use of K-means aims at space partitioning.

Figure 5 shows the outliers determined for ANTI, CORR and GAUSS by using the transformation T_3. In addition to the target space ((a), (b) and (c)), the position of the each outlier is also given in the original space ((d), (e) and (f)). As we have already mention, outliers may contain important information since they refer to points that are away from the competition, and therefore these points may require further investigation. In fact, as it is shown in Figures 5(a), (c), (e) and (f), some of the outliers are really important because their domination power is significant. It is up to the user to request the outliers towards further study.

Let us see now how K-means clustering can help. By applying K-means on the target space, the points are partitioned into groups. We apply clustering based on the cityblock distance of the points. Figure 6 gives the clustering results for GAUSS data set, for all transformations. Only the original space partitioning is shown. Clustering has been performed for 7 clusters. However, other values can be used as well. By inspecting Figure 6 we observe that clusters in T_1 are more crisp, in the sense that points are grouped together with respect to the diagonal they belong. This was expected since T_1 is based on the coordinates of each point only, and the image of each point is generated without taking into account the other points locations. On the other hand, T_2 and T_3 better express the notion of domination power. In both transformations, the shape of the clusters is quite different than that produced with T_1. For example, in T_2, the points indicated by arrows although they belong to the skyline, they receive a lower rank because the number of points they dominate is zero or very small. Moreover, notice that points in the same cluster does not necessarily share nearby positions in

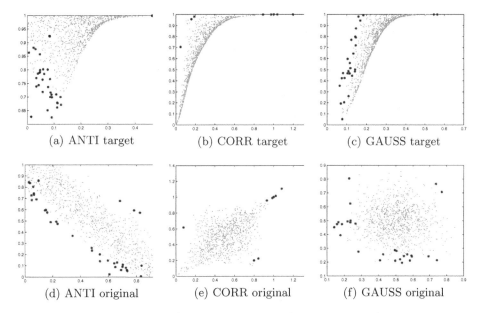

Fig. 5. Outlier detection for ANTI, CORR and GAUSS data sets (1,000 points in 2 dimensions)

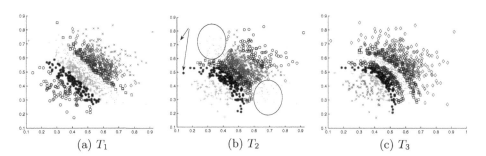

Fig. 6. Data partitioning with K-means for $K = 7$ (1,000 points, GAUSS data set)

the original space. The points enclosed by the two ellipses belong to the same cluster. T_3 produces similar results to those of T_2, but the emphasis is given on the number of points that dominate a particular point and the area that this point dominates (not the number of points dominated).

5   Concluding Remarks

Although the literature is rich in studying efficient algorithms for domination computation, only recently the issue of analyzing domination relationships has received some attention. Towards this direction, we have focused on applying well-established data mining techniques and query operations aiming at a more convenient scheme for determining the "best" items and extracting useful knowledge from domination relationships. Future research in the area may focus on: (i) mining domination relationships among different types of items, (ii) measuring the domination power in a set of items instead of individual ones, (iii) the evaluation of different metrics for domination power and distance and (iv) enhanced visualization methods for dominance relationships.

References

1. Beckmann, N., Kriegel, H.-P., Schneider, R., Seeger, B.: The R*-tree: an Efficient and Robust Access Method for Points and Rectangles. In: Proceedings of the ACM SIGMOD Conference, May 1990, Atlantic City, NJ, pp. 322–331 (1990)
2. Bentley, J.L., Clarkson, K.L., Levine, D.B.: Fast Linear Expected-Time Algorithms for Computing Maxima and Convex Hulls. In: Symposium on Discrete Algorithms, pp. 179–187 (1990)
3. Borzsonyi, S., Kossmann, D., Stocker, K.: The Skyline Operator. In: Proceedings of the International Conference on Data Engineering, pp. 421–430 (2001)
4. Chan, C.-Y., Jagadish, H.V., Tan, K.-L., Tung, A.K.H., Zhang, Z.: Finding k-Dominant Skylines in High Dimensional Space. In: Proceedings of the ACM SIGMOD Conference, pp. 503–514 (2006)
5. Guttman, A.: R-trees: a dynamic index structure for spatial searching. In: Proceedings of the ACM SIGMOD Conference, pp. 47–57 (1984)
6. Jin, W., Han, J., Ester, M.: Mining Thick Skylines over Large Databases. In: Boulicaut, J.-F., Esposito, F., Giannotti, F., Pedreschi, D. (eds.) PKDD 2004. LNCS (LNAI), vol. 3202, Springer, Heidelberg (2004)
7. Koltun, V., Papadimitriou, C.H.: Approximately Dominating Representatives. Theoretical Computer Science 371(3), 148–154 (2007)
8. Kung, H.T.: On Finding the Maxima of a Set of Vectors. Journal of the ACM 22(4) (1975)
9. Li, C., Ooi, B.C., Tung, A.K.H., Wang, S.: DADA: A Data Cube for Dominant Relationship Analysis. In: Proceedings of the ACM SIGMOD Conference, pp. 659–670 (2006)
10. Lin, X., Yuan, Y., Zhang, Q., Zhang, Y.: Selecting Stars: The k Most Representative Skyline Operator. In: Proceedings of the 23rd International Confernce on Data Engineering (2007)
11. Papadias, D., Tao, Y., Fu, G., Seeger, B.: Progressive Skyline Computation in Database Systems. ACM Transactions on Database Systems 30(1) (2005)

Semantic Knowledge Integration to Support Inductive Query Optimization

Nittaya Kerdprasop and Kittisak Kerdprasop

Data Engineering and Knowledge Discovery Research Unit,
School of Computer Engineering, Suranaree University of Technology,
111 University Avenue, Nakhon Ratchasima 30000, Thailand
{nittaya,kerdpras}@sut.ac.th

Abstract. We study query evaluation within a framework of inductive databases. An inductive database is a concept of the next generation database in that the repository should contain not only persistent and derived data, but also the patterns of stored data in a unified format. Hence, the database management system should support both data processing and data mining tasks. Having provided with a tightly-coupling environment, users can then interact with the system to create, access, and modify data as well as to induce and query mining patterns. In this paper, we present a framework and techniques of query evaluation in such an environment so that the induced patterns can play a key role as semantic knowledge in the query rewriting and optimization process. Our knowledge induction approach is based on rough set theory. We present the knowledge induction algorithm driven by a user's query and explain the method through running examples. The advantages of the proposed techniques are confirmed with experimental results.

1 Introduction

Since the emerging of knowledge discovery in databases (KDD), or data mining, as a new multi-disciplinary research area in the 1990's [9], it has been soon realized that the current database system should be extended or re-designed to support the KDD process. Imielinski and Mannila [13] have argued that existing KDD techniques are simply file mining because the inductive learning tools are built on top of the databases assuming a loose coupling between the two components. To fulfill a database mining concept, the mining engine has to be tightly coupled with the database system. In recent years, this idea has been realized and several research work along this line have been developed [2, 20, 23]. The tightly integration of databases with data mining gives rise to the new concept of inductive databases [5, 7, 19].

An inductive database is a database that contains not only data, but also patterns which are generalized information induced from data. By providing this tightly integration framework of data management system and pattern discovery engine, users can access patterns in the same manner as querying data. To achieve this aim a number of SQL-based inductive query languages, such as DMQL [11], MINE RULE [4], MSQL [14], have been proposed and implemented. Most of these languages are an SQL extension with some primitives to support the data mining task, that is, users can pose queries to induce, access and update patterns.

I.Y. Song, J. Eder, and T.M. Nguyen (Eds.): DaWaK 2007, LNCS 4654, pp. 157–169, 2007.

Besides the front-end functionalities, we propose that the induced patterns can also be useful in the back-end part of query answering. The induced patterns are viewed as a repository of semantic knowledge highly beneficial to the optimization process. The purpose of query optimization is to rewrite a given query into an equivalent one that uses less time and resources. Equivalence is defined in terms of identical answer sets. Query optimization utilizes syntactic and logic equivalence transformation of a given query expression. Semantic query optimization (SQO), on the contrary, uses not only syntactic transformations, but also semantic knowledge, such as integrity constraints and various forms of data generalization, to transform a query into an optimized one.

Early work on SQO [10, 15] transforms query by reasoning via heuristics such as index and restriction introduction, join elimination, contradiction detection. Since the introduction of SQO concept in 1981 [15], semantic-based transformation techniques have been developed constantly. Some proposed techniques in the literature are resolution refutation method [6], knowledge deduction [24], knowledge induction [25]. Recently, the interest on SQO has moved toward the setting of intelligent query answering [12, 17], which is defined as a procedure that can answer incorrect or incompletely specified query cooperatively and intelligently. The intelligence is obtained by analyzing the intent of a query and provide some generalized or associated answers. Necib and Freytag [21] propose an ontology-based optimization approach to rewrite a query into another one which is not necessary equivalent but can provide more meaningful result satisfying the user's intention.

Our research follows the direction of intelligent query answering with the emphasis on semantic-based optimization. We consider acquiring semantic knowledge using a rough set approach. By means of a rough set theory certain knowledge as well as rough (or vague) knowledge can be induced from the database content. The main purpose of this paper is to illustrate the idea of inducing and integrating certain and rough knowledge in the query rewriting and optimization process to produce an intelligent answer. We present the optimization process within the framework of inductive database systems that both data content and patterns are stored in the databases. Unlike previous work on inductive databases that express queries using a logic-based language [3, 4, 8], we formalize our idea based on a structured query language (SQL) as it is a typical format used extensively in most database systems.

The remainder of this paper is organized as follows. In section 2, we review the two important foundations of our work, that is, the relational inductive databases and rough set theory. We present our framework and algorithm of semantic knowledge induction using rough set concept in section 3. Section 4 illustrates the steps in query optimization with some experimental results. Section 5 concludes the paper and discusses the plausible extension of this research.

2   Preliminaries

2.1 Inductive Database Concept

Inductive databases can be viewed as an extension of the traditional database systems in that the databases do not only store data, but they also contain patterns of those data. Mannila [18] formalized a framework of inductive database $\mathcal{I}$ as a pair $(\mathcal{R}, \mathcal{P})$ where $\mathcal{R}$ is a database relation and $\mathcal{P}$ is a nested relation of the form $(\mathcal{Q}_{\mathcal{R}}, e)$ in which $\mathcal{Q}_{\mathcal{R}}$ is a set of patterns obtained from querying the base data and e is the evaluation function measuring some metrics over the patterns. As an example, consider the database (adapted from [3]) consisting of one base relation, $\mathcal{R}$. The induced patterns $\mathcal{P}$ are a set of rules represented as an implication LHS $\Rightarrow$ RHS; therefore, $\mathcal{Q}_{\mathcal{R}}$ = { LHS $\Rightarrow$ RHS | LHS, RHS $\subseteq \mathcal{R}$ } and the rule's quality metrics are support and confidence [1]. An inductive database $\mathcal{I} = (\mathcal{R}, \mathcal{P})$ containing one base relation $\mathcal{R}$ and a set $\mathcal{P}$ of all association patterns induced from $\mathcal{R}$ is shown in figure 1.

$\mathcal{R}$

X	Y	Z
1	0	0
1	1	1
1	0	1
0	1	1

$\mathcal{P}$

pattern	support	confidence
X $\Rightarrow$ Y	0.25	0.33
X $\Rightarrow$ Z	0.50	0.66
Y $\Rightarrow$ X	0.25	0.50
Y $\Rightarrow$ Z	0.50	1.00
Z $\Rightarrow$ X	0.50	0.66
Z $\Rightarrow$ Y	0.50	0.66
XY $\Rightarrow$ Z	0.25	1.00
XZ $\Rightarrow$ Y	0.25	0.50
YZ $\Rightarrow$ X	0.25	0.50

Fig. 1. An example of inductive database instance

Given the framework of an inductive database $\mathcal{I}$, users can query both the stored data (the part of $\mathcal{I}.\mathcal{R}$ in figure 1) as well as the set of patterns (the $\mathcal{I}.\mathcal{P}$ part). Formalization of inductive queries to perform data mining tasks has been studied by several research groups [5, 8]. We are, however, interested in the concept of inductive databases from a different perspective. Instead of using a sequence of queries and operations to create the induced patterns such as association rules [1], we shift our focus towards the induction of precise and rough rules and then deploy the stored information to support query answering. We unify the pattern representation to the relation format normally used in relational databases and call it relational inductive databases.

2.2 Rough Set Theory

The notion of rough sets has been introduced by Zdzislaw Pawlak in the early 1980s [22] as a new concept of set with uncertain membership. Unlike fuzzy set, uncertainty in rough set theory does not need probability or the value of possibility to deal with vagueness. It is rather formalized through the simple concepts of lower and upper approximation, which are in turn defined on the basis of set. Rough set concepts are normally explained within the framework of a decision system. The basic idea is partitioning universe of discourse into equivalence classes.

Definition 1. A *decision system* is any system of the form $\mathcal{A}$ = <U, A, d>, where U is a non-empty finite set of objects called the universe, A is a non-empty finite set of conditions, and d $\notin$ A is the decision attribute.

Definition 2. Given a decision system $\mathcal{A}$ = <U, A, d>, then with any B $\subseteq$ A there exists an *equivalence* or *indiscernibility relation* I_A(B) such that I_A(B) = {(x, x') $\in$ UxU | $\forall$a $\in$ B [a(x) = a(x')] }.

Table 1. A students' grading decision table

	Conditions			Decision
	age	score1	score2	grade
s1	19	0-20	0-20	fail
s2	19	0-20	21-40	fail
s3	20	0-20	41-60	fail
s4	20	0-20	41-60	fail
s5	19	0-20	81-100	pass
s6	19	41-60	41-60	pass
s7	19	21-40	61-80	pass
s8	20	21-40	21-40	pass

From the data samples in table 1, the followings are equivalent relations.

I(age) = {{s1, s2, s5, s6, s7}, {s3, s4, s8}}
I(score1) = {{ s1, s2, s3, s4, s5}, {s6}, {s7, s8}}
I(score2) = {{s1},{s2, s8},{s3, s4, s6}, {s5}, {s7}}
I(age, score1) = {{s1, s2, s5},{s3, s4},{s6},{s7},{s8}}
I(age, score2) = {{s1},{s2},{s3, s4},{s5},{s6},{s7},{s8}}
I(score1, score2) = {{s1},{s2},{s3, s4},{s5},{s6},{s7},{s8}}
I(age, score1, score2) = {{s1},{s2},{s3, s4},{s5},{s6},{s7},{s8}}

Equivalence relations partition the universe into groups of similar objects based on the values of some attributes. The question often arises is whether one can remove some attributes and still preserve the same equivalence relations. This question leads to the notion of reduct [16].

Definition 3. Let $\mathcal{A}$ = <U, A, d> be a decision system and P, Q $\subseteq$ A, P $\neq$ Q be two different sets of conditions. The set P is the *reduct* of set Q if P is minimal (i.e. no redundant attributes in P) and the equivalence relations defined by P and Q are the same.

It can be seen from the listed equivalence relations that I(age, score2) = I(score1, score2) = I(age, score1, score2). Therefore, (age, score1) and (score1, score2) are reducts of (age, score1, score2). The intersection of all reducts produces *core attributes*. According to our example, score2 is a core attribute. A reduct table of (score1, score2) and its partitions are shown in figure 2(a). If we are interested in the decision criteria for the pass grade, we can infer decision rules from the reduct table in figure 2(a) as follows.

IF (score1 = 0-20 $\wedge$ score2 = 81-100) THEN grade = pass
IF (score1 = 21-40 $\wedge$ score2 = 21-40) THEN grade = pass
IF (score1 = 21-40 $\wedge$ score2 = 61-80) THEN grade = pass
IF (score1 = 41-60 $\wedge$ score2 = 41-60) THEN grade = pass

	score1	score2	grade
s1	0-20	0-20	fail
s2	0-20	21-40	fail
s3	0-20	41-60	fail
s4	0-20	41-60	fail
s5	0-20	81-100	pass
s6	41-60	41-60	pass
s7	21-40	61-80	pass
s8	21-40	21-40	pass

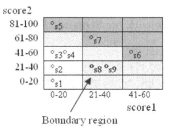

(a) equivalence relations on 8 students

	score1	score2	grade
s1	0-20	0-20	fail
s2	0-20	21-40	fail
s3	0-20	41-60	fail
s4	0-20	41-60	fail
s5	0-20	81-100	pass
s6	41-60	41-60	pass
s7	21-40	61-80	pass
s8	**21-40**	**21-40**	**pass**
s9	21-40	21-40	fail

(b) equivalence relations on 9 students

Fig. 2. A reduct table and (a) its partition into equivalence relations, each one is represented by a rectangular region, (b) equivalence relations with conflicting cases of s8 and s9

Suppose we are given additional information of the ninth student as shown in figure 2(b), then the above decision rules for the passing grade is no longer valid. It can be seen from figure 2(b) that s8 and s9 are in the same equivalence relation but their grades are different. It is such conflicting cases that inspires the rough set concept. Given the two decision sets of pass/fail, the uncertain cases such as s8 and s9 can be approximated their set membership by means of lower and upper approximation [16].

Definition 4. Let $\mathcal{A} = <U, A, d>$ be a decision system, $B \subseteq A$, $X \subseteq U$ be objects of interest and $[x]_B$ denote the equivalence class of $I_A(B)$. The *B-lower approximation* and *B-upper approximation* of X, denoted by bX and BX respectively, are defined by $bX = \{x \mid [x]_B \subseteq X\}$ and $BX = \{x \mid [x]_B \cap X \neq \varnothing\}$. The area between B-lower approximation and B-upper approximation is called *B-boundary region* of X, *BN*, and defined as $BN = BX - bX$.

The lower approximation of X is the set of all objects that certainly belong to X. This set is also called *B-positive region* of X. The *B-negative region* of X is defined as $U - BX$, or the set of all objects that definitely not belong to X. The *B-boundary region* of X is the set of all objects that cannot be classified as not belonging to X.

Given the information as shown in figure 2(b), B = {score1, score2} and X = {s5, s6, s7, s8} be set of students with passing grade, then bX = {s5, s6, s7} and BX = {s5, s6, s7, s8, s9}. The boundary region BN = {s8, s9}. *B-negative region* of X is {s1, s2, s3, s4} or the set of all students who definitely fail the exam.

If the boundary region is empty, it is a *crisp* (precise) set; otherwise, the set is *rough*. The set of passing students in figure 2(a) is a crisp set, whereas it is a rough set in figure 2(b). Decision rules generated from a rough set comprise of certain rules generated from the positive and negative regions, and possible rules generated from the boundary region.

Such method to generate decision rules is static because the decision attribute is defined in advance. Within the framework of query answering that decision attributes are usually not known in advance, the classical static rough set methodology is certainly impractical. We thus propose in the next section our method of dynamic rule induction driven by the query predicates.

3   Certain and Rough Knowledge Induction

3.1   A Framework for Semantic Knowledge Induction

In the typical environment of database systems, the size of data repository can be very large. With the classical rough set method that all prospective decisions have to be pre-specified, the number of generated rules can be tremendous. We thus propose a dynamic approach by taking predicate in the user's query to be a decision attribute at query processing time. By this scheme, we can limit the induction to only relevant decision rules and these rules are subsequently used as semantic knowledge in the process of query rewriting and optimization. The framework of our approach is shown in figure 3.

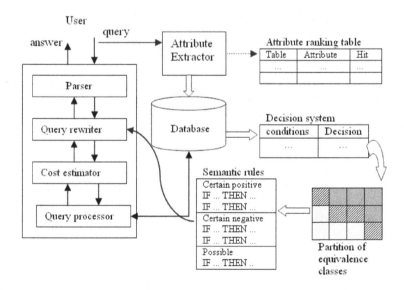

Fig. 3. A framework of query-driven induction for rough and precise knowledge

In our proposed framework, rule induction is invoked by user's query. Once the query has been posted, the component named *attribute extractor* has been called to extract the table's and attribute's name from the query. The *attribute ranking table* is also created to collect the history of attributes used in the queries. The column *hit* counts the number of times that attributes has been used. The approach of inducing rules based on query's attribute is described in the algorithm as shown in figure 4.

Algorithm. Query-driven semantic rule induction

Input: User's query, a database and background knowledge
Output: Certain and rough semantic rules

1. Call *Attribute Extractor* to extract table names T_i and attribute names A_j from the query
2. Access the *attribute ranking table* and update the hit counter identified by each T_i and A_j and sort the counter in descending order
3. Create a decision table $A = <U, A, d>$ where $d = A_j$, A = a set of attributes in T_i, U = a set of tuples in T_i
4. Pre-process A by
   - removing attributes with number of distinct values $= |A|$
   - discretizing attributes with real values
5. Partition U into equivalence classes and search for the first reduct R
6. From R, identify *bX, BX, BN* regions, then generate certain positive, certain negative, and possible rules
7. Generalize all three classes of rules using available background knowledge
8. Return the final rule set

Fig. 4. A query-driven semantic rule induction algorithm

3.2 Running Examples

We use the student data shown in table 1 with additional record <s9, 20, 21-40, 21-40, fail> as our running example. The information on interval order that 81-100 > 61-80 > 41-60 > 21-40 > 0-20 is used as background knowledge for rule generalization.

Example 1. Suppose there is a query asking whether the score1 = 55 is high enough for the passing grade.
Method:

(1) This query asks about grade with score1 as a condition. Hence, a reduct table as in figure 2(b) is constructed.
(2) Then, the following rules are generated.
   Certain positive rules: IF (score1=0-20 ∧ score2=81-100) THEN grade = pass
   　　　　　　　　　　　　　　　 IF (score1=21-40 ∧ score2=61-80) THEN grade = pass
   　　　　　　　　　　　　　　　 IF (score1=41-60 ∧ score2=41-60) THEN grade = pass
   Certain negative rules: IF (score1=0-20 ∧ score2=0-20) THEN grade = fail
   　　　　　　　　　　　　　　　 IF (score1=0-20 ∧ score2=21-40) THEN grade = fail
   　　　　　　　　　　　　　　　 IF (score1=0-20 ∧ score2=41-60) THEN grade = fail
   Possible rules: IF (score1=21-40 ∧ score2=21-40) THEN grade = pass
(3) The three classes of rules are generalized according to the given background knowledge. The final rules are as follow.

R1: IF (score1 > 20 ∧ score2 > 60) THEN grade = pass
R2: IF (score1 > 40 ∧ score2 > 40) THEN grade = pass
R3: IF (score1 > 20 ∧ score2 > 20) THEN grade = possibly pass

Notice that with the given information there is no matching rules from the negative class and R2 can be applied to answer this query.

Answer:

IF score2 > 40 THEN grade = pass.
IF score2 > 20 THEN grade = possibly pass.

Example 2. From the response of example 1, suppose the user wants to know further that based on the information of her first score, could her second score be predicted.
Method:

(1) The query asks for the value of score2, given the value of score1=55. Thus, a decision attribute is score2 and a decision table is as shown in table 2.

Table 2. A d ecision table with respect to query 2

	Conditions			Decision
	age	score1	grade	score2
s1	19	0-20	fail	0-20
s2	19	0-20	fail	21-40
s3	20	0-20	fail	41-60
s4	20	0-20	fail	41-60
s5	19	0-20	pass	81-100
s6	19	41-60	pass	41-60
s7	19	21-40	pass	61-80
s8	20	21-40	pass	21-40
s9	20	21-40	fail	21-40

(2) There is no reduct. So, all conditional attributes are used in the approximation of bX, BX, and BN regions. The decision objectives (X) are five sets of students whose score2 values are in the range 0-20, 21-40, 41-60, 61-80, and 81-100, respectively. From the approximation, these rules are induced:

Certain rules: IF (age=19 ∧ grade=pass ∧ score1 =0-20)   THEN score2 = 81-100
IF (age=19 ∧ grade=pass ∧ score1=21-20) THEN score2 = 61-80
IF (age=20 ∧ grade=fail ∧ score1 = 0-20)   THEN score2 = 41-60
IF (age=19 ∧ grade=pass ∧ score1 = 0-20) THEN score2 = 41-60
IF (age=20 ∧ score1 = 21-40)   THEN score2 = 21-40

Possible rules: IF (age=19∧score1=20∧grade=fail) THEN score2 = 0-20 ∨ 21-40

(3) Generalized rules are as follow.

R1: IF (score1 = 0-20)   THEN score2 = 81-100
R2: IF (score1 = 21-40)   THEN score2 = 61-80
R3: IF (score1 = 0-20 ∨ 41-60)   THEN score2 = 41-60
R4: IF (age=20 ∧ score1 = 21-40)   THEN score2 = 21-40
R5: IF (age=19 ∧ score1 = 20 ∧ grade=fail) THEN possibly score2 = 0-40

Answer:

IF score1 = 55   THEN score2 = 41-60.

4   Query Optimization in Inductive Databases

With the mechanism to induce semantic rules, we can then integrate the knowledge to rewrite and optimize queries in the framework of relational inductive database. We first explain the steps in query transformation, then show the results of our experimentation.

4.1   A Method of Query Rewriting and Optimization

We study intelligent query answering in inductive databases in a simplified framework of relational databases. We illustrate our idea through examples on a business database containing eight relations: customers, orders, products, categories, order_details, suppliers, shippers, and employees. To keep this section concise, we will consider only customers relation with schema as follows.

customers (*customerID, name, address, city, state, postalcode, country, phone, fax,*
        birthdate, marital, gender, education, member_card, total_children,
        occupation, houseowner, num_car)

Our objective is to turn this simple database $\mathcal{R}$ into an inductive database $\mathcal{I} = (\mathcal{R},$ $\mathcal{P})$ by inducing a set of patterns $\mathcal{P}$ from the base tables. The processes of inductive database creation (the $\mathcal{P}$ part of the database $\mathcal{I}$) and query rewriting composed of the following steps.

Step 1: Preprocess the base table by removing irrelevant attributes, i.e. those without inherent patterns such as phone number, customerID. After the attribute elimination step, the customers table contains information as shown in figure 5.

City	Marital	Gender	Education	Member card	Total children	Occupation	House owner	Num car
los angeles	s	m	bachelor	golden	0	management	yes	1
san diego	s	m	partial college	normal	0	skilled manual	no	1
nation city	m	f	bachelor	silver	3	skilled manual	yes	2
santa cruz	s	m	bachelor	silver	0	skilled manual	no	0

Fig. 5. Some examples from a set of customer instances

Step 2: Perform a semantic rule induction using the algorithm explained in figure 4. Suppose the query asks about *marital, gender, total_children,* and *houseowner* attributes, some of the induced patterns are shown in figure 6.

Step 3: Transform the induced rules into a tabular form (as shown in figure 7).

Step 4: Evaluate user's query for the possibility of null answer set detection. If the query contains unsatisfiable constraints that can be detected in an early stage using the induced patterns, then the subsequent processing is unnecessary. For example, given the query: *SELECT * FROM customers*

WHERE address = 'Bangkok' AND member_card = 'silver';

and the induced pattern: *IF address = 'Bangkok' THEN member_card = 'gold',*
then the answer 'No' can be returned instantly.

IF *gender* = *m* THEN *marital* = *s*
IF *total_children* = *0* THEN
        marital = *s*
IF *total_children* = *0* THEN
        gender = *m*
IF *gender* = *m* THEN
        total_children = *0*
IF *marital* = *m*   THEN
        houseowner = *yes*

tablename_1	column_1	value_1	tablename_2	column_2	value_2
customers	gender	m	customers	marital	s
customers	total_child	0	customers	marital	s
customers	total_child	0	customers	gender	m
customers	gender	m	customers	total_child	0
customers	total_child	0	customers	marital	s

Fig. 6. The patterns of customers data represented as semantic rules

Fig. 7. The pattern table with some sample values

Step 5: Rewrite query using the induced patterns. The conjunctive conditions C1∧C2∧C3∧... in the where clause of the SQL query are matched against the patterns in an iterative manner. In the first iteration C1 is matched against the antecedent part of the patterns to search for the one with the consequent part that is unifiable with either C2 or the rest of the conditions. If this is the case, the unified condition is considered redundant and thus, can be removed. The subsequent iterations are performed on the remaining conditions. Consider the following query Q, and the induced patterns P.

Q: SELECT * FROM customers
        WHERE total_children = '0'   AND marital = 's'   AND gender = 'm';

P: IF *total_childern* = *0 THEN marital* = *s*   (rule 1)
        IF *gender* = *m*   THEN *total_childern* = *0*   (rule 2)

First iteration: the antecedent and consequent parts of rule 1 can match with the first and second conditions of the query. These two conditions are redundant. Thus, the where clause can be simplified to: *WHERE total_children* = *'0' AND gender* = *'m';*

Second iteration: rule 2 states the fact regarding the association between gender = m and total_children = 0. Therefore, it can be applied to the query Q which can be finally rewritten as

                Q':   SELECT * FROM customers
                        WHERE gender = 'm';

4.2 Experimentation and Results

The proposed technique of query answering and refinement has been tested on a customer database implemented on MS SQL Server 2000. The patterns are induced from the customers table and four different queries pertaining to the customers have been tested on the database. We observe the returned answer set (number of tuples) as well as the query response time. We perform the experiments on the PC with CPU speed 3.2 GHz, 512 MB main memory and 80 GB HD. The query evaluation results are reported (in figure 8) comparatively between the original query processing and the answering from the query rewritten using the induced patterns.

Query Q1 illustrate the case of unsatisfiable query in which the condition of the query is found conflicting with the existing data content. The query asks for female customers who live in santa cruz. But the induced pattern states that there is no such

Query	Original query form	Pattern applied	Transformed query
Q1	SELECT * FROM customers WHERE city = 'santa cruz' AND gender = 'm' AND marital = 'm';	IF city = santa cruz THEN gender = m ∧ marital=s (9 patterns are induced)	None: detection of unsatisfiable condition
Q2	SELECT * FROM customers WHERE city = 'los angeles' AND houseowner = 'yes';	IF city = los angeles THEN houseowner = yes (2 patterns are induced)	SELECT * FROM customers WHERE city = 'los angeles';
Q3	SELECT * FROM customers WHERE gender = 'm' AND marital = 's' AND total_children = '0';	IF gender = m THEN marital= s ∧ total_children = 0 (4 patterns are induced)	SELECT * FROM customers WHERE gender = 'm';
Q4	SELECT * FROM customers WHERE city = 'santa cruz' AND gender = 'm' AND member_card = 'bronze';	IF city = santa cruz THEN gender = m (7 patterns are induced)	SELECT * FROM customers WHERE gender = 'm' AND member_card = 'bronze';

Size of answer sets (number of tuples) and response time (millisecond):

	Q1		Q2		Q3		Q4	
	size	time	size	time	size	time	size	time
Original query	0	104	27,660	1406	71,916	4665	16,596	1185
Transformed query	0	0	27,660	1268	71,916	3489	16,596	1074
Gain		100%		9.8%		25.2%		9.4%

Fig. 8. Experimental results of asking four queries on a customer database

customers; most customers in santa cruz are male. Therefore, this query can be answered instantly (i.e., the response time is 0). Queries Q2, Q3, and Q4 are the examples of queries with redundant predicates. Once redundancy has been removed, the query response time can be reduced.

5 Conclusions

Our query answering scheme presented in this paper is based on the setting of inductive databases. An inductive database is the concept proposed as the next generation of database systems. Within the framework of an inductive database system, data and patterns which are discovered from data are stored together as database objects. In such tightly coupling architecture patterns are considered first-class objects in that they can be created, accessed, and updated in the same manner as persistent data. We present the framework and techniques of query rewriting and answering that use stored patterns as semantic knowledge. Our knowledge induction process is based on the rough set theory. We propose the algorithm to induce rough and precise semantic rules. We limit the number of discovered rules by inducing only rules that are relevant to user's need. Relevancy is guided by query predicates. The intuitive idea of our knowledge induction algorithm is illustrated trough running examples.

In the query optimization process, we take into account two major techniques of transformations: semantically redundant predicate elimination and detection of unsatisfiable conditions, i.e. conditions that never been true. We plan to extend our work on additional rewriting techniques and experiments with different kinds of queries such as range queries, top-k queries. The test on effectiveness with real-world large database is also our future research.

Acknowledgements

This research has been supported by grants from the National Research Council. The second author is supported by the grant from Thailand Research Fund (TRF – grant number RMU5080026). The Data Engineering and Knowledge Discovery Research Unit is fully supported by the research grants from Suranaree University of Technology.

References

1. Agrawal, R., Imielinski, T., Swami, A.: Mining association rules between sets of items in large databases. In: Proc. ACM SIGMOD, pp. 207–216. ACM Press, New York (1993)
2. Agrawal, R., Shim, K.: Developing tightly-coupled data mining applications on a relational database system. In: Proc. KDD, pp. 287–290 (1996)
3. Bonchi, F.: Frequent pattern queries: Language and optimizations. Ph.D. Thesis, Computer Science Department, University of Pisa, Italy (2003)
4. Boulicaut, J.-F., Klemettinen, M., Mannila, H.: Querying inductive databases: A case study on the MINE RULE operator. In: Żytkow, J.M. (ed.) PKDD 1998. LNCS, vol. 1510, pp. 194–202. Springer, Heidelberg (1998)
5. Boulicaut, J.-F., Klemettinen, M., Mannila, H.: Modeling KDD processes within the inductive database framework. In: Mohania, M.K., Tjoa, A.M. (eds.) DaWaK 1999. LNCS, vol. 1676, pp. 293–302. Springer, Heidelberg (1999)
6. Charkravarthy, U.S., Grant, J., Minker, J.: Logic-based approach to semantic query optimization. ACM Transactions on Database Systems 15(2), 162–207 (1990)
7. De Raedt, L.: A perspective on inductive databases. SIGKDD Explorations 4(2), 69–77 (2002)
8. De Raedt, L., Jaeger, M., Lee, S., Mannila, H.: A theory of inductive query answering. In: Proc. IEEE ICDM, pp. 123–130. IEEE Computer Society Press, Los Alamitos (2002)
9. Fayyad, U.M., Piatetsky-Shapiro, G., Smyth, P., Uthurusamy, R.: Advances in Knowledge Discovery and Data Mining. AAAI Press (1996)
10. Hammer, M.M., Zdonik, S.B.: Knowledge base query processing. In: Proc. VLDB, pp. 137–147 (1980)
11. Han, J., Fu, Y., Wang, W., Koperski, K., Zaiane, O.: DMQL: A data mining query language for relational databases. In: Proc. ACM SIGMOD Workshop on Research Issues on Data Mining and Knowledge Discovery, pp. 27–34. ACM Press, New York (1996)
12. Han, J., Huang, Y., Cercone, N., Fu, Y.: Intelligent query answering by knowledge discovery techniques. IEEE Trans. on Knowledge and Data Engineering 8(3), 373–390 (1996)
13. Imielinski, T., Mannila, H.: A database perspective on knowledge discovery. Communications of the ACM 39(11), 58–64 (1996)

14. Imielinski, T., Virmani, A.: MSQL: A query language for database mining. Data Mining and Knowledge Discovery 2(4), 373–408 (1999)
15. King, J.: QUIST: A system for semantic query optimization in relational databases. In: Proc. VLDB, pp. 510–517 (1981)
16. Komorowski, J., Polkowski, L., Skowron, A.: Rough sets: A tutorial. In: Rough Fuzzy Hybridization: A New Trend in Decision-Making, pp. 3–98. Springer, Heidelberg (1999)
17. Lin, T., Cercone, N., Hu, X., Han, J.: Intelligent query answering based on neighborhood systems and data mining techniques. In: Proc. IEEE IDEAS, pp. 91–96. IEEE Computer Society Press, Los Alamitos (2004)
18. Mannila, H.: Inductive databases and condensed representations for data mining. In: Proc. Int. Logic Programming Symp., pp. 21–30 (1997)
19. Meo, R.: Inductive databases: Towards a new generation of databases for knowledge discovery. In: Proc. DEXA Workshop, pp. 1003–1007 (2005)
20. Meo, R., Psaila, G., Ceri, S.: A tightly-coupled architecture for data mining. In: Proc. IEEE ICDE, pp. 316–323. IEEE Computer Society Press, Los Alamitos (1998)
21. Necib, C., Freytag, J.: Semantic query transformation using ontologies. In: Proc. IEEE IDEAS, pp. 187–199. IEEE Computer Society Press, Los Alamitos (2005)
22. Pawlak, Z.: Rough sets. Int. Jour. Information and Computer Science 11(5), 341–356 (1982)
23. Sarawagi, S., Thomas, S., Agrawal, R.: Integrating association rule mining with relational database systems: Alternatives and implications. In: Proc. ACM SIGMOD, pp. 343–354. ACM Press, New York (1998)
24. Siegel, M., Sciore, E., Salveter, S.: A method for automatic rule derivation to support semantic query optimization. ACM Trans. on Database Systems 17(4), 563–600 (1992)
25. Sun, J., Kerdprasop, N., Kerdprasop, K.: Relevant rule discovery by language bias for semantic query optimization. Jour. Comp. Science and Info. Management 2(2), 53–63 (1999)

A Clustered Dwarf Structure to Speed Up Queries on Data Cubes*

Fangling Leng, Yubin Bao, Daling Wang, and Ge Yu

School of Information Science & Engineering,
Northeastern University, Shenyang 110004, P.R.China
{lengfangling,baoyb,dlwang,yuge}@mail.neu.edu.cn

Abstract. Dwarf is a highly compressed structure, which compresses the cube by eliminating the semantic redundancies while computing a data cube. Although it has high compression ratio, Dwarf is slower in querying and more difficult in updating due to its structure characteristics. So we propose two novel clustering methods for query optimization: the recursion clustering method for point queries and the hierarchical clustering method for range queries. To facilitate the implementation, we design a partition strategy and a logical clustering mechanism. Experimental results show our methods can effectively improve the query performance on data cubes, and the recursion clustering method is suitable for both point queries and range queries.

1 Introduction

Dwarf [1] is a highly compressed structure, which compresses the cube by eliminating the prefix and suffix redundancies while computing a data cube. What makes Dwarf practical is the automatic discovery of the prefix and suffix redundancies without requiring knowledge of the value distributions and without having to use sophisticated sampling techniques to figure them out. But the structure of Dwarf (see fig.1) has the characteristics: (1) It is a tree structure ignoring the suffix coalition, which has a common root and some nodes derived from the root. (2) The node cells store the node location information of the next dimension, which is similar to an index. (3) The nodes have different size. These characteristics make it inefficiently to store a Dwarf into a relational database. So Dwarf is usually stored in a flat file, which has good performance of the construction[1]. However, due to not considering the characteristics of queries and the high cost of maintaining files, it brings low performance of query and is hard to update. In order to solve the query performance problem, [1] gave a clustering algorithm, which clusters nodes according to the computational dependencies among the group-by relationships of the cube. This method can improve the query performance in a certain degree. For a cube, point queries and range queries are two important types of queries on it. We note that for point queries and range queries, the access order to nodes is

---

* Supported by the National Natural Science Foundation of China under Grant No.60473073, 60573090, 60673139.

I.Y. Song, J. Eder, and T.M. Nguyen (Eds.): DaWaK 2007, LNCS 4654, pp. 170–180, 2007.

according to the relationship of parent and child of the nodes from root to leaves in stead of the group-by relationships. Therefore, we propose two novel clustering algorithms to improve the query performance on Dwarf. As the data increases in complexity, the ability to refresh data in a modern data warehouse environment is currently more important than ever. The incremental update procedure in [1] expands nodes to accommodate new cells for new attribute values (by using overflow pointers), and recursively updates those sub-dwarfs which might be affected by one or more of the delta tuples. The frequent incremental update operations slowly deteriorate the original clustering of the Dwarf structure, mainly because of the overflow nodes created. Since Dwarf typically performs updates in periodic intervals, a process in the background periodically runs for reorganizing the Dwarf and transferring it into a new file with its clustering restored [1]. To avoid the reorganizing the Dwarf and maintain the clusters, we design a partition strategy and a logical cluster mechanism.

In [2], three algorithms are described for discovering tuples whose storage can be coalesced: MinCube guarantees to find all such tuples, but it is very expensive computation, while BU-BST and RBU-BST are faster, but only discover fewer coalesced tuples. Much research work has been done on approximating data cubes through various forms of compression such as wavelets [3], multivariate polynomials [4], or by sampling [5] [6] or data probability density distributions [7]. While these methods can substantially reduce the size of the cube, they do not actually store the values of the group-bys, but rather approximate them, thus not always providing accurate results. Relatively Dwarf is a promising structure to be discussed deeply.

The rest of this paper is organized as follows: Section 2 presents our two clustering methods for Dwarf, the recursion-clustering method and the hierarchical clustering method. Section 3 describes the physical structure of the clustered Dwarf, the paging partition strategy, and the logical clustering mechanism. Section 4 presents the performance experiments and the result analysis. The conclusion is given in Section 5.

2   Clustering Dwarf

For the following explanation, we give a sample data set in Table 1. Fig.1 shows the Dwarf structure of the sample data in Table 1 according to the method in [1], where node (5) is stored behind node (7) according to the computational dependencies, and the parent of node (5) is node (2).

Table 1. Sample Data Set

Store	Customer	Product	Price
S_1	C_2	P_2	70
S_1	C_3	P_1	40
S_2	C_1	P_1	90
S_2	C_1	P_2	50

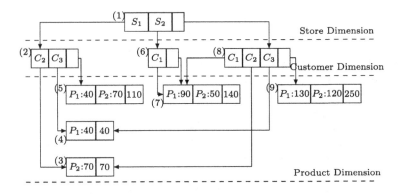

Fig. 1. The Dwarf Structure of Sample Data

When querying from node (2) to node 5, one more time I/O operation (node (7)) will be executed in order to read node (5). Therefore, we propose to cluster the Dwarf according to the relationship of parent and child of the nodes. Two clustering methods are designed to optimize Dwarf for improving the performance of point queries and range queries.

Suppose that the example query is :

> SELECT price
> FROM   example_cube
> WHERE (store=S_1) AND (customer=ALL) AND (product=P_1).

2.1   Recursion Clustering

When a point query is performed, the results must be in the sub-Dwarf pointed by the corresponding cell. That is a lengthways access manner. For the example query, when the condition $store = S_1$ is executed, the results must be in the sub-Dwarf pointed by cell S_1, i.e. node (2)(3)(4)(5). So if we can store the four nodes on the same cluster, the query performance will be improved.

In Fig.1, the root node has three sub-trees, rooted by (2), (6), (8). So they should be stored together. Furthermore, node (2) also has three sub-trees, rooted by (3), (4), (5), and they should be stored together too. These clustering steps are in the recursive manner. So we call our method as the recursion-clustering method (see Algorithm 1).

Apart from the leaf nodes, all internal nodes of a Dwarf contain *ALL* cells to indicate the summary information on the corresponding dimensions. [1] designed the recursion suffix coalition algorithm to compute the *ALL* cells. But it makes the clustering feature lost. We improve it by eliminating the recursive operations. Fig.2 gives an example of suffix coalition. Here the suffix coalition occurs at the *ALL* cell of node (1), and the following nodes are constructed with it. The node sequence writing to the disk in Fig.2 should be (5)(3)(6)(4)(2)(1). We use a stack to implement the recursion-clustering algorithm. The elements in the stack are the nodes to be closed and their sub-Dwarfs.

Algorithm 1. Recursion-Clustering Algorithm

1:   initialize the stack S
2:   push the begin node and its sub-Dwarf into S
3:   **while** S is not empty **do**
4:     pop the top element n in S
5:     **if** n has non-computed units **then**
6:       go to the first non-computed unit c of n
7:       compute the sub-Dwarf of c using the sub-Dwarf of n
8:       push n into S
9:       **if** c can be suffix coalition//not require creating new nodes
10:         write c's suffix coalition position into top element unit
11:       **else** //require creating new nodes
12:         create the new node n_1 using the sub-Dwarf of c
13:         push the n_1 and its sub-Dwarf into S
14:       **end if**
15:     **end if**
16: **end while**

2.2   Hierarchical Clustering

The range query often access several child nodes of a node. The sibling nodes will be accessed together. Because the sibling nodes are of the same dimension, we suggest the nodes of the same dimension should be clustered together.

Algorithm 2 shows the hierarchical clustering algorithm, which also supports the suffix coalition. Different from the recursion clustering, a queue is used in it. The elements of the queue are the nodes to be closed and the sub-Dwarfs rooted by these nodes. The process is similar to the breadth-first searching of graph. The storage sequence of the nodes in Fig.2 should be $(1)(2)(3)(4)(5)(6)$ according to the hierarchical clustering. Contrast with the recursion clustering, it has the following problem. Before the next layer nodes being written to the disk, i.e. the location of the next layer nodes is unknown, the node should finish writing to the disk. For example, if node (3) and node (4) have not been written to the disk, i.e. the value of cell B_1 and B_2 of node (2) are unknown, node (2) has to be written to the disk first to satisfy the cluster characteristics.

To solve this problem, we propose a node-indexing approach. It assigns a distinct index number to each node, and the cell in a node stores the index number but not the location on the disk of the next layer node. At the same time a node index table is built to maintain the mapping between the index number and the physical location of the node. For example, when writing node (2) to the disk, we only need to construct node (3) and node (4), and fill the index number of them into the cells of node (2). When the child nodes are written to the disk, their addresses will be filled into the node index table. Such index enables to write the parent node before its child nodes.

Restricting the impact of the update of a Dwarf is another advantage of the node-indexing approach. The update is very difficult because of the nodes with unequal size. When updating, some cells or nodes will be created or deleted. The Dwarf is usually stored in a flat file, and the frequent node insertion or deletion operations will produce many file fragments. Compared with deletion, inserting a new node or a new cell is very hard to deal with. If there is no node index, the operations will lead the parent node of the updated node to be updated passively because the corresponding cell of the parent node stores the offset of the updated child node in the file. Since using the node index, the problem on update operations is bounded within the child node itself.

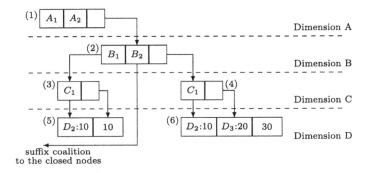

Fig. 2. An Example of Suffix Coalition

Algorithm 2. Hierarchical Clustering Algorithm
1:   initialize the queue Q, and put the begin node into Q
2:   **while** Q is not empty **do**
3:     get a node n from Q
4:     **if** n is the leaf node **then**
5:       compute all the units of n
6:       write n to the disk
7:       continue
8:     **end if**
9:     create the child nodes $n_1, n_2, \cdots, n_k$ of n and their sub-Dwarfs, and write nodes numbers into the corresponding units of n
10:    put the child nodes into Q by the creation time
11:    write n to the disk, and add n to the node index table
12: **end while**

According to the hierarchical clustering algorithm, the storage sequence of the nodes in Fig.1 should be (1)(2)(6)(8)(3)(4)(5)(7)(9). In hierarchical clustering, the suffix coalition of node (2) in Fig.1 is executed before node (1). If the nodes are always written to the end of the file, node (2) and node (8), which should be put together, will be separated by node (3)(4)(5). Hence, the physical organization structure should be considered further.

3   Physical Structure of Dwarf

3.1   Paging Partition Strategy

To solve the above clustering problem, we propose a partition organization strategy for Dwarf files. Firstly, allocate a certain space (called a dimension *partition*) on the disk for the node cluster of each dimension. Secondly, a dimension partition only stores the corresponding dimension nodes. When the remaining space cannot contain a new node, a new partition for that dimension will be appended to the end of the file. In this way, when the node is written to the disk, it may not be appended to the end of the file, but be written to its dimension partition. In this case, the hierarchical clustering feature is guaranteed.

Generally, we can extend the concept to the recursion clustering. Our partition management of Dwarf is designed as the paging storage management. Several pages consist of a partition, and a node is not allowed to span the pages. When clustering, a certain amount of pages is pre-allocated for each cluster, and a new partition is added at the end of the file only if a partition cannot contain the new nodes. Since a node is not allowed to span the pages, the fragments occur. But they can be used to save the broken cluster by updating operations. This will be addressed in the next section.

How many pages should we pre-allocate for a cluster? For the recursion clustering, a cluster is corresponding to a sub-Dwarf tree actually. In Fig.2, the cluster of nodes (2)(3)(4)(5)(6) is related with the sub-tree of ALL cell of node (1). Obviously, the cluster size is related with the height of the sub-tree. So when we pre-allocate the pages, we only need pre-allocate enough pages for each sub-tree of the root, and the low layer sub-trees recursively get the space through partitioning the pages. Suppose that there is a N-dimensional cube, we should pre-allocate $N' = 2^N$ pages for each sub-tree of the root. In the above expression, the number 2 indicates the qualitative analysis of the space. Its actual size should be studied further according to the data statistic features. For the hierarchical clustering, the nodes of the same dimension constitute a cluster. Obviously, the more close to the leaves the dimension is, the more amounts of the nodes the dimension has. For an N-dimensional cube, the page amount of the cluster of the k-th dimension should be $N_k = 2^k (1 \leq k \leq D)$.

3.2   Fragments in the Pages

As mentioned above, the fragments couldn't be avoided in the page-style partition clustering structure. The existence of the fragments wastes the storage space, and brings the side effect for compression. We propose a special mechanism, called *idle space manager*, to maintain the fragments space produced in the Dwarf system because of shrinking or deleting or enlarging nodes.

If the size of a node is reduced, there are two strategies to deal with it. (a) Only change the node and the node index table, which will produce new fragments, but the speed is very fast because of not moving other nodes. It only updates the node content, and adds a new record in the idle space table. But more and more such fragments will cause the total size of the idle space to be larger. (b) When shrinking the changed node, the fragments in a page is compacted and the following nodes are moved ahead within the page. At the same time, the node index table and the idle space table are updated. This method avoids the fragments, but the speed slows down. Our Dwarf system adopts this method.

If a node size is increased, and the increment is $\triangle S$, there are three cases. (1) If $\triangle S$ is less than the idle space of the page, the update will be restricted within the current page, and will be implemented by moving the following nodes forward. (2) If $\triangle S$ is greater than the idle space, but the total size of the node and the $\triangle S$ is less than a certain idle space or a certain idle page in the current cluster, the original node will be deleted and the current page will be shrunken. At the same time, the updated node is written on the new location of the partition, and the idle space table is updated. (3) When any idle space in a partition is not satisfied with the increased node, a new partition need to be appended to the end of the file, and the increased node will be written to the new partition. And the idle space table is updated. But the physical cluster feature is lost.

3.3   Logical Clustering Mechanism

To deal with the problem of breaking clusters, we propose a logical clustering mechanism, which is used to keep the physical cluster feature by tracking of the broken clusters. In fact, It cannot guarantee the pages in a logical cluster are stored in the sequential disk pages.

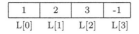

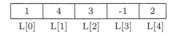

Fig. 3. The Records of Logical Clustering   **Fig. 4.** The Records after Updating

The logical cluster mechanism is similar to the disk chunk chain of FAT file system. We register a record for a page to hold the next page number in logic in a cluster. For example, there are 4 pages satisfied with the cluster characteristic before updating. Pages (0)(1) and pages (2)(3) are separately in the same cluster. Fig.3 shows the array of the logical cluster records. Here $L[i] = k$ indicates the logic page k is the neighbor of page i, and $L[i] = -1$ indicates page i is the file end in logic, but not physically. To the cluster having not been broken, it is always true that $L[i] = i + 1$, which is the basis of adjusting the broken cluster. Suppose that the node in page (1) is increased, and we have to append a new page after the file end to store the updated node. After updating the records of the logical cluster is shown in Fig.4. We can see that $L[1] = 4$, which tells us page (4) is the neighbor of page (1) in logic but not physically.

When the current storage layout seriously affects the query performance, the pages in a cluster in logic need to be adjusted together at once. At that time, the logical cluster record will work as a navigator. Since the node-indexing approach is used, the cell values of the nodes need not be changed when moving the pages, and we only need to modify the page numbers of the relative nodes in the node index table, which is another benefit of the node-indexing approach.

4   Performance Experiments

Contrast to [1], we focus on the query performance. We all know that the original intention of data cube is to speed up query performance. Since Dwarf already has high compression ratio, we consider that it is worthy to get better query performance at the cost of construction time and storage of Dwarf. The data set has 10 dimensions and 4×10^5 tuples. The cardinalities are 10, 100, 100, 100, 1000, 1000, 2000, 5000, 5000, and 10000, respectively. The aggregation operation is SUM. All the experiments are run on a Pentium IV with 2.6GHZ CPU and 512MB RAM, running Windows XP. We use ODBC connection to SQL Server 2000. In the following, non-clustering represents non-clustering Dwarf, Dwarf represents computational dependencies between the group-by relationships, R-Dwarf represents recursion clustering, and H-Dwarf represents hierarchical clustering.

4.1   Construction of Dwarf

Test 1: We test 7 to 10 dimensions with 4×10^5 tuples data set, respectively. The construction time costs are shown in Fig.5 and Fig.6. The construction time of our algorithm is about one time more than the original one. The storage space of our algorithm is about 0.3 times more than the original one. The main reason is that the idle space manager, the node-indexing, and the logical clustering mechanism need additional time and space.

4.2   Query on Dwarf

Test 2: The data cubes are the four Dwarfs used in Test 1 with 4×10^5 tuples. We create 1000 point queries randomly and run them continuously. Fig.7 shows the time cost. The recursion clustering outperforms the computational dependencies method by 10%. Hierarchical clustering behaves too badly, and it drops behind

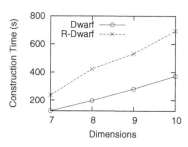

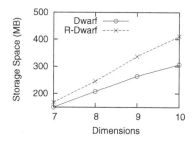

Fig. 5. Time vs Dims (4×10^5 Tuples)   **Fig. 6.** Space vs Dims (4×10^5 Tuples)

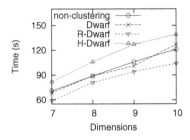

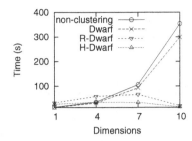

Fig. 7. Point Query Time vs Dims   **Fig. 8.** Range Query Time vs Dims

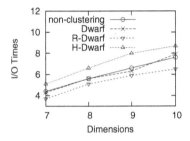

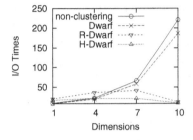

Fig. 9. Point Query I/O Times vs Dims   **Fig. 10.** Range Query I/O Times vs Dims

far from others, even the non-clustering method. Because it clusters the nodes of each dimension, and the query on each dimension will produce an I/O cost. To a point query, the average I/O number of it should be equal to the dimension number in theory. Here it is less than the dimension number because of the contributions of the high-speed buffer in Windows NT.

Test 3: The cube has 10 dimensions and 4×10^5 tuples. The sorting order is from small to big. To be fair, the query conditions are on the 1st, 4th, 7th, and 10th dimensions with the cardinalities are 10, 100, 2000, and 10000, respectively. Suppose each range query covers one dimension by the ratio 20%. We create 100 queries on each of the four dimensions, and use the four clustering methods mentioned in Test 2 to answer the queries. Fig.8 shows the response time vs dimension. The I/O times are shown in Fig.9 and Fig.10.

The former two methods sort the dimensions with the cardinalities changing from big to small when constructing the Dwarf, and the size of Dwarf is small. But when the conditions are pushed near to the root with the maximal cardinality, the time cost is growing fast. On the contrary, the sizes of the latter two methods are large, but the query response times do not fluctuate as much as the former two methods. And it seems to be similar to negative feedback. Because the limitation of space, the detail is not shown here.

It is obvious that for the query covering a dimension by a fixed proportion, the query time will increase with the increment of the cardinality, which is formed by the special tree structure of Dwarf. When querying at the 10th dimension with

10000 cardinality, the original method has to read about $10000 \times 20\% = 2000$ nodes of the next dimension to decide whether or not to go on, because the 10th dimension is the root. Since the data is sometimes sparse, the tuples matching the conditions are few. So most of the 2000 nodes read will be abandoned, which waste many I/O operations. Here the order is opposite to the original Dwarf, the 10th dimension is set to the leaf nodes. We only need to read a few leaf nodes to answer the queries. That is the reason why the I/O times using hierarchical clustering are close to the dimension number. In addition, we can see that in our experiments, when query on the root, the performance is still good, because the root has a small fan-out (only 10 cardinalities). To the query covering 20% range, we only need to judge 2 nodes but not the 2000 nodes of the next dimension at most. We call it query negative feedback of reversed dimension ordering. This negative feedback makes the performances of the range query using the recursion clustering and the hierarchical clustering fluctuates not much.

Generally, our clustering methods bring the construction and compression of Dwarf negative effect, but the effect is endurable. And our methods can remarkably improve the query performance on Dwarf. Since we have to maintain some additional structures, the performance of the update operations will be slow down, but it is endurable according to the simplicity of our structures. In the future, we will do some research on updating.

5   Conclusion

In this paper we propose two novel clustering algorithms of Dwarf to speed up querying and design a partition strategy and a logical clustering mechanism to facilitate updating and maintain the clusters. Experimental results and theoretic analysis show that the recursion clustering is the best for point queries, and the hierarchical clustering is the best for range queries. In general, the recursion clustering is suitable for both queries.

In the future, we will study the support of the iceberg cube [8] [9]. Compared with point queries and range queries, the iceberg query is a special type of query. The former searches aggregations from dimension values essentially, but the latter is opposite. Because the aggregations are all stored in the leaf nodes, an iceberg query will access almost all the nodes, which has no significance.

References

1. Sismanis, Y., Roussopoulos, N., Deligianannakis, A., Kotidis, Y.: Dwarf: Shrinking the Petacube. In: SIGMOD, pp. 564–475 (2002)
2. Wang, W., Feng, J., Lu, H., Yu, J.X.: Condensed Cube: An Effective Approach to Reducing Data Cube Size. In: ICDE, pp. 155–165 (2002)
3. Vitter, J.S., Wang, M., Iyer, B.: Data Cube Approximation and Histograms via Wavelets. In: CIKM, pp. 96–104 (1998)
4. Barbara, D., Sullivan, M.: A Space-Efficient Way to Support Approximate Multidimensional Databases. Technical report, ISSE-TR-98-03 (1998)

5. Gibbons, P.B., Matias, Y.: New Sampling-Based Summary Statistics for Improving Approximate Query Answers. In: SIGMOD, pp. 331–342 (1998)
6. Acharya, S., Gibbons, P.B., Poosala, V.: Congressional Samples for Approximate Answering of Group-By Queries. In: SIGMOD, pp. 487–498 (2000)
7. Shanmugasundaram, J., Fayyad, U., Bradley, P.S.: Compressed Data Cubes for OLAP Aggregate Query Approximation on Continuous Dimensions. In: KDD, pp. 223–232 (1999)
8. Beyer, K., Ramakrishnan, R.: Bottom-Up Computation of Sparse and Iceberg Cubes. In: SIGMOD, pp. 359–370 (1999)
9. Xin, D., Han, J., Li, X., Wah, B.W.: Star-Cubing: Computing Iceberg Cubes by Top-Down and Bottom-Up Integration. In: VLDB, pp. 476–487 (2003)

An OLAM-Based Framework for Complex Knowledge Pattern Discovery in Distributed-and-Heterogeneous-Data-Sources and Cooperative Information Systems

Alfredo Cuzzocrea

Department of Electronics, Computer Science, and Systems
University of Calabria
I-87036 Rende, Cosenza, Italy
cuzzocrea@si.deis.unical.it

Abstract. The problem of supporting advanced decision-support processes arise in many fields of real-life applications ranging from scenarios populated by distributed and heterogeneous data sources, such as conventional distributed data warehousing environments, to cooperative information systems. Here, data repositories expose very different formats, and knowledge representation schemes are very heterogeneous accordingly. As a consequence, a relevant research challenge is how to efficiently integrate, process and mine such distributed knowledge in order to make available it to end-users/applications in an integrated and summarized manner. Starting from these considerations, in this paper we propose an OLAM-based framework for *complex knowledge pattern* discovery, along with a formal model underlying this framework, called $\mathcal{M}ulti\text{-}Resolution\ \mathcal{E}nsemble\text{-}based\ Model\ for\ Advanced\ \mathcal{K}nowledge\ \mathcal{D}iscovery\ in\ Large\ \mathcal{D}atabases\ and\ Data\ Warehouses$ ($\mathcal{MRE}\text{-}\mathcal{KDD}^{+}$), and a reference architecture for such a framework. Another contribute of our work is represented by the proposal of KBMiner, a visual tool that supports the editing of even-complex KDD processes according to the guidelines drawn by $\mathcal{MRE}\text{-}\mathcal{KDD}^{+}$.

1 Introduction

The problem of supporting advanced decision-support processes arise in many fields of real-life applications ranging from distributed corporate database systems to *e-commerce* systems, and from supply-chain management systems to *cooperative information systems*, where huge, heterogeneous data repositories are cooperatively integrated in order to produce, process, and mine useful knowledge.

In such scenarios, intelligent applications run on top of enormous-in-size, heterogeneous data sources, ranging from transactional data to XML data, and from workflow-process log-data to sensor network data. Here, collected data are typically represented, stored and queried in large databases and data warehouses, which, without any loss of generality, define a collection of *distributed and heterogeneous data sources*, each of them executing as a singleton data-intensive software component (e.g., DBMS server, DW server, XDBMS server etc). Contrarily to this so-delineated

I.Y. Song, J. Eder, and T.M. Nguyen (Eds.): DaWaK 2007, LNCS 4654, pp. 181–198, 2007.

distributed setting, intelligent applications wish to extract *integrated, summarized knowledge* from data sources, in order to make strategic decisions for their business. Nevertheless, heterogeneity of data and platforms, and distribution of architectures and systems represent a serious limitation for the achievement of this goal. As a consequence, research communities have devoted a great deal of attention to this problem, with a wide set of proposals [13] ranging from *Data Mining* (DM) tools, which concern algorithms for extracting *patterns* and *regularities* from data, to *Knowledge Discovery in Databases* (KDD) techniques, which concern the overall process of discovering useful knowledge from data.

Among the plethora of techniques proposed in literature to overcome the above-highlighted gap between data and knowledge, *On-Line Analytical Mining* (OLAM) [16] is a successful solution that integrates *On-Line Analytical Processing* (OLAP) [14,4] with DM in order to provide an integrated methodology for extracting useful knowledge from large databases and data warehouses. The benefits of OLAM have been previously put-in-evidence [16]: (*i*) DM algorithms can execute on integrated, *OLAP-based multidimensional views* that are already pre-processed and cleaned; (*ii*) users/applications can take advantages from the interactive, exploratory nature of OLAP tools to decisively enhance the knowledge fruition experience; (*iii*) users/applications can take advantages from the flexibility of OLAP tools in making available a wide set of DM solutions for a given KDD task, so that, thanks to OLAP, different DM algorithms become easily *interchangeable* in order to decisively enhance the benefits coming from cross-comparative data analysis methodologies over large amounts of data.

Starting from these considerations, in this paper we propose an OLAM-based framework for *complex knowledge pattern* discovery, along with a formal model underlying this framework, called $\mathcal{Multi}$-$\mathcal{Resolution}$ $\mathcal{Ensemble}$-$based$ $Model$ for $Advanced$ $\mathcal{Knowledge}$ $\mathcal{Discovery}$ in $Large$ $\mathcal{Databases}$ and $Data$ $Warehouses$ ($\mathcal{MRE}$-$\mathcal{KDD}^+$), and a reference architecture for such a framework. On the basis of OLAP principles, $\mathcal{MRE}$-$\mathcal{KDD}^+$, which can be reasonably considered as an innovative contribution in this research field, provides a formal, rigorous methodology for implementing advanced KDD processes in data-intensive settings, but with particular regard to two specialized instances represented by (*i*) a general application scenario populated by distributed and heterogeneous data sources, such as a conventional distributed data warehousing environment (e.g., like those that one can find in B2B and B2C *e*-commerce systems), and (*ii*) the integration/data layer of cooperative information systems, where different data sources are integrated in a unique middleware in order to make KDD processes against these data sources transparent-to-the-user. The notion of complex patterns refers to patterns having a nature more advanced than that of traditional ones such as sequences, trees, graphs etc. Examples of complex patterns for KDD are: multidimensional domains, hierarchical structures, clusters etc.

Besides the widely-accepted benefits coming from integrating OLAM within its core layer [16], $\mathcal{MRE}$-$\mathcal{KDD}^+$ allows data-intensive applications adhering to the methodology it defines to take advantages from other relevant characteristics, among which we recall the following: (*i*) the *multi-resolution support* offered by popular OLAP operators/tools [21], which allow us to execute DM algorithms over integrated and summarized multidimensional views of data at different *level of granularity* and *perspective of analysis*, thus sensitively improving the quality of KDD processes; (*ii*) the *ensemble-based support*, which, briefly, consists in meaningfully combining results coming from different DM algorithms executed over a collection of multidimensional views in order to generate the final knowledge, and provide facilities at the knowledge fruition layer.

Another contribute of our work is represented by the proposal of **KBMiner**, a $\mathcal{MRE\text{-}KDD}^+$-based visual tool that allows us to edit the so-called *Knowledge Discovery Tasks* (KDT), which realize a *graphical formalism* for extracting useful knowledge from large databases and data warehouses according to the guidelines drawn by $\mathcal{MRE\text{-}KDD}^+$.

The remaining part of this paper is organized as follows. In Section 2, we survey principles and models of OLAM. In Section 3, we outline related work. In Section 4, we present in detail $\mathcal{MRE\text{-}KDD}^+$. In Section 5, we provide a reference architecture implementing the framework we propose, and the description of its key components. In Section 6, we illustrate the main functionalities of **KBMiner**, along with some running examples. Finally, in Section 7 we outline conclusions of our work, and further activities in this research field.

2 On-Line Analytical Mining (OLAM): Mixing Together OLAP and Data Mining

OLAM is a powerful technology for supporting knowledge discovery from large databases and data warehouses that mixes together OLAP functionalities for representing/processing data, and DM algorithms for extracting regularities (e.g., patterns, association rules, clusters etc) from data. In doing this, OLAM realizes a proper KDD process.

OLAM was proposed by Han in his fundamental paper [16], along with the OLAP-based DM system **DBMiner** [20], which can be reasonably considered as the practical implementation of OLAM. In order to emphasize and refine the capability of discovering useful knowledge from huge amounts of data, OLAM gets the best of both technologies (i.e., OLAP and DM). From OLAP, (*i*) the excellent capability of storing data, which has been of relevant interest during the last years (e.g., [1,22,34,31,24]), (*ii*) the support for *multidimensional and multi-resolution data analysis* [4]; (*iii*) the richness of OLAP operators [21], such as *roll-up*, *drill-down*, *slice-&-dice*, *pivot* etc; (*iv*) the wide availability of a number of query classes, such as *range-* [23], *top-k* [36] and *iceberg* [12] queries, which have been extensively studied during the last years, and can be used as baseline for implementing even-complex KDD tasks. From DM, the broad collection of techniques available in literature, each of them oriented to cover a specific KDD task; among these techniques, some are relevant for OLAM, such as: *mining association rules in transactional or relational databases* [2,3,18,26,29,30], *mining classification rules* [27,28,35,5,10], *cluster analysis* [25,11,33], *summarizing and generalizing data using data cube* [22,14,4] or *attribute-oriented inductive* [17,19] approaches.

According to the guidelines given in [16], there exist various alternatives to mix together the capabilities of OLAP and DM, mainly depending on the way of combining the two technologies. Among all, the most relevant ones are: (*i*) *cubing-then-mining*, where the power of OLAP operators generalized in the primitive *cubing* [16] (i.e., the universal operator for generating a new data cube from another one or from a collection of data cubes), is used to select, pre-process, and interactively-process the portion of (OLAP) data on which executing DM algorithms, and (*ii*) *mining-then-cubing*, where the power of DM is applied to the target data cube directly, and then OLAP operators/tools are used to further analyze results of DM in order to improve the overall quality and refinement of the extracted knowledge.

In our opinion, the most effective alternative is that provided by the cubing-then-mining mode, which is, in fact, the preferred solution in most application scenarios. In this mode, OLAP allows us to obtain summarized multidimensional views over large amounts of data, and DM algorithms successfully run on these views to extract knowledge. In other words, in this vest OLAP is recognized as a very efficient data-support/pre-processing-tool for DM algorithms, which also allows high performance in comparison to traditional *On-Line Transactional Processing* (OLTP – the technology of relational databases) operators/tools.

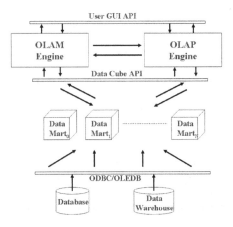

Fig. 1. A reference architecture for OLAM

In a data warehousing environment, subject-oriented multidimensional views of data sources are very often materialized into the so-called *Data Marts* [4], which can be generally intended as a sort of *specialized data cubes* built to support specific analysis requirements (e.g., sales, inventory, balance etc). Usually, given a collection of hetero-geneous data sources, a data warehouse application makes use of several data marts, each of them focused to support multidimensional, multi-resolution data analysis over a specific application context via OLAP operators/tools. In OLAM, DM algorithms are enabled to run over both a singleton data mart and, more interesting, *multiple data marts*, thus taking advantages from the amenity of extracting useful knowledge by means of complex JOIN-based OLAP operations over multiple sources.

A reference architecture for the cubing-then-mining OLAM mode is depicted in Figure 1 [21]. Here, the *OLAP Engine* and the *OLAM Engine* run in a combined manner in order to extract useful knowledge from a collection of subject-oriented data marts. Beyond the above-described OLAM features, this architecture also supports a leading OLAM functionality, the so-called *On-Line Interactive Mining* [21], which consists in iteratively executing DM algorithms over *different* views extracted from the *same* data mart via cubing. In this case, the effective "add-on" value given by OLAP is represented by a powerful *information gain* which cannot be easily supported by traditional OLTP operators/tools, without introducing excessive computational overheads.

3 Related Work

Starting from the first DM and KDD experiences, the idea of developing a framework for supporting advanced KDD processes from large databases and data warehouses via embedding complex data representation techniques (like OLAP) and sophisticated knowledge extraction methodologies (like DM) has been of relevant interest for the research community. The most important limitations to this ambition are represented by the difficult of integrating different-in-nature data representation models/methodologies, and the difficult of making heterogeneous DM algorithms intercommunicating.

While there are in literature a plethora of data representation techniques and DM algorithms, each of them developed for a particular application scenario, frameworks that integrate with-a-large-vision several techniques coming from different contexts via synthesizing data warehousing, DM and KDD principles are very few. Furthermore, while there exist an extremely wide set of DM and KDD tools (a comprehensive overview can be found in [15]), mainly focused to cover a specific KDD task (e.g., association rule discovery, classification, clustering etc), very few of them integrate heterogeneous KDD-oriented techniques and methodologies in a unique environment. Along these, the most significant experiences that have deeply influenced our work are DBMiner and WEKA [32]. In the following, we refer to both the environments in the vest of "realizations" of the respective underlying models.

DBMiner is a powerful OLAM-inspired system that allows us to (*i*) extract and represent knowledge from large databases and data warehouses, and (*ii*) mine knowledge via a wide set of very useful data analysis functionalities, mainly OLAP-inspired, such as data/patterns/results browse, exploration, visualization and intelligent querying. Specifically, at the representation/storage layer, DBMiner makes use of the popular *data cube* model (the foundation of OLAP), first proposed by Gray *et al.* [14], where relational data are aggregated on the basis of a multidimensional and multi-resolution vision of data. Based on the data cube model, DBMiner makes available to the user a wide set of innovative functionalities ranging from *time series analysis* to *prediction* of the data distribution of relational attributes to mining of complex objects (like those that one can find in a GIS); furthermore, DBMiner also offers a *data mining query language*, called *DMQL*, for supporting the standardization of DM functionalities and their integration with conventional DBMS. Finally, the graphical user interface of DBMiner supports various attracting, user-friendly forms implementing the above-listed features, and making them available to the user.

WEKA is a *Machine Learning* (ML) environment for efficiently supporting DM activities against large databases, and it has been designed to aid in decision-support processes in order to understand which information is relevant for the specific application context, and, consequently, make prediction faster. Similarly to DBMiner,

WEKA offers a graphical environment where users can (*i*) edit a ML technique, (*ii*) test it against external data sets, and (*iii*) study its performance under the stressing of various metrics. Moreover, WEKA users, just like DBMiner users, are allowed to mine the output knowledge of ML techniques by means of several advanced intelligent visualization components. Contrarily to DBMiner, WEKA does not make use of a particular data-representation/storage solution to improve data access/management/processing.

Finally, due to the nature and goals of both the outlined environments/models, we can claim that DBMiner is closer to our work rather than WEKA.

4 $\mathcal{MRE\text{-}KDD}^+$: A Multi-resolution, Ensemble-Based Model for Complex Knowledge Pattern Discovery from Large Databases and Data Warehouses

$\mathcal{MRE\text{-}KDD}^+$ is the innovative model underlying the framework we propose, and it has been designed to efficiently support complex knowledge pattern discovery from large databases and data warehouses according to a multi-resolution, ensemble-based approach. Basically, $\mathcal{MRE\text{-}KDD}^+$ follows the cubing-then-mining approach [16], which, as highlighted in Section 2, is the most promising OLAM solution for real-life applications. In this Section, we provide definition and main properties of $\mathcal{MRE\text{-}KDD}^+$.

4.1 $\mathcal{MRE\text{-}KDD}^+$ OLAP-Based Data Representation and Management Layer

Let $S = \{ S_0, S_1, \ldots, S_{K-1} \}$ be a set of K distributed and heterogeneous data sources, and $\mathcal{D} = \{ \mathcal{D}_0, \mathcal{D}_1, \ldots, \mathcal{D}_{P-1} \}$ be a set of P data marts defined over data sources in S. The first component of $\mathcal{MRE\text{-}KDD}^+$ is the so-called *Multidimensional Mapping Function* MMF, defined as a tuple MMF = $\langle \text{MMF}^{\mathcal{H}}, \text{MMF}^{\mathcal{F}} \rangle$, which takes as input a sub-set of M data sources in S, denoted by $S^M = \{ S_m, S_{m+1}, \ldots, S_{m+M-1} \}$, and returns as output a data mart $\mathcal{D}_k$ in $\mathcal{D}$, computed over data sources in S^M according to the construct $\text{MMF}^{\mathcal{H}}$ that models the *definition* of $\mathcal{D}_k$. $\text{MMF}^{\mathcal{H}}$ is in turn implemented as a conventional OLAP conceptual schema, such as *star* or *snowflake schemas* [6,21]. $\text{MMF}^{\mathcal{F}}$ is the construct of MMF that properly models the underlying function, defined as follows:

$$\text{MMF}^{\mathcal{F}}: S \to \mathcal{D} \qquad (1)$$

Given a MMF $\mathcal{G}$, we introduce the concept of *degree* of $\mathcal{G}$, denoted by $\mathcal{G}^{\Delta}$, which is defined as the number of data sources in S over which the data mart provided by $\mathcal{G}$ (i.e., $\mathcal{D}_k$) is computed, i.e. $\mathcal{G}^{\Delta} \equiv |S^M|$.

Due to the strongly "data-centric" nature of $\mathcal{MRE\text{-}KDD}^+$, management of OLAP data assumes a critical role, also with respect to performance issues, which must be taken in relevant consideration in data-intensive applications like those addressed by OLAM. To this end, we introduce the *Multidimensional Cubing Function* MCF, defined as a tuple MCF = $\langle MCF^{\mathcal{H}}, MCF^{\mathcal{F}} \rangle$, which takes as input a data mart $\mathcal{D}_k$ in $\mathcal{D}$, and returns as output a data mart $\mathcal{D}_h$ in $\mathcal{D}$, according to the construct $MCF^{\mathcal{H}}$ that models an OLAP operator/tool. In more detail, $MCF^{\mathcal{H}}$ can be one of the following OLAP operators/tools:

- *Multidimensional View Extraction* $\mathcal{V}$, which computes $\mathcal{D}_h$ as a multidimensional view extracted from $\mathcal{D}_k$ by means of a set of ranges $R_0, R_1, \ldots, R_{N-1}$ defined on the N dimensions of $\mathcal{D}_k$ $d_0, d_1, \ldots, d_{N-1}$, respectively, being each range R_j defined as a tuple $R_j = \langle L_l, L_u \rangle$, with $L_l < L_u$, such that L_l is the lower and L_u is the upper bound on d_j, respectively;
- *Range Aggregate Query* $\mathcal{Q}$, which computes $\mathcal{D}_h$ as a one-dimensional view with cardinality equal to 1 (i.e., an *aggregate value*) given by the application of a SQL aggregate operator (such as SUM, COUNT, AVG etc) applied to the collection of (OLAP) cells contained within a multidimensional view extracted from $\mathcal{D}_k$ by means of the operator $\mathcal{V}$;
- *Top-K Query* $\mathcal{K}$, which computes $\mathcal{D}_h$ as a multidimensional view extracted from $\mathcal{D}_k$ by means of the operator $\mathcal{V}$, and containing the (OLAP) cells of $\mathcal{D}_k$ whose values are the first K greatest values among cells in $\mathcal{D}_k$;
- *Drill-Down* $\mathcal{U}$, which computes $\mathcal{D}_h$ via decreasing the level of detail of data in $\mathcal{D}_k$;
- *Roll-Up* $\mathcal{R}$, which computes $\mathcal{D}_h$ via increasing the level of detail of data in $\mathcal{D}_k$;
- *Pivot* $\mathcal{P}$, which computes $\mathcal{D}_h$ via re-structuring the dimensions of $\mathcal{D}_k$ (e.g., changing the ordering of dimensions).

Formally, $MCF^{\mathcal{H}} = \{\mathcal{V}, \mathcal{Q}, \mathcal{K}, \mathcal{U}, \mathcal{R}, \mathcal{P}\}$. Finally, $MCF^{\mathcal{F}}$ is the construct of MCF that properly models the underlying function, defined as follows:

$$MCF^{\mathcal{F}}: \mathcal{D} \to \mathcal{D} \qquad (2)$$

It should be noted that the construct $MCF^{\mathcal{H}}$ of MCF operates on a singleton data mart to extract another data mart. In order to improve the quality of the overall KDD process, we also introduce the *Extended Multidimensional Cubing Function* MCF_E, defined as a tuple $MCF_E = \langle MCF_E^{\mathcal{H}}, MCF^{\mathcal{F}} \rangle$, which extends MCF by providing a different, complex OLAP operator/tool (i.e., $MCF_E^{\mathcal{H}}$) instead that the "basic" $MCF^{\mathcal{H}}$. $MCF_E^{\mathcal{H}}$ supports the amenity of executing $MCF^{\mathcal{H}}$ over multiple data marts, modeled as a sub-set of B data marts in $\mathcal{D}$, denoted by $\mathcal{D}^B = \{\mathcal{D}_b, \mathcal{D}_{b+1}, \ldots, \mathcal{D}_{b+B-1}\}$, being these data marts combined by means of the operator JOIN performed with respect to schemas of data marts. Specifically, $MCF_E^{\mathcal{H}}$ operates according to two variants: (*i*) in the first one, we first apply an instance of $MCF^{\mathcal{H}}$ to each data mart in $\mathcal{D}^B$, thus obtaining a set of *transformed* data marts $\mathcal{D}^T$, and then the operator JOIN to data marts in $\mathcal{D}^T$; (*ii*) in the second one, we first apply the operator JOIN to data marts in $\mathcal{D}^B$, thus obtaining a *unique* data mart $\mathcal{D}^u$, and then an instance of $MCF^{\mathcal{H}}$ to the data mart $\mathcal{D}^u$.

To give examples, let $\mathcal{D}^B = \{\mathcal{D}_0, \mathcal{D}_1, \mathcal{D}_2\}$ be the target sub-set of data marts, then, according to the first variant, a possible instance of $\mathsf{MCF}_E^{\mathcal{H}}$ could be: $\mathcal{V}(\mathcal{D}_0) \bowtie \mathcal{K}(\mathcal{D}_1) \bowtie \mathcal{U}(\mathcal{D}_2)$; contrarily to this, according to the second variant, a possible instance of $\mathsf{MCF}_E^{\mathcal{H}}$ could be: $\mathcal{U}(\mathcal{D}_0 \bowtie \mathcal{D}_1 \bowtie \mathcal{D}_2)$. Note that, in both cases, the result of the operation is still a data mart belonging to the set of data marts $\mathcal{D}$ of $\mathcal{MRE\text{-}KDD}^+$.

Formally, we model $\mathsf{MCF}_E^{\mathcal{H}}$ as a tuple $\mathsf{MCF}_E^{\mathcal{H}} = \langle \mathcal{D}^B, \mathcal{Y} \rangle$, such that (i) $\mathcal{D}^B$ is the sub-set of data marts in $\mathcal{D}$ on which $\mathsf{MCF}_E^{\mathcal{H}}$ operates to extract the final data mart, and (ii) $\mathcal{Y}$ is the set of instances of $\mathsf{MCF}^{\mathcal{H}}$ used to accomplish this goal.

4.2 $\mathcal{MRE\text{-}KDD}^+$ Data Mining Layer

DM algorithms defined in $\mathcal{MRE\text{-}KDD}^+$ are modeled by the set $\mathcal{A} = \{\mathcal{A}_0, \mathcal{A}_1, \ldots, \mathcal{A}_{T\text{-}1}\}$. These are classical DM algorithms focused to cover specific instances of con-solidated KDD tasks, such as discovery of patterns and regularities, discovery of association rules, classification, clustering etc, with the novelty of being applied to multidimensional views (or, equally, data marts) extracted from the data mart domain $\mathcal{D}$ of $\mathcal{MRE\text{-}KDD}^+$ via complex OLAP operators/tools implemented by the components MCF and MCF_E. Formally, an algorithm $\mathcal{A}_h$ of $\mathcal{A}$ in $\mathcal{MRE\text{-}KDD}^+$ is modeled as a tuple $\mathcal{A}_h = \langle I_h, \mathcal{D}_h, O_h \rangle$, such that: (i) I_h is the instance of $\mathcal{A}_h$ (properly, $\mathcal{A}_h$ models the *class* of the particular DM algorithm), (ii) $\mathcal{D}_h$ is the data mart on which $\mathcal{A}_h$ executes to extract knowledge, and (iii) O_h is the output knowledge of $\mathcal{A}_h$. Specifically, O_h representation depends on the nature of algorithm $\mathcal{A}_h$, meaning that if, for instance, $\mathcal{A}_h$ is a clustering algorithm, then O_h is represented as a collection of clusters (reasonably, modeled as sets of items) extracted from $\mathcal{D}_h$.

KDD process in $\mathcal{MRE\text{-}KDD}^+$ are governed by the component *Execution Scheme*, denoted by ES, which rigorously models *how* algorithms in $\mathcal{A}$ must be executed over multidimensional views of $\mathcal{D}$. To this end, ES establishes (i) how to combine multidimensional views and DM algorithms (i.e., which algorithm must be executed on which view), and (ii) the temporal sequence of executions of DM algo-rithms over multidimensional views. To formal model this aspect of the framework, we introduce the *Knowledge Discovery Function* KDF, which takes as input a collec-tion of R algorithms $\mathcal{A}^R = \{\mathcal{A}_r, \mathcal{A}_{r+1}, \ldots, \mathcal{A}_{r+R\text{-}1}\}$ and a collection of W data marts $\mathcal{D}^W = \{\mathcal{D}_w, \mathcal{D}_{w+1}, \ldots, \mathcal{D}_{w+W\text{-}1}\}$, and returns as output an execution scheme ES_p. KDF is defined as follows:

$$\mathsf{KDF}: \mathcal{A}^R \times \mathcal{D}^W \rightarrow \langle I^R \times \mathcal{D}^T, \varphi \rangle \tag{3}$$

such that: (i) I^R is a collection of instances of algorithms in $\mathcal{A}^R$, (ii) $\mathcal{D}^T$ is a collection of transformed data marts obtained from $\mathcal{D}^W$ by means of cubing operations provided by the components MCF or MCF_E of the framework, and (iii) φ is a collection deter-mining the temporal sequence of instances of algorithms in I^R over data marts in $\mathcal{D}^T$

in terms of ordered pairs $\langle I_r, \mathcal{D}^T_k \rangle$, such that the ordering of pairs indicates the temporal ordering of executions. From (3), we derive the formal definition of the component ES of $\mathcal{MRE\text{-}KDD}^+$ as follows:

$$ES = \langle I \times \mathcal{D}, \varphi \rangle \qquad (4)$$

Finally, the execution scheme ES_p provided by KDF can be one of the following alternatives:

- *Singleton Execution* $\langle I_r \times \mathcal{D}^T_k, \varphi \rangle$: execution of the instance I_r of the algorithm $\mathcal{A}_r$ over the transformed data mart $\mathcal{D}^T_k$, with $\varphi = \{\langle I_r, \mathcal{D}^T_k \rangle\}$.
- *$1 \times N$ Multiple Execution* $\langle I_r \times \{\mathcal{D}^T_k, \mathcal{D}^T_{k+1}, ..., \mathcal{D}^T_{k+N-1}\}, \varphi \rangle$: execution of the instance I_r of the algorithm $\mathcal{A}_r$ over the collection of transformed data marts $\{\mathcal{D}^T_k, \mathcal{D}^T_{k+1}, ..., \mathcal{D}^T_{k+N-1}\}$, with $\varphi = \{\langle I_r, \mathcal{D}^T_k \rangle, \langle I_r, \mathcal{D}^T_{k+1} \rangle, ..., \langle I_r, \mathcal{D}^T_{k+N-1} \rangle\}$.
- *$N \times 1$ Multiple Execution* $\langle \{I_r, I_{r+1}, ..., I_{r+N-1}\} \times \mathcal{D}^T_k, \varphi \rangle$: execution of the collection of instances $\{I_r, I_{r+1}, ..., I_{r+N-1}\}$ of the algorithms $\{\mathcal{A}_r, \mathcal{A}_{r+1}, ..., \mathcal{A}_{r+N-1}\}$ over the transformed data mart $\mathcal{D}^T_k$, with $\varphi = \{\langle I_r, \mathcal{D}^T_k \rangle, \langle I_{r+1}, \mathcal{D}^T_k \rangle, ..., \langle I_{r+N-1}, \mathcal{D}^T_k \rangle\}$.
- *$N \times M$ Multiple Execution* $\langle \{I_r, I_{r+1}, ..., I_{r+N-1}\} \times \{\mathcal{D}^T_k, \mathcal{D}^T_{k+1}, ..., \mathcal{D}^T_{k+M-1}\}, \varphi \rangle$: execution of the collection of instances $\{I_r, I_{r+1}, ..., I_{r+N-1}\}$ of the algorithms $\{\mathcal{A}_r, \mathcal{A}_{r+1}, ..., \mathcal{A}_{r+N-1}\}$ over the collection of transformed data marts $\{\mathcal{D}^T_k, \mathcal{D}^T_{k+1}, ..., \mathcal{D}^T_{k+M-1}\}$, with $\varphi = \{..., \langle I_{r+p}, \mathcal{D}^T_{k+q} \rangle, ...\}$, such that $0 \le p \le N - 1$ and $0 \le q \le M - 1$.

4.3 $\mathcal{MRE\text{-}KDD}^+$ Ensemble Layer

As stated in Section 1, at the output layer, $\mathcal{MRE\text{-}KDD}^+$ adopts an ensemble-based approach. The so-called *Mining Results* (MR) coming from the executions of DM algorithms over collections of data marts must be finally merged in order to provide the end-user/application with the extracted knowledge that is presented in the form of complex patterns. It should be noted that this is a relevant task in our proposed framework, as very often end-users/applications are interested in extracting useful knowledge by means of *correlated, cross-comparative* KDD tasks, rather than a singleton KDD task, according to real-life DM scenarios. Combining results coming from different DM algorithms is a non-trivial research issue, as recognized in literature. In fact, as highlighted in Section 4.2, the output of a DM algorithm depends on the nature of that algorithm, so that in some cases MR coming from very different algorithms cannot be combined directly.

In $\mathcal{MRE\text{-}KDD}^+$, we face-off this problematic issue by making use of OLAP technology again. We build multidimensional views over MR provided by execution schemes of KDF, thus giving support to a *unifying manner* of exploring and analyzing final results. It should be noted that this approach is well-motivated under noticing that usually end-user/applications are interested in analyzing final results based on a certain *mining metrics* provided by KDD processes (e.g., *confidence interval of association rules, density of clusters, recall of IR-style tasks* etc), and this way-to-do is perfectly suitable to be implemented within OLAP data cubes where (*i*) output of DM algorithms (e.g., item sets) is the data source, (*ii*) user-selected features of the output of DM algorithms are the (OLAP) dimensions, and (*iii*) the above-mentioned mining metrics are the (OLAP) measures. Furthermore, this approach also involves in the benefit of efficiently supporting the *visualization* of final results by mean of attracting user-friendly, graphical formats/tools such as multidimensional bars, charts, plots etc, similarly to the functionalities supported by DBMiner and WEKA.

Multidimensional Ensembling Function MEF is the component of $\mathcal{MRE}$-$\mathcal{KDD}^+$ which is in charge of supporting the above-described knowledge presentation/delivery task. It takes as input a collection of Q output results $O = \{O_0, O_1, ..., O_{Q-1}\}$ provided by KDF-formatted execution schemes and the definition of a data mart Z, and returns as output a data mart $\mathcal{L}$, which we name as *Knowledge Visualization Data Mart* (KVDM), built over data in O according to Z. Formally, MEF is defined as follows:

$$\text{MEF: } \langle O,Z \rangle \to \mathcal{D} \tag{5}$$

It is a matter to note that the KVDM $\mathcal{L}$ becomes part of the set of data marts $\mathcal{D}$ of $\mathcal{MRE}$-$\mathcal{KDD}^+$, but, contrarily to the previous data marts, which are used to knowledge processing purposes, it is used to knowledge exploration/visualization purposes.

5   A Reference Architecture for Supporting Complex Knowledge Pattern Discovery from Large Databases and Data Warehouses

Figure 2 shows a reference architecture implementing the framework we propose. This architecture is suitable to be implemented on top of any distributed software platform, under the design guidelines given by component-oriented software engineering best practices. Despite being orthogonal to any distributed data-intensive environment, as stated in Section 1, this architecture is particularly useful in a general application scenario populated by distributed and heterogeneous data sources, and in the integration/data layer of cooperative information systems. As we will put in evidence throughout the remaining part of this Section, components of the reference architecture implement constructs of the underlying model $\mathcal{MRE}$-$\mathcal{KDD}^+$, according to a meaningful abstraction between formal constructs and software components.

As shown in Figure 2, in the proposed architecture, distributed and heterogeneous data sources, located at the Data Source Layer, are first processed by means of *Extraction, Transformation and Loading* (ETL) tasks implemented by the ETL Engine, and then integrated in a common Relational Data Layer, in order to ensure flexibility at the next data processing/transformation steps, and take advantages from mining correlated knowledge. The Data Mart Builder, which implements the component MMF of $\mathcal{MRE}$-$\mathcal{KDD}^+$, is in charge of constructing a collection of subject-oriented data marts, which populate the Data Mart Layer, via accessing data at the Relational Data Layer, and in-dependence-on specific requirements of the target data-intensive application running on top of the proposed architecture. The OLAP Engine, which implements the components MCF and MCF_E of $\mathcal{MRE}$-$\mathcal{KDD}^+$, provides conventional and complex OLAP operators/tools over data marts of the

Data Mart Layer, thus originating a collection of multidimensional views located at the OLAP View Layer. These views constitute the input of the OLAM Engine, which, by accessing a set of conventional DM algorithms stored in the DM Algorithm Repository, implements the component KDF of $\mathcal{MRE}$-$\mathcal{KDD}^{+}$ via combining views and algorithms to execute even-complex KDD processes. Finally, the Mining Result Merging Component, which implements the component MEF of $\mathcal{MRE}$-$\mathcal{KDD}^{+}$, combines different MR to obtain the final knowledge, and meaningfully support the knowledge fruition experience via complex patterns such as multidimensional domains, hierarchical structures and clusters.

Another integrated component of the proposed architecture is the $\mathcal{MRE}$-$\mathcal{KDD}^{+}$-based visual tool KBMiner, which is not depicted in Figure 2 for the sake of simplicity. KBMiner can connect to the architecture in order to efficiently support the editing of KDT (see Section 1) for discovering useful knowledge from large databases and data warehouses. However, being completely independent of the particular implementation, the proposed architecture can also be realized as core-, inside-, stand-alone-platform within distributed data-intensive environments, where KDD functionalities

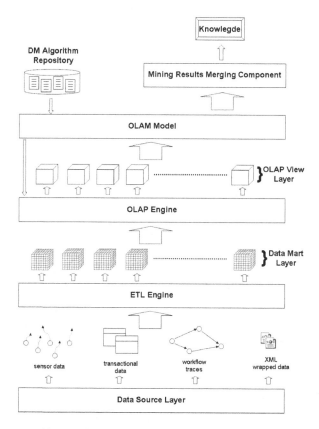

Fig. 2. A reference architecture implementing the proposed complex knowledge pattern discovery framework

are available in the vest of component-oriented API to intelligent applications executing KDD processes rather than in the vest of plug-in components of **KBMiner** to *knowledge workers* authoring KDD processes (similarly to the former one, in the latter case the edited KDT must be executed at-the-very-end by an ad-hoc software component implementing a "general-purpose" $\mathcal{MRE}\text{-}\mathcal{KDD}^+$ engine).

5.1 Performance Issues

Despite OLAP technology has already reached relevant performance, accessing and processing OLAP data still remain a problematic bottleneck for data-intensive applications embedding an OLAP engine inside their core layer. Basically, this is due to the fact that, in real-life repositories, data size grows with an exponential trend, and TB/PB is the typical data magnitude achieved.

A very efficient solution to this problem is represented by *data-cube-compression/approximate-query-answering* techniques, which allow us to sensitively speed-up data access activities and query evaluation tasks against OLAP data by (*i*) reducing the size of data cubes, thus obtaining *compressed representations* of them, and (*ii*) implementing efficient algorithms capable of evaluating queries over these representations, thus obtaining *approximate answers* that are perfectly tolerable in OLAP [7]. In an OLAM architecture such that proposed in Figure 2, which is *intrinsically* OLAP-based, these techniques can be easily integrated (specifically, with regard to our proposed architecture, inside the **OLAP Engine**) and successfully exploited in order to improve the overall performance of even-complex KDD processes, since they allow us to reduce the complexity of resource-intensive operations (i.e., multidimensional data access and management). Among the plethora of data-cube-compression/approximate-query-answering techniques proposed in literature, we recall *analytical synopses* [8] and *sampling-based approaches* (e.g., [9]), which represent relevant results in this research field.

6 KBMiner: A $\mathcal{MRE}\text{-}\mathcal{KDD}^+$-Based Visual Tool

It is widely recognized that knowledge discovery is intrinsically a *semi-automatic* process, meaning that it requires the interaction of system author, which usually is also an expert of the investigated application scenario, in order to: (*i*) define the goals of the target KDD process, (*ii*) define the parameters of the target KDD process, (*iii*) set the default/input values of such parameters, (*iv*) check the alignment and the correctness of intermediate results generated by the execution of the target KDD process, (*v*) compose the final results, (*vi*) understand and mine the final results. On the other hand, very often system author is not an ICT expert; as an example, this is a common case in the *Business Intelligence* (BI) context, where OLAP/OLAM technology has been widely applied, and knowledge workers are typically non-ICT-expert business managers and administrators. In consequence of both aspects, in real-life systems/applications there is the straightly needed of *visual authoring tools* able to efficiently support the editing of even-complex KDD processes according to a transparent-for-user manner. In other words, these tools must be capable of allowing system authors to design KDD processes in a simple, intuitive and interactive manner, by means of meaningful metaphors offered by visual programming. It should be noted that **DBMiner** and **WEKA** adhere to this evidence.

Following similar considerations, **KBMiner** is a $\mathcal{MRE}\text{-}\mathcal{KDD}^+$-based visual tool supporting the editing of KDT for the discovery of useful knowledge from large databases and data warehouses according to the $\mathcal{MRE}\text{-}\mathcal{KDD}^+$ guidelines. As highlighted in Section 1, KDT allow system author to "codify" KDD processes, being the underlying "programming language" based on the constructs of $\mathcal{MRE}\text{-}\mathcal{KDD}^+$.

A KDT is a directed graph where nodes represent algorithms/tasks, and arcs represent data flows or inter-algorithm/tasks operations. Furthermore, a KDT also includes the multidimensional-views/data-marts on which the previous algorithms/tasks execute, and other components modeling the composition of MR according to the ensemble-based approach defined by $\mathcal{MRE}\text{-}\mathcal{KDD}^+$. Furthermore, in **KBMiner**, system author is also allowed to associate the so-called *Mining ruLes* (ML), which are *logical rules* defined on $\mathcal{MRE}\text{-}\mathcal{KDD}^+$ entities, to KDT arcs, in order to codify on top of the target KDT a sort of *control algorithm* that is in charge of *driving* the overall KDD process via controlling the values (TRUE or FALSE) given by the evaluation of ML across intermediate tasks of the process. As an example, given the domain of $\mathcal{MRE}\text{-}\mathcal{KDD}^+$ entities $\{\{\mathcal{A}_h, \mathcal{A}_k, \mathcal{A}_z\}, \{\mathcal{D}_m, \mathcal{D}_n, \mathcal{D}_p\}\}$, such that $\mathcal{A}_i$ with $i \in \{h, k, z\}$ is a clustering algorithm, and $\mathcal{D}_j$ with $j \in \{m, n, p\}$ is a data mart, a ML r, after the execution of $\mathcal{A}_h$ on $\mathcal{D}_m$ producing the output O_h in terms of a collection of clusters, could decide to run $\mathcal{A}_k$ on $\mathcal{D}_n$ or, alternatively, $\mathcal{A}_z$ on $\mathcal{D}_p$ on the basis of the fact that densities of clusters in O_h are greater than a given threshold V or not.

Other two relevant features supported by **KBMiner** are the following: (*i*) *interface operations* between DM algorithms, which establish how the output O_i of an algorithm $\mathcal{A}_i$ must be provided as input $\mathcal{D}_j$ to another algorithm $\mathcal{A}_j$ – the simplest way of implementing this operation is to directly transfer O_i to $\mathcal{D}_j$, but more complex models can be devised, such as generating $\mathcal{D}_j$ from O_i via a given OLAP operator/tool; (*ii*) *merging operations* between MR, which establish how MR must be combined at both the intermediate tasks of the target KDD process and the output layer in order to produce the final knowledge – some examples of operations used to this end are: union, intersection, OLAP-query-based projection etc.

6.1 Running Example

A meaningful example of KDT that can be edited in **KBMiner** is depicted in Figure 3. The KDD process modeled by the KDT of Figure 3 is composed by the following tasks:

1. DM algorithm $\mathcal{A}_i$ is executed over the multidimensional view originated by the OLAP query Q_t against the data marts $\mathcal{D}_k$ and $\mathcal{D}_h$, and it produces a multidimensional view representing the output O_i.
2. DM algorithm $\mathcal{A}_j$ is executed over O_i, and it produces the output O_j.
3. ML R_p is evaluated against O_j: if the value of R_p is TRUE, then DM algorithm $\mathcal{A}_k$ is executed over O_j, thus producing the output O_k; otherwise, if the value

of R_p is FALSE, then O_j is partitioned in two multidimensional views by means of the OLAP query Q_l. The first view, given by the result of Q_l, denoted by $DS(Q_l)$, constitutes the input of the DM algorithm $\mathcal{A}_f$, which produces the output O_f. The second view, given by the complementary set of the result of Q_l, denoted by $\neg DS(Q_l)$, constitutes, along with the data mart $\mathcal{D}_w$, the input of DM algorithm $\mathcal{A}_g$, which produces the output O_g.

4.   The final result of the KDD process modeled by the KDT of the running example is obtained in dependence on the value of the ML R_p against O_j: if that value is TRUE, then the final result corresponds to O_k; otherwise, if that value is FALSE, then the final result corresponds to the multidimensional view given by $O_f \cap O_g$.

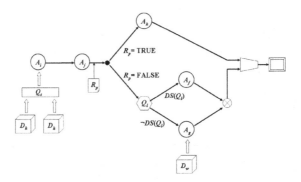

Fig. 3. A KDT in KBMiner

As shown by the running example above, **KBMiner** is able to efficiently model and execute KDD processes according to a simple and intuitive metaphor, which, however, allows us to author even-complex KDD tasks via meaningful ML; the combined effects of these two aspects contribute to make **KBMiner** a very useful tool for real-life data-intensive applications.

6.2 $\mathcal{MRE\text{-}KDD^+}$ UML Class Diagram

Finally, Figure 4 shows the **KBMiner** UML class diagram, which puts in evidence the various components of **KBMiner**, and how these components interact in order to discovery useful knowledge from large databases and data warehouses according to the guidelines of $\mathcal{MRE\text{-}KDD^+}$. As shown in Figure 4, **KBMiner** components are the following:

*   **Main Menu**, which coordinates all the **KBMiner** components;
*   **Data Mart Editor**, which allows users to edit and build subject-oriented data marts (it corresponds to the construct MMF of $\mathcal{MRE\text{-}KDD^+}$);
*   **OBDC Explorer**, which provides data access and data source linkage functionalities;

- **ETL Tool**, which supports ETL tasks;
- **OLAP Data Browser**, which allows users to access, explore and query OLAP-data-cubes/data-marts;
- **KDT Editor**, which allows users to edit KDT (it corresponds to the construct KDF of $\mathcal{MRE}\text{-}\mathcal{KDD}^+$);
- **OLAM Instance Editor**, which builds an $\mathcal{MRE}\text{-}\mathcal{KDD}^+$-based mining model starting from an input KDT;
- **OLAP Operation Editor**, which supports the editing of conventional and complex OLAP operators/tools (it corresponds to the constructs MCF and MCF_E of $\mathcal{MRE}\text{-}\mathcal{KDD}^+$);
- **DM Algorithm Repository Explorer**, which allows users to access and explore the DM algorithm repository;
- **Mining Rule Editor**, which allows users to edit ML;
- **Interface Operation Composer**, which allows users to edit interface operations between DM algorithms;
- **Merging Operation Composer**, which allows users to edit merging operations between MR at intermediate tasks of a KDD process as well as at the output layer (it corresponds to the construct MEF of $\mathcal{MRE}\text{-}\mathcal{KDD}^+$);

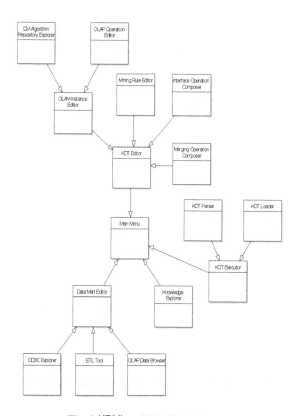

Fig. 4. KBMiner UML Class Diagram

- **KDT Executor**, which executes KDT;
- **KDT Parser**, which parses KDT and builds the corresponding $\mathcal{MRE}$-$\mathcal{KDD}^+$-based mining models;
- **KDT Loader**, which loads KDT in the vest of $\mathcal{MRE}$-$\mathcal{KDD}^+$-based mining models to be executed;
- **Knowledge Explorer**, which allows users to load and explore a previously-edited $\mathcal{MRE}$-$\mathcal{KDD}^+$-based mining model, and, if needed, re-execute it.

7 Conclusions and Future Work

Starting from successful OLAM technologies, in this paper we have presented a complete framework for supporting advanced knowledge discovery from large databases and data warehouses, which is useful for any data-intensive setting, but with particular emphasis for a general application scenario populated by distributed and heterogeneous data sources, and the integration/data layer of cooperative information systems. To this end, we have formally provided principles, definitions and properties of $\mathcal{MRE}$-$\mathcal{KDD}^+$, the model underlying the framework we propose. Other contributions of our work are: (*i*) a reference architecture implementing the framework, which can be realized in any distributed software platform, under the design guidelines given by component-oriented software engineering best practices; (*ii*) **KBMiner**, a visual tool that allows users to edit even-complex KDD processes according to the $\mathcal{MRE}$-$\mathcal{KDD}^+$ guidelines in a simple, intuitive and interactive manner. Future work is oriented along two main directions: (*i*) testing the performance of our framework against real-life scenarios such as those drawn by distributed corporate data warehousing environments in B2B and B2C *e*-commerce systems, and (*ii*) extending the actual capabilities of $\mathcal{MRE}$-$\mathcal{KDD}^+$ as to embed novel functionalities for supporting *prediction of events* in new DM activities edited by users/applications on the basis of the "history" given by logs of previous KDD processes implemented in similar or correlated application scenarios.

References

1. Agarwal, S., et al.: On the Computation of Multidimensional Aggregates. In: VLDB, pp. 506–521 (1996)
2. Agrawal, R., et al.: Mining Association Rules between Sets of Items in Large Databases. In: ACM SIGMOD, pp. 207–216 (1993)
3. Agrawal, R., et al.: Fast Algorithms for Mining Association Rules. In: VLDB, pp. 487–499 (1994)
4. Chaudhuri, S., et al.: An Overview of Data Warehousing and OLAP Technology. SIGMOD Record 26(1), 65–74 (1997)
5. Cheeseman, P., et al.: Bayesian Classification (AutoClass): Theory and Results. In: Fayyad, U., et al. (eds.) Advances in Knowledge Discovery and Data Mining, pp. 153–180. AAAI/MIT Press, Menlo Park, CA, USA (1996)

6. Colliat, G.: OLAP, Relational, and Multidimensional Database Systems. SIGMOD Record 25(3), 64–69 (1996)
7. Cuzzocrea, A.: Overcoming Limitations of Approximate Query Answering in OLAP. In: IEEE IDEAS, pp. 200–209. IEEE Computer Society Press, Los Alamitos (2005)
8. Cuzzocrea, A.: Improving Range-Sum Query Evaluation on Data Cubes via Polynomial Approximation. Data & Knowledge Engineering 56(2), 85–121 (2006)
9. Cuzzocrea, A., et al.: Approximate Range-Sum Query Answering on Data Cubes with Probabilistic Guarantees. Journal of Intelligent Information Systems 28(2), 161–197 (2007)
10. Elder IV, J., et al.: A Statistical Perspective on Knowledge Discovery in Databases. In: Fayyad, U., et al. (eds.) Advances in Knowledge Discovery and Data Mining, pp. 83–115. AAAI/MIT Press, Menlo Park, CA, USA (1996)
11. Ester, M., et al.: Knowledge Discovery in Large Spatial Databases: Focusing Techniques for Efficient Class Identification. In: SSD, pp. 67–82 (1995)
12. Fang, M., et al.: Computing Iceberg Queries Efficiently. In: VLDB, pp. 299–310 (1998)
13. Fayyad, U., et al.: From Data Mining to Knowledge Discovery: An Overview. In: Fayyad, U., et al. (eds.) Advances in Knowledge Discovery and Data Mining, pp. 1–35. AAAI/MIT Press, Menlo Park, CA, USA (1996)
14. Gray, J., et al.: Data Cube: A Relational Aggregation Operator Generalizing Group-By, Cross-Tab, and Sub-Totals. Data Mining and Knowledge Discovery 1(1), 29–54 (1997)
15. Goebel, M., et al.: A Survey of Data Mining and Knowledge Discovery Software Tools. SIGKDD Explorations 1(1), 0–33 (1999)
16. Han, J.: OLAP Mining: An Integration of OLAP with Data Mining. In: IFIP 2.6 DS, pp. 1–9 (1997)
17. Han, J., et al.: Data-driven Discovery of Quantitative Rules in Relational Databases. IEEE Transactions on Knowledge and Data Engineering 5(1), 29–40 (1993)
18. Han, J., et al.: Discovery of Multiple-Level Association Rules from Large Databases. In: VLDB, pp. 420–431 (1995)
19. Han, J., et al.: Exploration of the Power of Induction in Data Mining. In: Fayyad, U., et al. (eds.) Advances in Knowledge Discovery and Data Mining, pp. 399–421. AAAI/MIT Press, Menlo Park, CA, USA (1996)
20. Han, J., et al.: DBMiner: A System for Mining Knowledge in Large Relational Databases. In: KDD, pp. 250–255 (1996)
21. Han, J., et al.: Data Mining: Concepts and Techniques. Morgan Kaufmann Publishers, San Francisco, CA, USA (2000)
22. Harinarayan, V., et al.: Implementing Data Cubes Efficiently. In: ACM SIGMOD, pp. 205–216 (1996)
23. Ho, C.-T., et al.: Range Queries in OLAP Data Cubes. In: ACM SIGMOD, pp. 73–88 (1997)
24. Karayannidis, N., et al.: SISYPHUS: the Implementation of a Chunk-Based Storage Manager for OLAP. Data & Knowledge Engineering 45(2), 155–180 (2003)
25. Ng, R., et al.: Efficient and Effective Clustering Method for Spatial Data Mining. In: VLDB, pp. 144–155 (1994)
26. Park, J.S., et al.: An Effective Hash-based Algorithm for Mining Association Rules. In: ACM SIGMOD, pp. 175–186 (1995)
27. Piatetsky-Shapiro, G.: Discovery, Analysis, and Presentation of String Rules. In: Piatetsky-Shapiro, G., et al. (eds.) Knowledge Discovery in Databases, pp. 229–238. AAAI/MIT Press, Menlo Park, CA, USA (1991)

28. Quinlan, J.R.: C4.5: Programs for Machine Learning. Morgan Kaufmann Publishers, San Francisco, CA, USA (1993)
29. Savasere, A., et al.: An Efficient Algorithm for Mining Association Rules in Large Databases. In: VLDB, pp. 432–443 (1995)
30. Srikant, R., et al.: Mining Generalized Association Rules. In: VLDB, pp. 407–419 (1995)
31. Vitter, J.S., et al.: Approximate Computation of Multidimensional Aggregates of Sparse Data Using Wavelets. In: ACM SIGMOD, pp. 194–204 (1999)
32. Witten, I., et al.: Data Mining: Practical Machine Learning Tools and Techniques, 2nd edn. Morgan Kaufmann Publishers, San Francisco, CA, USA (2005)
33. Zhang, T., et al.: BIRCH: An Efficient Data Clustering Method for Very Large Databases. In: ACM SIGMOD, pp. 103–114 (1996)
34. Zhao, Y., et al.: An Array-based Algorithm for Simultaneous Multidimensional Aggregates. In: ACM SIGMOD, pp. 159–170 (1997)
35. Ziarko, W.: Rough Sets, Fuzzy Sets and Knowledge Discovery. Springer, New York, NY, USA (1994)
36. Xin, D., et al.: Answering Top-k Queries with Multi-Dimensional Selections: The Ranking Cube Approach. In: VLDB, pp. 463–475 (2006)

Integrating Clustering Data Mining into the Multidimensional Modeling of Data Warehouses with UML Profiles

Jose Zubcoff[1], Jesús Pardillo[2], and Juan Trujillo[2]

[1] Departament of Sea Sciences and Applied Biology,
University of Alicante, Spain
Jose.Zubcoff@ua.es
[2] Department of Software and Computing Systems,
University of Alicante, Spain
{jesuspv,jtrujillo}@dlsi.ua.es

Abstract. Clustering can be considered the most important unsupervised learning technique finding similar behaviors (clusters) on large collections of data. *Data warehouses* (DWs) can help users to analyze stored data, because they contain preprocessed data for analysis purposes. Furthermore, the *multidimensional* (MD) model of DWs, intuitively represents the system underneath. However, most of the clustering data mining are applied at a low-level of abstraction to complex unstructured data. While there are several approaches for clustering on DWs, there is still not a conceptual model for clustering that facilitates modeling with this technique on the *multidimensional* (MD) model of a DW. Here, we propose (i) a conceptual model for clustering that helps focusing on the data-mining process at the adequate abstraction level and (ii) an extension of the *unified modeling language* (UML) by means of the UML *profiling* mechanism allowing us to design clustering data-mining models on top of the MD model of a DW. This will allow us to avoid the duplication of the time-consuming preprocessing stage and simplify the clustering design on top of DWs improving the discovery of knowledge.

Keywords: Conceptual Modeling, Multidimensional Modeling, UML Extension, KDD, Data Warehouse, Data Mining, Clustering.

1 Introduction

Clustering [1] is the process of organizing objects into groups whose members are similar in some way. This data-mining technique helps to discover behavior patterns in large amount of data. While data mining is an instance of the *knowledge discovery in databases* (KDD) [2], the whole KDD process is the "nontrivial extraction of implicit, previously unknown, and potentially useful information from data". Main drawbacks of viewing KDD process as single isolated operations instead of an integrated process are (i) the duplicity of time-consuming processes and that (ii) every step in the KDD process increases the knowledge

I.Y. Song, J. Eder, and T.M. Nguyen (Eds.): DaWaK 2007, LNCS 4654, pp. 199–208, 2007.

of data under analysis, then analyst is missing opportunities of learning from previous knowledge. Data-mining algorithms require input data that are coherent with a specific data schema, however, they are applied on complex unstructured flat files, and thus preprocessing steps have become the most time-consuming phase of the analysis. On the other hand, a *data warehouse* (DW) is a "subject-oriented, integrated, time-variant and non-volatile collection of data in support of management's decision making process" [3]. Then, it is consistent that a DW is the previous step in KDD. Moreover, integrating clustering data mining into a DW project allows analysts to carry out the KDD process establishing business objectives from the early steps of project development, which ensures having the needed data in an adequate schema.

It is widely accepted that the development of DWs is based on the *multidimensional* (MD) modeling. This semantic rich structure represents data elements using facts, dimensions, and hierarchy levels, describing the system domain. This kind of representation is completely intuitive for analysts, in contrast to the usage of complex unstructured flat files. Main advantages of the conceptual modeling are abstracting from specific issues (platform, algorithms, parameters, etc.), the flexibility, and reusability. Moreover, a conceptual model remains the same when it is implemented on different platforms, or whenever the platform is updated. It also provides users with common notation and further semantics that improve the understanding of the system.

All considering, in this paper we propose a conceptual model for clustering at the needed level of abstraction to focus only in the main concepts of clustering and taking advantage of the previous domain knowledge. In order to avoid the learning of new languages, we use a well-known modeling language to accomplish this objective: the *unified modeling language* (UML) [4], that allows us to extend it for specific domain by means of UML *profiles*. Previously, we proposed in [5] this kind of extension for designing MD models in DWs. We also used UML profiles to design data-mining models for association rules [6,7] and classification [8] on top of MD models.

Outline. The remainder of this paper is structured as follows: Section 2 explains the clustering domain model. Section 3 proposes a new UML profile for designing the clustering data-mining models in the context of the MD modeling. Section 4 applies our UML profile to the design of a credit-card purchase clustering, and implements the resulting model in a commercial database system. Section 5 explains other works related to data mining, DWs, and modeling. Finally, section 6 presents the main conclusions and introduces immediate and future work.

2   Conceptualizing Clustering Algorithms

Data-mining can be represented as a domain model from which we can derive a modeling language to design data-mining models. In this section, we introduce the domain model of the MD-data clustering. In the next section, we will translate this model to a well-known modeling language like UML.

2.1 Clustering on MD Data

Clustering data mining identifies common behaviors in a data set that users might not logically derive through casual observation. A cluster is therefore a collection of objects that are *similar* among them and *dissimilar* to the objects belonging to other clusters. The result from cluster analysis is a number of groups (clusters) that form a partition, or a structure of partitions, of the data set. The suitable MD model of a DW easily represents data under study.

As shown in Fig. 1, the *fact* under analysis contains *measures* that are contextualized by *dimensions*. Each dimension used as *input* represents an axis, and each *case* corresponds to partitions of these axes. A clustering algorithm [9,10] differs from other algorithms (such as classification or association rules) because there is not a specific predict attribute that builds a clustering model. Then, clustering algorithms use input attributes to build a MD space on which we can measure similarities existing in the data to be clustered.

Fig. 1. Clustering data mining on the MD data of DWs

2.2 Domain Model of MD-Data Clustering

The clustering data-mining technique can take advantage of the ease-to-see structured MD data (Fig. 2, leyend: ∘— shared aggregation, ◆— composite aggregation, ◁— generalization, ←-- dependency). In a MD model, data is organized in terms of *facts* and *dimensions*. Whilst facts are the focus of analysis, modeling collections of *measures* (typically numerical data); the associated dimensions provide a context-of-analysis *descriptions* (like textual data) forming *hierarchies* on which measures can be aggregated at different granularity *levels*.

Data mining with a clustering *technique* will allow analysts to discover groups of similar behaviors upon the *fact* under study along *dimensions* at any *level* of granularity in their *hierarchies*. In this way, the MD structure can help to understand in deep the underlying data, since it represents the system domain close to the way of analysts' thinking.

The *algorithm*'s *parameters* (not shown in Fig. 2) are: the *maximum number of iterations* to build clusters, the *maximum number of clusters* built, the *number of clusters* to preselect, the *minimum support* to specify the number of cases that are needed to build a cluster, the *minimum error tolerance* to stop the building

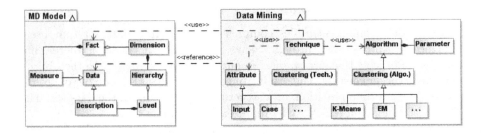

Fig. 2. Overview of the MD-data clustering domain model

of clusters, the *maximum number of input attributes* used to build a cluster, the *sample size* controlling the number of cases used to build the clusters, the *maximum number of classes* to specify the number of categories in an attribute value, and the *sensitivity* to detect smaller or larger density variation as clusters.

3   An Approach for Conceptual Modeling of Clustering

Once we have a model of our domain, we can translate it to other languages by using some mapping mechanism. Whilst the source model (modeling our domain) can be represented under some kind of formalism or simply in natural language, the target model is typically hosted by existing formal languages (instantiating or extending them) providing designers with tools developed to support these languages. One of the mapping mechanisms that we can use for modeling is a UML profile, the well-formalized extension mechanism of UML.

The *unified modeling language* (UML) [4] is general-purpose modeling language to model systems with a visual notation. Although it is a widely accepted language for specifying software systems, it also incorporates a light-weight mechanism (UML profiles) for extending the language to model specific domains. Profiling UML is mainly based on the concepts of *stereotypes, tag definitions*, and *constraints*. We easily extend UML and model with the profiled UML, reusing existing modeling tools and designing portable models interchangeable among them. Additionally, we support our UML profiles on restricting languages like the *object constraint language* (OCL) [11] to specify some semantics from translated domains that are not directly representable.

A UML profile (the extending model) extends UML (the metamodel; i.e., the modeling language) by stereotyping the existing UML modeling elements. Stereotypes are entities that translate specific domains into the UML modeling elements (metaclasses); the closer the domain-concept semantics are to the selected metaclass, the better the mapping will be. Moreover, UML provides us with a standard visual notation for defining UML profiles. Thus, we can present UML profile specifications by means of this visual notation, easily visualizing the extension concepts. However, there are some elements in the definition of our models (such as constraints or descriptions about semantics) that are difficult to present in the provided visual notation. For this reason, other complementary specification mechanisms (such as enumerating lists, tables, or textual notations) are also commonly used[1].

3.1   UML Profile for Conceptual Modeling of MD-Data Clustering

All considering, we propose a UML extension for conceptual modeling of clustering data mining based on the UML profiling mechanism. Our UML profile reuses a previous profile for MD modeling of DWs [5]. An excerpt of its specification is presented in Fig. 3. Visually, we define stereotypes and the extended metaclasses as boxes labeled with «*stereotype*» and «*metaclass*», respectively. MD concepts like facts, dimensions, and aggregation hierarchies are mapped to the UML *Class* modeling element by defining *Fact*, *Dimension*, and *Base* stereotypes. In this case, a *Class* is the closest concept of UML for mapping on, since it models structural abstractions from domain entities. It is a main modeling resource in UML, widely used in almost every UML model. Moreover, this profile also translates MD data such as fact measures (*FactAttribute* stereotype) or level descriptions, like dimension attributes (*DimensionAttribute*) or object identity specifications (*OID*), into the UML *Property* metaclass that typically models attributes of UML metaclasses.

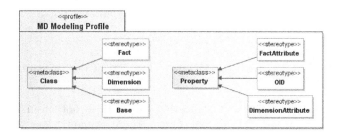

Fig. 3. Excerpt of the UML profile mapping MD-modeling concepts into UML

Reusing the UML profile for MD modeling of DWs [5], we present our proposed conceptual model for clustering by using the same UML extending mechanism (Fig. 4). Thus, we import the former mappings for MD concepts. For mapping clustering-related concepts into UML, we define four stereotypes: *Clustering* (the generalization of clustering algorithms including their parameters), *Input* (input attributes of the data-mining technique referring MD data), *Case* (case ones), and the abstract[2] *Attribute* (data-mining attributes referring MD data).

---

[1] These mechanisms do not substitute the well-defined standard UML visual notation, and we should not use them whereas we can clearly specify our extending elements with the UML visual notation.

[2] An abstract stereotype (the name represented in italics) can not be attached to any modeling element, but it can be useful for helping in taxonomy definitions.

The input and case mappings are implemented in the profiling mechanism by specializing the *Attribute* stereotype that extends the UML *Class* metaclass to model references from data-mining attributes to MD data (with its *reference* tag definition). The algorithm mapping is implemented by means of the *Clustering* stereotype extending the UML *InstanceSpecification* metaclass (it models specific objects that conform to previously defined classes) and the *Settings* class modeling the algorithm settings (it is not a stereotype, but it can also be considered a mapping resource). Each settings parameter is represented indicating its domain and default value in the UML visual notation. A UML profile also allows us to carry out notational changes. The simplest one, the addition of an icon to the standard visual notation of the extended metaclass, is applied to inputs ($\nearrow$→) and cases ($\blacktriangleleft$). Otherwise (as with the *Clustering* stereotype), the default UML notation represents stereotypes textually («*clustering*»).

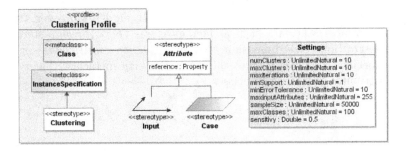

Fig. 4. The UML profile for MD-data clustering at the conceptual level

Constraint Specification. The hosting model does not often have enough expressivity to translate domain concepts without semantic loss. For instance, the assertion *one input is needed for clustering* is not directly representable in the previous UML profile specification. To solve it, UML profile specifications can be enriched with constraint specifications powered with other languages such as OCL. With OCL, we can define formal constraints from our domains not directly representable in UML. Due to space constraints, we only show one of the constraints for the clustering domain that is not contemplated with profiles:

- One input is needed for clustering.
  ```
  context Clustering inv:
  Usage.allInstances()->exists(supplier->includes(self) and
      client->includes(oclIsKindOf(Class) and
          not oclAsType(Class).extension_Input->isEmpty()))
  ```

Each constraint has a context (referring by the **self** keyword) from which we can navigate the UML metamodel (consulting the associated metaclasses with the notation `<class>.<property>`) to check boolean conditions such as `forAll()` or `isEmpty()`, imposed on the mapped metaclasses.

4   Applying Clustering on Data Warehouses

In this section, we will show the suitability of our conceptual model for clustering on MD data. The case study is the analysis of the credit-card purchases. Decision-makers wishes a customer segmentation (groups of customer having similar spending and purchasing behavior), information that serves as a useful tool for selective marketing. Analysts integrate data sources from operational databases, demographic or census data, etc. The time-consuming task of preprocessing data is done when the DW is built. For the sake of simplicity, we will sketch a data-mining session with clustering, showing only the required data.

Modeling the Data Warehouse Repository. Credit-card purchasing data is stored in a DW, modeling with the MD approach (Fig. 5). We model the MD data of the DW applying the MD-modeling profile [5]. A *Purchase* is the *Fact* (shown as «*fact*» textual notation) of analysis, with the purchase *amount* as the only measure (*FactAttribute*, represented with *FA* iconic notation). Providing context of analysis, we model three *Dimensions* (shown as «*dimension*»): *Location*, *Date*, and *Cardholder*. These dimensions aggregate (diamond arrowheads) purchase amount through aggregation hierarchies (each granularity level is shown as «*base*»). For instance, a purchase location is modeled with the *Street–City–State–Region* hierarchy. In our case, each level attribute is labeled as a *DimensionAttribute* (represented with *DA*) or an *OID*.

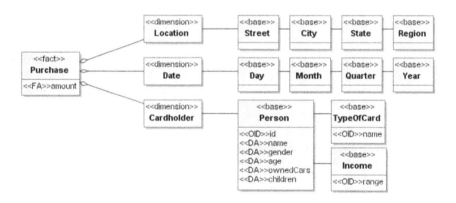

Fig. 5. MD data model of the credit-card purchases

Modeling Clustering on the Data Warehouse. We wish to cluster the credit-card purchases into some cardholder attributes (Fig. 6). We model the purchase clustering (*clustered purchases*) specifying an instance of the *Settings* class from the *Clustering Profile* stereotyped as a *Clustering* settings (shown as «*clustering*»). For this cluster, we set the values of the parameters *numClusters*, *maxInputAttributes*, and *minSupport* (the rest of parameters has their default values). The clustering process is carry out on purchase *amount* (an

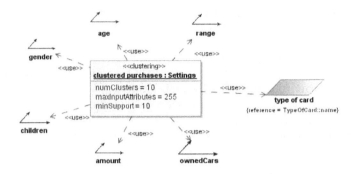

Fig. 6. The model for clustering credit-card purchases in our case tudy

input attribute), some person attributes as inputs (*gender, age, ownedCars*, and *children*), and the income *range*. In addition, the *type of card* is used as a case.

For the sake of simplicity, we hide the *reference* tag values from the *Attribute* stereotype (except for illustrating the type-of-card one) since these tags are only used to solve references to MD data, and they are not needed in the visualization of our models (that use attribute names as identification of the references).

Implementation. The clustering data-mining model designed in the previous lines is a platform-independent model that can be easily transformed to a specific platform. For instance, we have implemented it on Microsoft SQL Server 2005 platform. Specifically, in the Microsoft Analysis Services tool, which allow us to implement the MD modeling of DWs as well as data-mining techniques. Based on the previous model (Fig. 6), we have defined the clustering structure in this tool; a cluster resulting from this implementation is also shown in Fig. 7. It identifies a high-probability behavior occurrence of having less than two children at home,

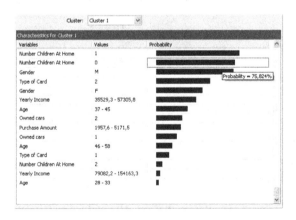

Fig. 7. Implementation of the purchases clustering model in Microsoft Analysis Services

a yearly income between 35, 500 and 57, 300 euros, between 37 and 58 years old, and using a `Gold` card spending between 2, 000 and 5, 000.

5   Related Work

Related to the conceptual modeling, the KDD process, there are few works in the literature, mainly focused on describing at a low-level detail the data mining step. Current approaches for modeling data mining from different perspectives are: (i) the *common warehouse model* (CWM) [12], that models data mining focusing in the metadata interchange; (ii) the *predictive model markup language* (PMML) [13], that provides a vendor-independent format for defining learning algorithms; and (iii) the *pattern base management system* (PBMS) [14,15] that proposes a repository for patterns.

CWM is centered on low-level metadata, and thus it is too complex to be used by analysts for designing purposes. PMML is focused on sharing learning models among tools, however it is neither a design-oriented nor process-oriented standard. PBMS is defined as a platform for storing and managing patterns, then it is not oriented to design data-mining models, but to model its results.

Finally, none of these approaches: (i) proposes a conceptual model for clustering, (ii) facilitates the design of clustering data-mining on MD data, (iii) takes advantage of the previous phases (in the KDD process) to design the data-mining models, (iv) assures data quality integrating clustering data mining into the KDD process, or (v) considers business goals from the early stages of a DW project. Here, we propose a conceptual model for clustering data mining on top of a MD model of the DW that covers this gap.

6   Conclusion

In this paper, we have presented a UML profile that allow data-mining analysts to design clustering models on top of the MD model of a DW. The main benefits of our proposal are that it (i) proposes a conceptual model for clustering, (ii) facilitates the design of clustering data-mining on MD data, (iii) takes advantage of the previous phases (in the KDD process) to design the data-mining models, (iv) assures data quality integrating clustering data-mining into the KDD process, and (v) considers business goals from the early stages of the DW project. In addition, these objectives are achieved using a widely accepted visual modeling language, the UML. To show the feasibility of our approach, we have applied it to design a credit-card purchase clustering, implemented on a commercial database management server with data-mining facilities. In this way, analysts can easily address data-mining issues improving the whole KDD process.

Future Work. Our immediate future work is to formalize the mapping from our conceptual model for clustering to specific-platform physical models. We are also working on including privacy-preserving capabilities integrating this issue into the clustering data-mining models from the early stages of a DW project.

Acknowledgments. This work has been partially supported by the METASIGN (TIN2004-00779) project from the Spanish Ministry of Education and Science, and by the DADS (PBC- 05-012-2) project from the Castilla-La Mancha Ministry of Education and Science (Spain). Jesús Pardillo is funded by the Spanish Ministry of Education and Science under a FPU grant (AP2006-00332).

References

1. Jain, A.K., Murty, M.N., Flynn, P.J.: Data Clustering: A Review. ACM Comput. Surv. 31(3), 264–323 (1999)
2. Frawley, W.J., Piatetsky-Shapiro, G., Matheus, C.J.: Knowledge Discovery in Databases: An Overview. In: Knowledge Discovery in Databases, pp. 1–30. AAAI/MIT Press (1991)
3. Inmon, W.H.: Building the Data Warehouse, 2nd edn. John Wiley & Sons, Inc., New York, NY, USA (1996)
4. Object Management Group: Unified Modeling Language (UML), version 2.1.1. (February 2007) http://www.omg.org/technology/documents/formal/uml.htm
5. Luján-Mora, S., Trujillo, J., Song, I.-Y.: A UML profile for multidimensional modeling in data warehouses. Data Knowl. Eng. 59(3), 725–769 (2006)
6. Zubcoff, J.J., Trujillo, J.: Extending the UML for Designing Association Rule Mining Models for Data Warehouses. In: Tjoa, A.M., Trujillo, J. (eds.) DaWaK 2005. LNCS, vol. 3589, pp. 11–21. Springer, Heidelberg (2005)
7. Zubcoff, J., Trujillo, J.: A UML 2.0 profile to design Association Rule mining models in the multidimensional conceptual modeling of data warehouses. Data Knowl. Eng. (In Press, 2006), doi:10.1016/j.datak.2006.10.007
8. Zubcoff, J.J., Trujillo, J.: Conceptual modeling for classification mining in data warehouses. In: Tjoa, A.M., Trujillo, J. (eds.) DaWaK 2006. LNCS, vol. 4081, pp. 566–575. Springer, Heidelberg (2006)
9. Rasmussen, E.M.: Clustering Algorithms. Information Retrieval: Data Structures & Algorithms, 419–442 (1992)
10. Jain, A.K., Dubes, R.C.: Algorithms for Clustering Data. Prentice-Hall, Englewood Cliffs (1988)
11. Object Management Group: Object Constraint Language (OCL), version 2.0. (May 2006), http://www.omg.org/technology/documents/formal/ocl.htm
12. Object Management Group: Common Warehouse Metamodel (CWM), version 1.1 (March 2003), http://www.omg.org/technology/documents/formal/cwm.htm
13. Data Mining Group: Predictive Model Markup Language (PMML), version 3.1 (Visited April 2007), http://www.dmg.org/pmml-v3-1.html
14. Rizzi, S., Bertino, E., Catania, B., Golfarelli, M., Halkidi, M., Terrovitis, M., Vassiliadis, P., Vazirgiannis, M., Vrachnos, E.: Towards a Logical Model for Patterns. In: Song, I.-Y., Liddle, S.W., Ling, T.-W., Scheuermann, P. (eds.) ER 2003. LNCS, vol. 2813, pp. 77–90. Springer, Heidelberg (2003)
15. Rizzi, S.: UML-Based Conceptual Modeling of Pattern-Bases. In: PaRMa (2004)

A UML Profile for Representing Business Object States in a Data Warehouse*

Veronika Stefanov and Beate List

Women's Postgraduate College for Internet Technologies
Vienna University of Technology
{stefanov,list}@wit.tuwien.ac.at

Abstract. Data Warehouse (DWH) systems allow to analyze business objects relevant to an enterprise organization (e.g., orders or customers). Analysts are interested in the states of these business objects: A customer is either a potential customer, a first time customer, a regular customer or a past customer; purchase orders may be pending or fullfilled.

Business objects and their states can be distributed over many parts of the DWH, and appear in measures, dimension attributes, levels, etc.

Surprisingly, this knowledge – how business objects and their states are represented in the DWH – is not made explicit in existing conceptual models. We identify a need to make this relationship more accessible.

We introduce the *UML Profile for Representing Business Object States in a DWH*. It makes the relationship between the business objects and the DWH conceptually visible. The UML Profile is applied to an example.

1 Introduction

Data Warehouse (DWH) systems are used by decision makers for performance measurement and decision support. The data offered by a DWH describes business objects relevant to an enterprise organization (e.g., accounts, orders, customers, products, and invoices) and allows to analyze their status, development, and trends. *Business objects*, also known as domain objects, represent entities from the "real world enterprise" and should be recognizable to business people, as opposed to implementation objects (e.g., menu items, database tables).

Business objects can be characterized by the states they have during their lifecycle. For example, a customer can have different states: There are potential customers, first time customers, regular customers, customers who pay their bills on time, fraudulent customers, high volume customers, past customers, etc.

In many organizations today, DWHs have grown over time and become large and complex. Business objects and their states can be found all over the DWH: Data relevant to analyzing e.g. customers may be distributed over many parts of the DWH, and the different states of the customer may appear in many different ways,

---

* This research has been funded by the Austrian Federal Ministry for Education, Science, and Culture, and the European Social Fund (ESF) under grant 31.963/46-VII/9/2002.

I.Y. Song, J. Eder, and T.M. Nguyen (Eds.): DaWaK 2007, LNCS 4654, pp. 209–220, 2007.

i.e. as measures, dimensional attributes, levels, etc. A business object can easily have 20 states, which might be hidden in just as many facts or even more: Potential customers in a contacts fact of the marketing data mart, regular customers in a sales fact, customers who pay on time in a payment transaction fact, etc.

Surprisingly, this knowledge - how business objects and their states are represented in the data warehouse - is not made explicit in existing conceptual models. When new analysis requirements appear, it is often difficult, time consuming and costly to find out whether the information about a certain state of a business object is already contained somewhere in the DWH, or whether the data model of the DWH has to be extended or changed.

Our goals therefore are to

- make the relationship between the business objects that the uses want to analyze and the DWH visible and accessible
- show where the business objects and their states can be analyzed in the DWH
- offer a new perspective on DWHs, which emphasizes business objects and their states instead of solely fact and dimension tables
- show how the business object states relate to the data model

We achieve these goals by introducing the *UML*[1] *Profile or Representing Business Object States in a DWH* (Section 4), which makes the connection between data warehouses and business object states visible. State machines (Section 2) describe how objects change states in reaction to events, and are suitable for capturing business logic. We relate state machines of business objects to DWH elements and use them as a new viewpoint on data warehouses. Additionally, using the UML Profile, we define 14 correspondence patterns between state machines and DWH elements (Section 5).

The contributions of the UML Profile for Representing Business Object States in a DWH along with the correspondence patterns are:

- It provides a straightforward way to make visible where a business object such as a customer is available in the DWH and can be analyzed, and how its various states correspond to facts, dimensions and measures.
- Through the business objects, it offers a bigger picture as it links the needs of the business organization to the DWH that stores data on business objects for analytical purposes.
- The Profile for Representing Business Object States in a DWH allows modeling on several levels of detail and thus enables the modeler to choose the right level of detail for different purposes or target audiences. A high level overview model shows only the whole business objects and how they are related to facts and dimensions, where as a more detailed model can show measures, dimension attributes and hierarchy levels for the individual states of the object.
- By extending standard UML 2.0 state machines, the UML profile offers reuse of a well-known notation as well as tool reuse, avoiding costs of learning a new notation or additional tools.

---

[1] Unified Modeling Language.

- The models allow locating business object states in the DWH, and thus recognizing cases where business object states are not available at all, which may indicate business requirements that are not yet addressed. If the model shows that a business object state is available at several locations at the same time, it can be checked whether this was a deliberate design decision or whether this indicates a problematic situation.
- Together with the correspondence patterns the models can support the design phase of a DWH project, as they provide hints on possible facts, measures and dimensions that can be derived from the business object states.

2 UML State Machines

We use UML 2.0 state machines to model business object states. State machines in general describe "the possible life histories of an object" [1]. A state machine comprises a number of states, interconnected by transitions. Events trigger the transitions. There may be several transitions for one event, and they may be guarded by conditions (see Fig. 1).

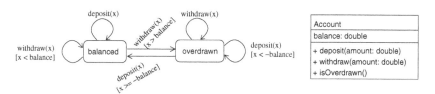

Fig. 1. State machine for Account objects

State machines are used for example in Software Engineering to achieve higher quality software. During the analysis phase of a software development process, state machines for the main business objects (account, order, etc.) are modeled. Such a state machine cannot be transformed directly to code, but is very useful for designers as it allows to recognize "illegal" or conceptually impossible state transitions and thus assess correctness of the final software product. The UML Profile for Representing Business Object States in a DWH presented in this paper aims at bringing the power of state machines to Data Warehousing.

Figure 2 gives an overview of state chart elements and their use in UML 2.0. States may be nested in other states (c), and the so-called sub-machines can be divided into regions (d). This allows access to the model on various levels of detail. Figure 3 shows an example state machine. The object modeled here is the contract between a telecommunications provider and the user of a post-paid mobile phone.

Before it is signed, the contract is in the state "potential". After the customer has signed it, it is registered and a credit check is performed. During this phase, it may become "cancelled", if the check fails. Otherwise, it the credit is OK, the overall state is now called "current", which is a composite state that contains a lot of detail in terms of substates. If the contract is never signed, it becomes "past" after one year.

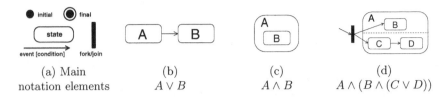

(a) Main
notation elements

(b)
$A \lor B$

(c)
$A \land B$

(d)
$A \land (B \land (C \lor D))$

Fig. 2. UML State Machines: Syntax and combinations of states

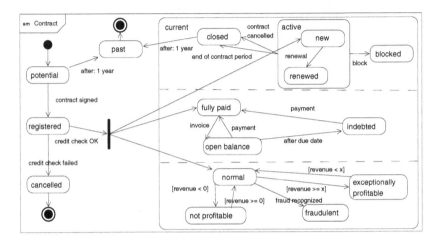

Fig. 3. State Machine diagram for a post-paid mobile phone contract

"Current" contracts are initially "normal", "fully paid", "active", and "new", as indicated by the transition arcs entering the composite state. The three regions within "current" are concurrent and orthogonal, meaning that at each point in time the object has a valid state in each of the three regions, and that what happens in one region does not influence the other regions.

In the first region, the contract starts as "active" and can become "closed", either because the contract period ends or because the contract is cancelled beforehand. If a contract is blocked for some reason, it is neither active nor closed. Active contracts start as "new" and can become "renewed". Closed contracts are moved from "current" to "past" after one year.

The middle region relates to the financial aspects of the contract: Each time an invoice is issued for a contract, it moves from "fully paid" to "open balance", and when the payment for the invoice arrives, it moves back again. If an open balance is not paid until its due date, the contract is "indebted". Issues regarding the amounts of the payments are not shown for sake of clarity (cf. Fig. 1).

In the last region, the movements of the contract object between the states depend mostly on certain conditions becoming true: If the revenue gained from this contract supercedes the amount x, it becomes "exceptionally profitable", meaning that this customer is very valueable to the company. Whereas, if the revenue falls below 0, keeping this contract is actually causing financial damage (i.e. by producing higher costs than monthly payments). The fourth state in this region is "fraudulent": It indicates that fraud connected to this contract has been discovered.

3   Business Objects in the DWH

The aim of our approach is to create conceptual models for making visible where business objects and their states are available in the DWH. Before we can construct a UML Profile, we have to identify the DWH elements we will use. This section contains a metamodel of these elements, an overview of the correspondences between the DWH elements and business object states, the user groups that the UML Profile is aimed at, and finally a tabular overview of the correspondence patterns.

3.1   The Metamodel

Figure 4 shows a metamodel of the elements that may represent business objects in a DWH. There are different kinds of *data repositories*, one of which may be a *data mart*. Other subtypes of data repositories [2] are not used in this paper for sake of simplicity. Data repositories contain *data objects*, which are *facts* or *entities*, depending on the type of the repository. Entities have *entity attributes*, whereas facts may have *measures*. Each fact consists of at least two *dimensions*, which may have several *levels* connected to each other via *roll-up* relationships. Dimensions are described by *dimension attributes*. Facts come in different types: *Transaction facts* or *snapshot facts*, of which there are two sub-types, *periodic* and *accumulating*.

The elements data repository, data mart, fact and entity of this metamodel (and also the UML Profile) are based on previous work by the authors. See [2] for a more detailed description.

3.2   Representation of Business Objects and Their States in a DWH

Not all elements of this metamodel are useful for all conceptual modeling needs. We have identified a number of correspondence patterns along with the main usage scenarios of our approach. They are grouped into three levels and one additional category: the *overview* level, the *fact* level, the *attribute* level, and finally correspondences related to *aggregation and classification*. This subsection gives an overview; details and examples are given in Section 5.

Overview. To show at a glance, where a business object can be found in the DWH, it can be linked as a whole to several data marts, fact tables, dimension tables or entities.

Fact Level. The relationship of transitions and states of business objects to different types of fact tables can be shown on this level. As Kimball described in [3], there are three main types of measurements, which are the

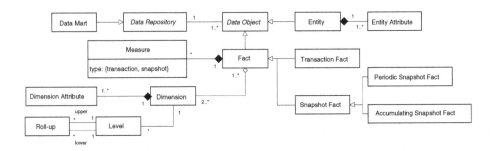

Fig. 4. A metamodel of DWH elements used to represent states of business objects

fundamental grains of fact tables: *transaction, periodic snapshot*, and *accumulating snapshot*. A *transaction fact* corresponds to a *transition* between states, whereas *snaphot facts* correspond to *states*.

Attribute Level. Showing even more detail, a state of a business object may be represented by dimensions or measures, whereas the transactions and their guard conditions are found in measures or dimension attributes.

Aggregation and classification. Guard conditions on state transitions may also appear in aggregation and classification hierarchies. Nested states correspond directly to roll-up relationships.

3.3   User Groups

The conceptual model presented here is designed to answer the needs of two separate types of users:

DWH managers and *business users* are looking for the big picture, an overview of what to find where. They will use models showing business objects with correspondences mainly on the *overview level* (see Section 5.1 below). This allows them to find answers to questions such as "Which business object is in which data mart?", "Is there a fact that corresponds to this business object?", "If I have this dimension, does it represent this business object?"

DWH designers and developers need to understand the details of how business objects states and transactions between them can be represented in a DWH. They will use a fine-grained state machine model of a business object mapped to a data model. The *correspondence patterns* identified by us give hints on which elements can be used in the DWH model to represent the characteristics of the business objects, thus improving the creation and evolution of DWH models.

3.4   Correspondence Patterns

Based on an analysis of the main requirements of the user groups, Table 1 gives an overview of which elements of the UML Profile may be linked to which in the conceptual model to show how business objects (elements on horizontal lines)

Table 1. Correspondence patterns between elements describing business object states (horizontal) and elements of the DWH (vertical)

	Fact	Transaction F.	Snapshot F.	Measure	Dimension	D. Attribute	Level	Roll-up	Entity	E. Attribute	Data Mart
Object of State Machine	X				X				X		X
State			X	X		X	X				
Transition		X							X		
Guard condition				X		X	X				
Nested States								X			

are represented in the DWH, and which elements of the DWH (vertical columns) can be used to represent the states of the business objects (further details of the correspondence patterns in Section 5).

4   The UML Profile for Representing Business Object States in a Data Warehouse

We introduce the UML Profile for Representing Business Object States in a DWH. It provides an easy to use yet formally founded way to model the correspondences between business object states and DWHs.

The Unified Modeling Language (UML) can be extended and adapted to a specific application area through the creation of profiles [4]. UML profiles are special UML packages with the stereotype ≪profile≫. A profile adds elements while preserving the syntax and semantic of existing UML elements. It contains stereotypes, constraints and tag definitions.

A *stereotype* is a model element defined by its name and by the base class(es) to which it is assigned. Base classes are usually metaclasses from the UML metamodel, for instance the metaclass Class. A stereotype can have its own notation, e.g. a special icon.

Constraints are applied to stereotypes in order to enforce restrictions. They specify pre- or postconditions, invariants, etc., and must comply with the restrictions of the base class [4]. We use the Object Constraint Language (OCL) [5] which is widely used in UML profiles to define constraints in our profile, but any language, such as a programming language or natural language, may be used.

Tag definitions are additional attributes assigned to a stereotype, specified as name-value pairs. They have a type and can be used to attach arbitrary information to model elements.

The UML Profile for Representing Business Object States in a DWH makes it possible to model how the DWH used to analyze business objects is related to the lifecycle and states of these objects. Its stereotypes are described in detail in Table 2. These stereotypes can be used directly in UML State Machine diagrams [4]. In Fig. 5 we show a part of the UML 2 metamodel (light) to illustrate how the stereotypes we designed (dark) fit into to the existing UML meta-model.

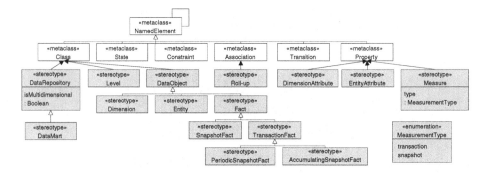

Fig. 5. UML stereotypes for representing business object states in a DWH

The relationships between the stereotypes correspond to the metamodel presented in Section 3.

We use `Class`, `Association`, and `Property` as base classes for the stereotypes. This allows us to model their relationships with elements used in state machines such as `State`, `Transition` and `Constraint`.

Table 2. Stereotype Definitions of the UML Profile for Representing Business Object States in a DWH (cf. diagrams in Fig. 4 and 5)

Name	**DataRepository**	
Base class	Class	
Description	A data repository represents a type of database used in data warehouse environments.	
Tag Definition	isMultidimensional	
	Type: AuxiliaryConstructs::PrimitiveTypes::Boolean	
	Multiplicity: 1	
	Description: Indicates whether the data model of the DataRepository	
	is a multidimensional data model	
Constraints	A DataRepository must be related to at least one DataObject:	
	`self.dataObject->size() >= 1`	
Name	**DataMart**	DM
Generalization	DataRepository	
Description	A data mart is a departmental subset of a DWH focused on a single subject area.	
Name	**DataObject**	
Base class	Class	
Description	A data object is part of the data model contained in a data repository. The stereotypes Fact and Entity are derived from DataObject.	
Constraints	A DataObject must belong to exactly one DataRepository:	
	`self.dataRepository.size() = 1`	
Name	**Fact**	
Generalization	DataObject	
Description	A fact is a data object of a multidimensional data model.	
Constraints	The DataRepository containing a fact must have a multidimensional data model:	
	`self.oclAsType(DataObject).dataRepository.isMultidimensional = true`	
	The Fact must have at least two Dimensions:	
	`self.dimension->size() >= 2`	
Name	**Entity**	
Generalization	DataObject	
Description	An entity is a data object of an E/R model.	«Entity»
Constraints	The DataRepository containing an entity must not have a multidimensional data model:	
	`not self.oclAsType(DataObject).dataRepository.isMultidimensional`	
	The Entity has at least one EntityAttribute:	
	`self.entityAttribute->size() >= 1`	
Name	**TransactionFact**	
Generalization	Fact	
Description	A TransactionFact contains measures of transactions.	T
Constraints	A TransactionFact may only contain transaction measures.	
	`self.allAttributes->forAll(a	a.oclIsTypeOf(Measure) implies a.type=transaction)`	
Name	**SnapshotFact**	
Generalization	Fact	
Description	A Snapshotfact contains snapshot measures, e.g. inventories.	S
Constraints	A SnapshotFact may only contain snapshot measures.	
	`self.allAttributes->forAll(a	a.oclIsTypeOf(Measure) implies a.type=snapshot)`	

Table 2. *(continued)*

Name	**AccumulatingSnapshotFact**	
Generalization	SnapshotFact	aS
Description	The measures of an entry in the AccumulatingSnapshotFact are gathered over time	
Name	**PeriodicSnapshotFact**	
Generalization	SnapshotFact	pS
Description	The measures of a PeriodicSnapshotFact are acquired periodically for all instances	
Name	**Measure**	
Base class	Attribute	
Description	A numeric Measure is the object of analysis.	
Tag Definition	measurementType	«Measure»
	Type: MeasurementType (Enumeration)	
	Multiplicity: 1	
	Description: Indicates the type of measurement: *transaction* or *snapshot*	
Constraints	A Measure must belong to exactly one Fact:	
	`self.fact.size() = 1`	
Name	**EntityAttribute**	
Base class	Attribute	«Entity–
Description	An attribute to an Entity.	Attribute»
Constraints	An EntityAttribute must belong to exactly one Entity:	
	`self.entity.size() = 1`	
Name	**Dimension**	
Base class	Class	
Description	Dimensions provide context for the measures and together are assumed to uniquely determine them.	«Dimension»
Constraints	A Dimension is used by at least one Fact:	
	`self.fact->size() >= 1`	
	A Dimension has at least one DimensionAttribute:	
	`self.dimensionAttribute->size() >= 1`	
Name	**DimensionAttribute**	
Base class	Attribute	«Dimension–
Description	DimensionAttributes describe Dimensions.	Attribute»
Constraints	A Dimension Attribute belongs to exactly one Dimension:	
	`self.dimension->size() = 1`	
Name	**Level**	
Base class	Class	«Level»
Description	Levels of Dimensions are used to aggregate Measures.	
Constraints	A Level belongs to exactly one Dimension:	
	`self.dimension->size() = 1`	
Name	**Roll-up**	
Base class	Association	
Description	A Roll-up-Association between two Levels indicates that Measures can be aggregated from the lower to the upper Level.	⟶
Constraints	A Roll-up-relationship has exactly one upper and one lower Level:	
	`self.upper->size() = 1 and self.lower->size() = 1`	

5   Correspondence Patterns and Examples

To create conceptual models for making visible where business objects and their states are available in the DWH, we link elements from DWH conceptual data models with state machines of business objects. In this section, we apply the correspondence patterns already introduced in Section 3 to examples.

5.1   Overview Level

We can link the business object to one or more *data marts* to provide an overview (Figure 6a). Then we can identify where each business object can be analyzed

Fig. 6. The business object "Contract" in the fact "Sales" (left) and data marts (right)

in the DWH model. A business object may correspond to a *fact table* (e.g., order, shipment, account; Figure 6b), a *dimension table* (e.g., product, customer, account), or in the case of a pure E/R-model to an *entity*.

5.2   Fact Level

There are three main types of measurements or fundamental grains of fact tables: *transaction, periodic snapshot*, and *accumulating snapshot* [3]. Looking at state charts, we find that a *transaction fact* corresponds to a *transition* between states (Fig. 7 (a)), whereas *snaphot facts* correspond to *states*. Periodic snapshots record the current state for all instances of a business object at regular intervals (b), whereas accumulating snapshots "follow" each instance as it passes from state to state, adding values to the record over time (c). In the latter case, the order in which the values are added must conform to the state chart.

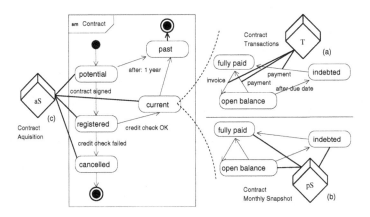

Fig. 7. Transitions in transaction facts (a), states in snapshot facts (b, c)

5.3   Attribute Level

To provide a more detailed conceptual model, i.e. for DWH design and development, this level contains correspondence patterns related to all kinds of attributes (measures, dimension attributes, entity, attributes, guard conditions).

States. A state in a state chart may play several roles in the DWH model. The most straightforward case is when there is an explicit "status" dimension (e.g., for account facts, insurance policies, etc.). In this case, several states of a business object are modeled by the dimension, i.e. as dimension attributes. Second, a state may be modeled as a measure. For a single state, this measure is of type boolean (either the object is in this state of not), and for several states, the measure would be an enumeration with the values corresponding to the states (which may be seen as a degenerate type of the status dimension).

Transitions. Transitions in state charts may be guarded by conditions. These conditions often are found in the DWH as measures or dimension attributes. If a guard condition contains a recognizable variable (e.g. "revenue" in Fig. 8(a)), this variable can turn into a measure in the DWH model.

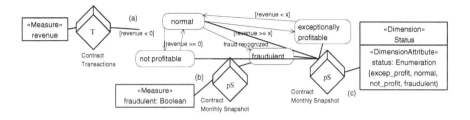

Fig. 8. Guard condition as measure (a), states as measure (b) or dimension (c)

5.4 Aggregation and Classification

State charts of business objects may also provide hints regarding aggregation and classification hierarchies. First, nested states correspond to roll-up relationships, the inner state being the special case and the outer state the more general. Also, guard conditions may define levels in the hierarchy, as they separate instances into groups. In the case of a specific "status" dimension mentioned above, the states covered by this dimension then may also correspond to levels.

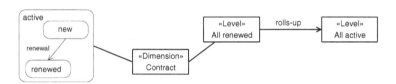

Fig. 9. Nested states of a business object correspond to a roll-up relationship

6 Related Work

Using UML Profiles to model the structure and behavior of Data Warehouses as well as related aspects such as ETL processes has become increasingly popular, also due to the rise of the Model-Driven Architecture (MDA [6]).

Trujillo and Luján-Mora introduced a Data Warehouse design framework in [7], which is supported among others by the UML Profile for Modeling ETL processes [8], a Profile for Physical Modeling of Data Warehouses [9], and a Profile for Multidimensional Modeling [10].

UML has also been applied to aspects such as DWH security. Fernandez-Medina et al. have extended UML for Designing Secure Datawarehouses [11].

In [12], Mazon et al. show how the MDA can be applied to Data Warehousing. Our contribution widens the scope of conceptual modelling in Data Warehousing to include more than structural models.

7 Conclusion

In this work, we have addressed that fact that there are no existing models for describing how business objects and their states are represented in a DWH. Business object states can be distributed over many parts of the DWH, and appear in measures, dimension attributes, levels, etc.

We introduced the *UML Profile for Representing the Lifecycle of Business Objects in a Data Warehouse*. It makes the relationship between the business objects and the DWH visible and accessible, as it allows to model various elements of DWHs and their data models in combination with UML 2.0 state machines. Using the UML Profile, we identified 14 correspondence patterns between state machines and DWH. The UML Profile and the correspondences are applied to an example.

References

1. Rumbaugh, J., Jacobson, I., Booch, G.: The Unified Modeling Language Reference Manual. Addison-Wesley, Reading (2004)
2. Stefanov, V., List, B., Korherr, B.: Extending UML 2 Activity Diagrams with Business Intelligence Objects. In: Tjoa, A.M., Trujillo, J. (eds.) DaWaK 2005. LNCS, vol. 3589, pp. 53–63. Springer, Heidelberg (2005)
3. Kimball, R.: Fundamental Grains. Intelligent Enterprise 2(5) (1999)
4. Object Management Group, Inc.: UML 2.0 Superstructure (2005),
   http://www.omg.org/cgi-bin/apps/doc?formal/05-07-04.pdf
5. Object Management Group, Inc.: UML 2.0 Object Constraint Language (OCL) Specification (2005),
   http://www.omg.org/cgi-bin/apps/doc?ptc/05-06-06.pdf
6. Object Management Group, Inc.: Model Driven Architecture (MDA) (2004),
   http://www.omg.org/cgi-bin/doc?formal/03-06-01
7. Luján-Mora, S., Trujillo, J.: A Data Warehouse Engineering Process. In: Yakhno, T. (ed.) ADVIS 2004. LNCS, vol. 3261, pp. 14–23. Springer, Heidelberg (2004)
8. Trujillo, J., Luján-Mora, S.: A UML Based Approach for Modeling ETL Processes in Data Warehouses. In: Song, I.-Y., Liddle, S.W., Ling, T.-W., Scheuermann, P. (eds.) ER 2003. LNCS, vol. 2813, pp. 307–320. Springer, Heidelberg (2003)
9. Luján-Mora, S., Trujillo, J.: Physical modeling of data warehouses using UML. In: DOLAP 2004, Proceedings, pp. 48–57. ACM Press, New York (2004)
10. Luján-Mora, S., Trujillo, J., Song, I.Y.: A UML profile for multidimensional modeling in data warehouses. Data Knowl. Eng. 59(3), 725–769 (2006)
11. Fernández-Medina, E., Trujillo, J., Villarroel, R., Piattini, M.: Extending UML for Designing Secure Data Warehouses. In: Atzeni, P., Chu, W., Lu, H., Zhou, S., Ling, T.-W. (eds.) ER 2004. LNCS, vol. 3288, pp. 217–230. Springer, Heidelberg (2004)
12. Mazon, J.N., Trujillo, J., Serrano, M., Piattini, M.: Applying MDA to the development of data warehouses. In: Proceedings DOLAP '05, pp. 57–66. ACM Press, New York, NY, USA (2005)

Selection and Pruning Algorithms for Bitmap Index Selection Problem Using Data Mining

Ladjel Bellatreche[1], Rokia Missaoui[2], Hamid Necir[3], and Habiba Drias[3]

[1] Poitiers University - LISI/ENSMA France
bellatreche@ensma.fr
[2] University of Quebec in Outaouais (UQO) - Canada
rokia.missaoui@uqo.ca
[3] Institut National d'Informatique (INI) - Algerie
ncrhmd@yahoo.fr, h_drias@ini.dz

Abstract. Indexing schemes are redundant structures offered by DBMSs to speed up complex queries. Two types of indices are available: mono-attribute indices (B-tree, bitmap, hash, etc.) and multi-attribute indices (join indices, bitmap join indices). In relational data warehouses, bitmap join indices (BJIs) are bitmap indices for optimizing star join queries through bit-wise operations. They can be used to avoid actual joins of tables, or to greatly reduce the volume of data that must be joined, by executing restrictions in advance. BJIs are defined using non-key dimension attributes and fact key attributes. Moreover, the problem of selecting these indices is difficult because there is a large number of candidate attributes (defined on dimension tables) that could participate in building these indices. To reduce this complexity, we propose an approach which first prunes the search space of this problem using data mining techniques, and then based on the new search space, it uses a greedy algorithm to select BJIs that minimize the cost of executing a set of queries and satisfy a storage constraint. The main peculiarity of our pruning approach, compared to the existing ones (that use only appearance frequencies of indexable attributes appearing in queries as a pruning metric), is that it uses others parameters such as the size of their dimension tables, the length of each tuple and the size of a disk page.

Keywords: Physical design, bitmap join index, pruning algorithms, data mining.

1 Introduction

Relational data warehouses are usually modelled using relational schemas like star and snow flake schemas. On the top of these schemas, complex queries are executed and are characterized by their restrictions on the dimension non-key attributes and join operations between dimension tables and the fact table. The major bottleneck in evaluating such queries is the join between a large fact table and the surrounding dimension tables [14]. To optimize join operations, many structures have been proposed in the literature. We can classify them into two main categories: (1) *non redundant structures* and (2) *redundant structures*.

I.Y. Song, J. Eder, and T.M. Nguyen (Eds.): DaWaK 2007, LNCS 4654, pp. 221–230, 2007.
© Springer-Verlag Berlin Heidelberg 2007

The first category concerns different implementations of the join operation (e.g., hash join). These structures are efficient when (a) the size of joined tables is reasonable and (b) the join concerns two tables [7]. Redundant structures, like materialized views [13] and join indices [16], are more efficient to speed up join operation involving many tables [11]. Their main drawbacks are the extra storage requirement and the maintenance overhead.

In this paper, we concentrate on one redundant structure called *bitmap join index* (BJI). The join index has been proposed in 80's [16] and it consists in pre-computing the join operation usually involving *two relations*. In the context of relational data warehouses, a join index has been extended to a multi-table join index by concatenating columns from different dimension tables and lists RIDs (row identifiers) in the fact table from each concatenated value [15]. This index allows multidimensionality to be explicitly represented [7]. Another variant of the join index, called *bitmap join index*, is proposed. It is defined as a bitmap index on the fact table F based on one or several column(s) of dimension table(s). Besides disk saving (due to their binary representation and potential compression [9]), such index speeds up queries having Boolean operations (such as AND, OR and NOT) and COUNT operations.

The task of index selection is to automatically select an appropriate set of indices for a data warehouse and a workload under resource constraints (storage, maintenance, etc.). It is challenging for the following reasons [3]: The size of a relational data warehouse schema may be large (many tables with several columns), and indices can be defined on a set of columns. Therefore, the search space of indices that are relevant to a workload can be very huge [17]. To deal with this problem, most of approaches use two main phases: (1) *generation of candidate attributes* and (2) *selection of a final configuration*. The first phase prunes the search space of index selection problem by eliminating some *non relevant* attributes. In the second phase, the final indices are selected using greedy algorithms [4], linear programming algorithms [3], etc. The *quality of the final set of indices depends essentially on the pruning phase*. To prune the search space of index candidates, many approaches were proposed [4,17,3,10,2], that can be classified into two categories: *heuristic enumeration-driven approaches* and *data mining driven approaches*. In heuristic enumeration-driven approaches, heuristics are used. For instance, in [4], a greedy algorithm is proposed that uses optimizer cost of SQL Server to decide the goodness of a given configuration of indices. The weakness of this work is that it *imposes the number of generated candidates*. DB2 Advisor is another example belonging to this category [17], where the query parser is used to pick up selection attributes used in workload queries. The generated candidates are obtained by a few *simple combinations* of selection attributes [17].

In data mining driven approaches, the pruning process is done using data mining techniques, like in [2], where CLOSE algorithm [12] is used to generate BJI candidates, without imposing the number of index candidates as in the first category. The basic idea is to generate frequent closed itemsets representing groups of attributes that could participate in selecting the final configuration of BJIs.

Our proposed work belongs to the second category (data mining driven approaches). By examining the unique work done on selecting BJIs using data mining techniques [2], we figure out that it only uses *the appearance frequencies of attributes* to generate frequent closed itemsets, where each one represents a BJI candidate. Since in the vertical partitioning of databases the frequency parameter is not sufficient in getting a good fragmentation schema [5], we believe that a similar observation holds for the BJI selection problem. Indeed, Fun et al. [5] showed the weakness of affinity-based algorithms (that use only query frequencies) in reducing the query processing cost. To get a better result, they recommend the use of other parameters like the size of tables, the length of tuples, the size of disk page, etc. Based on that observation, we propose two pruning algorithms, called DynaClose and DynaCharm (based on CLOSE and CHARM, respectively) that take into account the above parameters as part of the pruning metric, since the cost of join operations depends heavily on the size of joined tables [6]. Once the pruning phase is processed, a greedy algorithm is executed to select a set of BJIs that reduces the query processing cost and satisfies the storage constraint.

This paper is organized into five sections: Section 2 presents a formalization of the problem of selecting BJIs as an optimization problem, and shows the limitation of the unique related work. Section 3 presents our approach by describing pruning algorithms to generate BJI candidates and a greedy algorithm for selecting the final BJIs. Section 4 validates our work using experimental studies. Section 5 concludes the paper and suggests future directions.

2   Bitmap Join Indices

A BJI is defined on one or several non key dimension attributes with low cardinality (that we call indexable column) joining their tables with the fact table. An indexable attribute A_j of a given dimension table D_i for a BJI is a column $D.A_j$, such that there is a condition of the form $D_i.A_j \ \theta \ Expression$ in the WHERE clause. The operator θ must be among $\{=, <, >, <=, >=\}$. An example of SQL statement for creating a BJI defined on dimension table *Customer* and fact table *Sales* based on *gender* attribute (indexable column) is given below:

CREATE BITMAP INDEX Sales_c_gender_bjix ON (Customers.cust_gender) FROM Sales, Customers WHERE Sales.cust_id = Customers.cust_id;

This index may speed up the following query: SELECT count(*) FROM Sales, Customers WHERE gender='F' and Sales.cust_id = Customers.cust_id;
by counting only the number of 1 in the bitmap corresponding to the gender ='F', *without any access* to the fact table Sales and dimension table Customer.

Since a BJI can be built on one or several attributes to speed up join operations between several tables, the number of all possible BJIs may be very large. Moreover, selecting an optimal configuration of BJIs for a given data warehouse schema and a workload is an NP-complete problem [3,17].

BJI selection problem can be formalized as follows: given a data warehouse with a set of dimension tables $D = \{D_1, D_2, ..., D_d\}$ and a fact table F, a workload Q

of queries $Q = \{Q_1, ..., Q_i, ..., Q_m\}$, where each query Q_i has an access frequency, and storage constraint S, the aim of BJI selection problem is to find a set of BJIs minimizing the query processing cost and satisfying the storage requirement S.

2.1   Basic Concepts on Data Mining

In APRIORI-like algorithms [1], rule mining is conducted in two steps: detection of all frequent (large) itemsets (i.e., groups of items that occur in the database with a support $\geq$ a minimal support called *minsup*), and utilization of frequent itemsets to generate association rules that have a confidence $\geq$ *minconf*. The first step may lead to a set of frequent itemsets (FIs) that grows exponentially with the whole set of items. To reduce the size of the FI set, some studies were conducted on frequent closed itemsets (FCIs) and maximal frequent itemsets (i.e., itemsets for which every superset is infrequent).

The CLOSE algorithm [12] is one of the first procedures for FCI generation. Like APRIORI, it performs a level-wise computation within the powerset lattice. However, it exploits the notion of generator of FCIs to compute closed itemsets. CHARM [18] is another procedure which generates FCIs in a tree organized according to inclusion. The computation of the closure and the support is based on a efficient storage and manipulation of *TID-sets* (i.e., the set of transactions per item). Closure computation is accelerated using *diffsets*, the set difference on the TID-sets of a given node and its unique parent node in the tree. CLOSET [8] generates FCIs as maximal branches of a *FP-tree*, a structure that is basically a prefix tree (or *trie*) augmented with transversal lists of pointers.

In the following we will exploit CLOSE and CHARM to identify frequent closed itemsets (in our case, the attributes to be indexed).

2.2   Weakness of Existing Pruning Approaches

To show the limitations of one of the existing approaches for index selection, we consider the following example: suppose we have a star schema with two dimension tables *Channels* (denoted by Ch) and *Customer* (C) and a fact table *Sales* (S). The cardinality of these tables (number of instances) is: $||Sales|| = 46521, ||channels|| = 5$ and $||customers|| = 30000$. On the top of this schema, five queries are executed (see Table 1). The CLOSE algorithm used by [2] to select index candidates with a *minsup* = 3 (in absolute value) returns a BJI (called, *sales_desc_bjix*) defined on *Channels* and *Sales* using *channel_desc* attribute. This is due to the fact that there are three occurrences of the same selection predicate defined on that attribute in the five queries. Since the *cust_gender* attribute is not more frequent than *C.channel_desc* attribute, no index is proposed. However, the presence of such index may reduce the query processing cost due to the large size of the *Customer* table (30 000 instances), while the table *Channels* that participates to the *sales_desc_bjix* has only five instances. This example shows the limitations of Aouiche's approach that uses only the frequency of attributes to prune the search space for the BJI selection problem.

Table 1. Query Description

(1) select S.channel_id, sum(S.quantity_sold) from S, C
where S.channel_id=C.channel_id and *C.channel_desc='Internet'* group by S.channel_id
(2) select S.channel_id, sum(S.quantity_sold), sum(S.amount_sold) from S, C
where S.channel_id=C.channel_id and *C.channel_desc ='Catalog'* group by S.channel_id
(3) select S.channel_id, sum(S.quantity_sold),sum(S.amount_sold) from S, C
where S.channel_id=C.channel_id and *C.channel_desc ='Partners'* group by S.channel_id
(4) select S.cust_id, avg(quantity_sold) from S, C
where S.cust_id=C.cust_id and C.cust_gender='M' group by S.cust_id
(5) select S.cust_id, avg(quantity_sold) from S, C
where S.cust_id=C.cust_id and *C.cust_gender='F'* group by S.cust_id

To overcome this limitation, we enrich the pruning function by considering other parameters like the size of tables, tuples, and pages.

3   Our Proposed Approach

In order to select a set of BJIs that minimizes the query processing cost and satisfies the storage constraint S, we consider an approach with two steps commonly found in the existing index selection approaches: (1) generating candidate attributes and (2) selecting the final configuration and BJIs.

3.1   Generation of Candidate Attributes

The input of this step is the context matrix. It is constructed using the set of queries $Q = \{Q_1, Q_2, ..., Q_m\}$ and a set of indexable attributes $A = \{A_1, A_2, ..., A_l\}$. The matrix has rows and columns that represent queries and indexable attributes, respectively. A value of this matrix is given by:

$$Uses(Q_i, A_j) = \begin{cases} 1 \text{ if query } Q_i \text{ uses a selection predicate defined on } A_j \\ 0 \text{ otherwise} \end{cases}$$

Example 1. Assume we have five queries (see Table 1). To facilitate the construction of the context matrix, we rename the indexable attributes as follows: $Sales.cust\_id = A_1$, $Customers.cust\_id = A_2$; $Customers.cust\_gender = A_3$, $Channels.channel\_id = A_4$; $Sales.channel\_id = A_5$; $Channels.channel\_desc = A_6$. The matrix is given below.

	A_1	A_2	A_3	A_4	A_5	A_6
Q_1	0	0	0	1	1	1
Q_2	0	0	0	1	1	1
Q_3	0	0	0	1	1	1
Q_4	1	1	1	0	0	0
Q_5	1	1	1	0	0	0
Support	$\frac{2}{5}$	$\frac{2}{5}$	$\frac{2}{5}$	$\frac{3}{5}$	$\frac{3}{5}$	$\frac{3}{5}$

To prune the search space of the BJI selection problem, we propose two algorithms *DynaClose* and *DynaCharm* which are an adaptation of CLOSE [12] and CHARM [18], respectively. We used CLOSE because it has been exploited by Aouiche et al. [2] for the same purpose (i.e., selecting BJIs), and hence will allow us to better compare the performance of our approach with Aouiche's approach. Our adaptation concerns especially the pruning metric which is different from the one (i.e., the support) used in CLOSE and CHARM. The new metric, called fitness metric, penalizes the FCIs related to small dimension tables. It is expressed by: $\frac{1}{n} \times (\sum_{i=1}^{n} sup_i \times \alpha_i)$, where n represents the number of non-key attributes in each FCI and α_i is equal to $\frac{|D_i|}{|F|}$, where $|D_i|$ and $|F|$ represent the number of pages needed to store the dimension table D_i and the fact table F, respectively [1].

Example 2. The itemset $\{A_1, A_2, A_3\}$ ($\{A_4, A_5, A_6\}$ respectively) has a support equal to 0.4 (0.6 resp.) and a fitness equal to 0.12896 (0.0001 resp.).

The set of the FCIs generated by DynaClose must be purified to avoid non coherent BJIS. Let us recall that a BJI is built between a fact table and dimension table(s) based on non-key attributes. Three scenarios of purification are considered: (1) a generated FCI contains only keys attributes (of dimension tables) or foreign keys of the fact table, (2) a given FCI has a number of key attributes higher than the number of non-key attributes. For one non-key attribute, we need two key attributes for building BJIs. Generally, for N selection attributes, $2 \times N$ key-attributes are required, and (3) the FCI contains only non-key attributes.

3.2   Selection of the Final Configuration

Once the candidate attributes (purified FCIs) are computed, the final configuration of BJIs is calculated. To select this configuration, we propose a greedy algorithm. The input of this algorithm includes: (a) a star schema, (b) a set of queries: $Q = \{Q_1, Q_2, \cdots, Q_m\}$, (c) $PURBJI$ (representing the purified FCIs), and (d) a storage constraint S. Our algorithm starts with a configuration having a BJI defined on an attribute (of $PURBJI$) with smallest cardinality (let say, I_{min}), and iteratively improves the initial configuration by considering other attributes of $PURBJI$ until no further reduction in total query processing cost is possible or the constraint S is violated. Our greedy algorithm is based on a mathematical cost model, an adaptation of an existing one [2] that computes the number of disk page accesses (I/O cost) when executing the set of queries.

4   Experimental Studies

To evaluate our approach, we first implemented in Java five algorithms, namely CLOSE, CHARM, DynaClose, DynaCharm and the greedy algorithm on a PC

---

[1] The number of pages occupied by a table T is calculated as follows: $|T| = \left\lceil \frac{||T|| \times LT}{PS} \right\rceil$, where $||T||$, LT and PS represent the number of instances of T, the length of an instance of T and the page size, respectively.

Algorithm 1. Greedy Algorithm for BJIs Selection

Let: BJI_j be a BJI defined on attribute A_j. $Size(BJI_j)$ be the storage cost of BJI_j
Intput: Q: a set of queries, $PURBJI$: purified FCIs, and a storage constraint S.
Output: $Config_{finale}$: set of selected BJIs.
begin

    $Config_{finale} = BJI_{min}$;
    $S := S - Size(BJI_{min})$;
    $PURBJI := PURBJI - A_{min}$; A_{min} is the attribute used to defined BJI_{min}
    WHILE $(Size(Config_{finale}) \leq S)$ DO
        FOR each $A_j \in BJISET$ DO
            IF $(COST[Q, (Config_{finale} \cup BJI_j))] < COST[Q, Config_{finale}])$
            AND $((Size(Config_{finale} \cup BJI_j) \leq S))$ THEN
                $Config_{finale} := Config_{finale} \cup BJI_j$;
                $Size(Config_{finale}) := Size(Config_{finale}) + Size(BJI_j)$;
                $PURBJI := PURBJI - A_j$;
end

Pentium IV and 256Mo of memory, where the first two procedures rely on the support to compute FCIs while DynaClose, DynaCharm make use of the fitness. We then conducted a set of experiments using the same dataset exploited by Aouiche *et al.* [2]. The star schema of data has one fact table *Sales* (16 260 336 tuples), five dimension tables: *Customers* (50 000 tuples), *Products* (10 000 tuples), *Promotions* (501 tuples), *Time* (1 461 tuples) and *Channels* (5 tuples). The proposed benchmark includes also forty OLAP queries.

The experiments were conducted according to four scenarios: (1) identification of the values of *minsup* that give an important number of FCIs, (2) evaluation of different approaches (CLOSE, DynaClose, CHARM and DynaCharm) by executing the 40 queries on a non indexed data warehouse without considering storage constraint, (3) evaluation of different approaches by considering the storage constraint and (4) measurement of CPU bound of different approaches.

First, we carried out experiments to set the appropriate value of *minsup* that allows the generation of a large set of FCIs. The results show that the appropriate *minsup* value should be set to 0.05.

4.1   Evaluation Without Storage Constraint

Figure 1 shows how different indexing approaches reduce the cost of executing the 40 queries with an increasing number of minimum support. The main result is that Dynaclose outperforms approaches for almost all values of *minsup*. However, its performance deteriorates (in the sense that no candidate indices can be generated) when the *minsup* value becomes high. We notice that for *minsup* values exceeding 0.475, CLOSE, CHARM and DynaCharm stop generating new FCIs, and hence the query processing cost remains stable.

A comparison between DynaCharm and DynaClose shows that they have a similar performance for small *minsup* values (ranging between 0.05 and 0.175). These results coincide with the experimental study of Zaki et al. [18]. However, as we

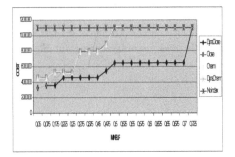

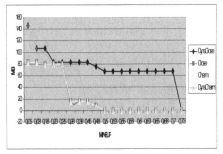

Fig. 1. Performance of five alternatives

Fig. 2. Behavior of procedures according to space storage

increase the *minsup*, the performance gap between DynaClose and DynaCharm becomes larger. This is due to the fact that DynaCharm processes branches in a depth-first fashion, and FCIs are formed only at the end of an iteration.

4.2   Evaluation of DynaClose et Dynacharm with Storage Constraint

The set of BJIs generated by DynaClose and DynaCharm for a *minsup*=0.05 requires storage of 147 MB (see Figure 3). This value is very high if we compare it with the size of the fact table which is 372 MB. Consequently, we execute our greedy algorithm for selecting BJIs by considering various storage values with a fixed value of *minsup* equal to 0.05. This value allows the generation of a large number of index candidates. Figure 2 shows that DynaClose and DynaCharm improve the performance by 43% (compared to the solution without indexing) for 44 MB of storage (almost 3 times smaller than the initial space of 147 MB). With the same storage, CHARM and CLOSE give a 33,56% gain. Our proposed variants Dynaclose and Dynacharm provide a better performance than the traditional approaches (i.e., using support rather than fitness) for all the values of the considered storage, except for the storage space 84 MB (where all approaches provide the same gain of 58,19%).

4.3   Evaluation of Different Approaches Based on CPU Bound

We have conducted experiments about the execution time (in microseconds) of each algorithm according to a varying value of *minsup*. Figure 4 shows that DynaCharm and DynaClose need more execution time to prune the search space compared to CHARM and CLOSE. This result was foreseeable since contrary to traditional approaches (CLOSED and CHARM) which prune according to the *minsup*, our two algorithms take more time since the pruning phase is done in two stages: it enumerates first the infrequent itemsets according to *minsup* and then calculates the fitness function for each generated FCI. The results show that DynaCharm is the procedure which requires the highest time for almost all *minsup* values that we consider, but it remains stable for high *minsup*. This is due to the pruning phase of DynaCharm since it is carried out only when we get maximized closed itemsets.

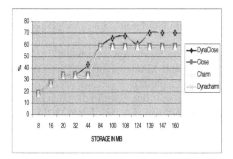
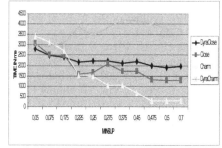

Fig. 3. Storage according to *minsup*

Fig. 4. Execution time (in ms) of the four approaches

5   Conclusion

In this paper, we proposed an approach that automatically selects bitmap join indices by computing frequent closed itemsets (attribute sets). Our approach proceeds in two main steps: (1) it uses data mining procedures, called *DynaClose* and *DynaCharm* (adaptations of CLOSE and CHARM algorithms) to prune the search space of the bitmap join index selection problem, and (2) it uses a greedy algorithm to select the final configuration of indices. The main peculiarity of our pruning approach, compared to the existing ones where only frequency of dimension attributes is used, lies in the utilization of not only the frequency of an attribute in a query but also other parameters like the size of tables, pages, and tuples. Once the pruning phase is executed, we exploit our greedy algorithm that uses a cost model to generate the appropriate set of indices. Our approach was validated through experimentations.

We plan to extend this work into two directions: (i) dynamic change of BJIs when significant changes in the workload and table sizes occur, and (ii) handle the problem of view materialization in data warehouses using a similar approach.

References

1. Agrawal, R., Srikant, R.: Fast algorithms for mining association rules. In: 20th International Conference on Very Large Data Bases (VLDB'94), pp. 487–499 (1994)
2. Aouiche, K., Boussaid, O., Bentayeb, F.: Automatic Selection of Bitmap Join Indexes in Data Warehouses. In: Tjoa, A.M., Trujillo, J. (eds.) DaWaK 2005. LNCS, vol. 3589, Springer, Heidelberg (2005)
3. Chaudhuri, S.: Index selection for databases: A hardness study and a principled heuristic solution. IEEE Transactions on Knowledge and Data Engineering 16(11), 1313–1323 (2004)
4. Chaudhuri, S., Narasayya, V.: An efficient cost-driven index selection tool for microsoft sql server. In: Proceedings of the International Conference on Very Large Databases, August 1997, pp. 146–155 (1997)
5. Fung, C.-H., Karlapalem, K., Li, Q.: Cost-driven vertical class partitioning for methods in object oriented databases. VLDB Journal 12(3), 187–210 (2003)

6. Getoor, L., Taskar, B., Koller, D.: Selectivity estimation using probabilistic models. In: Proceedings of the ACM SIGMOD International Conference on Management of Data, pp. 461–472 (2001)

7. Golfarelli, M., Rizzi, E., Saltarelli, S.: Index selection for data warehousing. In: Proceedings 4th International Workshop on Design and Management of Data Warehouses (DMDW'2002), Toronto, Canada, pp. 33–42 (2002)

8. Han, J., Pei, J., Yin, Y.: Mining Frequent Patterns without Candidate Generation. In: Proceedings of the ACM-SIGMOD 2000 Conference, Dallas, Texas, USA, pp. 1–12 (2000)

9. Johnson, T.: Performance measurements of compressed bitmap indices. In: Proceedings of the International Conference on Very Large Databases, pp. 278–289 (1999)

10. Labio, W., Quass, D., Adelberg, B.: Physical database design for data warehouses. In: Proceedings of the International Conference on Data Engineering (ICDE) (1997)

11. Oneil, P.: Multi-table joins through bitmapped join indioces. SIGMOD 24(03) (1995)

12. Pasquier, N., Bastide, Y., Taouil, R., Lakhal, L.: Discovering frequent closed itemsets. In: ICDT, pp. 398–416 (1999)

13. Rizzi, S., Saltarelli, E.: View materialization vs. indexing: Balancing space constraints in data warehouse design. In: Eder, J., Missikoff, M. (eds.) CAiSE 2003. LNCS, vol. 2681, Springer, Heidelberg (2003)

14. Stöhr, T., Märtens, H., Rahm, E.: Multi-dimensional database allocation for parallel data warehouses. In: Proceedings of the International Conference on Very Large Databases, pp. 273–284 (2000)

15. Red Brick Systems. Star schema processing for complex queries. White Paper (July 1997)

16. Valduriez, P.: Join indices. ACM Transactions on Database Systems 12(2), 218–246 (1987)

17. Valentin, G., Zuliani, M., Zilio, D.C., Lohman, G.M., Skelley, A.: Db2 advisor: An optimizer smart enough to recommend its own indexes. In: ICDE 00, pp. 101–110 (2000)

18. Zaki, M.J., Hsiao, C.J.: Charm: An efficient algorithm for closed itemset mining. In: Proceeding of the 2nd SIAM International Conference on Data Mining (ICDM 02) (2002)

MOSAIC: A Proximity Graph Approach for Agglomerative Clustering

Jiyeon Choo[1], Rachsuda Jiamthapthaksin[1], Chun-sheng Chen[1],
Oner Ulvi Celepcikay[1], Christian Giusti[2], and Christoph F. Eick[1]

[1] Computer Science Department, University of Houston, Houston, TX 77204-3010, USA
{jchoo,rachsuda,lyons19,onerulvi,ceick}@cs.uh.edu
[2] Department of Mathematics and Computer Science, University of Udine, Via delle Scienze,
33100, Udine, Italy
giusti@dimi.uniud.it

Abstract. Representative-based clustering algorithms are quite popular due to their relative high speed and because of their sound theoretical foundation. On the other hand, the clusters they can obtain are limited to convex shapes and clustering results are also highly sensitive to initializations. In this paper, a novel agglomerative clustering algorithm called MOSAIC is proposed which greedily merges neighboring clusters maximizing a given fitness function. MOSAIC uses Gabriel graphs to determine which clusters are neighboring and approximates non-convex shapes as the unions of small clusters that have been computed using a representative-based clustering algorithm. The experimental results show that this technique leads to clusters of higher quality compared to running a representative clustering algorithm stand-alone. Given a suitable fitness function, MOSAIC is able to detect arbitrary shape clusters. In addition, MOSAIC is capable of dealing with high dimensional data.

Keywords: Post-processing, hybrid clustering, finding clusters of arbitrary shape, agglomerative clustering, using proximity graphs for clustering.

1 Introduction

Representative-based clustering algorithms form clusters by assigning objects to the closest cluster representative. k-means is the most popular representative-based clustering algorithm: it uses cluster centroids as representatives and iteratively updates clusters and centroids until no change in the clustering occurs. k-means is a relatively fast clustering algorithm with a complexity of $O(ktn)$, where n is the number of objects, k is the number of clusters, and t is the number of iterations. The clusters generated are always contiguous. However, when using k-means the number of clusters k has to be known in advance, and k-means is very sensitive to initializations and outliers. Another problem of k-means clustering algorithm is that it cannot obtain clusters that have non-convex shapes [1]: the shapes that can be obtained by representative-based clustering algorithms are limited to convex polygons.

I.Y. Song, J. Eder, and T.M. Nguyen (Eds.): DaWaK 2007, LNCS 4654, pp. 231–240, 2007.

In theory, agglomerative hierarchical clustering (AHC) [2] is capable of detecting clusters of arbitrary shape. However, in practice, it performs a very narrow search, merging the two closest clusters without considering other merging candidates and therefore often misses high quality solutions. Moreover, its time complexity is $O(n^2)$ or worse. Finally, many variations of AHC obtain non-contiguous clusters. [3]

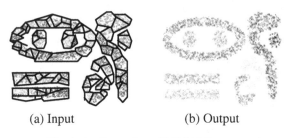

(a) Input   (b) Output

Fig. 1. An illustration of MOSAIC's approach

This paper proposes a hybrid clustering technique that combines representative-based with agglomerative clustering trying to maximize the strong points of each approach. A novel agglomerative clustering algorithm called MOSAIC is proposed, which greedily merges neighboring clusters maximizing a given fitness function and whose implementation uses Gabriel graphs [4] to determine which clusters are neighboring. Non-convex shapes are approximated as the union of small convex clusters that have been obtained by running a representative-based clustering algorithm, as illustrated in Fig. 1. Creating mosaics in art is the process of assembling small pieces to get a sophisticated design. Similarly, the proposed MOSAIC algorithm pieces convex polygons together to obtain better clusters.

1.   Run a representative-based clustering algorithm to create a large number of clusters.
2.   Read the representatives of the obtained clusters.
3.   Create a merge candidate relation using proximity graphs.

4.   WHILE there are merge-candidates (C_i, C_j) left
    BEGIN
        Merge the pair of merge-candidates (C_i, C_j), that enhances fitness function q the most, into a new cluster C'
        Update merge-candidates:
        $\forall C$ Merge-Candidate$(C', C) \Leftrightarrow$
            Merge-Candidate$(C_i, C) \lor$ Merge-Candidate(C_j, C)
    END
RETURN the best clustering X found.

Fig. 2. Pseudo code for MOSAIC

Relying on proximity graphs the MOSAIC conducts a much wider search which, we claim, results in clusters of higher quality. Moreover, the expensive, agglomerative clustering algorithm is only run for usually less than 1000 iterations; therefore, the impact of its high complexity on the overall run time is alleviated, particularly for very large data sets. Furthermore, the proposed post-processing technique is highly generic in that it can be used with any representative-based clustering algorithm, with any proximity graph and with any cluster evaluation function. Fig. 2 gives the pseudo code of the proposed MOSAIC algorithm.

In summary, MOSAIC merges pairs of neighboring clusters maximizing an externally given fitness function q, and this process is continued until only one cluster is left. Finally, the best clustering is determined and returned. Using cluster representatives obtained from a representative-based clustering algorithm as an input, a proximity graph is generated to determine which of the original clusters are neighboring and a merge-candidate relation is constructed from this proximity graph. When clusters are merged, this merge-candidate relation is updated incrementally without any need to regenerate proximity graphs.

The main contributions of the paper are;

- It introduces a hybrid algorithm that combines strong features of representative-based clustering and agglomerative clustering.
- The algorithm provides flexibility by enabling to plug-in any fitness functions and is not restricted to any specific cluster evaluation measure.
- The algorithm conducts a much wider search, compared to traditional agglomerative clustering algorithms, by considering neighboring clusters as merge candidates.

The organization of our paper is as follows. Section 2 describes MOSAIC in more detail. Then the performance study of MOSAIC and the comparative study with DBSCAN and k-means are explained in section 3. Related work is reviewed in Section 4, and a conclusion is given in Section 5 respectively.

2   Post-processing with MOSAIC

This section discusses MOSAIC in more detail. First, proximity graphs are introduced and their role in agglomerative clustering is discussed. Next, an internal cluster evaluation measure will be discussed that will serve as a fitness function in the experimental evaluation. Finally, MOSAIC's complexity is discussed

2.1   Using Gabriel Graphs for Determining Neighboring Clusters

Different proximity graphs represent different neighbor relationships for a set of objects. There are various kinds of proximity graphs [5], with Delaunay graphs [6] (DG) being the most popular ones. The Delaunay graph for a set of cluster representatives tells us which clusters of a representative-based clustering are neighboring and the shapes of representative-based clusters are limited to Voronoi cells, the dual to Delaunay graphs.

Delaunay triangulation (DT) [7] is the algorithm that constructs the Delaunay graphs for a set of objects. Unfortunately, using DT for high dimensional datasets is impractical since it has a high complexity of $O(n^{\frac{d}{2}})$ (when d>2), where d is the number of dimensions of a data set. Therefore, our implementation of MOSAIC uses another proximity graph called Gabriel graphs (GG) [4] instead, which is a sub-graph of the DG. Two points are said to be Gabriel neighbors if their diametric sphere does not contain any other points. The pseudo code of an algorithm that constructs the GG for a given set of objects is given in Fig. 3. Constructing GG has a time complexity of $O(dn^3)$ but faster, approximate algorithms ($O(dn^2)$) to construct GG exist [8].

Let $R = \{r_1, r_2,..., r_k\}$, be a set of cluster representatives
FOR each pair of representatives (r_i, r_j),
   IF for each representative r_p, the following inequality

$$d(r_i, r_j) \leq \sqrt{d^2(r_i, r_p) + d^2(r_j, r_p)}$$

   where $p \neq i, j$ and $r_p \in R$, is true,
THEN r_i and r_j are neighboring.
$d(r_i, r_j)$ denotes the distance of representatives r_i and r_j.

Fig. 3. Pseudo code for constructing Gabriel graphs

Gabriel graphs are known to provide good approximations of Delaunay graphs because a very high percentage of the edges of a Delaunay graph are preserved in the corresponding Gabriel graph [9]. MOSAIC constructs the Gabriel graph for a given set of representatives, e.g. cluster centroids in the case of k-means, and then uses the Gabriel graph to construct a boolean merge-candidate relation that describes which of the initial clusters are neighboring. This merge candidate relation is then updated incrementally when clusters are merged. The illustration of the Gabriel graph construction in MOSAIC is shown in Fig. 4 in which cluster representatives are depicted as squares, objects in a cluster a represented using small blue circles, and clusters that have been obtained by a representative algorithm are visualized using doted lines.

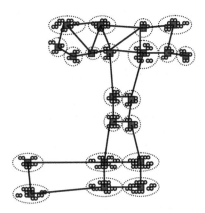

Fig. 4. Gabriel graph for clusters generated by a representative-based clustering algorithm

2.2 Cluster Evaluation Measures for Traditional Clustering

Many evaluation measures have been proposed in the literature [2]. In this paper, we use Silhouettes [10] which is an internal evaluation measure that has been widely used to assess cluster quality. Silhouettes is a popular cluster evaluation technique that takes into consideration both cohesion and separation. The definition of Silhouettes is as follows:

$$s_i = \frac{(a_i - b_i)}{\max(a_i, b_i)} \tag{1}$$

where,

$$a_i = \min_m \left(\frac{1}{|C_m|} \sum_{o_j \in m} d_{ij} \right), \quad m \neq k \text{ and} \tag{2}$$

$$b_i = \frac{1}{|C_k|} \sum_{o_j \in C_k} d_{ij} . \tag{3}$$

In the formula (1), b_i is average dissimilarity of an object o_i to all other objects o_j in the same cluster. a_i is minimum of average dissimilarity of an object o_i to all objects o_j in another cluster (the closest cluster). To measure quality not for one object but entire clustering, we use average of Silhouettes over whole dataset. The fitness function $q(X)$ is defined as follows:

$$q(X) = \frac{1}{n} \sum_{i=1}^{n} s_i \tag{4}$$

where n is the number of objects in a dataset.

2.3 Complexity

The time complexity of our proposed hybrid clustering algorithm depends on two factors: the complexity of the representative-based clustering algorithm and of the MOSAIC algorithm itself. Analyzing MOSAIC's complexity, we already discussed that the cost for constructing the Gabriel Graph is $O(k^3)$. After that, we have to merge the k vertices of the Gabriel Graph. Basically, a Delaunay Graph is a planar graph; since a Gabriel Graph is a connected subset of a Delaunay Graph, we have that the number e of edges of our GG is $k-1 \leq e \leq 3k-6$. This means that the number of edges e in the graph is always linear with respect to the number of vertices: $e=O(k)$. Thus, at the ith iteration, $O(k)-i$ new merge-candidates are created for the newly created cluster that have to be evaluated, which adds up to $O(k^2)$ fitness function evaluations: $(O(k-1) + O(k-2) + \ldots + 1)$. Putting this all together, we obtain a time complexity of the MOSAIC algorithm of: $O(k^3 + k^2*(O(q(X)))$ where $O(q(X))$ is the time complexity of the fitness function. A lower complexity for MOSAIC can be obtained if the fitness of a particular clustering can be computed incrementally during the merging stages based on results of previous fitness computations.

3 Experiments

We compare MOSAIC using the Silhouettes fitness function with DBSCAN and k-means[1]. Due to space limitations we are only able to present a few results; a more detailed experimental evaluation can be found in [3]. We conducted our experiments on a Dell Inspiron 600m laptop with a Intel(R) Pentium(R) M 1.6GHz processor with 512 MB of RAM. We set up three experiments that use the following datasets: an artificial dataset called **9Diamonds**[11] consisting of 3,000 objects with 9 natural clusters, **Volcano**[11] containing 1,533 objects, **Diabetes**[12] containing 768 objects, **Ionosphere**[12] containing 351 objects, and **Vehicle**[12] containing 8,469 objects.

Experiment 1: The experiment compares the clustering results generated by running k-means with k=9 with MOSAIC for the 9Diamonds dataset.

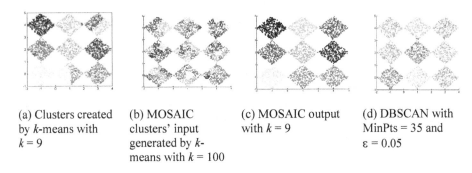

(a) Clusters created by k-means with $k = 9$

(b) MOSAIC clusters' input generated by k-means with $k = 100$

(c) MOSAIC output with $k = 9$

(d) DBSCAN with MinPts = 35 and ε = 0.05

Fig. 5. Experimental results for the 9Diamonds dataset

Discussion: As shown in Fig. 5 (a), k-means is not able to discover the natural clusters. MOSAIC, on the other hand, is able to discover the natural clusters by iteratively merging the sub-clusters that have been depicted in Fig. 5 (b) by maximizing the Silhouettes fitness function: the clustering with the highest fitness value is displayed in Fig. 5 (c).

Experiment 2: The experiment compares the clustering results generated by MOSAIC and DBSCAN for two two-dimensional datasets: 9Diamonds and Volcano.

Discussion: To use DBSCAN, we have to choose values for two parameters: MinPts and ε. One challenge of this experiment is to find proper values for those parameters. First, we used the procedure that has been proposed in the original paper [13] to select values for MinPts and ε. Unfortunately, this procedure did not work very well: DBSCAN just created a single cluster for both datasets tested. Therefore, relying on human intuition, we employed a manual, interactive procedure that generated 80 pairs of parameters for DBSCAN. The parameters selected by the second procedure lead to

---

[1] In general, we would have preferred to compare our algorithm also with CHAMELEON and DENCLUE. However, we sadly have to report that executable versions of these two algorithms no longer exist.

much better results. We observed that ε values that produce better clustering results are much smaller than those suggested by analyzing the sorted k-dist graph. Fig. 5 (d) depicts one of the best clustering results obtained for DBSCAN for the 9Diamonds dataset. MOSAIC correctly clusters the dataset while DBSCAN reports a small number of outliers in the left corner of the bottom left cluster.

Volcano is a real world dataset that contains chain-like patterns with various densities. In general, DBSCAN and MOSAIC produced results of similar quality for this dataset. Fig. 6 depicts a typical result of this comparison: MOSAIC does a better job in identifying the long chains in the left half of the display (Fig. 6 (a)), whereas DBSCAN correctly identifies the long chain in the upper right of the display (Fig. 6. (b)). DBSCAN and MOSAIC both fail to identify all chain patterns.

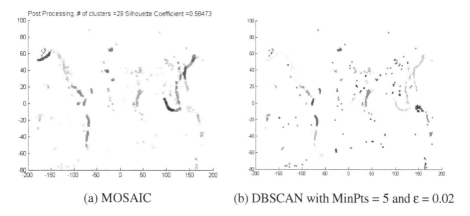

(a) MOSAIC   (b) DBSCAN with MinPts = 5 and ε = 0.02

Fig. 6. Experimental results of MOSAIC and DBSCAN on Volcano dataset

Experiment 3: This experiment compares MOSAIC and k-means on three high dimensional datasets: Vehicle, Ionosphere, and Diabetes. The quality of clustering results is compared using the Silhouettes cluster evaluation measure. MOSAIC's input were 100 clusters that have been created by running k-means. Next, MOSAIC was run and its Silhouette values were averaged over its 98 iterations. These Silhouette values were compared with the average Silhouette values obtained by running k-means with $k = 2 - 99$. Table 1 summarizes the findings of that experiment.

Table 1. Information for the high dimensional datasets and experimental results

Dataset	Number of objects	Number of dimensions	Average Silhouette coefficient of k-means	Average Silhouette coefficient of MOSAIC
Vehicle	8,469	19	0.20013	0.37157
Ionosphere	351	34	0.2395	0.26899
Diabetes	768	8	0.23357	0.24373

Discussion: MOSAIC outperforms k-means quite significantly for the Vehicle dataset and we see minor improvements for the Ionosphere and Diabetes datasets.

4 Related Work

Discovering arbitrary shape clusters is very important in many domains such as hot spot detection, region discovery and spatial data mining. Jiang [1] proposes spatial clustering techniques that employ hierarchical clustering accompanied by tree-like diagrams and claim that this is a beneficiary for visualizing cluster hierarchies at different levels of detail. Anders [14] developed an unsupervised graph-based clustering algorithm, called Hierarchical Parameter-free Graph Clustering (HPGCL) for spatial data analysis. Various proximity graphs are used to define coarse-to-fine hierarchical segmentation of data.

Density-based clustering methods [13, 15, 16, and 17] have been found to be efficient for discovering dense arbitrary shaped clusters, such as the ones in the 9Diamonds dataset. The main drawbacks of density-based clustering algorithms are their need for parameters tuning, and their usually poor performance for datasets with varying density. Moreover, they do not seem to be suitable for high dimensional data.

There has been significant research centering on hybrid clustering. CURE is a hybrid clustering algorithm that integrates a partitioning algorithm with an agglomerative hierarchical algorithm [18]. CURE agglomeratively merges the two clusters that have the closest pair of representatives, and updates mean and a set of representative points. CHAMELEON [19] provides a sophisticated two-phased clustering algorithm. In the first phase, it uses a multilevel graph partitioning algorithm to create an initial set of clusters and in the second phase it iteratively merges clusters maximizing relative inter-connectivity and relative closeness. MOSAIC also relies on two-phase clustering but it has a major advantage over CHAMELEON and CURE by being able to plug-in any fitness function and not being restricted to evaluate clusters based on inter-connectivity and closeness. Lin and Zhong [20 and 21] propose hybrid clustering algorithms that combine representative-based clustering and agglomerative clustering methods. However they employ different merging criteria and perform a narrow search that only considers a single pair of merge candidates. Surdeanu [22] proposes a hybrid clustering approach that combines agglomerative clustering algorithm with the Expectation Maximization (EM) algorithm.

5 Conclusion

This paper proposes a novel approach that approximates arbitrary-shape clusters through unions of small convex polygons that have been obtained by running a representative-based clustering algorithm. An agglomerative clustering algorithm called MOSAIC is introduced that greedily merges neighboring clusters maximizing an externally given fitness function. Gabriel graphs are used to determine which clusters are neighboring. We claim that using proximity graphs increases the number of merge candidates considerably over traditional agglomerative clustering algorithms that only consider "closest" clusters for merging, resulting in clusters of higher quality. MOSAIC is quite general and can be used with any representative-based clustering algorithm, any proximity graph, and any fitness function. Moreover, we claim that MOSAIC can be effectively applied to higher dimensional data.

MOSAIC also has some similarity with agglomerative grid-based clustering algorithms; both approaches employ micro-clusters which are grid-cells in their approach and convex polygons in our approach and greedily merge neighboring clusters. However, our approach is much more general by supporting more variety of shapes and it allows for convex polygons of different sizes. On the other hand, for a given grid structure it is easy to determine which clusters are neighboring, which is not true for our approach.

We conducted experiments whose results suggest that using MOSAIC in conjunction with k-means can significantly improve cluster quality. Using Silhouettes function as a fitness function we also compared MOSAIC with DBSCAN; both algorithms obtained results of similar quality for most datasets tested. However, before using DBSCAN we had to spend significant efforts for parameter tuning which is not the case when using MOSAIC which only requires a single input parameter: the fitness function.

However, based on our initial experimental results, we don't believe that the Silhouettes function is the best possible fitness function to find arbitrary shape clusters. Consequently, in our current research we investigate to find more suitable fitness functions for this purpose.

References

1. Jiang, B.: Spatial Clustering for Mining Knowledge in Support of Generalization Processes in GIS. In: ICA Workshop on Generalisation and Multiple representation (2004)
2. Tan, M., Steinbach, M., Kumar, V.: Introduction to Data Mining, 1st edn. Addison-Wesley, Reading (2005)
3. Choo, J.: Using Proximity Graphs to Enhance Representative-based Clustering Algorithms. Master Thesis, Department of Computer Science, University of Houston, TX (2007)
4. Gabriel, K., Sokal, R.: A New Statistical Approach to Geographic Variation Analysis. Systematic Zoology 18, 259–278 (1969)
5. Toussaint, G.: The Relative Neighborhood Graph of A Finite Planar Set. In: Int. Conf. Pattern Recognition, vol. 12, pp. 261–268 (1980)
6. Kirkpatrick, D.: A note on Delaunay and Optimal Triangulations. Information Processing Letters 10, 127–128 (1980)
7. Okabe, A., Boots, B., Sugihara, K.: Spatial Tessellations: Concepts and Applications of Voronoi Diagrams. Wiley, New York (1992)
8. Bhattacharya, B., Poulsen, R., Toussaint, G.: Application of Proximity Graphs to Editing Nearest Neighbor Decision Rule. In: Int. Sym. on Information Theory (1981)
9. Asano, T., Imai, H., Ibaraki, T., Nishizeki, T.: SIGAL 1990. LNCS, vol. 450, pp. 70–71. Springer, Heidelberg (1990)
10. Rousseeuw, P.J., Silhouettes, A.: Graphical Aid to The Interpretation and Validation of Cluster Analysis. Int. J. Computational and Applied Mathematics 20, 53–65 (1987)
11. Data Mining and Machine Learning Group website, University of Houston, Texas, http://www.tlc2.uh.edu/dmmlg/Datasets
12. UCI Machine Learning Repository, http://www.ics.uci.edu/~mlearn/MLRepository.html
13. Ester, M., Kriegel, H.P., Sander, J., Xu, X.: Density-Based Spatial Clustering of Applications with Noise. In: Int. Conf. Knowledge Discovery and Data Mining (1996)

14. Anders, K.H.: A Hierarchical Graph-Clustering Approach to Find Groups of Objects. Technical Paper. In: ICA Commission on Map Generalization, 5th Workshop on Progress in Automated Map Generalization (2003)
15. Sander, J., Ester, M., Kriegel, H.P., Xu, X.: Density-Based Clustering in Spatial Databases: The Algorithm GDBSCAN and its Applications. In: Inf. Conf. Data Mining and Knowledge Discovery, pp. 169–194 (1998)
16. Kriegel, H.P., Pfeifle, M.: Density-Based Clustering of Uncertain Data. In: Int. Conf. Knowledge Discovery in Data Mining, pp. 672–677 (2005)
17. Hinneburg, A., Keim, D.: An Efficient Approach to Clustering in Large Multimedia Databases with Noise. In: Conf. Knowledge Discovery in Data Mining (1998)
18. Guha, S., Rastogi, R., Shim, K.: CURE: An Efficient Clustering Algorithm for Large Databases. In: Int. Conf. ACM SIGMOD on Management of data, pp. 73–84. ACM Press, New York (1998)
19. Karypis, G., Han, E.H., Kumar, V.: CHAMELEON: A Hierarchical Clustering Algorithm Using Dynamic Modeling. IEEE Computer 32, 68–75 (1999)
20. Lin, C., Chen, M.: A Robust and Efficient Clustering Algorithm based on Cohesion Self-Merging. In: Inf. Conf. 8th ACM SIGKDD on Knowledge Discovery and Data Mining, pp. 582–587. ACM Press, New York (2002)
21. Zhong, S., Ghosh, J.: A Unified Framework for Model-based Clustering. Int. J. Machine Learning Research 4, 1001–1037 (2003)
22. Surdeanu, M., Turmo, J., Ageno, A.: A Hybrid Unsupervised Approach for Document Clustering. In: Int. Conf. 11h ACM SIGKDD on Knowledge Discovery in Data Mining, pp. 685–690. ACM Press, New York (2005)

A Hybrid Particle Swarm Optimization Algorithm for Clustering Analysis

Yannis Marinakis[1], Magdalene Marinaki[2], and Nikolaos Matsatsinis[1]

[1] Decision Support Systems Laboratory, Department of Production Engineering and Management, Technical University of Crete, 73100 Chania, Greece
marinakis@ergasya.tuc.gr, nikos@ergasya.tuc.gr
[2] Industrial Systems Control Laboratory, Department of Production Engineering and Management, Technical University of Crete, 73100 Chania, Greece
magda@dssl.tuc.gr

Abstract. Clustering is a very important problem that has been addressed in many contexts and by researchers in many disciplines. This paper presents a new stochastic nature inspired methodology, which is based on the concepts of Particle Swarm Optimization (PSO) and Greedy Randomized Adaptive Search Procedure (GRASP), for optimally clustering N objects into K clusters. The proposed algorithm (Hybrid PSO-GRASP) for the solution of the clustering problem is a two phase algorithm which combines a PSO algorithm for the solution of the feature selection problem and a GRASP for the solution of the clustering problem. Due to the nature of stochastic and population-based search, the proposed algorithm can overcome the drawbacks of traditional clustering methods. Its performance is compared with other popular stochastic/metaheuristic methods like genetic algorithms and tabu search. Results from the application of the methodology to a survey data base coming from the Paris olive oil market and to data sets from the UCI Machine Learning Repository are presented.

Keywords: Particle Swarm Optimization, GRASP, Clustering Analysis.

1 Introduction

Clustering analysis identifies clusters (groups) embedded in the data, where each cluster consists of objects that are similar to one another and dissimilar to objects in other clusters ([5], [13], [17]). The typical cluster analysis consists of four steps (with a feedback pathway) which are the *feature selection* or *extraction*, the *clustering algorithm design* or *selection*, the *cluster validation* and the *results interpretation* [17].

The *basic feature selection problem (FSP)* is an optimization one, where the problem is to search through the space of feature subsets to identify the optimal or near-optimal one with respect to a performance measure. In the literature many successful feature selection algorithms have been proposed ([6], [10]). *Feature extraction* utilizes some transformations to generate useful and novel

I.Y. Song, J. Eder, and T.M. Nguyen (Eds.): DaWaK 2007, LNCS 4654, pp. 241–250, 2007.

features from the original ones. The *clustering algorithm design* or *selection* step is usually combined with the selection of a corresponding proximity measure ([5], [13]) and the construction of a clustering criterion function which makes the partition of clusters a well defined optimization problem. However, it should be noted that the problem is NP-hard as the clustering objective functions are highly non-linear and multi-modal functions and as a consequence it is difficult to investigate the problem in an analytical approach. Many heuristic, metaheuristic and stochastic algorithms have been developed in order to find a near optimal solution in reasonable computational time. An analytical survey of the clustering algorithms can be found in [5], [13], [17]. *Cluster validity* analysis is the assessment of a clustering procedure's output. Effective evaluation standards and criteria are used in order to find the degree of confidence for the clustering results derived from the used algorithms. External indices, internal indices, and relative indices are used for cluster validity analysis ([5], [17]). In the *results interpretation* step, experts in the relevant fields interpret the data partition in order to guarantee the reliability of the extracted knowledge.

In this paper, a new hybrid metaheuristic algorithm based on an Particle Swarm Optimization (PSO) [7] algorithm for the solution of the feature selection problem and on a Greedy Randomized Adaptive Search Procedure (GRASP) [2] for the solution of the clustering problem is proposed. Such an algorithm that combines a nature inspired intelligence technique like PSO and a stochastic metaheuristic like GRASP is applied for the first time for the solution of this kind of problems. In order to assess the efficacy of the proposed algorithm, this methodology is evaluated on datasets from the UCI Machine Learning Repository and to a survey data set from Paris olive oil market. Also, the method is compared with the results of three other metaheuristic algorithms for clustering analysis that use a Tabu Search Based Algorithm [3], a Genetic Based Algorithm [4] and an Ant Colony Optimization algorithm [1] for the solution of the feature selection problem [10]. The rest of this paper is organized as follows: In the next section the proposed Hybrid PSO-GRASP Algorithm is presented and analyzed in detail. In section 3, the analytical computational results for the datasets used in this study are presented while in the last section conclusions and future research are given.

2   The Proposed Hybrid PSO-GRASP Algorithm for Clustering

2.1   Introduction

The proposed algorithm (Hybrid PSO-GRASP) for the solution of the clustering problem is a two phase algorithm which combines a Particle Swarm Optimization (PSO) [7] algorithm for the solution of the feature selection problem and a Greedy Randomized Adaptive Search Procedure (GRASP) for the solution of the clustering problem. In this algorithm, the activated features are calculated by the Particle Swarm Optimization algorithm (see 2.2) and the fitness (quality) of each particle is calculated by the clustering algorithm (see 2.3).

The problem of clustering N objects (patterns) into K clusters is considered. In particular the problem is stated as follows: Given N objects in R^n , allocate each object to one of K clusters such that the sum of squared Euclidean distances between each object and the center of its belonging cluster (which is also to be found) for every such allocated object is minimized. The clustering problem can be mathematically described as follows:

$$\text{Minimize } J(w, z) = \sum_{i=1}^{N} \sum_{j=1}^{K} w_{ij} \parallel x_i - z_j \parallel^2 \tag{1}$$

Subject to

$$\sum_{j=1}^{K} w_{ij} = 1, \qquad\qquad i = 1, ..., N \tag{2}$$

$$w_{ij} = 0 \text{ or } 1, \qquad\qquad i = 1, ..., N, j = 1, ..., K$$

where K is the number of clusters (given or unknown), N is the number of objects (given), $x_i \in R^n, (i = 1, ..., N)$ is the location of the ith pattern (given), $z_j \in R^n, (j = 1, ..., K)$ is the center of the jth cluster (to be found), $(z_j = \frac{1}{N_j} \sum_{i=1}^{N} w_{ij} x_i$, where N_j is the number of objects in the jth cluster), and w_{ij} is the association weight of pattern x_i with cluster j, (to be found), where w_{ij} is equal to 1 if pattern i is allocated to cluster j, $\forall i = 1, ..., N, j = 1, ..., K$ and is equal to 0, otherwise.

Initially in the first phase of the algorithm a number of features are activated, using the Particle Swarm Optimization Algorithm. In order to find the clustering of the samples (fitness or quality of the PSO algorithm) a GRASP algorithm is used. The clustering algorithm has the possibility to solve the clustering problem with known or unknown number of clusters. When the number of clusters is known, the equation (1), denoted as SSE, is used in order to find the best clustering. In the case that the number of clusters is unknown, the selection of the best solution of the feature selection problem cannot be performed based on the sum of squared Euclidean distances because when the features are increased (or decreased) a number of terms are added (or subtracted) in equation (1) and the comparison of the solutions is not possible, using only the SSE measure. Thus the minimization of a validity index ([12], [14]) is used, given by:

$$validity = \frac{SSE}{SSC}. \tag{3}$$

where $SSC = \sum_{i}^{K} \sum_{j}^{K} (\parallel z_i - z_j \parallel)^2$ is the distance between the centers of the clusters.

2.2 Particle Swarm Optimization for the Feature Selection Problem

Particle Swarm Optimization (PSO) is a population-based swarm intelligence algorithm. It was originally proposed by Kennedy and Eberhart as a simulation of the social behavior of social organisms such as bird flocking and fish schooling [7]. PSO uses the physical movements of the individuals in the swarm and has a flexible and well-balanced mechanism to enhance and adapt to the global and local exploration abilities. Most applications of PSO have concentrated on the optimization in continuous space while recently, some work has been done to the discrete optimization problem [8].

The PSO algorithm first randomly initializes a swarm of particles. The position of each individual (called particle) is represented by a d-dimensional vector in problem space $s_i = (s_{i1}, s_{i2}, ..., s_{id})$, $i = 1, 2, ..., M$ (M is the population size), and its performance is evaluated on the predefined fitness function. Thus, each particle is randomly placed in the d-dimensional space as a candidate solution (in the feature selection problem d corresponds to the number of activated features). One of the key issues in designing a successful PSO for Feature Selection Problem is to find a suitable mapping between Feature Selection Problem solutions and particles in PSO. Every candidate feature in PSO is mapped into a binary particle where the bit 1 denotes that the corresponding feature is selected and the bit 0 denotes that the feature is not selected. The velocity of the i-th particle $v_i = (v_{i1}, v_{i2}, ..., v_{id})$ is defined as the change of its position. The flying direction of each particle is the dynamical interaction of individual and social flying experience. The algorithm completes the optimization through following the personal best solution of each particle and the global best value of the whole swarm. Each particle adjusts its trajectory toward its own previous best position and the previous best position attained by any particle of the swarm, namely p_{id} and p_{gd}. In the discrete space, a particle moves in a state space restricted to zero and one on each dimension where each v_i represents the probability of bit s_i taking the value 1. Thus, the particles' trajectories are defined as the changes in the probability and v_i is a measure of individual's current probability of taking 1. If the velocity is higher it is more likely to choose 1, and lower values favor choosing 0. A sigmoid function is applied to transform the velocity from real number space to probability space:

$$sig(v_{id}) = \frac{1}{1 + exp(-v_{id})} \tag{4}$$

In the binary version of PSO, the velocities and positions of particles are updated using the following formulas:

$$v_{id}(t + 1) = wv_{id}(t) + c_1 rand1(p_{id} - s_{id}(t)) + c_2 rand2(p_{gd} - s_{id}(t)) \tag{5}$$

$$s_{id}(t + 1) = \begin{cases} 1, & \text{if } rand3 < sig(u_{id}) \\ 0, & \text{if } rand3 >= sig(u_{id}) \end{cases} \tag{6}$$

where $p_{id} = (p_{i1d}, ..., p_{ind})$ is the best position encountered by i-th particle so far; p_{gd} represents the best position found by any member in the whole swarm population; t is iteration counter; s_{id} is the valued of the d-th dimension of particle s_i, and $s_{id} \in \{0, 1\}$; v_{id} is the corresponding velocity; $sig(v_{id})$ is calculated according to the Equation (4), c_1 and c_2 are acceleration coefficients; $rand1$, $rand2$ and $rand3$ are three random numbers in $[0, 1]$. Acceleration coefficients c_1 and c_2 control how far a particle will move in a single iteration. Low values allow particles to roam far from target regions before being tugged back, while high values result in abrupt movement towards, or past, target regions. Typically, these are both set to a value of 2.0, although assigning different values to c_1 and c_2 sometimes leads to improved performance. As in basic PSO, a parameter V_{max} is incorporated to limit the v_{id} so that $sig(v_{id})$ does not approach too closely 0 or 1 [7]. Such implementation can ensure that the bit can transfer between 1 and 0 with a positive probability. In practice, V_{max} is often set at ± 4. The proposed algorithm is established based on standard PSO, namely basic PSO with inertia weight developed by Shi and Eberhart in [15], where w is the inertia weight that controls the impact of previous histories of velocities on current velocity. The particle adjusts its trajectory based on information about its previous best performance and the best performance of its neighbors. The inertia weight w is also used to control the convergence behavior of the PSO. In order to reduce this weight over the iterations, allowing the algorithm to exploit some specific areas, the inertia weight w is updated according to the following equation:

$$w = w_{max} - \frac{w_{max} - w_{min}}{iter_{max}} \times iter \qquad (7)$$

where w_{max}, w_{min} are the maximum and minimum values that the inertia weight can take, and $iter$ is the current iteration (generation) of the algorithm while the $iter_{max}$ is the maximum number of iterations (generations).

2.3   Greedy Randomized Adaptive Search Procedure for the Clustering Problem

As it was mentioned earlier in the clustering phase of the proposed algorithm a **Greedy Randomized Adaptive Search Procedure (GRASP)** ([2], [9]) is used which is an iterative two phase search algorithm. Each iteration consists of two phases, a **construction phase** and a **local search phase**. In the construction phase, a randomized greedy function is used to built up an initial solution which is then exposed for improvement attempts in the local search phase. The final result is simply the best solution found over all iterations. In the first phase, a **randomized greedy technique** provides feasible solutions incorporating both greedy and random characteristics. This phase can be described as a process which stepwise adds one element at a time to the partial (incomplete) solution. The choice of the next element to be added is determined by ordering all elements in a candidate list (**Restricted Candidate List - RCL**) with respect to a greedy function. The heuristic is adaptive because the benefits associated with every element are updated during each iteration of the construction phase to reflect the changes brought on by the selection of the previous element. The probabilistic component of a **GRASP** is characterized by randomly choosing one of the best candidates in the list but not necessarily the top candidate. The greedy algorithm is a simple one pass procedure for solving the clustering problem. In the second phase, a **local search** is initialized from the solution of the first phase, and the final result is simply the best solution found over all searches. In the following the way the GRASP algorithm is applied for the solution of the clustering problem is analyzed in detail. An initial solution (i.e. an initial clustering of the samples in the clusters) is constructed step by step and, then, this solution is exposed for development in the local search phase of the algorithm. The first problem that we have to face was the selection of the number of the clusters. Thus, the algorithm works with two different ways.

If the number of clusters is known a priori, then a number of samples equal to the number of clusters are selected randomly as the initial clusters. In this case, as the iterations of GRASP increased the number of clusters do not change. In each iteration, different samples (equal to the number of clusters) are selected as initial clusters. Afterwards, the RCL is created. The RCL parameter determines the level of greediness or randomness in the construction. In our implementation, the best promising candidate samples are selected to create the RCL. The samples in the list are ordered taking into account the distance of each sample from all centers of the clusters and the ordering is from the smallest to the largest distance. From this list, the first D samples (D is a parameter of the problem) are selected in order to form the final RCL. The candidate sample for inclusion in the solution is selected randomly from the RCL using a random number generator. Finally, the RCL is readjusted in every iteration by recalculated all the distances based on the new centers and replacing the sample which has been included in the solution by another sample that does not belong to the RCL, namely the $(D + t)$th sample where t is the number of the current iteration. When all the samples have been assigned to clusters two measures are calculated (see section 2.1) and a local search strategy is applied in order to improve the solution. The local search works as follows: For each sample the probability of its reassignment in a different cluster is examined by calculating the distance of the sample from the centers. If a sample is reassigned to a different cluster the new centers are calculated. The local search phase stops when in an iteration no sample is reassigned. If the number of clusters is unknown then initially a number of samples are selected randomly as the initial clusters. Now, as the iterations of GRASP increased the number of clusters changes but cannot become less than two. In each iteration a different number of clusters can be found. The creation of the initial solutions and the local search phase work as in the previous case. The only difference compared to the previous case concerns the use of the validity measure in order to choose the best solution because as we have different number of clusters in each iteration the sum of squared Euclidean distances varies significantly for each solution.

3   Computational Results

3.1   Data and Parameter Description

The performance of the proposed methodology is tested on 9 benchmark instances taken from the UCI Machine Learning Repository and a survey data set coming from the Paris olive oil market [16]. The datasets from the UCI Machine Learning Repository were chosen to include a wide range of domains and their characteristics are given in Table 1. The data varies in term of the number of observation from very small samples (Iris with 80 observations) up to larger data sets (Spambase with 4601 observations). Also, there are data sets with two and three clusters. In one case (Breast Cancer Wisconsin) the data set is appeared with different size of observations because in this data set there is a number of missing values. The problem of missing values was faced with two different ways. In the first way where all the observations are used we took the mean values of all the observations in the corresponding feature while in the second way where we have less values in the observations we did not take into account the observations that they had missing values. Some data sets involve only numerical features, and the remaining include both numerical and categorical features. Concerning the survey that was conducted in Paris, France, interviews took place in the households and the sample was equal to 204 olive oil consumers. The questionnaire employed (total 115 questions) was designed in such a way in order to collect information of the sample's demographic characteristics, determine the oils generally used, frequency of olive oil use, type of use of the different oils and consumer perceptions with respect to attributes and factors such as health, taste, purity and pricing, the size and type of packaging, brand loyalty, branding color, odor, taste, packaging, labeling and finally which one of six olive oil brands the consumer would purchase in a ranking order. For each data set, Table 1 reports the total number of features and the number of categorical features in parentheses. The parameter settings for Hybrid PSO-GRASP based metaheuristic algorithm were selected after thorough empirical testing and they are: The number of swarms is set equal to 1, the number of particles is set equal to 50, the number of generations is set equal to 50, the size of RCL varies between 30 and 150, the number of GRASP's iterations is equal to 100 and the coefficients are $c_1 = 2, c_2 = 2$, $w_{max} = 0.9$ and $w_{min} = 0.01$. The algorithm was implemented in Fortran 90 and was compiled using the Lahey f95 compiler on a Centrino Mobile Intel Pentium M 750 at 1.86 GHz, running Suse Linux 9.1.

3.2 Results of the Proposed Algorithm

The objective of the computational experiments is to show the performance of the proposed algorithm in searching for a reduced set of features with high clustering of the data. The purpose of feature variable selection is to find the smallest set of features that can result in satisfactory predictive performance. Because of the curse of dimensionality, it is often necessary and beneficial to limit the number of input features in order to have a good predictive and less computationally intensive model. In general there are $2^{number of features} - 1$ possible feature combinations and, thus, in our cases the problem with the fewest number of feature combinations is the Iris (namely $2^4 - 1$), while the most difficult problem is the Olive Oil where the number of feature combinations is $2^{115} - 1$.

A comparison with other metaheuristic approaches for the solution of the clustering problem is presented in Table 2. In this Table, three other metaheuristic

Table 1. Data Sets Characteristics

Data Sets	Observations	Features	Clusters
Australian Credit (AC)	690	14(8)	2
Breast Cancer Wisconsin 1 (BCW1)	699	9	2
Breast Cancer Wisconsin 2 (BCW2)	683	9	2
Heart Disease (HD)	270	13(7)	2
Hepatitis 1 (Hep1)	155	19 (13)	2
Ionosphere (Ion)	351	34	2
Spambase(spam)	4601	57	2
Iris	150	4	3
Wine	178	13	3
Olive Oil	204	115	unknown

algorithms are used for the solution of the feature subset selection problem. The first one is a classic Tabu Search algorithm [3] running for 1000 iterations and with size of the Tabu List equal to 10, the second is a genetic algorithm [4] running for 20 generations, having a population size equal to 500, and using a single 1-point crossover operator with probability equal to 0.8 and a mutation operator with a probability equal to 0.25 and the third is an Ant Colony Optimization algorithm [1] with number of ants equal to the number of features and number of iterations equal to 100. In the clustering phase of these three algorithms the GRASP algorithm is used. The results of these three algorithms are explained in [11]. From this table it can be observed that the Hybrid PSO-GRASP algorithm performs better (has the largest number of correct classified samples) than the other three algorithms in all instances. It should be mentioned that in some instances the differences in the results between the Hybrid PSO-GRASP algorithm and the other three algorithms are very significant. Mainly, for the two data sets that have the largest number of features compared to the other data sets, i.e. in the Ionosphere data set the percentage for the Hybrid PSO-GRASP algorithm is 85.47%, for the Hybrid ACO-GRASP algorithm is 82.90%, for the Genetic-GRASP algorithm is 75.78% and for the Tabu-GRASP algorithm is 74.92% and in the Spambase data set the percentage for the Hybrid PSO-GRASP algorithm is 87.13%, for the Hybrid ACO-GRASP algorithm is 86.78%, for the Genetic-GRASP algorithm is 85.59% and for the Tabu-GRASP algorithm is 82.80%. These results prove the significance of the solution of the feature selection problem in the clustering algorithm as when a more sophisticated method (PSO) for the solution of this problem was used the performance of the clustering algorithm was improved.

It should, also, be mentioned that the algorithm was tested with two options: with known and unknown number of clusters. In case that when the number of clusters was unknown and thus in each iteration of the algorithm different initial values of clusters were selected the algorithm always converged to the optimal number of clusters and with the same results as in the case that the number of clusters was known. Concerning the Olive Oil data set, where the number of

Table 2. Results of the Algorithm

Instance	PSO		ACO		Genetic		Tabu	
	Selected Features	Correct Classified	Sel. Feat.	Correct Classified	Sel. Feat.	Correct Classified	Sel. Feat.	Correct Classified
BCW2	5	662(96.92%)	5	662(96.92%)	5	662(96.92%)	6	661(96.77%)
Hep1	7	135(87.09%)	9	134(86.45%)	9	134(86.45%)	10	132(85.16%)
AC	8	604(87.53%)	8	603(87.39%)	8	602(87.24%)	9	599(86.81%)
BCW1	5	676(96.70%)	5	676(96.70%)	5	676(96.70%)	8	674(96.42%)
Ion	11	300(85.47%)	2	291(82.90%)	17	266(75.78%)	4	263(74.92%)
spam	51	4009(87.13%)	56	3993(86.78%)	56	3938(85.59%)	34	3810(82.80%)
HD	9	232(85.92%)	9	232(85.92%)	7	231(85.55%)	9	227(84.07%)
Iris	3	145(96.67%)	3	145(96.67%)	4	145(96.67%)	3	145(96.67%)
Wine	7	176(98.87%)	8	176(98.87%)	7	175(98.31%)	7	174(97.75%)

clusters is unknown, the algorithm found in the optimum solution the number of clusters equal to 3 and the number of selected features equal to 105.

4   Conclusions and Future Research

In this paper a new metaheuristic algorithm, the Hybrid PSO-GRASP, is proposed for solving the Clustering Problem. This algorithm is a two phase algorithm which combines an Particle Swarm Optimization (PSO) algorithm for the solution of the feature selection problem and a Greedy Randomized Adaptive Search Procedure (GRASP) for the solution of the clustering problem. Three other metaheuristic algorithms for the solution of the feature selection problem were also used for comparison purposes. The performance of the proposed algorithms is tested using various benchmark datasets from UCI Machine Learning Repository and a survey data set coming from the Paris olive oil market. The objective of the computational experiments, the desire to show the high performance of the proposed algorithms, was achieved as the algorithms gave very efficient results. The significance of the solution of the clustering problem by the proposed algorithm is proved by the fact that the percentage of the correct clustered samples is very high and in some instances is larger than 96%. Also, the focus in the significance of the solution of the feature selection problem is proved by the fact that the instances with the largest number of features gave better results when the PSO algorithm was used. Future research is intended to be focused in using different algorithms both to the feature selection phase and to the clustering algorithm phase.

References

1. Dorigo, M., Stutzle, T.: Ant Colony Optimization. In: A Bradford Book, The MIT Press, Cambridge, Massachusetts, London, England (2004)
2. Feo, T.A., Resende, M.G.C.: Greedy randomized adaptive search procedure. Journal of Global Optimization 6, 109–133 (1995)
3. Glover, F.: Tabu Search I. ORSA Journal on Computing 1(3), 190–206 (1989)

4. Goldberg, D.E.: Genetic Algorithms in Search, Optimization, and Machine Learning. Addison-Wesley Publishing Company, INC, Massachussets (1989)

5. Jain, A.K., Murty, M.N., Flynn, P.J.: Data Clustering: A Review. ACM Computing Surveys 31(3), 264–323 (1999)

6. Jain, A., Zongker, D.: Feature Selection: Evaluation, application, and Small Sample Performance. IEEE Transactions on Pattern Analysis and Machine Intelligence 19, 153–158 (1997)

7. Kennedy, J., Eberhart, R.: Particle swarm optimization. In: Proceedings of 1995 IEEE International Conference on Neural Networks, vol. 4, pp. 1942–1948 (1995)

8. Kennedy, J., Eberhart, R.: A discrete binary version of the particle swarm algorithm. In: Proceedings of 1997 IEEE International Conference on Systems, Man, and Cybernetics, vol. 5, pp. 4104–4108 (1997)

9. Marinakis, Y., Migdalas, A., Pardalos, P.M.: Expanding Neighborhood GRASP for the Traveling Salesman Problem. Computational Optimization and Applications 32, 231–257 (2005)

10. Marinakis, Y., Marinaki, M., Doumpos, M., Matsatsinis, N., Zopounidis, C.: Optimization of Nearest Neighbor Classifiers via Metaheuristic Algorithms for Credit Risk Assessment. Journal of Global Optimization (accepted, 2007)

11. Marinakis, Y., Marinaki, M., Doumpos, M., Matsatsinis, N., Zopounidis, C.: A New Metaheuristic Algorithm for Cluster Analysis. Journal of Global Optimization (submitted, 2007)

12. Ray, S., Turi, R.H.: Determination of Number of Clusters in K-means Clustering and Application in Colour Image Segmentation. In: Proceedings of the 4th International Conference on Advances in Pattern Recognition and Digital Techniques (ICAPRDT99), Calcutta, India (1999)

13. Rokach, L., Maimon, O.: Clustering Methods. In: Maimon, O., Rokach, L. (eds.) Data Mining and Knowledge Discovery Handbook, pp. 321–352. Springer, New York (2005)

14. Shen, J., Chang, S.I., Lee, E.S., Deng, Y., Brown, S.J.: Determination of Cluster Number in Clustering Microarray Data. Applied Mathematics and Computation 169, 1172–1185 (2005)

15. Shi, Y., Eberhart, R.: A modified particle swarm optimizer. In: Proceedings of 1998 IEEE World Congress on Computational Intelligence, pp. 69–73 (1998)

16. Siskos, Y., Matsatsinis, N.F., Baourakis, G.: Multicriteria analysis in agricultural marketing: The case of French olive oil market. European Journal of Operational Research 130, 315–331 (2001)

17. Xu, R., Wunsch II, D.: Survey of Clustering Algorithms. IEEE Transactions on Neural Networks 16(3), 645–678 (2005)

Clustering Transactions with an Unbalanced Hierarchical Product Structure

MinTzu Wang[1], PingYu Hsu[2], K.C. Lin[3], and ShiuannShuoh Chen[4]

[1] Department of Business Administration, National Central University
Jhongli City, Taoyuan County 32001, Taiwan, R.O.C.
93441024@cc.ncu.edu.tw
[1] Department of Information Management, Technology and Science Institute of Northern Taiwan
Taipei 112, Taiwan, R.O.C.
mtwang@tsint.edu.tw
[2,4] Department of Business Administration, National Central University
Jhongli City, Taoyuan County 32001, Taiwan, R.O.C.
pyhsu@mgt.ncu.edu.tw,
shuoh@yahoo.com
[3] Department of Management Information Systems, National Chung Hsing University
Taichung 402, Taiwan R.O.C.
kclin@nchu.edu.tw

Abstract. The datasets extracted from large retail stores often contain sparse information composed of a huge number of items and transactions, with each transaction only containing a few items. These data render basket analysis with extremely low item support, customer clustering with large intra cluster distance and transaction classifications having huge classification trees. Although a similarity measure represented by counting the depth of the least common ancestor normalized by the depth of the concept tree lifts the limitation of binary equality, it produces counter intuitive results when the concept hierarchy is unbalanced since two items in deeper subtrees are very likely to have a higher similarity than two items in shallower subtrees. The research proposes to calculate the distance between two items by counting the edge traversal needed to link them in order to solve the issues. The method is straight forward yet achieves better performance with retail store data when concept hierarchy is unbalanced.

Keywords: data mining, clustering, similarity (distance) measure, hierarchy.

1 Introduction

The availability of inexpensive and quality information technology has in many retail organizations resulted in an abundance of sales data. Retailers have developed their model from a product-oriented to a customer-oriented business. Since superior customer service comes from superior knowledge in understanding customer purchase behavior revealed through their sales transaction data, the transaction data can be applied with various popular data mining algorithms, such as basket analysis [1] [5] [15], customer segmentation (clustering) or classification [2]. All of these algorithms

I.Y. Song, J. Eder, and T.M. Nguyen (Eds.): DaWaK 2007, LNCS 4654, pp. 251–261, 2007.
© Springer-Verlag Berlin Heidelberg 2007

assume that transactions are a composition of items and some repeatedly appear in a significant portion of transactions so that similar transactions or subsets of transactions can be discovered, clustered, and classified.

However, the assumption may not be valid with sparse datasets. In these datasets, the number of items is huge and yet each transaction contains only a few items. The result is that each item only appears in a small portion of the transactions. More often than not, the transaction data derived from large retail stores is sparse. These data render basket analysis with extremely low item support, customer clustering with large intra cluster distance and transaction classifications with huge classification trees.

To alleviate the issue, several researchers have proposed to take into consideration the concept of hierarchy to increase the counts of items. Attribute Oriented Induction [7] is proposed to reduce the number of items by replacing low level items in concept hierarchy with higher level items to increase the number of occurrences. Multiple-Level Association rule [5] is proposed to count the occurrence of items at a higher level with the summation of item counts at a lower level to increase the counts of items at higher levels. Although some patterns or rules of higher level items can be derived with these approaches, a detailed level of knowledge is discarded.

The limitation imposed by binary equality relationships between items in transactions, [12] is proposed to change the relationships to a similarity represented by counting the depth of the least common ancestor normalized by the depth of the concept tree. Although this approach lifts the limitation of binary equality, it produces counter intuitive results when the concept hierarchy is unbalanced since two items in deeper subtrees are very likely to have higher similarity then two items in the shallower subtrees even though the two pairs have to climb the same number of nodes to reach to the least common ancestors. The research proposes to calculate the distances between two items by counting the edge traversal needed to link the two items to solve the issues. The method is straight forward yet reaches better performance with retail store data when concept hierarchy is unbalanced.

The rest of the paper is organized as follows: section 2 presents the related similarity and distance measures, section 3 proposes the model and methodology, section 4 compares the proposed method against traditional similarity computation models by showing several experiment results, and section 5 summarizes the contribution and future work of the research.

2 Related Works

The similarity of instances can be measured by distance, the higher the similarity, the less the distance. There are a variety of ways to measure the inter-object similarity where it plays an important role [11] [12] [13]. The notion of similarity is used in many contexts to identify objects having common "characteristics." Among them, the most dominant are correlation measures and distance measures. The correlation measure of similarity does not look at the magnitude but instead at the pattern of the values, while the distance measures focus on the magnitude of values which may have different patterns across attributes. There are several types of distance measures where Euclidean distance and Manhattan distance are the most popular mechanisms to calculate distance between items.

In other cases, the distances are calculated based on itemsets. The itemsets are modeled as vectors in an n-dimensional space. The Vector Model is a popular model in the information retrieval domain [8]. One advantage of this model is that we can now weigh the components of the vectors, by using schemes such as TF-IDF [14]. The Cosine-Similarity Measure (CSM) defines the similarity between two vectors to be the cosine of the angle between them. This measurement (equation 1) is very popular for query-document and document-document similarity in text retrieval [9] [14].

Some other studies use the Pearson Correlation Coefficient as a similarity measure; GroupLens[4] pertains a vector model with a "vote" of the user for a particular item. A collaborative filtering system looks for people sharing common interests [3]. This similarity is given by the formula (equation 2):

$$S_{\cos}(x_i, x_j) = \frac{\sum_{k=1}^{m}(x_{ik} \times x_{jk})}{\sqrt{(\sum_{k=1}^{m} x_{ik}^2 \times \sum_{k=1}^{m} x_{jk}^2)}} \tag{1}$$

$$c(x, y) = \frac{\sum_j (x_j - \bar{x})(y_j - \bar{y})}{\sqrt{\sum_j (x_j - \bar{x})^2 \sum_j (y_j - \bar{y})^2}} \tag{2}$$

In other cases, the similarity between two itemsets is often determined by set intersection: the more the items have in common, the similar the itemsets are. Jaccard's Coefficient is one of them [6] [9] [10] [11].

Few existing similarity calculations take product hierarchy into consideration. In this research, a bottom-up distance calculation is proposed and various similarity estimations between sets of leaves are compared, and evaluated for different applications.

3 The Proposed Model and the Methodology

3.1 Unbalanced Hierarchy

A product hierarchy is a tree structure where each internal node represents a summary category of products. The nodes beneath an internal node represent either sub-categories of the category denoted by the internal node or products classified into the denoted category. Products are marked as leaf nodes, and categories are denoted by internal nodes. Except for the root node, each internal node should represent a category. The root node is just a dummy node to link all categories and the nodes below into a tree. Figure 1 shows an example of a Product Hierarchy. The products (leaves) can be at different levels. In the figure, internal nodes and leaf nodes are marked with strings prefixed with capital letters C and D, respectively. Readers can find that D_1 and D_2 are beneath the same category of C1111. In the figure, D1 and D2 are products and C1111 is a category.

Even though most researches computing distances with the auxiliary of hierarchies assume domain hierarchy is balanced [5] [12], we would like to point out that in many applications unbalanced hierarchies are more natural. For example, a Deli emphasizing healthy food sells an abundant variety of milk and only a limited selection of beer may have a product hierarchy that classifies milk in three levels of categories according to the percentage of fat, flavor, calcium, and supplemented Vitamins and classify beer with a single layer of categories, namely, domestic and foreign brands.

On the other hand, if a balanced product hierarchy is imposed on the store, the managers have to face the unpleasant dilemma of either shrinking the category hierarchy under milk or meaninglessly enlarge the hierarchy beneath beer. Shrinking the milk category hierarchy may obscure important business signals and enlarging the beer hierarchy may render the discovered beer information meaningless with few numbers of items in each category.

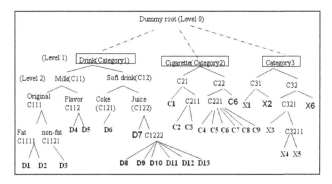

Fig. 1. Part of product hierarchy with 6 levels in a retailing store

3.2 Computing Distances on Unbalanced Hierarchy

To cluster transactions into groups, a distance calculation mechanism based on unbalanced hierarchies is proposed. With a hierarchy, distance can be computed in top-down or bottom-up manners. Defining the corresponding distance for the product items (leaf nodes) in the concept hierarchy helps us to define the distance between transactions which we will apply to clustering.

Definition 1: *The depth of a node, d(n), is defined as the number of edges needed to be traversed from the root to the node.*

For example, d(root) = 0, and d(C1) =3

Definition 2: *The Lowest Common Ancestor of two nodes, lca(n_i,n_j), is defined as the node that is a common ancestor of n_i and n_j and has the greatest depth among all the common ancestor nodes of n_1 and n_2.* $lca(n_1, n_2) = arg(max_{\{n \text{ is ancestor of } n_1 \text{ and } n_2\}} d(n))$

For example, lca(D_1,D_2)=C1111, lca(D_1,C_2)=root.

Definition 3: *Minimum Traversal Length of two nodes mtl(n_i,n_j)= the number of the edges on the shortest path from node n_i to n_j.*

For example, mtl(D_1,D_2)=2, mtl(D_1,D_7)=7.

Definition 4: *The top-down distance between 2 leaf nodes*

$$tdd(n_i, n_j) = \begin{cases} \dfrac{mtl(n_i, n_j)}{d(n_i) + d(n_j)} & , if\ lca(n_i, n_j) <> root. \\ 1 & otherwise. \end{cases} \quad (3)$$

For example, tdd(D1, D2)=2/(5+5)=0.2, tdd(C2,C3)=2/(4+4)=0.25, tdd(D1, D1)=0, and tdd(C1, D2)=1. On the other hand, from a customers' perspective, they usually don't know how deep the products are in the hierarchy level, but products of different brands seem quite similar to them, i.e., same distance, for example, tdd(D1, D2) should be almost equivalent to tdd(C2,C3). We use the branches at level 1 as a different sub-tree. In figure 1, we have 3 sub-trees.

Definition 5

- n_i *is said to be in a subtree rooted by* n_j *if there exist a path originating from* n_i *to* n_j
- *The sub-tree depth of a node* n_i, $std(n_i) = \max_{\{n\ is\ a\ descendent\ of\ n_i\}} d(n)$
- *The level difference of a node* n_i *is defined as* $ld(n_i) = \max_{n\ is\ a\ node\ in\ the\ tree} std(n) - std(n_i)$

For example, std(C1)= std(C2) = std(C3)=4, std(D1)= std(D2) = std(D4)=5, ld(C1)=1, and ld(D1)=0.

Definition 6: *The bottom-up distance between 2 nodes*

$$bud(n_i, n_j) = \begin{cases} \dfrac{mtl(n_i, n_j)}{(d(n_i) + ld(n_i)) + (d(n_j) + ld(n_j))}, & if\ lca(n_i, n_j) <> root \\ 1 & otherwise \end{cases} \quad (4)$$

For example, bud(C2,C3) = 2/(5+5)=0.2, bud(C1,C2)=3/(4+5)=0.33.

Now we may proceed to define the similarity of objects (transactions). The distance between two transactions is the average of the pair wise minimal distance from the 2 items. This idea is derived from Jaccard's Coefficient measure with introduced hierarchy, and it is very intuitive and more general.

Definition 7: *Let* T_a, T_b *be two transactions.* $T_a=\{i_1, i_2, ...i_m\}$, $T_b =\{j_1, j_2, ...j_n\}$, *where* i_i, j_j *are items.*

- *The distance between* T_a *and* T_b *measured with top-down product hierarchy traversal is defined as* $dt(T_a,T_b)$

$$dt(T_a, T_b) = \frac{\sum_{i=1}^{m} \min(tdd(i_1,j_1), tdd(i_1,j_2),...,tdd(i_1,j_n)) + \sum_{j=1}^{n} \min(tdd(i_1,j_j), tdd(i_2,j_j),...,tdd(i_m,j_j))}{n+m}. \quad (5)$$

- *The distance between* T_a *and* T_b *measured with bottom-up product hierarchy traversal is defined as* $bdt(T_a,T_b)$.

$$bdt(T_a, T_b) = \frac{\sum_{i=1}^{m} \min(bud(i_1,j_1), bud(i_1,j_2),...,bud(i_1,j_n)) + \sum_{j=1}^{n} \min(bud(i_1,j_j), bud(i_2,j_j),...,bud(i_m,j_j))}{n+m}. \quad (6)$$

For example, $T_1 =< D_4, D_8>$, $T_2 =< D_5, D_9>$, $T_3 =< D_4, D_5, D_7>$, $T_5 =< D_1, C_2>$, $T_6 =< D_2, C_3>$,then bdt(T_1, T_2)=0.225, bdt(T_1, T_3)=0.183, bdt(T_5, T_6)=0.2, and dt(T_5, T_6)=0.225.

Table 1 is a sample transaction database with Figure 1 as the corresponding concept hierarchy. The top-down and bottom up distances of products D1, D3, D4, C1, C2, X1, and X6 are shown in Table 2. The distance matrix of T5, T6, T7, T8, T9, and T15 are listed in Table 3.

Table 1. A Sample Transaction Database

Tran#	Products	Tran#	Products	Tran#	Products
T1	D4, D8	T11	C1, C3, C3	T21	D9, D11
T2	D5, D9	T12	C4, X5	T22	D12, D13
T3	D4, D5, D7	T13	C5, X6	T23	C4, C6
T4	D4, D6	T14	C3, X5	T24	C5, C7
T5	D1, C2	T15	X5, X6	T25	C8, C9
T6	D2, C3	T16	X3, X6		
T7	D2, C1, C2	T17	D1		
T8	D3, C5	T18	X1		
T9	C2, C3	T19	D1, D3		
T10	C1, C2	T20	D8, D10		

Table 2. Distance Matrix of Products with top-down and bottom up methods

Product	D1		D3		D4		C1		C2		X1		X6	
	tdd	bud	tdd	bud	tdd	bud	tdd	bud	tdd	bud	tdd	bud	tdd	bud
D1	0	0	0.4	0.4	0.56	0.56	1	1	1	1	1	1	1	1
D3	0.4	0.4	0	0	0.56	0.56	1	1	1	1	1	1	1	1
D4	0.56	0.56	0.56	0.56	0	0	1	1	1	1	1	1	1	1
C1	1	1	1	1	1	1	0	0	0.43	0.33	1	1	1	1
C2	1	1	1	1	1	1	0.43	0.33	0	0	1	1	1	1
X1	1	1	1	1	1	1	1	1	1	1	0	0	0.67	0.67
X6	1	1	1	1	1	1	1	1	1	1	0.67	0.67	0	0

Table 3. Distance Matrix of Transactions with top-down and bottom up methods

Tran #	T05		T06		T07		T08		T09		T15	
	dt	bdt	dt	bdt	dt	bdt	dt	bdt	dt	bdt	dt	bdt
T05	0	0	0.225	0.2	0.166	0.147	0.575	0.5	0.313	0.3	1	1
T06	0.225	0.2	0	0	0.186	0.147	0.575	0.5	0.313	0.3	1	1
T07	0.166	0.147	0.186	0.147	0	0	0.596	0.502	0.336	0.307	1	1
T08	0.575	0.5	0.575	0.5	0.596	0.502	0	0	0.813	0.7	1	1
T09	0.313	0.3	0.313	0.3	0.336	0.307	0.813	0.7	0	0	1	1
T15	1	1	1	1	1	1	1	1	1	1	0	0

3.3 Algorithm of Computing Transaction Distance with an Unbalanced Hierarchy

After defining the distance of the product items in the concept hierarchy, we then apply the distance between transactions to clustering and a brief example

```
Algorithm: Computing Transaction Distance with an Unbalanced Hierarchy
```

```
Input:
1.   UProdH - an unbalanced product hierarchy
2.   TranDB  -  transactions  database  where  each  transaction  contains
     products in UProdH
Output:
1.   dt(Ta,Tb)and bdt(Ta,Tb) - Distance between two transactions  with top-
     down and bottom-up method
Method:
1.   For each pair of products pi, pj  in UProdH,
         d(pi)= depth of pi, d(pj)= depth of pj ,
         Find lowest common ancestor of pi, pj , lca(pi, pj)
         If lca(pi, pj)= root
             Then top-down distance tdd(pi, pj)= bud(pi, pj)=1
         Else
             compute minimum traversal length mtl(pi, pj)
             Compute level difference ld(pi), ld(pj)
             tdd(pi, pj)= mtl(pi, pj)/(d(pi)+d(pj))
             bud(pi, pj)= mtl(pi, pj)/(d(pi)+ld(pi)+d(pj)+ ld(pj))
2.   For each pair of transactions ta:<p1, p2, ....pm>, tb:<p1, p2, ....pn> in
     TranDB,
         For i= 1 to m
             Mintdd=1, Minbud=1
             For j= 1 to n
                 If Mintdd > tdd(pi, pj) then Mintdd = tdd(pi, pj)
                 If Minbud > bud(pi, pj) then Minbud = bud(pi, pj)
             Next j
             dt(Ta,Tb)=dt(Ta,Tb)+Mintdd, bdt(Ta,Tb)=bdt(Ta,Tb)+Minbud
         Next i
         For j= 1 to n
             Mintdd=1, Minbud=1
             For i= 1 to m
                 If Mintdd > tdd(pi, pj) then Mintdd = tdd(pi, pj)
                 If Minbud > bud(pi, pj) then Minbud = bud(pi, pj)
             Next i
             dt(Ta,Tb)=dt(Ta,Tb)+Mintdd, bdt(Ta,Tb)=bdt(Ta,Tb)+Minbud
         Next j
         dt(Ta,Tb)= dt(Ta,Tb)/(n+m), bdt(Ta,Tb)= bdt(Ta,Tb)/(n+m)
 Return dt((Ta,Tb), bdt(Ta,Tb)
```

Fig. 2. Algorithm: Computing Transaction Distance with an Unbalanced Hierarchy

demonstration, this section describes the algorithm of computing transaction distance with hierarchy.

4   Experimental Results

To verify the usefulness of the proposed distance measurement, the measurement is applied to a transaction set derived from a retail chain store, which has over 37,593 products organized into an unbalanced hierarchy and a sparse transaction set. The experiment shows that with traditional distance measuring tools, neither association rules nor clustering can render reasonable knowledge, whereas, with the proposed methods, reasonable clustering can be grouped.

4.1 Data Description

The dataset contains 37,593 products organized into a five-level product hierarchy with 1 root, 12 level-1 categories, 144 level-2 categories, 602 level-3 categories and level-4 as products. On average, each level-3 category has 62 products. The dataset contains 48,886 transactions. Each transaction contains 1 to 98 products with an average of 5 products. The top two most popular products are purchased in 854 transactions (4.06%) and 737 transactions (3.5%), respectively. When applying Apriori [1][15] to the dataset, we found that no patterns can be formed with a support threshold at 0.3%.

When trying to cluster the transactions with AHC (Agglomerative Hierarchical Clustering) with single link distance calculation, we found that if the threshold of intra distance of clusters is set at 0.2, only around 100 clusters have more than one transaction with most of these transactions being singleton. To make the experiment result interesting, we kept only transactions that contained 2 to 5 products and deleted the rest of data from the dataset. The number of transactions and transaction details that remained were 4,041 and 11,243, respectively.

4.2 Comparisons of the Different Distance Measures

A good clustering method should not only produce high quality clusters with high intra-cluster similarity (low intra-cluster distance) and low inter-cluster similarity (high inter-cluster distance) but also consequential clusters for users.

In this experiment, we compared 4 similarity measures including JC, balanced hierarchy, top-down, and bottom-up methods with an unbalanced hierarchy. Single link AHC is used to perform the clustering where two clusters with the nearest distances among the grouped transactions are merged. The distances are normalized to between 0 and 1.

The quality of the clustering results is measured by the number of transactions in the largest cluster, and average intra and average inter distances. The first indicator shows the effectiveness of the proposed distance calculation in clustering transactions. Fig. 3 shows that regardless of increased category levels, the bottom-up distance method yields the largest clusters whereas the Jaccard method yields the tiniest clusters and the top-down distance method's performance is between the two. It is important because in a large sparse dataset, it is hard to find any cluster with sufficient amount of transactions, but only lots of small clusters which cannot be representative for user to form any strategies. In the experiment, increased category level is achieved by inserting a dummy node between each of the top five hundred products and the corresponding concept. Figure 4 shows that bottom-up method has better intra distances than the top down methods while also having more transactions in the clusters. Figure 5 shows that given the threshold of intra cluster distances, the clusters grouped by the bottom up method have larger clusters compared with the other two methods.

From Figure 7 and 8, the reader can find that the bottom up method has a smaller intra distance at every step of the clustering merge and the benefit is more significant when the concept hierarchy is skewed. Figure 9 shows that the bottom-up method doesn't have a better inter cluster distance at each merge step until clusters are forced to be merged into a few clusters at the latter stage of the algorithm execution; however, it still denotes discrimination quality between clusters.

The top-down similarity measure and bottom-up similarity measure is not like JC and SC measures, they are not proportional. They will not yield exactly the same clusters; they will suck different transactions into the cluster by way of top-down viewpoint or bottom-up viewpoint.

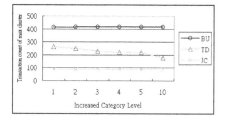

Fig. 3. Transaction counts of the maximal clusters

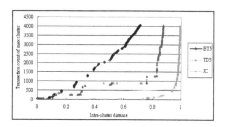

Fig. 6. Transaction counts of the largest clus-ters with category level increased by 5

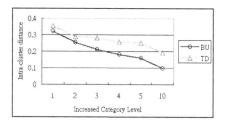

Fig. 4. Intra-cluster distance with BU and TD

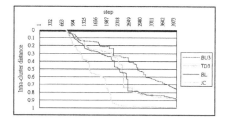

Fig. 7. Intra distance of each merge step with category level increased by 3

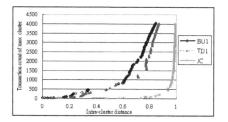

Fig. 5. Transaction counts of the largest clusters with category level increased by 1

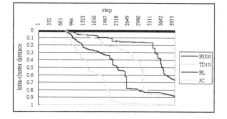

Fig. 8. Intra distance of each merge step with category level increased by 10

With all the data, readers can find that the bottom up method yields the best clustering result while the traditional Jaccard method yields the worst clustering result, and the performance of the top-down method lies between. The more significant the difference the more skewed the concept hierarchy is.

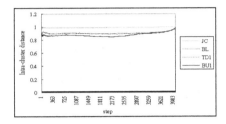

Fig. 9. Inter-cluster distance of each merge step with category level increased by 1

5 Conclusions

This study presents a novel clustering scheme based on a similarity (distance) measure of transactions in an Unbalanced Hierarchical Product Structure. We also provide an experimental comparison of our measures against traditional similarity measures, and evaluate how well our measures match human intuition. Clustering via this similarity measure offers valuable information for the decision maker.

In summary, the main contributions of this research are as follows.

1. We introduce concise similarity (distance) measures that can exploit unbalanced hierarchical domain structure, leading to similarity scores that are more intuitive than the ones generated by traditional similarity measures.
2. We analyze the differences between various measures, compare them empirically, and show that most of them are very different from measures that don't exploit the domain hierarchy.

The study applies the unbalanced concept hierarchy to clusters only. The line of research can be extended to discovering other knowledge, such as frequent patterns, knowledge summary and classifications. Therefore, the research presents a possibility for many other related researches.

References

1. Agrawal, R.S.: Fast Algorithms for Mining Association Rules. In: Proc. of the 20th Int'l Conference on Very Large Databases, Santiago, Chile (1994)
2. Ball, G.H., Hall, D.J.: ISODATA: a novel technique for data analysis and pattern classification. Standford Res. Inst., Menlo Park, CA (1965)
3. Golsberg, D., Nichols, D., Oki, B.M., Terry, D.: Using collaborative filtering to weave an information tapestry. Commun. ACM 35(12), 61–70 (1992)
4. Herlocker, et al.: An Algorithmic Framework for Performing Collaborative Filtering. In: Proceedings of the 1999 Conference on Research and Development in Information Retrieval (1999)
5. Han, J., Fu, Y.: Mining multiple-level association rules in large databases. IEEE Transactions on Knowledge and Data Engineering 11(5), 798–805 (1999)
6. Han, J., Kamber, M.: Data Mining: Concepts and Techniques. Morgan Kaufmann, San Francisco (2000)

7. Han, J., Cai, Y., Cercone, N.: Knowledge Discovery in Databases: An Attribute-Oriented Approach. In: VLDB, pp. 547–559 (1992)
8. McGill, M.J.: Introduction to Modern Information Retrieval. McGraw-Hill, New York (1983)
9. Kantardzic, M.: Data Mining: concepts, models, methods, and algorithms. John Wiley, Chichester (2002)
10. Tan, P.-N., Steinbach, M., Kumar, V.: Introduction to Data Mining. Pearson International Edition (2005)
11. Tan, P.-N., Kumar, V., Srivastava, J.: Selecting the right objective measure for association analysis. Information Systems 29, 293–313 (2004)
12. Ganesan, P., Garcia-Molina, H., Widom, J.: Exploiting hierarchical domain structure to compute similarity. ACM Transactions on Information Systems 21(1) (2003)
13. Chen, S., Han, J., Yu, P.S.: Data Mining: An Overview from a Database Perspective. IEEE Transactions on Knowledge and Data Engineering 8(6), 866–883 (1996)
14. Salton, G., Buckley, C.: Term-weighting approaches in automatic text retrieval. Inf. Process. Manage. 24(5), 513–523 (1988)
15. Srikant, R., Agrawal, R.: Mining generalized association rules. In: Proceedings of VLDB '95, pp. 407–419 (1995)

Constrained Graph b-Coloring Based Clustering Approach

Haytham Elghazel, Khalid Benabdeslem, and Alain Dussauchoy

LIESP Laboratory, Claude Bernard University of Lyon I, 43 Bd du 11 novembre 1918,
69622 Villeurbanne cedex, France
{elghazel,kbenabde,dussauchoy}@bat710.univ-lyon1.fr

Abstract. Clustering is generally defined as an unsupervised data mining process which aims to divide a set of data into groups, or clusters, such that the data within the same group are similar to each other while data from different groups are dissimilar. However, additional background information (namely constraints) are available in some domains and must be considered in the clustering solutions. Recently, we have developed a new graph b-coloring clustering algorithm. It exhibits more important clustering features and enables to build a fine partition of the data set in clusters when the number of clusters is not pre-defined. In this paper, we propose an extension of this method to incorporate two types of Instance-Level clustering constraints (must-link and cannot-link constraints). In experiments with artificial constraints on benchmark data sets, we show improvements in the quality of the clustering solution and the computational complexity of the algorithm.

1 Introduction

Clustering is an important task in the process of data analysis. Informally, this task consists on the division of instances (or synonymously, data points, objects, patterns, etc.) into homogenous groups. Each group, called cluster, contains objects that are similar between themselves and dissimilar to objects of the other groups. The clustering problem has been addressed in many contexts and by researchers in many disciplines, including machine learning, data mining, pattern recognition, image analysis, etc. [1].

Traditionally clustering techniques are broadly divided in hierarchical and partitioning. While hierarchical algorithms build clusters gradually and then give a cluster hierarchy whose leaves are the instances and whose internal nodes represent nested clusters of various sizes, partitioning algorithms learn clusters directly. In fact, they represent each cluster by its centroid or by one of its objects and try to discover cluster using an iterative control strategy to optimize an objective function. When the dissimilarities among all pairs of data in $X = \{x_1,...,x_n\}$ are specified, these can be summarized as a weighed dissimilarity matrix D in which each element $D(x_i,x_j)$ stores the corresponding dissimilarity. Based on D, the data can also be conceived as a graph $G=(V,E)$ where each vertex v_i in V corresponds to a data x_i in X and each edge in E corresponds to a pair of vertices (v_i,v_j) with label $D(x_i,x_j)$.

I.Y. Song, J. Eder, and T.M. Nguyen (Eds.): DaWaK 2007, LNCS 4654, pp. 262–271, 2007.

Recently, we have proposed a new clustering approach [2] based on the concept of *b-coloring* of a graph [3]. A *graph b-coloring* is the assignment of colors (clusters) to the vertices of the graph such that

(i)   no two adjacent vertices have the same color (*proper coloring*),
(ii)   for each color there exists at least one *dominating vertex* which is adjacent to all the other colors.

The *b-coloring based clustering method* in [2] enables to build a fine partition of the data set (*numeric* or *symbolic*) in clusters when the number of clusters is not specified in advance. Such a partition possesses several properties that are desirable for clustering.

While clustering is an unsupervised learning process, users require sometimes incorporating some background information about the data set (named constraints) in these algorithms. These latter vary from the user and the domain but we are usually interested to the use of background information in the form of instance-level *must-link* and *cannot-link* constraints. A *must-link constraint* enforces that two instances must be placed in the same cluster while a *cannot-link constraint* enforces that two instances must not be placed in the same cluster.

Setting these constraints requires some modifications in the clustering algorithms which is not always feasible. In that case, we deal with a semi-supervised clustering problem. Many authors investigated the use of constraints in clustering problem. In [4], the authors have proposed a modified version of COBWEB clustering algorithm that uses background information about pairs of instances to constrain their cluster placement. Equally, a recent work [5] has looked at extending the ubiquitous k-Means algorithm to incorporate the same types of *instance-level hard constraints* (*must-link* and *cannot-link*).

In this paper, we are interested in ways to integrate background information into the *b-coloring based clustering* algorithm. The proposed algorithm is evaluated against benchmark data sets and the results of this study indicate the effectiveness of the *instance-level hard constraints* to offer real benefits (accuracy and runtime) for clustering problem.

In the next section, we briefly describe the *b-coloring* based clustering method. Section 3 is devoted to the modified *graph b-coloring based clustering algorithm* which we will refer to as COP-*b-coloring* (for *constraint portioning b-coloring*). Next, some experiments using relevant benchmarks data set are presented in Section 4. This section includes the evaluation method and the experimental results. Section 5 summarizes our contribution.

2   A b-Coloring Based Clustering Method

In this section, we provide some background on the *b-coloring* based clustering method that was recently introduced in [2].

Consider the data $X = \{x_1,...,x_n\}$ to be clustered as an undirected complete edge-weighted graph $G = (V, E)$, where $V = \{v_1,...,v_n\}$ is the vertex set and $E = V \times V$ is the edge set. Vertices in G correspond to instances (vertex v_i for instances x_i), edges represent neighborhood relationships, and edge-weights reflect dissimilarity between pairs of linked vertices. The graph G is traditionally represented with the corresponding weighted dissimilarity matrix, which is the $n \times n$ symmetric matrix $D=\{d_{j,j'}| x_j, x_{j'} \in X\}$. A common informal definition states that "a cluster is a set of entities which are *similar*, and entities from different clusters are not *similar*". Hence, the edges between two vertices within one cluster should be small weighted (denoting weak dissimilarity), and those between vertices from two clusters should be large weighted (high dissimilarity). The clustering problem is hence formulated as a graph *b-coloring* problem. The *b-coloring* of such a *complete graph* is not interesting for the clustering problem. Indeed, the *trivial partition* is returned where each cluster (*color*) is assumed to contain one and only one instance (*vertex*). Consequently, our clustering approach requires to construct a *superior threshold graph*, which is a partial graph of the original one $G=(V,E)$. Let $G_{>\theta}=(V,E_{>\theta})$ be the superior threshold graph associated with threshold value θ chosen among the dissimilarity table D. In other words, $G_{>\theta}$ is given by $V = \{v_1,...,v_n\}$ as vertex set and $\{(v_i, v_j)| D(x_i, x_j)=d_{ij} >\theta\}$ as edge set.

Table 1. A dissimilarity Matrix

x_i	A	B	C	D	E	F
A	0					
B	0.20	0				
C	0.10	0.20	0			
D	0.20	0.20	0.25	0		
E	0.10	0.20	0.10	0.05	0	
F	0.40	0.075	0.15	0.15	0.15	0

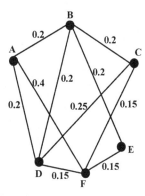

Fig. 1. A threshold graph with $\theta =0.1$ for data in Table 1

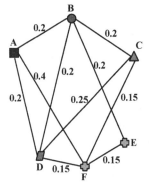

Fig. 2. Initializing the colors of vertices with maximal colors in Fig. 1

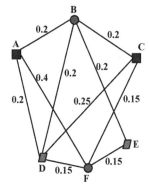

Fig. 3. A *b-coloring graph* constructed from Fig. 2 by removing colors without any dominating vertex

The data to be clustered are now depicted by a *non-complete edge-weighted graph* $G=(V,E_{>\theta})$. In order to divide the vertex set V into a partition $P=\{C_1,C_2,..,C_k\}^1$ where for $\forall\ C_i,C_j \in P$, $C_i \cap C_j=\varnothing$ for $i\neq j$ (when the number of clusters k is not pre-defined), our *b-coloring based clustering algorithm* performed on the graph G consists of two steps: 1) initializing the colors of vertices with maximum number of colors, and 2) removing colors without any dominating vertex using a *greedy procedure*.

As an illustration, let suppose the data set related to the weighted dissimilarity table D in Table 1. Fig. 1 shows the *superior threshold graph* $\theta=0.1$ for Table 1. The edges are labeled with the corresponding dissimilarities. The figures Fig. 2 and Fig. 3 illustrate the two stages of the *b-coloring algorithm* performed on the superior threshold graph $\theta=0.1$. The vertices with the same color (shape) are grouped into the same cluster. Therefore, {A,C}, {B,F}, {D,E} are the clusters, and the nodes with bold letter are the dominating vertices.

The clustering algorithm is iterative and performs multiple runs, each of them increasing the value of the dissimilarity threshold θ. Once all threshold values passed, the algorithm provides the optimal partitioning (corresponding to one threshold value θ_o) which maximizes *Dunn's generalized index* ($Dunn_G$) [6]. $Dunn_G$ is designed to offer a compromise between the *intercluster separation* and the *intracluster cohesion*. So, it is the more appropriated to partition data set in *compact* and *well-separated* clusters. As an illustration, successive threshold graphs are constructed for each threshold θ selected from the dissimilarity Table 1, and our approach is used to give the *b-coloring* partition of each graph. The value of the *Dunn's generalized index* is computed for the obtained partitions. We conclude that the partition $\theta=0.15$ has the maximal $Dunn_G$ among other ones with different θ.

3   The Constrained b-Coloring Clustering Algorithm

This section is devoted to discuss our modification to the *b-coloring* based clustering algorithm. In the sequel, we define two forms of instance-level hard constraints we are using. We then show our investigation to incorporate this kind of constraints into a *b-coloring clustering algorithm*.

3.1   Constraints

Let $X=\{x_1,...,x_n\}$ denotes the given set of instances which must be partitioned such that the number of clusters is not given beforehand. In the context of clustering algorithms, instance-level constraints are a useful way to express a priori knowledge that constrains a placement of instances into clusters. In general, constraints may be derived from partially labeled data or from background knowledge about the domain of real data set. We consider the clustering problem of the data set X under the following types of constraints.

– *Must-Link* constraints denoted by $ML(x_i,x_j)$ indicates that two instances x_i and x_j must be in the same cluster.

---

[1] The notation of P is used to both represent a set of clusters as well as a set of colors, because each cluster $C_i \in P$ corresponds to a color in our *b-coloring based clustering algorithm*.

- *Cannot-Link* constraints denoted by $CL(x_i,x_j)$ indicates that two instances x_i and x_j must not be in the same cluster.
- *Transitively derived* Instance-Level constraints from:
  - $ML(x_i,x_j)$ and $ML(x_j,x_k)$ imply $ML(x_i,x_k)$,
  - $ML(x_i,x_j)$, $ML(x_k,x_l)$ and $CL(x_i,x_k)$ imply both $CL(x_i,x_l)$ and $CL(x_j,x_k)$.

3.2 The Constraint Algorithm

In the remainder of this section, we describe the constrained b-coloring clustering approach called COP-*b-coloring* (for *constraint portioning b-coloring*). The algorithm takes in a *data set* $X=\{x_1,...,x_n\}$, a *pairwise dissimilarity table* $D=\{d_{j,j'}| x_j,x_{j'} \in X\}$, a full set of *must-link constraints* (both directly and transitively) denoted by $Con_=$, and a full set of *cannot-link constraints* (both directly and transitively) denoted by $Con_{\neq}$. It returns a partition P of the data set X that satisfies all specified constraints (the number of clusters in not given beforehand).

As mentioned above, the *b-coloring clustering approach* requires a *non-complete edge-weighted graph* to return a partition P of $X=\{x_1,...,x_n\}$ data set. In order to incorporate the *instance-level constraints* into the clustering problem, our changes concern the construction of this *non-complete graph* that will be presented to the *b-coloring* algorithm. For that, we need to introduce the following definition:

Definition 1. A *composite vertex* v' is a subset of instances such that all pairs among these instances appear together in $Con_=$. As an illustration, the two *must-link constraints* $ML(x_i,x_j)$ and $ML(x_j,x_k)$ transitively imply $ML(x_i,x_k)$. Thus, both must-link constraints can be viewed as a single one namely $ML(x_i,x_j,x_k)$. A composite vertex v' is hence identified as a subset $\{x_i,x_j,x_k\}$.

The construction of the *non-complete edge-weighted graph* $G'=(V',E')$ is now formulated using the following instructions:

- Transform the full set of *must-link constraints* $Con_=$ on a *composite vertex* set $V_1' = \{v'_1,...,v'_m\}$. The composite vertices in V_1' are pairwise disjointed. In the other hand, the remaining r instances $(X - \cup_{i=1..m} v'_i)$ which are not involved in any *must-link constraint* are affected to new r composite vertices $V_2' = \{v'_{m+1},...,v'_r\}$. Finally V_1' and V_2' are combined into $V'=\{v'_1,...,v'_r\}$ where $r<n$.
- Using the full set of *cannot-link constraints*, for any two composite vertices v'_i and v'_j in V', the edge (v'_i, v'_j) is in E' if there is at least one instance x_i in v'_i and another x_j in v'_j appear together in a *cannot-link constraint* (i.e. $\exists x_i \in v'_i \exists x_j \in v'_j$ *such that* $CL(x_i,x_j)$).
- Using the threshold value θ chosen among the dissimilarity table D, for any two composite vertices v'_i and v'_j in V', the edge $(v'_i, v'_j) \in E'$ if there $\exists x_i \in v'_i$ $\exists x_j \in v'_j$ *such that* $D(x_i,x_j)=d_{ij} >\theta$.

The data set X, the full set of *must-link constraints* $Con_=$, and the full set of *cannot-link constraints* $Con_{\neq}$ are now summarized within an *undirected non-complete graph* $G'=(V',E')$. The main idea consists to apply the *b-coloring based clustering algorithm* [2] on G'. Indeed, the goal is to give an assignment of colors (clusters) to the vertices (*i.e.* a composite vertices) of *G'* so that no two adjacent vertices have the same color, and that for each color class, at least one *dominating vertex* is adjacent to at least one vertex of every other color sets. The color of each composite vertex is then assigned to its members.

Proposition 1. The partition returned by the *b-coloring based clustering algorithm* satisfies all specified constraints (both *must-link* and *cannot-link*).

Proof. Each composite vertex consists of instances such that all pairs among these instances appear together in $Con_=$. The b-coloring algorithm affects a color for each composite vertex which is then assigned to its members. Consequently, the vertices appear in a *must-link constraint* will be placed in the same cluster (color). Thus, the given partition satisfies all *must-link constraints*. Each *cannot-link constraint* (among $Con_{\neq}$) between two instances (x_i, x_j) is transformed as an edge between their composite vertices in *G'*. According to the property of the b-coloring of *G'*, the colors of two adjacent vertices are different. Thus, x_i and x_j can never appear in the same cluster. Therefore, the given partition satisfies all *cannot-link constraints*.

Proposition 2. The incorporation of the instances-level constraints decreases the runtime of the clustering algorithm.

Proof. The *b-coloring based clustering algorithm* in [2] generates the *b-coloring* of any graph *G* (associated with a threshold value θ) in $O(n^2\Delta)$ where *n* is the number of vertices (instances). In our case, the incorporation of the all instance-level constraints transforms the initial graph *G* (with *n* vertices) into a graph *G'* (with *r* vertices) where $r < n$. Therefore, the incorporation of the *instances-level constraints* allow to decrease the complexity of the clustering algorithm to $O(r^2\Delta)$ (*c.f.* Section 4 for illustrations).

4   Experimental Results

In this section, we illustrate our algorithm's performance on several standard UCI data sets [7] and show that incorporating background information in the form of instance-level *must-link* and *cannot-link* constraints improves the clustering accuracy while decreasing runtime.

4.1   Evaluation Metrics

The evaluation of clustering algorithms is always a challenge. In our case, the used UCI data set includes *class information* (label) for each data instance. Since the objective was to perform a semi-supervised classification that correctly identifies the underlying *classes* in the given data when the number of clusters is not pre-defined, we use the predefined *class labels* for the evaluation step. Consequently, as used in [4,5], our evaluation will be based on a label matching scheme called *Rand index* which concerns the *clustering accuracy*.

The returned partition (the *b-coloring* one) will be considered as a relation on the instances: for each pair of instances, they have either the same label or different ones. For a data set with *n* instances, there are $n(n-1)/2$ unique pairs of instances (x_i, x_j), and thus there are $n(n-1)/2$ pairwise decisions reflected in our returned partition. The *Rand index* [8] is defined as:

$$Rand(P, P') = Accuracy = \frac{a+b}{n \times (n-1)/2} \tag{1}$$

where

- P is the correct partition which is produced using the predefined class label.
- P' is the returned partition through the COP-*b-coloring* algorithm.
- a and b are the correct decisions. a is the number of decisions where x_i and x_j are in the same cluster for the partitions P and P'. b is the number of decisions where x_i and x_j are placed in different cluster for P and P'.

In order to examine the effectiveness of the COP-*b-coloring* algorithm, our experimental methodology follows the same principle used in [4,5]. For each data set, the main idea is to generate a number of artificial constraints and compute the *accuracy improvement* as more constraints are included into the COP-*b-coloring* algorithm. The constraints generation is given as follows: for each constraint, we randomly select two instances from the data set and check their labels. If they have the same label, we generate a must-link constraint. Otherwise, we generate a cannot-link constraint.

For an interesting assess of the learning improvements gained with the *COP-b-coloring algorithm*, we try to examine its ability to generalize the constraint information to the unconstrained instances. Thus, we propose to compute, aside the overall accuracy, the one on a *hold-out test set* (a subset of data set composed of instances that are not directly or transitively affected by the constraints). As used in [4,5], this is obtained using 10-fold cross-validation: we generate constraints on nine folds and evaluate performance on the tenth. Both evaluation metrics (*overall accuracy* and *hold-out test set accuracy*) are determined by averaging the results given from 100 trials conducted on each used data set where a trial is one 10-fold cross-validation run.

We note that the Euclidian distance is applied to define the dissimilarity level between two instances characterized with m features a_f ($f \in \{1...m\}$) as given by the following formula:

$$d_{i,j} = D(x_i, x_j) = \left(\sum_{f=1}^{m} \left(\frac{g_f(a_{i,f}, a_{j,f})}{m_f} \right)^2 \right)^{1/2} \tag{2}$$

where m_f is the *normalized coefficient* for the attribute a_f and g_f is the comparative dissimilarity function between the two attribute values $a_{i,f}$ and $a_{j,f}$ corresponding respectively to the instances x_i and x_j.

For *numeric* items, g_f is: For *categorical* items, g_f is:

$$g_f(a_{i,f}, a_{j,f}) = |a_{i,f} - a_{j,f}| \qquad g_f(a_{i,f}, a_{j,f}) = \begin{cases} 1 & \text{iff } a_{i,f} \neq a_{j,f} \\ 0 & \text{iff } a_{i,f} = a_{j,f} \end{cases} \tag{3}$$

4.2 Results

We report here our experiments using four relevant benchmark data sets chosen from UCI database [7]. We suppose that the number of cluster k is not given beforehand. Thus, the clustering algorithm was required to select the best value of k using the $Dunn_G$.

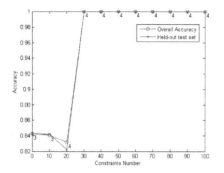

Fig. 4. *COP-b-coloring* results on soybean

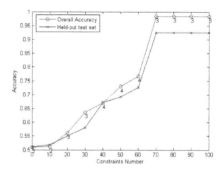

Fig. 5. *COP-b-coloring* results on mushroom

4.2.1 soybean data set contains 47 instances with 35 features and 4 output classes. The Figure 4 provides the clustering accuracy with a varying number of directly added constraints from 5 to 100. It gives also the number of cluster identified at each step. Without any constraints, the *graph b-coloring clustering algorithm* achieves an accuracy of 84%. The overall accuracy reaches 100% after 30 random constraints which attain 5.4% in isolation. Likewise, hold-out accuracy improves and achieving 100% with 30 constraints. Then, we deduce that incorporating 30 constraints achieves a 16% increase in accuracy. Moreover, our algorithm produces best results than COP-COBWEB and COP-KMEANS which achieve a held accuracy respectively of 96% and 98% for 100 random constraints[2].

4.2.2 mushroom data set contains a sample of 50 selected instances (using a *proportionate stratified sampling*) with 21 *categorical* attributes. A record also contains a *poisonous* or *edible* label for the mushroom. Lacking constraints, the *graph b-coloring clustering algorithm* achieves an accuracy of 51%. After incorporating 100 constraints (which attain 71% in isolation) the overall accuracy reaches 98%. Hold-out accuracy climbs to 93% (82% for COP-KMEANS and 83% for COP-COBWEB) yielding an improvement of 42% over the baseline (*c.f.* Fig. 5).

4.2.3 tic-tac-toe data set contains 100 instances, each described by 9 categorical attributes. The instances were also classified into two classes. COP-b-coloring starts at 48% accuracy with no constraints, reaching an overall accuracy of 95% and a hold-out of 82% with 500 random constraints (56% for COP-KMEANS and 49% for COP-COBWEB) and yielding an improvement of 34% over the baseline (*c.f.* Fig. 6). The set of 500 random constraints achieves 70% accuracy before any clustering occurs.

---

[2] We note that the results of COP-COBWEB and COP-KMEANS algorithms are given from [5] and not be reproduced in this paper.

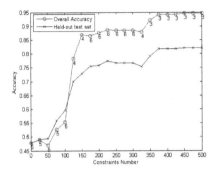 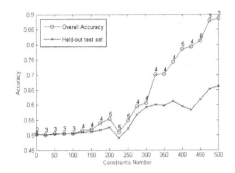

Fig. 6. *COP-b-coloring* results on tic-tac-toe   **Fig. 7.** *COP-b-coloring* results on cleve

4.2.4   cleve data set is very interesting due to its real and mixture appearance. It consists of 303 instances of heart disease (generated at the Cleveland Clinic on 1988) with 13 features. There are 5 numeric and 8 categorical attributes. The instances were also classified into two classes each class is either healthy (buff) or with heart-disease (sick). In the absence of constraints, the *graph b-coloring clustering algorithm* achieves an accuracy of 50%. After incorporating 500 constraints the overall accuracy reaches 89%. Here, 500 random constraints achieve 54% accuracy before any clustering occurs. Hold-out accuracy climbs to 66% yielding an improvement of 16% over the baseline (*c.f.* Fig. 7).

The following Table 2 provides a comparison on runtime of COP-b-coloring in the absence of constraints to the runtime when including 10, 50, and 100 constraints. As viewed in Section 3, because constraints transforms the initial graph G into G' with a smaller number of vertices, the runtime of COP-b-coloring decreases significantly as more constraints are incorporated. Note that the runtime depends also on the pairwise dissimilarity matrix D from the data set due to the multiple runs of the algorithm, each of them increasing the value of the dissimilarity threshold θ selected among D.

Table 2. Runtime comparison (in seconds) in the presence of 0, 10, 50, and 100 constraints

Data set	# instances	# of thresholds θ	0	10	50	100
Soybean	47	77	0.08	0.07	0.05	0.04
Mushroom	50	18	0.05	0.04	0.02	0.01
tic-tac-toe	100	7	0.06	0.05	0.03	0.02
cleve	303	2355	3.57	3.43	3.18	2.86

5   Conclusion

This paper has proposed an extension of the *b-coloring based clustering approach* to incorporate two types of *Instance-Level constraints*. We have shown significant improvements in accuracy and runtime as demonstrated by the results obtained over four *UCI data sets* in the form of *overall* and *hold-out accuracy criterions*. We have concluded that combining the power of clustering with background information achieve better performance than either in isolation. A real advantage of this method is that it performs a semi-supervised classification that correctly satisfies all specified constraints when the number of clusters is not pre-defined and without any exception on the type of data (as long as a dissimilarity table can be constructed). The obtained results have also illustrated the efficiency of our COP-b-coloring algorithm to generate better results than COP-KMEANS and COP-COBWEB algorithms.

References

[1] Jain, A.K., Murty, M.N., Flynn, P.J.: Data Clustering: A Review. ACM Computing Surveys 31, 264–323 (1999)
[2] Elghazel, H., Deslandres, V., Hacid, M-S., Dussauchoy, A., Kheddouci, H.: A new clustering approach for symbolic data and its validation: Application to the healthcare data. In: Esposito, F., Raś, Z.W., Malerba, D., Semeraro, G. (eds.) ISMIS 2006. LNCS (LNAI), vol. 4203, pp. 473–482. Springer, Heidelberg (2006)
[3] Irving, W., Manlove, D.F.: The b-chromatic number of a graph. Discrete Applied Mathematics 91, 127–141 (1999)
[4] Wagsta, K., Cardie, C.: Clustering with instance-level constraints. In: Proceedings of the 17th International Conference on Machine Learning, pp. 1103–1110 (2000)
[5] Wagsta, K., et al.: Constrained K-means Clustering with Background Knowledge. In: Proceedings of the 18th International Conference on Machine Learning, pp. 577–584 (2001)
[6] Kalyani, M., Sushmita, M.: Clustering and its validation in a symbolic framework. Pattern Recognition Letters 24(14), 2367–2376 (2003)
[7] Blake, C.L., Merz, C.J.: UCI repository of machine learning databases. In: University of California, Irvine, Dept. of Information and Computer Sciences (1998), Available from http://www.ics.uci.edu/ ̃mlearn/MLRepository.html
[8] Rand, W.M.: Objective criteria for the evaluation of clustering methods. Journal of the American Statistical Association 66, 846–850 (1971)

An Efficient Algorithm for Identifying the Most Contributory Substring

Ben Stephenson

Department of Computer Science, Room 355 Middlesex College, University of
Western Ontario, 1151 Richmond Street, London, Ontario, Canada N6A 5B7
ben@csd.uwo.ca

Abstract. Detecting repeated portions of strings has important applications to many areas of study including data compression and computational biology. This paper defines and presents a solution for the
Most Contributory Substring Problem, which identifies the single substring that represents the largest proportion of the characters within a
set of strings. We show that a solution to the problem can be achieved
with an $O(n)$ running time (where n is the total number of characters in
all of the input strings) when overlapping occurrences of the most contributory substring are permitted. Furthermore, we present an extended
algorithm that does not permit occurrences of the most contributory substring to overlap. The expected running time of the extended algorithm
is $O(n \log n)$ while its worst case performance is $O(n^2)$.

1 Introduction

This paper considers the problem of determining the most contributory substring
of a set of strings. We define the most contributory substring to be the string
of characters w such that the number of occurrences of w times its length, $|w|$,
is maximal. Put another way, the most contributory substring is the string w
such that removing all occurrences of w from the set of strings reduces the size
of the set by the greatest number of characters. Being able to identify the most
contributory substring is useful because it can aid in pattern matching tasks
such as analyzing profile data and identifying strings that should be replaced
with short code words in compression algorithms.

The remainder of this paper is organized in the following manner. Sections 2
and 3 outline related problems and define the terms used in the remainder of the
paper. Our initial algorithm is presented in 4. It is followed by an extended version of the algorithm which ignores overlapping substrings in Section 5. Section
6 discusses applications for this algorithm. We briefly identify areas of future
work and summarize in Section 7.

2 Related Work

The most contributory substring problem is related to two other well known
problems: the Longest (or Greatest) Common Substring Problem and the All
Maximal Repeats Problem (AMRP). However it is distinct from each of these.

I.Y. Song, J. Eder, and T.M. Nguyen (Eds.): DaWaK 2007, LNCS 4654, pp. 272–282, 2007.

While the Longest Common Substring Problem identifies the longest substring common across all strings in a set, the Most Contributory Substring Problem identifies the string that contributes the greatest number of characters. One important distinction between these problems is that there is no guarantee that the most contributory substring will occur in every string in the set while the longest common substring will. The longest common substring problem also ignores multiple occurrences of the same substring within each string in the set while the Most Contributory Substring Problem counts every occurrence.

A variation on the Longest Common Substring Problem has also been presented [4] that identifies the longest common substring to k strings in a set. This problem is also distinct from the Most Contributory Substring Problem because it fails to consider the value of multiple occurrences of a substring within one string in the input set.

The All Maximal Repeats Problem identifies occurrences of identical substrings α and β in a string s such that the characters to the immediate left and right of α are district from the characters to the immediate left and right of β. In contrast, the Most Contributory Substring Problem is concerned with identifying strings that occur many times. Each occurrence of the string may or may not have distinct characters to its immediate left and right.

3   Definitions

A substring w of a string s is defined to be a string for which there exist strings (possibly empty) p and q such that $s = pwq$. A suffix, w of a string s is defined to be a non empty string meeting the constraint $s = pw$ for some (possibly empty) string p.

Let $\mathcal{L}$ represent the set of m strings under consideration. We will denote a specific string within the set as $\mathcal{L}_i$ such that $0 \leq i < m$. Let $\Sigma_0...\Sigma_{m-1}$ represent the alphabet employed by each of the m strings in $\mathcal{L}$. The alphabet used across all strings is constructed as $\Sigma_0 \cup ... \cup \Sigma_{m-1}$ and will be denoted by Σ.

A generalized suffix tree is a data structure constructed from a collection of strings. An equivalent tree can be constructed for a single string that is the concatenation of all of the strings in the set provided that a unique *sentinel* character is placed after each string. The set of sentinel characters is denoted by $\mathcal{S}$. In our examples we will represent sentinel characters with the punctuation marks # and $. We will use the symbol s to represent the string generated by concatenating the strings in $\mathcal{L}$ with a distinct sentinel character between each string. Because a sentinel character is introduced after each input string during the construction of s, $|\mathcal{L}|$ is equal to $|\mathcal{S}|$.

Several linear time and space suffix tree construction algorithms have been developed [6, 7, 8, 9] for alphabets that are a small, constant size compared to the length of the input string. Another algorithm was developed subsequently that provided the ability to construct a suffix tree in linear time and space over integer alphabets [3]. Assuming that the string ends in a sentinel character, the

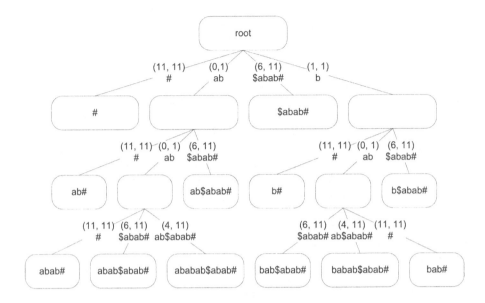

Fig. 1. Suffix Tree for the string *ababab$abab#*

following properties hold for a suffix tree independent of the algorithm used to construct it:

1. The tree contains n leaves where each leaf corresponds to one of the n suffixes of a string of length n.
2. Each branch in the tree is labeled with one or more characters from the string. The label is represented by two integers which specify the starting and ending position as indices in the concatenated string. Collecting the characters as one traverses from the root of the tree to a leaf will give one of the suffixes of the string.
3. The first character on each branch leaving a node is distinct. As a result, no node has more than $|\Sigma| + |\mathcal{S}|$ children.
4. Each non-leaf node has a minimum of two children.
5. The string depth of a leaf node is the length of the suffix of s represented by that node. The string depth of an interior node is the length of the prefix that is common to all of the suffixes represented by leaf nodes below it.

Figure 1 shows a suffix tree constructed for the strings *ababab* and *abab* after they have been merged into a single string and the sentinel characters have been added. The suffix generated by traversing the tree is shown in each leaf node.

4   Initial Algorithm

This section presents an initial algorithm for solving the most contributory substring problem which permits overlapping occurrences of the identified string.

It is subsequently extended to consider only non-overlapping occurrences of the most contributory substring in Section 5.

Our algorithm for determining the most contributory substring begins by creating a suffix tree for s using one of the linear time construction algorithms published previously. Once the suffix tree is constructed it is transformed so that all of the strings in $\mathcal{L}$ are represented by an interior node, and so that all strings containing a sentinel character are represented by a leaf node. Any string containing a sentinel character occurs due to the construction of s. Consequently, the identification process can disregard any string represented by a leaf node while ensuring that every substring of a string in $\mathcal{L}$ is considered. The following steps are taken to transform the tree to meet these constraints:

- Any leaf node that is reached by a branch label that begins with a sentinel character is left untouched.
- Any leaf node that is reached by a branch labeled with a string that begins with a letter in Σ is split into two nodes. A new interior node with only one child is inserted between the leaf and its parent. The branch to the new node is labeled with all characters before the first sentinel character in the original branch label. The branch from the new node to the leaf node is labeled with the remaining characters.

Once this transformation has been performed the number of nodes in the tree has increased by at most n, retaining a total number of nodes that is $O(n)$. The algorithm employed is shown in Figure 2. The effect of this transformation can be seen in the bottom-most nodes in Figure 3.

The SplitLeaves transformation can be performed in $O(n \log |\mathcal{S}|)$ time if the positions of the sentinel characters are recorded when the concatenated string is formed. Then finding the position of the first sentinel character can be implemented as a binary search for the first value in the sentinel character position list that is greater than the starting index of branch b. This strategy requires $O(s)$ additional space to store the list of sentinel characters.

An alternative strategy is to build a table which maps each position in s to the position of the next sentinel character. This table can be constructed in linear time during the construction of s and requires additional space that is $O(n)$. With this table, finding the position of the first sentinel character in the label be can be accomplished in constant time. This results in overall time complexity for SplitLeaves which is $O(n)$. Which of these strategies is superior depends on the size of $\mathcal{S}$ and the trade-off between space and time costs for the specific context in which the algorithm is employed.

A depth first traversal of the tree is performed once the leaf nodes have been split. This traversal assigns a score to each node, determined by computing the product of the string depth of the node and the number of leaf nodes below it. Depending on the application, this score can be recorded in the node and utilized later or two variables can be used to record the best string and score encountered so far.

Figure 3 shows Figure 1 after scoring has been completed. It shows the values used to compute the score in addition to the score awarded for each node. From

```
Algorithm SplitLeaves(node n)

For each branch, b, leaving n
  If the node, m, at the end of b is a leaf and the label on b starts with a
  character in Sigma
    Create a new node, p
    Create a branch, c, from p to m
    Change the target of branch b to node p

    Find the position, x, of the first sentinel character in the label on b

    Label c with all characters from x to the end of the string
    Change the label on b so that it only includes those characters before x

  Else
    SplitLeaves(b.Target)
End For
```

Fig. 2. Algorithm SplitLeaves

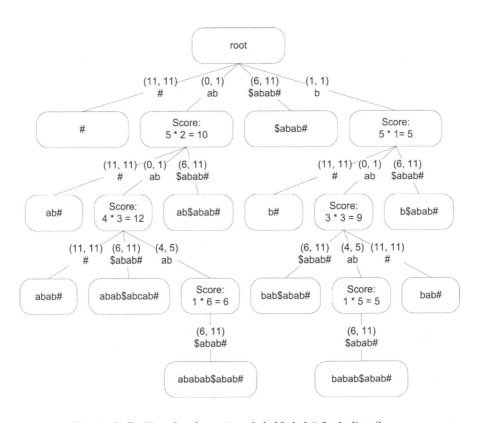

Fig. 3. Suffix Tree for the string *ababab$abab#* Including Scores

this diagram we can see that the node corresponding to the string *abab* received the highest score with the value 12. As a result, we would conclude that the string *abab* is the most contributory substring for the input strings *ababab* and *abab*. The occurrences of the string are indicated below using underlining and over-lining.

$$a\overline{ba}bab\$\underline{abab}\#$$

5   Handling Overlapping Occurrences

It is important to observe that the single underlined and single over-lined occurrences of *abab* are overlapping. While this may not be a problem in some situations, any application that 'uses up' the string once it is identified will destroy the subsequent overlapping occurrence(s). Consequently the current identification technique will over estimate the value of some strings if overlapping should be prohibited. This section extends the preceding algorithm so that it detects and ignores overlapping occurrences of the string.

Two occurrences of a substring are known to overlap if the absolute value of the difference in their starting positions within *s* is less than the length of the substring. We annotate each interior node in the tree with the score based on the number of non-overlapping occurrences of the substring represented by the node using the depth first traversal algorithm shown in Figure 4.

During this traversal, a balanced binary search tree is constructed for each node in the suffix tree. It contains the starting indexes of the substring represented by the node in the suffix tree. For leaf nodes, the binary search tree consists of a single node. At each interior node, the balanced binary search trees from its children are merged to form a new larger tree. Each tree from a child is merged in sequence, with the smaller trees being merged into the largest tree.

In the pathological case where the height of the suffix tree is n, the amount of time required to merge the binary search trees is $O(\log n)$ for each level in the tree because only one new node is merged, giving an overall complexity of $O(n \log n)$. A similar situation results in the pathological case where $n = |\Sigma| + |\mathcal{L}|$. In this case, the height of the suffix tree is 1, but n leaf nodes must be merged into the tree at a cost that is $O(\log n)$ for each merge. As a result, this case also has a complexity which is $O(n \log n)$.

A myriad of suffix trees exist with heights between the pathological cases described in the previous paragraph. However, the amount of time required to construct the balanced binary search trees is also $O(n \log n)$ for these cases. The number of merges is minimized by merging the smaller trees into the largest tree at each merge point, resulting in each merge still having a cost that is $O(\log n)$. Due to the construction of a suffix tree, it is known that the size of the largest binary search tree will grow by at least one at each level in the tree (since each interior node in the suffix tree has at least two children except for the split leaf nodes). When the binary search tree grows by only one element at each level, the tree has height n and the overall time required to build the binary search trees is $O(n \log n)$ as described previously. If the size of the largest binary search tree grows by more than 1 at any point, then the height of the suffix tree is known to be less than n. In fact, if the increase in size of the largest binary search tree is denoted by Δ_j at level j then the maximum height of the suffix tree is $n - \sum(\Delta_j - 1)$, where $\sum$ indicates summation in this instance. While any level that has a value of Δ_i that is greater than 1 requires multiple merges that each cost $O(\log n)$, the additional merges performed at that level are offset by the decreased height of the tree, resulting in an overall running time that remains $O(n \log n)$.

Once the binary search tree for a node is constructed, it is traversed using algorithm CountUniqueOccurrences, shown in Figure 5. This algorithm is responsible for determining if two or more of the occurrences of the substring represented by suffix tree node **n** overlap. This is accomplished by traversing the balanced binary search tree using an in-order traversal, counting only those nodes whose value differs by at least **length**. This value is returned to the NonOverlappingScore algorithm, which uses it to compute the score for the node.

Since each starting position can only exist in one binary search tree at each level in the suffix tree, the total time required to traverse all of the binary search trees at a level is $O(n)$. Unfortunately, this leads to worst case running time for CountUniqueOccurrences which is $O(n^2)$ in those rare cases when the height of the suffix tree approaches n. Studies have been conducted that show, contrary to the worst case height, the expected height of a suffix tree is $O(\log n)$ [1, 2, 5]. Consequently, while it is possible that CountUniqueOccurrences may cause the overall running time of NonOverlappingScore to reach $O(n^2)$, its expected performance should not exceed the $O(n \log n)$ performance of CountUniqueOccurrences. This expected performance exceeds the cost associated with constructing the suffix tree initially $(O(n))$, and splitting its leaf nodes $(O(n \log |\mathcal{S}|))$. Thus we conclude that the overall cost of determining the Most Contributory Substring when overlapping occurrences are prohibited is $O(n^2)$ in the worst case, with an expected performance of $O(n \log n)$.

The NonOverlappingScore algorithm is $O(n)$ in space. In a degenerate suffix tree with height n, it is necessary to allocate the ChildTrees array for every level in the tree before any call to NonOverlappingScore returns. Initially, this may appear to require $O((|\Sigma| + |\mathcal{L}|)n)$ space. However, it is impossible for every node in the tree to to have $|\Sigma| + |\mathcal{L}|$ children unless $|\Sigma| + |\mathcal{L}|$ is three or less. This constraint exists because the total number of nodes in the modified suffix tree is bounded by $3n$. Thus, even if we must allocate space to hold a pointer to every child's tree before any calls to NonOverlappingScore return, the total amount of space allocated with still be $3n$ pointers or less.

The amount of space required to store the binary trees used to identify the unique occurrences is also $O(n)$. The total number of nodes in the binary trees at any level in the suffix tree is bounded by n. Furthermore, binary trees only exist at two levels in the suffix tree at any time. Once the binary tree for the current node in the suffix tree is constructed, the binary trees for all of its children are deallocated. This means that a total of no more than $2n$ nodes will be present

```
Algorithm NonOverlappingScore(node n)

ChildTrees: an array of pointers to binary search trees
Retval: a pointer to a balanced binary search tree

If n.NumChildren == 0  // it's a leaf node
  Retval = new binary search tree with one node
  Set the value in Retval's root node to |s| - n.StringDepth
  Return Retval
End If


Allocate n.NumChildren pointers for ChildTrees  // Proceed depth first
For i = 0 to n.NumChildren - 1                   // through the suffix tree
  ChildTrees[i] = NonOverlappingScore(n.Child[i])
End For


For i = 1 to n.NumChildren - 1 // Ensure ChildTrees[0] points to largest tree
  If ChildTrees[0].num_nodes < ChildTrees[i].num_nodes
    swap(ChildTrees[0], ChildTrees[n])
  End If
End For


Retval = ChildTrees[0]
For i = 1 to n.NumChildren - 1 Retval = BSTreeMerge(Retval, ChildTrees[i])
n.Score = CountUniqueOccurrences(Retval,n.StringDepth) * n.StringDepth


Deallocate the binary tree for each child and the ChildTrees array
return Retval
```

Fig. 4. Algorithm NonOverlappingScore

```
Algorithm CountUniqueOccurrences(BinarySearchTree t, Integer str_len)

count = 0                 // number of non-overlapping occurrences
last_position = MIN_INT   // last_position is initialized to the
                          // negative value of largest magnitude
InOrderTraversal(t.root, count, last_position)
return count



Algorithm InOrderTraversal(BSTNode b, Ref Integer count,
                           Ref Integer last_position)

If (b.left != NULL) InOrderTraversal(b.left, count, last_position)
If ((last_position + str_len) <= b.value)
  count++
  last_position = b.value
End If
If (b.right != NULL) InOrderTraversal(b.right, count, last_position)
```

Fig. 5. Algorithm CountUniqueOccurrences

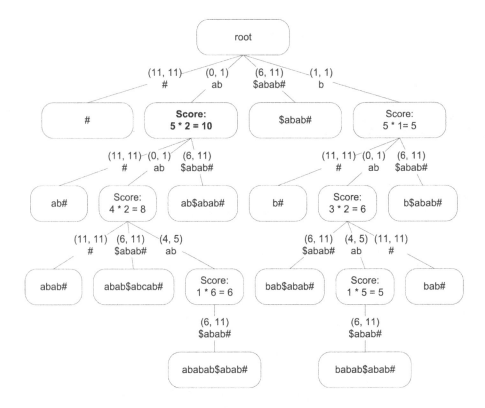

Fig. 6. Suffix tree for the string *ababab$abab#* Including Score Computed Using Only Non-Overlapping Occurrences

in the binary trees before the children's binary trees are deallocated, giving a space requirement that is $O(n)$.

After algorithm NonOverlappingScore executes the number of disjoint occurrences of each substring has been determined and the resulting score has been computed. Figure 6 shows the tree after scoring has been performed using the NonOverlappingOccurrences algorithm. It shows that the best substring is *ab* with a score of 10 – a different result than the *abab* achieved when overlapping strings were permitted.

6   Applications

This algorithm was originally developed to analyze profile data collected during the execution of an application. The profile data consisted of approximately 200 distinct events which occurred millions of times. While determining which *distinct* event occurred with greatest frequency was straightforward, determining which *sequence* of events contributed the most to the execution of the application was more challenging. Analyzing this data using an inefficient algorithm took a prohibitively large amount time. As a result, the most contributory substring problem was identified and solved. Using the solution to the most contributory substring problem made it possible to quickly identify what sequence or sequences of events contributed the most to the execution of the application.

We also believe that using this algorithm can improve the level of compression achieved by Huffman coding. In its simplest form, Huffman coding uses variable length bit sequences to represent each symbol. Shorter sequences of bits are used to represent those symbols that occur with greater frequency while longer sequences are used to represent symbols that occur less frequently.

A variation on Huffman coding has also been developed which considers fixed length groups such as 2 symbols at a time. The frequency of each sequence of 2 symbols is determined and bit sequences are assigned based on the frequencies of the sequences. Using the most contributory substring algorithm can extend this idea further by using variable length sequences. The most contributory substring can be identified repeatedly in order to determine what sequence (possibly consisting of only one symbol) represents the largest portion of the string being compressed. This sequence would then be encoded using the shortest code word. Using this strategy degenerates to standard Huffman encoding when each most contributory substring identified has length one. We have yet to implement this variation of Huffman coding, so we are presently unsure how much of an improvement is achieved for real data sets.

Other applications for this algorithm may also exist. In particular, we believe that it may have utility in the realm of computational biology.

7   Conclusion

An algorithm is presented which solves the Most Contributory Substring Problem. This problem identifies a substring of its input that represents the largest proportion of the characters in the input string. In its first presentation, the algorithm identified potentially overlapping occurrences of the most contributory substring in a running time that was $O(n)$. An extended version of the algorithm was also presented that discounted overlapping occurrences of the most contributory substring. While the extended algorithm may require a running time as large as $O(n^2)$ for pathological input cases, previous studies have shown the expected height of a suffix tree is $O(\log n)$ rather than $O(n)$, reducing the complexity of the Most Contributory Substring Algorithm to $O(n \log n)$ in the expected case.

While our algorithm gives an optimal result when one substring is identified, it does not necessarily give an optimal result when applied iteratively to identify the k most contributory substrings of s. We hope to extend our algorithm to efficiently solve this problem in the future.

References

[1] Apostolico, A., Szpankowski, W.: Self-alignments in words and their applications. J. Algorithms 13(3), 446–467 (1992)

[2] Devroye, L., Szpankowski, W., Rais, B.: A note on the height of suffix trees. SIAM J. Comput. 21(1), 48–53 (1992)

[3] Farach, M.: Optimal suffix tree construction with large alphabets. In: FOCS '97: Proceedings of the 38th Annual Symposium on Foundations of Computer Science (FOCS '97), Washington, DC, USA, 1997, p. 137. IEEE Computer Society Press, Los Alamitos (1997)

[4] Gusfield, D.: Algorithms on strings, trees, and sequences: computer science and computational biology. Cambridge University Press, New York, NY, USA (1997)

[5] Jacquet, P., Szpankowski, W.: Autocorrelation on words and its applications: Analysis of suffix trees by string-ruler approach. JCTA: Journal of Combinatorial Theory, Series A, 66 (1994)

[6] McCreight, E.M.: A space-economical suffix tree construction algorithm. J. ACM 23(2), 262–272 (1976)

[7] Ukkonen, E.: Constructing suffix trees on-line in linear time. In: Proceedings of the IFIP 12th World Computer Congress on Algorithms, Software, Architecture - Information Processing '92, vol. 1, pp. 484–492, North-Holland (1992)

[8] Ukkonen, E.: On-line construction of suffix trees. Algorithmica 14(3), 249–260 (1995)

[9] Weiner, P.: Linear pattern matching algorithms. In: Proceedings of the 14th IEEE Symp on Switching and Automata Theory, pp. 1–11. IEEE Computer Society Press, Los Alamitos (1973)

Mining High Utility Quantitative Association Rules

Show-Jane Yen and Yue-Shi Lee

Department of Computer Science and Information Engineering, Ming Chuan University
5 The-Ming Rd., Gwei Shan District, Taoyuan County 333, Taiwan
{sjyen,leeys}@mcu.edu.tw

Abstract. Mining *weighted association rules* considers the profits of items in a transaction database, such that the association rules about important items can be discovered. However, high profit items may not always be high revenue products, since purchased quantities of items would also influence the revenue for the items. This paper considers both profits and purchased quantities of items to calculate utility for the items. Mining *high utility quantitative association rules* is to discover that when some items are purchased on some quantities, the other items on some quantities are purchased too, which have high utility. In this paper, we propose a data mining algorithm to find *high utility itemsets* with purchased quantities, from which high utility quantitative association rules also can be generated. Our algorithm needs not generate candidate itemsets and just need to scan the original database twice. The experimental results show that our algorithm is more efficient than the other algorithms which only discovered high utility association rules.

1 Introduction

The previous approaches [3, 8, 9, 10] which discovered high utility association rules cannot generate high utility quantitative association rules. Mining *quantitative association rules* [1, 5, 7] is to discover that most of the customers purchased some items on some quantities, also purchased the other items on some quantities. For an item, different purchased quantities for the item are regarded as different *quantitative items*. For example, three cakes and four cakes are different quantitative items. There may be many different purchased quantities for each item in the database. It is very costly to process a large number of quantitative items. In order to reduce and speed up the computation time and generate useful information, the purchased quantities for each item need to be combined to some ranges. However, the combinations for quantities of items may lead to information loss. Besides, mining quantitative association rules just considered frequently purchased quantities of items, but not considered profits or importance of items, such that association rules about high profit items cannot be discovered.

This paper considers both profit and purchased quantities for each item to compute utilities for quantitative itemsets, and proposes an algorithm to find high utility quantitative itemsets. High utility quantitative association rules can be discovered from high utility quantitative itemsets. An example of such a rule is (bread, 3, 4) $\Rightarrow$ (milk, 2, 3), which means that most of customers purchased three or four breads, also

I.Y. Song, J. Eder, and T.M. Nguyen (Eds.): DaWaK 2007, LNCS 4654, pp. 283–292, 2007.
© Springer-Verlag Berlin Heidelberg 2007

purchased two or three milk. We can use this information to package products with quantities that have high utility and estimate the number of items which need to be reserved according to the number of the other items.

For mining high utility quantitative association rules, a criterion of high utility need to be specified. If the utility for a quantitative itemset satisfy the criterion, then the quantitative itemset is a high utility quantitative itemset. However, there may be different utilities for the items with the same profit, since the purchased quantities for the items may be different. For a quantitative itemset which does not satisfy the criterion, it may satisfy the criterion by combining the quantities with the adjacent quantities of the quantitative itemset. Therefore, we propose a combination method to combine the quantities for an itemset, such that useful information can be discovered.

Besides, the subsets for high utility quantitative itemsets may not be high utility, that is, there is no downward closure property for the problem of mining high utility quantitative itemsets. We also propose a *support bound method* to predict if a quantitative itemset can be joined with the other quantitative itemsets to generate high utility quantitative itemsets. Our algorithm just needs to scan original database twice to generate small datasets, and use the support bound method to reduce the number of datasets. High utility quantitative itemsets can be discovered from these small datasets without generating candidate quantitative itemsets. Hence, our algorithm is very efficient than the other approaches [3, 8, 9, 10] which just discovered high utility itemsets without quantity information.

2   An Algorithm for Mining Quantitative Association Rules

In this section, we present our algorithm HUQA (High Utility Quantitative Association Rules) for mining high utility quantitative association rules. Table 1 is a transaction database, in which TID denotes transaction identifier. Each transaction contains items purchased in this transaction and their purchased quantities. The weights for each item are shown in Table 2.

Table 1. A transaction database

TID	Transaction	TID	Transaction
1	(A,2)(B,5)(C,2)(D,1)	7	(A,3)(C,2)(D,1)
2	(B,4)(C,1)(D,1)	8	(B,6)(C,1)
3	(A,2)(B,6)(C,2)	9	(A,3)(B,6)(C,1)(D,1)
4	(B,5)(C,1)	10	(B,5)
5	(B,4)(C,2)	11	(A,2)(B,6)(C,2)(D,1)
6	(A,2)(B,6)(C,1)(D,1)	12	(A,3)(B,4)(C,2)(D,1)

Before introducing our algorithm, we first define and describe the problem of mining high utility quantitative association rules: A quantitative item (q_item) x = (i_p, l, u) is an item with purchased quantities, in which i_p is an item, l and u are lower bound and upper bound of purchased quantities, respectively. For example, q_item x = (B, 2, 4) represents that the purchased quantities of item B are between 2 and 4. If $l = u$, then x can be denoted as (i_p, l), which means that the purchased quantity of item i_p is l. A quantitative itemset (q_itemset) is a set of q_items. *The length of a q_itemset* is the number of q_items contained in the q_itemset. A q_itemset with length k is called k_q_itemset. For example, q_itemset X = {(A, 1)(B, 5)} means that the item A and item B were purchased together, and the purchased quantities for A and B are 1 and 5, respectively. The length of q_itemset {(A, 1)(B, 5)} is 2, which is a 2_q_itemset.

Table 2. The weight for each item

Item	Weight
A	0.8
B	0.3
C	0.2
D	0.4

For any two q_items x = (i_1, l_1, u_1) and y = (i_2, l_2, u_2), if $l_1 \le l_2$ and $u_1 \ge u_2$, then x contains y. For any two q_itemsets X = {$(x_1, l_1, u_1)(x_2, l_2, u_2) \dots (x_n, l_n, u_n)$} and Y = {$(y_1, p_1, q_1)(y_2, p_2, q_2) \dots (y_m, p_m, q_m)$}, if there exists $1 \le i_1, \dots, i_m \le n$ such that $y_1 = x_{i1}$, $y_2 = x_{i2}, \dots, y_m = x_{im}$, $l_{i1} \le p_1, l_{i2} \le p_2, \dots, l_{im} \le p_m, u_{i1} \ge q_1, u_{i2} \ge q_2, \dots, u_{im} \ge q_m$, then q_itemset X contains q_itemset Y. If a q_itemset X contains a subset of a trans- action, then the transaction *supports* X. The *support for q_itemset* X is the ratio of the number of the transactions which support X to the total number of transactions in the database. The *support count for q_itemset* X is the number of the transactions which support X, which is denoted as SC(X).

The utility u(x) of a q_item x = (i, j) is the product of the weight W(i) of item i and the quantity j. For example, u(A,2) = W(A) × 2 = 0.8 × 2 = 1.6. If a transaction T_q supports q_item x and x contains q_item (i_p, j) in transaction T_q, then the utility of q_item x in T_q is u (x , T_q) = u(i_p, j) = W(i_p) × j. *The utility of a q_itemset in a transaction* is the sum of the utilities for all the q_items contained in the q_itemset in the transaction. For example, the utility of q_itemset {(A,1,3)(B,4,6)} in the first transaction in Table 1 is (0.8 × 2)+(0.3 × 5) = 3.1. *The utility of a q_item in a transac- tion database* is the total sum of the utilities of the q_item in all the transactions in the database. The total utility of the transaction database is defined in expression (1), in which D is the set of all the transactions in the database. *The utility support of a q_item* x = {i_p, l, u} is the ratio of the utility of x in the database to the total utility of the transaction database, which is shown in expression (2).

$$TU = \sum_{Tq \in D} \sum_{x \in Tq} u(x, Tq) \tag{1}$$

$$US(i_p, l, u) = \frac{\sum_{j=l}^{u} [SC(i_p, j) \times u(i_p, j)]}{TU} \tag{2}$$

$$US(X) = \frac{\sum_{j_1=l_1}^{u_1} \dots \sum_{j_k=l_k}^{u_k} [SC((i_1, j_1) \dots (i_k, j_k))] \times [u(i_1, j_1) + u(i_2, j_2) + \dots + u(i_k, j_k)]}{TU} \tag{3}$$

The utility of a q_itemset in a transaction database is the sum of the utilities of the q_itemset in all the transactions in the database. *The utility support of a q_itemset* X = $\{(i_1, l_1, u_1)(i_2, l_2, u_2)...(i_k, l_k, u_k)\}$ in a database is the ratio of the utility of X in the database to the total utility of the database, which is shown in expression (3). For example, the utility support of q_itemset {(A,2)(B,5,6)} in Table 1 is computed as follows. Because SC((A,2)(B,5)) = 1 and SC((A,2)(B,6)) = 3; W(A) = 0.8 and W(B) = 0.3 in Table 2, the utilities of q_itemsets {(A,2)(B,5)} and {(A,2)(B,6)} are u((A,2)(B,5)) = 0.8×2 + 0.3×5 = 3.1 and u((A,2)(B,6)) = 0.8×2 + 0.3×6 = 3.4, respectively. The total utility of transaction database in Table 1 is 36.9. Hence, the utility support   of q_itemset {(A,2)(B,5,6)} is US((A,2)(B,5,6)) = [SC((A,2)(B,5)) × u((A,2)(B,5)) + SC((A,2)(B,6)) × u((A,2)(B,6))] / TU = [1×3.1 + 3×3.4] / 36.9 = 0.36. If the utility support for a q_itemset is no less than *user-defined minimum utility support (MUS)* threshold, then the q_itemset is a *high utility q_itemset*.

2.1   Mining High Utility Quantitative Itemsets

Our algorithm HUQA includes two steps for mining high utility q_itemsets: The first step is to find the q_itemsets whose utility supports satisfy minimum utility support. Because the subsets of high utility q_itemsets may not be high utility, we need to find *weak utility q_itemsets* that can be extended to generate high utility q_itemsets by joining with the other q_itemsets. For the first database scan, the support count and the utility support for each q_item are computed, and the high utility q_itemsets and weak utility q_itemsets can be generated. For each transaction in the database, the q_items are sorted decreasingly according to the utilities of the q_items in the first database scan. For the second step, the conditional database for each high (weak) utility q_itemset can be constructed. After constructing conditional databases, high (weak) utility q_itemsets are generated from the conditional databases, which is similar the first step. Hence, HUQA executes the two steps recursively.

Because there are different q_items for an item, there may be too many q_items in the database to generate high utility q_itemsets. We propose a combining method to combine a set of adjacent q_items, which these q_items with adjacent quantities and the same item. For a q_item with single quantity, if the utility support for the q_item satisfies minimum utility support (MUS), then this q_item is a high utility 1_q_itemset. Otherwise, this q_item is considered to be combined with the other q_items with adjacent quantity. If all the q_items with adjacent quantities are combined to form another q_item, then the range of the quantities for the q_item will be large. Hence, a combined threshold needs to be specified to avoid combining the q_item with low utility support. For a set of adjacent q_items $(i, l_1), (i, l_2), ...$ and (i, l_k), if their utility supports are less than MUS and greater than or equal to MUS/r, in which r is a user-specified *quantitative related coefficient* and r≥1, then these q_items can be combined to form another q_item (i, l_1, l_k). If the utility support for the combined q_item is no less than MUS, then this q_item is a high utility q_itemset. If the utility support for a q_itemset is less than MUS/r, then this q_itemset cannot be combined with the other q_itemsets.

For example, suppose that the minimum utility support (MUS) and quantitative related coefficient r are 20% and 2, respectively. Since the utility support for (B, 6) satisfies MUS, (B, 6) is a high utility q_itemset. Because utility supports for (A, 2),

(A, 3), (B, 4) and (B, 5) are no less than MUS/r, these q_items can be combined as (A, 2, 3) and (B, 4, 5). Since the utility supports for (A, 2, 3) and (B, 4, 5) are 0.37 and 0.23, respectively, which are greater than MUS (0.2), the two q_itemset (A, 2, 3) and (B, 4, 5) are also high utility q_itemsets.

A q_itemset X which is not high utility may be extended to generate high utility q_itemsets by joining with the other q_itemsets, because the utilities for the extended q_itemsets are larger than that of q_itemset X. Hence, we construct conditional database for q_itemset X in order to find high utility q_itemsets which contain X. However, if conditional databases are constructed for all q_itemsets, then many conditional databases need to be generated. In order to generate conditional databases as few as possible, we propose a method to find the q_itemsets without high utilities, which can be extended to generate high utility q_itemsets. Expression (4) represents that the utility support for a high utility q_itemset (i, j) is greater than or equal to MUS, which can be transformed into expression (5).

Hence, for a q_itemset which can be extended to generate high utility q_itemsets, the support count for the q_itemset must be no less than the *support bound (SB)* for the q_itemset. The support bound for a q_item (i, j) is shown in expression (6), in which $MaxU (i, j)$ is the largest utility among the utilities of the q_itemsets after q_item (i, j) in the transactions. If the support count for a q_itemset is less than its support bound, then the q_itemset cannot be joined with the other q_itemsets to generate high utility q_itemsets. Hence, in the first database scan, we compute and record the largest utility of all the q_itemsets which can be joined with (i, j) in the transactions of the database. For a q_itemset whose utility support is less than MUS, if support count for the q_itemset is no less than its support bound, then this q_itemset is called *weak utility q_itemset*.

$$\frac{SC(i, j) \times u(i, j)}{TU} \geq MUS \tag{4}$$

$$SC(i, j) \geq \frac{MUS \times TU}{u(i, j)} \tag{5}$$

$$SB(i, j) = \left\lceil \frac{MUS \times TU}{u(i, j) + [MaxU(i, j)]} \right\rceil \tag{6}$$

2.2 Conditional Database Generation

For the first database scan, HUQA finds all the high and weak q_items. For the second database scan, the conditional databases for all the high and weak q_items are constructed. The conditional database for a q_item x is constructed as follows: For each transaction T in the database, if x is in transaction T, then x and all the q_items after x in T are put into conditional database for x. If x is not in T, then T is ignored. After scanning all the transactions, the conditional databases for all the high and weak utility q_items can be constructed.

After constructing conditional databases for high and weak utility k_q_itemsets ($k \geq 1$), HUQA finds all the high and weak (k+1)_q_itemsets from the conditional databases according to the first step. Hence, HUQA just need to scan original data-base twice to generate high (weak) utility 1_q_itemsets and construct conditional databases for them, and generate high (weak) utility k_q_itemsets and conditional databases for high (weak) utility k_q_itemsets from conditional databases for (k-1)_q_itemsets recursively.

3 Experimental Results

Because there is no algorithms for discovering high utility quantitative association rules, we evaluate the performance of our HUQA algorithm by comparing with Two-Phase association rule algorithm [10], which is the most efficient algorithm for min-ing high utility itemsets. Owing to real data is hard to be obtained, we generate synthetic datasets to simulate real world datasets. Table 3 shows the parameters used in the synthetic datasets.

Table 3. Parameters for synthetic datasets

|D|	Number of transactions
|T|	Average number of items per transaction
|L|	Number of maximal potentially high utility quantitative itemsets
|I|	Average length of maximal potentially high utility quantitative itemsets
N	Number of items
MQ	The maximum purchased quantity of items

It is different between the utility defined in [10] and the utility support defined in this paper. In addition, HUQA algorithm is to find high utility quantitative itemsets. Liu et al. [10] proposed Two-Phase algorithm just to find high utility itemsets. The differences between Two-Phase algorithm and HUQA algorithm are shown in Table 4. Hence, the two algorithms cannot specify the same threshold. We compare our HUQA algorithm with Two-Phase algorithm under the same number of high utility (quantita-tive) itemsets generated by the two algorithms.

We generate synthetic dataset T6.I4.D10K, and the number of items is set to 1000. In the following two experiments, the numbers of high utility (quantitative) itemsets generated by the two algorithms are 1000, 2000, 3000 and 4000, respec-tively. The execution times for the two algorithms are shown in Figure 1. From

Table 4. Comparisons of Two-Phase and HUQA

	Two-Phase	HUQA
Threshold	$u(X) \geqq MU$	$us(X) \geqq MUS$
Pattern	High utility itemset {A, C}	High utility q_itemset {(A,2)(C,2,3)}

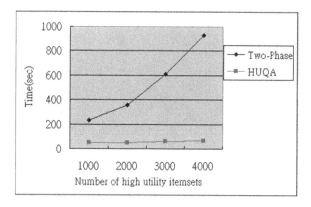

Fig. 1. Execution times for Two-Phase HUQA algorithms

Figure 1, we can see that the performance gap between Two-Phase algorithm and HUQA algorithm increases significantly as the number of generated high utility (quantitative) itemsets increases, since the number of candidate itemsets and the number of database scans also increases for Two-Phase algorithm. However, our HUQA algorithm needs not generate candidate itemsets and the longer the lengths of high utility q_itemsets are, the smaller the sizes of conditional databases are. Hence, HUQA algorithm outperforms Two-Phase algorithm significantly.

We also use dataset T6.I4.D10K to compare the size of memory space used by Two-Phase and HUQA algorithms based on the same configuration as the above experiment. Figure 2 shows that the memory space used by Two-Phase algorithm is more than that of HUQA algorithm. Because a lot of candidate itemsets, the total utility of all the transactions which contains each candidate itemset, and support count and utility of each candidate itemsets need to be stored for Two-phase algorithm, a large number of storage space need to be used. However, comparing with Two-Phase algorithmm, HUQA has to take memory space to store conditional databases, but all the conditional databases are small.

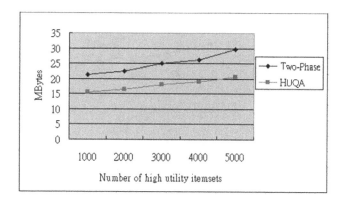

Fig. 2. The memory usages for Two-Phase and HUQA algorithms

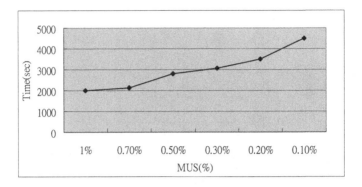

Fig. 3. The executing times for different minimum utility support threshold

In the following, we evaluate the scalability for our HUQA algorithm. The dataset used in the experiment is T6.I4.D1000K, and the number of items is 1000. Figure 3 shows that the execution times increase smoothly as minimum utility support decreases, since the number of high utility q_itemsets increases and the number of conditional databases increases when minimum utility support decreases.

Figure 4 shows the execution times when the number of items is from 1000 to 5000. The dataset used in the experiment is T6.I4.D100K, and minimum utility support is set to 0.5%. From Figure 4, we can see that the execution times slightly decrease as the number of items increases, since support count for each q_item decreases and the number of high utility q_itemsets decreases when the number of items increases.

Finally, for the dataset T6.I4.D1000K, the number of transactions is set to 300K, 500K, 700K and 1000K, respectively, and minimum utility support is set to 0.5%. The execution times of HUQA algorithm for different number of transactions are shown in Figure 5, in which the execution times increase gradually as the number of transactions increases. This is because the number of transactions which need to be processed increases and the sizes of conditional databases increases when the number of transactions increases.

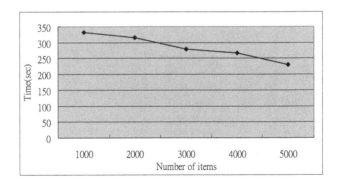

Fig. 4. Execution times for different number of items

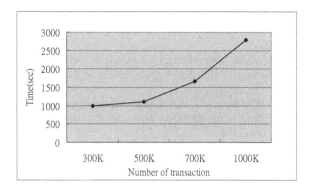

Fig. 5. The executing times for different number of transactions

4 Conclusions

Mining weighted association rules [2, 4, 6] only considers purchased frequency and profit for an itemset; however, a high profit item may not always be a high utility item, since the purchased quantities for the itemset also needs to be considered. Therefore, this paper defines utility support of an itemset, which is calculated by weights, purchased quantities and purchased frequency for the itemset. We also propose an efficient algorithm HUQA to discover high utility quantitative itemsets from a transaction database. High utility quantitative association rules can be generated from the high utility quantitative itemsets. For HUQA algorithm, a method for combining adjacent quantities for an item and a method for reducing the number of conditional databases generated by HUQA are proposed. HUQA algorithm does not generate candidate q_itemsets and just needs to scan original database twice, which is very efficient. For the experiments, we compare our algorithm with Two-Phase algorithm which is the most efficient algorithms for mining high utility itemsets. The experimental results show that our algorithm outperforms Two-Phase algorithm [10] under the same number of generated high utility (quantitative) itemsets.

References

1. Agrawal, R., Srikant, R.: Mining quantitative association rules in large relational tables. In: Proc. ACM SIGMOD, pp. 1–12 (1996)
2. Cai, C.H., Fu, A.W.C., Cheng, C.H., Kwong, W.W.: Mining Association Rules with Weighted Items. In: IDEAS, pp. 68–77 (1998)
3. Chan, R., Yang, Q., Shen, Y.-D.: Mining high utility itemsets. In: Proceedings of the 2003 IEEE International Conference on Data Mining, Melbourne, FL, November 2003, pp. 19–26 (2003)
4. Tao, F., Murtagh, F., Farid, M.: Weighted Association Rule Mining using Weighted Support and Significance Framework. In: Proceedings of the ACM SIGKDD Conference on Knowledge Discovery and Data Mining, pp. 661–666 (2003)
5. Tsai, P.S.M., Chen, C.-M.: Mining Quantitative Association Rules in a Large Database of Sales Transactions. J. Inf. Sci. Eng. 17(4), 667–681 (2001)

6. Wang, W., Yang, J., Yu, P.: Efficient mining of weighted association rules (WAR). In: Proceedings of the ACM SIGKDD Conference on Knowledge Discovery and Data Mining, pp. 270–274 (2000)

7. Yen, S.J., Lee, Y.S., Chen, S.W.: Mining Quantitative Association Rules from Transaction Database. In: Proceedings of 10th National Conference on Fuzzy Theory and Its Applications, pp. D520–D525 (2002)

8. Yao, H., Hamilton, H.J., Butz, C.J.: A Foundational Approach to Mining Itemset Utilities from Databases. In: Proc. of the 4th SIAM International Conference on Data Mining, Florida, USA, pp. 482–486 (2004)

9. Shen, Y.D., Yang, Q., Zhang, Z.: Objective-oriented utility-based association mining. In: Proc. of the IEEE Int. Conf. on Data Mining, Japan, December 2002, pp. 426–433 (2002)

10. Liu, Y., Liao, W.-K., Choudhary, A.: A Fast High Utility Itemsets Mining Algorithm. In: Proc. ACM Press Conference on Knowledge Discovery in Data, pp. 90–99 (2005)

Extraction of Association Rules Based on Literalsets

G. Gasmi[1,2], S. Ben Yahia[1], E.M. Nguifo[2], and S. Bouker[1]

[1] Department of computer science
Faculty of Science of Tunis
[2] Research center of computer science of Lens
Rue de l'Université SP 16, 62307 Lens Cedex
ghada_gasmi@yahoo.fr, sadok.benyahia@fst.rnu.tn, mephunguifo@gmail.com,
slimkrm@yahoo.fr

Abstract. In association rules mining, current trend is witnessing the emergence of a growing number of works toward bringing negative items to light in the mined knowledge. However, the amount of the extracted rules is huge, thus not feasible in practice. In this paper, we propose to extract a subset of generalized association rules (*i.e.,* association rules with negation) from which we can retrieve the whole set of generalized association rules. Results of experiments carried out on benchmark databases showed important profits in terms of compactness of the introduced generic basis.

1 Introduction

The problem of association rules mining is one of the most important problems of data mining which was intensively studied since its definition [1]. This problem concerned the market basket analysis. Indeed, association rules analysis is the task of discovering correlations between itemsets witch are frequently occurring together (*e.g.,* 70% of costumers who buy chips and drinks buy also pizza with a probability equal to 90%). Nevertheless, in many domains, one might be interested in discovering assocation rules taking into account the absence of some items. For example, *e.g.,* 60% of customers who do not buy chocolate do not buy also toys in 80% of cases. However, incorporating negation into association rule framework is problematic. Indeed, in an average transaction, only a small fraction of items occurs. At the same time, almost all possible items are not present in each transaction. Therefore, each transaction supports a huge number of *litteralsets* (*i.e.,* a generalization of itemsets which takes into account the absence of items) containing negation. Consequently, the number of association rules with negation (*generalized association rules*) is huge. It is nearly impossible for the end users to comprehend or validate such large number of complex generalized association rules, thereby limiting the usefulness of the data mining results.

In this paper, we propose an adaptation of PRINCE algorithm [2], called GENGBR, to extract a subset of generalized association rules (*generic basis of generalized association rules*) from which we can retrieve the remaining rules.

I.Y. Song, J. Eder, and T.M. Nguyen (Eds.): DaWaK 2007, LNCS 4654, pp. 293–302, 2007.

Thus, we help user to explore better the presented knowledge. Results of experiments carried out on benchmark databases showed a significant compactness of the generalized association rules raging from 87% to 99%.

The remainder of the paper is organized as follows. Section 2 overviews the related work on the extraction of generalized association rules. In section 3, we present key notions used throughout the paper. In section 4, we describe the algorithm GenGbr permitting to extract generic basis of generalized association rules. In section 5, we present the results of experiments carried out on benchmark databases. Section 6 concludes the paper with conclusion and future work.

2   Related Work

The idea of mining generalized association rules was first presented in [3] where the authors introduce the concept of excluding associations. They present a versatile method for finding associations of the form $AB\overline{C} \Rightarrow D$, where $AB \Rightarrow D$ is not valid due to an insufficient confidence value. Such a rule represents the fact that A and B imply D when C does not occur. Their solution is to transform the database into a trie structure and extract both positive association rules and excluding association rules directly from the trie.

In [4], the authors mentioned the notion of negative relationships based on chi-square. Indeed, this statistical metric is used to verify the independence between two variables. To determine the nature of the relationship (*i.e.*, positive or negative), a correlation metric was used. In [5], the authors presented a new idea to mine strong negative rules. They combined positive frequent itemsets with domain knowledge in the form of taxonomy to mine negative associations. However, it is not obvious to generalize their algorithm it is domain dependent and requires a predefined taxonomy. In [6], the authors proposed a new algorithm for generating both positive and negative association rules. They add on top of the support-confidence framework another measure called *mininterest* for a better pruning of the frequent litteralsets generated.

Another approach was presented in [7] permitting to find confined generalized association rules of the form $\overline{X} \Rightarrow Y$, $\overline{X} \Rightarrow \overline{Y}$, or $X \Rightarrow \overline{Y}$, where the entire antecedent or consequent is a conjunction of only negative or a conjunction of only positive items. However, it is noteworthy that this approach is not general enough to capture all types of generalized association rules. However, limiting the algorithm to the discovery of confined negative association rules only allows the authors to develop an efficient method based on the correlation coefficient analysis.

Also, the problem of mining generalized association rules has been attacked in [8]. Itemsets are divided into derivable and non-derivable based on the existence of certain rules (functional dependencies) in the dataset. The authors present an efficient method permitting to extract a concise representation of a huge number of patterns with negation using negative border and rule generators, called *generalized disjunction-free literal set representation* ($\mathcal{GDFLR}$). In [9], the authors adapted proposed $\mathcal{GDFLR}$ to propose a novel concise representation which represents all frequent litteralsets with at most k negative items.

In [10], the authors defined the dissociation rules allowing users to find negatively associated sets of items. However, this set of rules does not capture all types of generalized association rules.

3   Key Notions

In the following, we present some key notions which will be used in the remaining of the paper.

Definition 1. *(literal) A litteral l is an item i (i.e., positive litteral) or the negation of an item $\bar{i}$ (i.e., negative litteral).*

For example, the first transaction of the extraction context depicted by Figure 1 contains the following litterals: A, $\overline{B}$, $\overline{C}$, D and $\overline{E}$.

Definition 2. *(Extraction context) An extraction context is a triplet $\mathcal{K} = (\mathcal{O}, \mathcal{I}, \mathcal{R})$, such that $\mathcal{O}$ represents a finite set of objects (or transactions), $\mathcal{I}$ represents a finite set of attributes (or items) and $\mathcal{R}$ is a binary relation (i.e., $\mathcal{R} \subseteq \mathcal{O} \times \mathcal{I}$). Each couple $(o,i) \in \mathcal{R}$ expresses that the transaction $o \in \mathcal{O}$ contains the item $i \in \mathcal{I}$.*

Definition 3. *(literalset) A literalset L noted by $X\,\overline{Y}$ is a conjunction of:*

- *a positive literalset X which is a conjunction of positive litterals.*
- *a negative literalset $\overline{Y}$ which is a conjunction of negative litterals.*

A literalset L is said to be frequent if and only if its relative support given by $supp(L) = \frac{|\{T \in \mathcal{D}, X \subseteq T \wedge \forall l \in \overline{Y} l \notin T|\}}{|T|}$ is at least equal to a minimal threshold minsup fixed by the user.

For $minsup = \frac{2}{5}$, we can extract from the database of Figure 1 the frequent literalset $\overline{C}$ D which has a support equal to $\frac{4}{5}$. Indeed, from a context containing n items, we can derive 3^n literalsets. Thus, this number is usually huge. A solution consists on extracting a condensed representation of the whole set of literalsets which consists on the set of closed literalsets.

Definition 4. *(Variation)*
Let L be a positive literalset. A variation of L is obtained after negating any number of positive litterals in L.

	A	B	C	D	E
1	×			×	
2			×	×	
3			×	×	
4		×	×		×
5	×			×	×

Fig. 1. An extraction context $\mathcal{K}$

Definition 5. *(Positive variation) Let L noted by $X\overline{Y}$ be a literalset. P is the positive variation of literalset L if and only if P is equal to XY such that $Y = \{\overline{l}, l \in \overline{Y}\}$.*

A positive variation of the literalset $A \, \overline{B} \, \overline{C}$ is $A \, B \, C$.

Definition 6. *(Galois connection) Let an extraction context $\mathcal{K} = (\mathcal{O}, \mathcal{I}, \mathcal{R})$. Let ϕ be the application from the power set $\mathcal{O}$ (i.e. the set of all subsets of $\mathcal{O}$), noted by $P(\mathcal{O})$, to the power set of literals $\mathcal{L}$, noted by $P(\mathcal{L})$. ϕ associates to a set of objects $O \subseteq \mathcal{O}$, the set of literals $L \in \mathcal{I}$ shared by all objects $o \in O$.*

$$\phi : P(\mathcal{O}) \rightarrow P(\mathcal{L})$$
$$\phi(O) = \{l \in \mathcal{L} | \forall o \in O, (o, l) \in \mathcal{R}\}$$

Let ψ be the application of the power set of $\mathcal{L}$ to the power set of $\mathcal{O}$. It associates to a set of literals $L \subseteq \mathcal{L}$, the set of objects $O \in \mathcal{O}$ containing items $i \in L$:

$$\psi : P(\mathcal{L}) \rightarrow P(\mathcal{O})$$
$$\psi(L) = \{o \in \mathcal{O} | \forall i \in L, (o, l) \in \mathcal{R}\}$$

Definition 7. *(Closure operators) The couple of applications (ϕ, ψ) defines a Galois connection between the power set of $\mathcal{O}$ and that of $\mathcal{L}$. Applications $\omega = \phi \circ \psi$ and $\gamma = \psi \circ \phi$ are called closure operators of Galois connection [11].*

Definition 8. *(Closed literalset) A literalset $L \subseteq \mathcal{L}$ is closed if and only if $L = \omega(L)$. L is the maximal set of items shared by a set of objects.*

Definition 9. *(Minimal generator) Let g be a literalset. It is said a minimal generator of a closed literalset L if and only if $\omega(g) = L$ and $\nexists \, g' \subseteq g$ such that $\omega(g') = L$.*

4   Extraction of Generic Basis of Generalized Association Rules

Incorporating negation into association rule framework is very problematic. Indeed, in an average transaction, only a small fraction of items is present. At the same time, almost all possible items are not present in every transaction. Therefore, each transaction supports a huge number of literalsets containing negation. Consequently, the number of association rules with negation (*generalized association rules*) is huge. It is nearly impossible for the end users to comprehend or validate such large number of complex generalized association rules, thereby limiting the usefulness of the data mining results. To palliate this limit, a solution consists on extracting a condensed representation of the whole set of generalized association rules called a *generic basis of generalized association rules*. In the following, we present an adaptation of **Prince** algorithm [2], called GENGBR, permitting to:

1. extract the set of frequent minimal generators and their corresponding closed literalsets;
2. deriving a generic basis of generalized association rules;

4.1 GENGBR Algorithm

GENGBR algorithm takes as input the extraction context $\mathcal{K}$, *minsup* and *minconf* measures and returns a generic basis of generalized association rules. Indeed, GENGBR operates in three steps. 1) extraction of the set of frequent minimal generators. 2) ordering partially this set of minimal generators 3) extraction of generic basis of generalized association rules.

Extraction of Minimal Generators. GENGBR initializes the set of 1-minimal generators with items and their respective negations. The algorithm computes the support of positive 1-minimal generators after a scan to the extraction context. Then, it deduces the support of negative 1-minimal generators. 1-minimal generators having a support equal to that of the empty set are pruned since they can not be minimal.

The algorithm generates k-minimal generators by joining pairs of k-1-minimal generators. In order to compute the supports of minimal generators, we use the algorithm FASTERIE [12] permitting to compute the support of literalsets efficiently. Indeed, for each k-minimal generator g, we check if all its k-1-subsets belong to the set of k-1-minimal generators. The estimated support of g is equal to the smallest support of its k-1-subsets. Notice that k-1-minimal generators are stored in a prefix tree. Hence, according to the proposition 1, we can compute the support of a minimal generator g by using supports of its subsets. Then, we calculate positive variation of all retained candidates in one pass over context, in order to obtain their real supports. If the estimated support of a minimal generator g is equal to its real support, then g is pruned.

Proposition 1. *Let* $L = \{i_1, ..., i_k, ..., i_p, \overline{j}_1, ..., \overline{j}_z, ..., \overline{j}_n\}$ *be a literalset, where* $i_k, j_z \in \mathcal{I}$, $\{i_1, ..., i_k, ..., i_p\}$ *is the set of positive items and* $\{\overline{j}_1, ..., \overline{j}_z, ..., \overline{j}_n\}$ *is the set of negative items. The support of* L *is given by*

$$supp(\{i_1, ..., i_k, ..., i_p, \overline{j}_1, ..., \overline{j}_z, ..., \overline{j}_n\}) = (-1)^n supp(\{i_1, ..., i_k, ..., i_p, j_1, ..., j_z, ..., j_n\}) +$$

$$\sum_{S \subset \{\overline{j}_1, ..., \overline{j}_n\}} (-1)^{|S'|} supp(\{i_1, ..., i_k, ..., i_p\} \cup \{\overline{j}_z | z \in S\})$$

Where $|S'|=|S|$ *if* **n** *is* **odd** *and* $|S'|=|S|+1$ *if* **n** *is* **even**.

Ordering the Set of Minimal Generators Partially. In this step, GENGBR orders the set of minimal generators by grouping them into equivalence classes (i.e., minimal generators having the same support).

Extraction of Generic Basis of Generalized Association Rules. In this step, the algorithm computes the closure of each equivalence class. Hence, we can extract generic basis of generalized association rules. In the following we present an extension of the generic basis of exact association rules $\mathcal{GBE}$ and the generic basis of approximative association rules $\mathcal{TGBA}$ which are introduced in [13].

$\mathcal{GBE}$	
$R_1: B \overset{1}{\Rightarrow} E \overline{D}$	$R_6: C \overline{D} \overset{1}{\Rightarrow} B E$
$R_2: E \overset{1}{\Rightarrow} B \overline{D}$	$R_7: \overline{A} \overset{1}{\Rightarrow} B E \overline{D}$
$R_3: \overline{D} \overset{1}{\Rightarrow} B E$	$R_8: A \overset{1}{\Rightarrow} C$
$R_4: B C \overset{1}{\Rightarrow} E \overline{D}$	$R_9: C \overline{A} \overset{1}{\Rightarrow} B E \overline{D}$
$R_5: C E \overset{1}{\Rightarrow} B \overline{D}$	

$\mathcal{TGBA}$	
$R_{10}: \emptyset \overset{0.8}{\Rightarrow} B E \overline{D}$	$R_{15}: E \overset{0.75}{\Rightarrow} B C \overline{D}$
$R_{11}: \emptyset \overset{0.8}{\Rightarrow} C$	$R_{16}: E \overset{0.75}{\Rightarrow} B \overline{A} \overline{D}$
$R_{12}: B \overset{0.75}{\Rightarrow} C E \overline{D}$	$R_{17}: \overline{D} \overset{0.75}{\Rightarrow} B C E$
$R_{13}: B \overset{0.75}{\Rightarrow} E \overline{A} \overline{D}$	$R_{18}: \overline{D} \overset{0.75}{\Rightarrow} B E \overline{A}$
$R_{14}: C \overset{0.75}{\Rightarrow} B E \overline{D}$	

Fig. 2. Left: $\mathcal{GBE}$ and **Right:** $\mathcal{GBE}$ for $minsup=2$ and $minconf=0.7$

Definition 10. *Let $\mathcal{FCL}$ be the set of frequent closed literalsets and $\mathcal{G}_c$ be the set of minimal generators of a frequent closed literalset c. $\mathcal{GBE}$ basis is defined as the following:*

$$\mathcal{GBE} = \{R : g \Rightarrow (c - g) \mid c \in \mathcal{FCL} \wedge g \in \mathcal{G}_c \wedge g \neq c\}.$$

Definition 11. $\mathcal{TGBA} = \{R: g \Rightarrow (c\text{-}g), \ c \in \mathcal{FCL} \wedge g \in \mathcal{G} \wedge \omega(g) \subset c, \nexists \ c'$ $\mathcal{FCL} \wedge c' \subset c \wedge confidence(R) \geq minconf\}.$

Example 1. From the extraction context 1, we can extract the generic basis of generalized association rules shown by Figure 2.

5   Experimental Results

All programs was developed in the C language. Experiments were conducted on a PentiumM PC with 1.73 GHz and 1GB of main memory, running Fedora Linux. In the following, we present the results of experiments carried out on Benchmark datasets which are taken from the UCI Machine Learning Repository [1].

Table 1 summarizes the characteristics of these datasets. It shows the number of transactions, the number of items and the average transaction length.

In Table 3, we report, respectively, in the second and the third column the number of frequent literalsets and the number of frequent closed literalsets. According to the fourth column which shows the compactness degree, we can note that frequent closed literalsets represents an efficient reduction of the whole set of frequent literalsets. Indeed, the compactness degree ranges from 53% to 99%. Especially, we can notice that the more $minsup$ measure decreases, the more the compactness degree increases.

Table 3 presents the number of minimal generators and closed frequent literalset for a value of $minsup$ and reports the evolution of the size of $(\mathcal{GBE}, \mathcal{TGBA})$ following the variation of $minconf$ threshold. As we see, we can notice that the size of $(\mathcal{GBE}, \mathcal{TGBA})$ is sensitive to the variation of $minconf$. Indeed, the more $minconf$ decreases, the more the size of $(\mathcal{GBE}, \mathcal{TGBA})$ increases. This is can be explained by the fact that the more $minconf$ decreases, the more number of minimal generators which satisfy $minconf$ increases (c.f., definition 11).

---

[1] http://www.ics.uci.edu/~mlearn/MLRepository.html

Table 1. Datasets characteristics

Datasets	Number of transactions	Number of items	Average size
Nursery	12960	31	9
Monks	432	19	7
Flare	323	32	10
Zoo	101	28	9

Table 2. Variation of closed literalsets vs frequent litterasets

Zoo			
$minsup$ (%)	|closed literalsets|	|literalsets|	compactness
50	4496	85976	94%
45	6534	175138	96%
40	9440	411597	97%
35	13287	3475290	99%
30	18546	12005931	99%

Flare			
$minsup$ (%)	|closed literalsets|	|literalsets|	compactness
70	785	10179	92%
60	2697	55138	95%
50	8222	291967	97%
40	22673	1376138	98%
30	63309	7834112	99%

Monks			
$minsup$ (%)	|closed literalsets|	|literalsets|	compactness
5	20091	402562	95%
4	23787	670078	96%
3	31152	1069381	97%
2	52119	2952838	98%
1	65647	9858007	99%

Nursery			
$minsup$ (%)	|closed literalsets|	|literalsets|	compactness
30	7820	16639	53%
25	23865	52915	54%
20	55189	155093	64%
15	148782	559223	73%
10	527181	3591330	85%

Table 3 present the evolution of redundant association rules, extracted by APRI-ORI algorithm [1], and the evolution of the size of $(\mathcal{GBE}, \mathcal{TGBA})$. According to this table, we can notice that the extraction of $(\mathcal{GBE}, \mathcal{GBA})$ reduces significantly

Table 3. Evolution of generic basis vs *minconf* variation

Zoo				
minsup	|minimal generators|	|closed literalsets|	*minconf*	|($\mathcal{GBE}, \mathcal{TGBA}$)|
40%	14965	9440	100%	13878
			80%	76511
			60%	80665
			40%	82141
10%	160369	48335	100%	158961
			70%	809708
			40%	894402
			10%	902803

Flare				
minsup	|minimal generators|	|closed literalsets|	*minconf*	|($\mathcal{GBE}, \mathcal{TGBA}$)|
30%	409346	63309	90%	1639699
			70%	2420995
			50%	2621381
			30%	2627627
20%	1300799	158743	80%	6961732
			60%	8673651
			40%	9039247
			20%	9049337

Monks				
minsup	|minimal generators|	|closed literalsets|	*minconf*	|($\mathcal{GBE}, \mathcal{TGBA}$)|
2%	406048	52119	100%	405344
			50%	2186412
			10%	2190424
			2%	2190424
1%	742120	65547	100%	741416
			60%	2449336
			20%	4107928
			1%	4107928

Nursery				
minsup	|minimal generators|	|closed literalsets|	*minconf*	|($\mathcal{GBE}, \mathcal{TGBA}$)|
20%	67814	55189	80%	117015
			60%	337823
			40%	346969
			20%	346985
10%	831917	527181	100%	720666
			70%	3199069
			40%	5295733
			10%	5299004

Table 4. Evolution of redundant association rules vs variation of $(\mathcal{GBE}, \mathcal{TGBA})$

Zoo				
$minsup$	$minconf$	redundant rules	$|(\mathcal{GBE}, \mathcal{TGBA})|$	compactness
	100%	1757107	5726	99%
	90%	9395726	26804	99%
50%	70%	21927278	32986	99%
	50%	25707122	33305	99%
	100%	36633849	13878	99%
	80%	126906052	76511	99%
40%	60%	214911116	80665	99%
	40%	243781052	82141	99%

Flare				
$minsup$	$minconf$	redundant rules	$|(\mathcal{GBE}, \mathcal{TGBA})|$	compactness
	90%	3197382	38740	98%
	80%	7005934	44881	99%
60%	70%	9479458	47580	99%
	60%	9992856	47677	99%
	100%	3016819	37275	98%
	90%	26490103	153579	99%
50%	70%	86872737	197179	99%
	50%	113240360	199535	99%

Monks				
$minsup$	$minconf$	redundant rules	$|(\mathcal{GBE}, \mathcal{TGBA})|$	compactness
	100%	2562376	78152	97%
	60%	7486234	282724	97%
5%	20%	27239400	377524	98%
	5%	34459908	377524	98%
	100%	5606560	108368	98%
	60%	14327608	377224	98%
4%	20%	55183848	526600	99%
	4%	73640844	526600	99%

Nursery				
$minsup$	$minconf$	redundant rules	$|(\mathcal{GBE}, \mathcal{TGBA})|$	compactness
	80%	898992	117015	87%
	60%	2829756	337823	88%
20%	40%	5187964	346969	93%
	20%	7059822	346985	95%
	100%	5606560	149291	97%
	75%	14327608	598750	95%
15%	45%	55183848	1134618	98%
	15%	73640844	1135112	98%

the amount of knowledge presented at the user. Indeed, the compactness degree ranges from 87% to 99%.

6 Conclusion

In this paper, we presented an approach permitting to extract a subset of generalized association rules. Results of experiments carried out on benchmark databases showed important profits in terms of compactness of the introduced generic basis. Our future work agenda includes investigation of new metric different from $minsup$ and $minconf$ permitting to reduce the number of literalsets.

References

1. Agrawal, R., Imielinski, T., Swami, A.: Mining association rules between sets of items in large databases. In: Proceedings of the ACM SIGMOD Intl. Conference on Management of Data, Washington, USA, pp. 207–216. ACM Press, New York (1993)
2. Hamrouni, T., Yahia, S.B., Slimani, Y.: PRINCE: Extraction optimisée des bases génériques de régles sans calcul de fermetures. In: Proceedings of the Intl. INFORSID Conference, Hermés, Grenoble, France, pp. 353–368 (2005)
3. Amir, A., Feldman, R., Kashi, R.: A new versatile method for association generation. In: Princ. of Data Mining and Knowledge Disc., pp. 221–231 (1997)
4. Brin, S., Motawni, R., Silverstein, C.: Beyond market baskets: Generalizing association rules to correlation. In: Proceedings of the SIGMOD, Tucson, Arizona (USA), pp. 265–276 (1997)
5. Savasere, A., Omiecinski, E., Navathe, S.B.: Mining for strong negative associations in a large database of customer transactions. In: ICDE, Orlando, Florida, USA, pp. 494–502 (1998)
6. Wu, X., Zhang, C., Zhang, S.: Efficient mining of both positive and negative association rules. ACM Transactions on Information Systems, 381–405 (2004)
7. Antonie, M.L., Zaiane, O.R.: Mining positive and negative association rules: An approach for confined rules. Technical report, University of Alberta, Department of Computing Science (2004)
8. Kryszkiewicz, M., Cichon, K.: Support oriented discovery of generalized disjunction-free representation of frequent patterns with negation. In: Ho, T.-B., Cheung, D., Liu, H. (eds.) PAKDD 2005. LNCS (LNAI), vol. 3518, pp. 672–682. Springer, Heidelberg (2005)
9. Kryszkiewicz, M.: Generalized disjunction-free representation of frequents patterns with at most k negations. In: Ng, W.-K., Kitsuregawa, M., Li, J., Chang, K. (eds.) PAKDD 2006. LNCS (LNAI), vol. 3918, pp. 468–472. Springer, Heidelberg (2006)
10. Morzy, M.: Efficient mining of dissociation rules. In: Tjoa, A.M., Trujillo, J. (eds.) DaWaK 2006. LNCS, vol. 4081, Springer, Heidelberg (2006)
11. Ganter, B., Wille, R.: Formal Concept Analysis. Springer, Heidelberg (1999)
12. Bouker, S., Gasmi, G., Yahia, S.B., Nguifo, E.M.: Extraction des bases génériques de régles dŠassociation généralisées. In: EGC 2007, Namur, Belgium (2007)
13. Bastide, Y., Pasquier, N., Taouil, R., Lakhal, L., Stumme, G.: Mining minimal non-redundant association rules using frequent closed itemsets. In: Proceedings of the Intl. Conference DOOD'2000. LNCS, pp. 972–986. Springer, Heidelberg (2000)

Cost-Sensitive Decision Trees Applied to Medical Data

Alberto Freitas[1,2], Altamiro Costa-Pereira[1,2], and Pavel Brazdil[3]

[1] CINTESIS – Center for Research in Health Information Systems and Technologies,
Al. Prof. Hernâni Monteiro; 4200-319 Porto, Portugal
[2] Department of Biostatistics and Medical Informatics, Faculty of Medicine,
University of Porto, Portugal
{alberto,altamiro}@med.up.pt
[3] Artificial Intelligence and Data Analysis Group, Artificial Intelligence and Computer
Science Laboratory, Faculty of Economics, University of Porto, Portugal
pbrazdil@liacc.up.pt

Abstract. Classification plays an important role in medicine, especially for medical diagnosis. Health applications often require classifiers that minimize the total cost, including misclassifications costs and test costs. In fact, there are many reasons for considering costs in medicine, as diagnostic tests are not free and health budgets are limited. Our aim with this work was to define, implement and test a strategy for cost-sensitive learning. We defined an algorithm for decision tree induction that considers costs, including test costs, delayed costs and costs associated with risk. Then we applied our strategy to train and evaluate cost-sensitive decision trees in medical data. Built trees can be tested following some strategies, including group costs, common costs, and individual costs. Using the factor of "risk" it is possible to penalize invasive or delayed tests and obtain decision trees patient-friendly.

Keywords: Classification, Costs and Cost Analysis, Data Mining in Medicine, Cost-Sensitive Learning.

1 Introduction

In medical care, as in other areas, knowledge is crucial for decision making support, biomedical research and health management [1]. Data mining and machine learning can help in the process of knowledge discovery. Data mining is the non-trivial process of identifying valid, novel, potentially useful and ultimately understandable patterns in data [2]. Machine learning is concerned with the development of techniques which allow computers to "learn" [3].

Classification methods can be used to build models that describe classes or predict future data trends. It generic aim is to build models that allows predicting the value of one categorical variable from the known values of other variables. Classification is a common, pragmatic tool in clinical medicine. It is the basis for finding a diagnosis and, therefore, for the definition of distinct strategies of therapy. In addition, classification plays an important role in Evidence-Based Medicine. Machine learning systems can be used to enhance the knowledge bases used by expert systems as it can

I.Y. Song, J. Eder, and T.M. Nguyen (Eds.): DaWaK 2007, LNCS 4654, pp. 303–312, 2007.
© Springer-Verlag Berlin Heidelberg 2007

produce a systematic description of clinical features that uniquely characterize clinical conditions. This knowledge can be expressed in the form of simple rules, or decision trees [4].

The majority of existing classification methods was designed to minimize the number of errors. Nevertheless, real-world applications often require classifiers that minimize the total cost, including misclassifications costs (each error has an associated cost) and test (attribute) costs. In medicine a false negative prediction, for instance failing to detect a disease, can have fatal consequences; while a false positive prediction can be, in many situations, less serious (e.g. giving a drug to a patient that does not have a certain disease). Each diagnostic test has also a cost and to decide if is worthwhile pay the costs of tests it is necessary to know both misclassification and tests costs. There are many reasons for considering costs in medicine. Diagnostic tests, as other health interventions, are not free and budgets are limited.

Misclassification and test costs are on the most important costs, but there are also other types of costs [5]. Cost-sensitive learning (also known as cost-sensitive classification) is the area of machine learning that deals with costs in inductive learning.

Our aim with this work was to study and implement a strategy for learning and testing cost-sensitive decision trees, considering several costs, and with application to medical data.

The rest of this paper is organized as follows. In the next section we expose the main types of costs. Then we review related work. After that, we explain our cost-sensitive decision tree strategy. Next, we present and compare some experimental results. And finally, we conclude and point out some future work.

2   Types of Costs

Turney [5] presents a taxonomy for many possible types of costs that may occur in classification problems. From his enumeration, misclassification and test are on the most important costs. Costs can be measured in many distinct units as, for instance, money (euros, dollars), time (seconds, minutes) or other types of measures (e.g., quality of life).

2.1   Misclassification Costs

A problem with n classes is normally associated with a matrix $n*n$, where the element in line i and column j represent the cost of classifying a case in class i being from class j. Usually the cost is zero when $i = j$. Typically, misclassification costs are constant, that is, the cost is the same for any instance classified in class i but belonging to class j. The traditional error rate measure occurs when the cost is 0 for $i = j$ and 1 for all the other cells.

In some cases, the cost of misclassification errors can be conditional, that is, it can be dependent of specific features or dependent on a moment in time. The cost of prescribing a specific drug to an allergic patient may be different than prescribing that drug to a non allergic patient.

The cost of misclassification can be associated with the moment it occurs. A medical device can issue an alarm when a problem occurs and, in this situation, the cost is dependent simultaneously on the correctness of the classification and on the time the alarm is issued, that is, the alarm will only be useful if there is time for an adequate action [6]. Misclassification costs can also be dependent on the classification of other cases. In the previous example, if an alarm is correctly and consecutively issued for the same problem, then the benefit of the first alarm should be greater than the benefit of the others.

2.2 Cost of Tests

In medicine, the majority of diagnostic tests have an associated cost (e.g., an echography or a blood test). These costs can be highly distinct between different tests (attributes).

The costs of tests may be constant for all patients or may change according to specific patient features. A bronchodilatation test, for instance, has a higher cost for children less than 6 years, which means that the feature age has influence in the cost of the test.

Medical tests can also be very distinct when considering their influence in the "quality of life". A range of tests are completely harmless for patients (e.g., obstetric echography), others can be dangerous and put patient life at risk (e.g., cardiac catheterism), and some can be (only) uncomfortable (e.g., digestive endoscopy).

Some tests can be cheaper (and faster) when ordered together (in group) than when ordered individually and sequentially (e.g., renal, digestive and gynecological echography). Some tests can also have common costs that can be priced only once. Blood tests, for instance, share a common cost of collecting the blood sample. There is not only an economic reduction but also a non-economical reduction in the cost of "worry" the patient.

A number of tests might depend of the results of other tests. The test "age", for instance, may influence the cost of the bronchodilatation test. Some tests can have an increased price as result of secondary effects. Other tests can have patient specific, time dependent or emergency dependent costs.

In general, tests should only be ordered if their costs are not superior to the costs of classification errors.

3 Current Cost-Sensitive Approaches

In inductive learning (learning by examples), the majority of the work is concerned with the error rate (or success rate).

Nevertheless, some work has been done considering non-uniform misclassification costs, that is, different costs for different types of errors [7], [8]. Other literature is concerned with the cost of tests, without taking into account misclassification costs [9], [10].

And there is also some work concerned simultaneously with more than one type of costs, including the work of Turney [11], Zubek and Dieterich [12], Greiner et al. [13], Arnt and Zilberstein [14], and Ling et al. [15], [16], [17], [18], [19], [20]. At this level, the work of Turney [11] was the first to consider both test and misclassification costs. Next, we give a brief overview of this work considering both costs.

Turney [11] implemented a system, the ICET system, that uses a genetic algorithm for building a decision tree that minimizes test and misclassification costs. The ICET system was robust but very time consuming.

Several authors associated the cost problem to a Markov decision process that has the disadvantage of being computationally expensive. Zubek and Dietterich [12] used an optimal search strategy, while Arnt and Zilberstein [14] included a utility cost for the time needed to obtain the result of a test.

Greiner et al. [13] analyzed the problem of learning cost-sensitive optimal active classifiers, using a variant of the probably-approximately-correct (PAC) model.

Chai et al. [15] proposed a cost-sensitive naïve Bayes algorithm for reducing the total cost. Ling et al. [16] proposed a decision tree algorithm that uses a cost reduction splitting criteria during training, instead of minimum entropy. After that, Sheng et al. [17], presented an approach where a decision tree is built for each new test case. In another paper, Sheng et al. [18] proposed a hybrid model that results from the integration of a decision tree sensitive to costs with a naïve Bayes classifier. Zhang et al. [19] compared strategies for checking if missing values should be or not obtained, and stated that, for tests with high costs or high risk, it should be more cost-effective not to obtain their values. Recently, Ling et al. [20], updated their strategy for building cost-sensitive decision trees, with the inclusion of sequential test strategies.

4   A Cost-Sensitive Strategy: Taking Risk into Account

Next, we briefly describe our strategy for implementing a cost-sensitive decision tree. Our aim was to implement a tool for building decision models sensitive to costs, namely test costs, misclassification costs and other types of costs.

We opted to use decision trees because they have several interesting characteristics for health professionals. They are easy to understand and use, and present an intuitive and appealing structure. Their structure is congruent with decision making methods that physicians normally use in daily routine, when they try to understand which is the best diagnostic test or the best treatment for one patient [21].

We modified the C4.5 algorithm [22] to contemplate costs and, consequently, to generate cost-sensitive decision trees. Specifically we used de j48 class, implemented in the open source package for data mining Weka [3].

4.1   Cost Function

We adapted the decision tree splitting criteria to contemplate costs, through the following cost function (1):

$$\frac{\Delta I_i}{\left(C_i \phi_i\right)^\omega}. \tag{1}$$

Where ΔI_i is the information gain (or gain ratio) for attribute i, C_i is the cost of attribute (test) i, ϕ_i is the factor of "risk" associated with attribute (test) i, and ω is the cost scale factor.

When building a tree, for each node, the algorithm will select the attribute that maximizes the defined heuristic for the cost function. The cost function does not consider misclassification costs, as these costs do not have influence in the decision tree splitting criteria [23].

Attributes without cost, as age and gender, are assigned the value 1. Attributes with higher costs lead to lower results in the cost function and have, consequently, fewer chances to be selected.

The factor of "risk" is an influent piece in the cost function. This factor was introduced to penalize attributes that might be invasive, may cause discomfort and disturb, or could somehow contribute to low patient quality of life (e.g., the invasive test "coronary angiography"). The value 1 means absence of influence and is equivalent to a completely innocuous test, while values higher than 1 means that exists influence. Higher factors leads to lower results in the cost function and, therefore, to lower possibilities for an attribute to be selected.

If both test cost and factor of "risk" are equal to one, then their participation is neutral and, thus, the cost function is basically the traditional information gain function. The cost of the attribute can be adjusted in two ways. It can increase or decrease considering the inoffensiveness of the test (factor of "risk"), or it can be modified by the cost scale factor. The cost scale factor is a general parameter that is equally applied to all tests, reducing or increasing the influence of costs in attribute selection.

The factor of "risk" can also be used to penalize delayed tests. A longer test can have consequences in the patient quality of life and may increase other costs, as those related to staff, facilities and increased length of stay. Between two similar tests, but with one longer than the other, it makes sense that the faster test should be preferred. Hence, considering the average length of tests, an adjustment can be made through the factor of "risk".

The cost scale factor regulates the influence of costs, as it can make trees more (or less) sensitive to costs. This factor may also be used to adequate the cost function to the used scale. An inexpensive test, such as the "age" test, has the same cost for any scale. But, tests with real costs have clearly different values in different scales. If a test "A" cost 10 in the Euro scale then it costs 1000 in the centime scale, that is, the ratio between the tests "age" and "A" are 1 to 10 in the Euro scale and 1 to 1000 in the other. To avoid that a change in the scale could benefit some attributes, it is important to adjust it with the cost scale factor.

For a cost scale factor of 0 the costs are not considered as the denominator of the cost function became 1. This situation is equivalent to consider the original C4.5 information gain. With an increase in the cost scale factor, the costs of tests will have more influence and less expensive tests will be preferred. The cost scale factor typically will assume a value between 0 and 1.

For making our model sensitive to misclassification costs, we used a meta-learner implemented in Weka, CostSensitiveClassifier [3].

4.2 Decision Tree Test and Usage Strategies

After building cost-sensitive models, we want to test and use them. For that, we have a test strategy where we consider the cost of a test individually or in a group. Within groups of tests it is possible to distinguish between (i) tests ordered simultaneously or (ii) tests with a common cost.

Sometimes (i), there is a big difference between the cost of a group of tests and the sum of individual tests costs. Many medical tests do not have an immediate result and therefore it also important to consider the length of time for the group against the total delay of the individual tests. When a medical doctor is at a node of the decision tree, and have the first of a group of tests, he must decide if he will ask for a group of tests or only for the node test. If he orders a group of tests, then the cost considered will be the group cost, even if some tests in the group are not used.

In the other situation (ii), it is possible to separate a common cost for a group of tests. In a group of blood tests, for instance, the cost of collecting blood is a common cost for all tests in the group. Only the first test of the group will be priced for that common cost.

Delayed tests are considered in the training phase of the decisions tree. Slower tests will, therefore, have tendency to be tested only after faster ones.

4.3 Individual Costs

Our strategy also allows (a) to consider specific patient characteristics, (b) to modify test costs in situations where their values are already known (tested previously), and (c) to consider availability and slowness of some tests.

(a) The cost of each test can be conditional on the characteristics of the patient, that is, it is possible to have a variable cost associated with specific features of the patient. As seen before, the age can change the cost of the test "bronchodilatation" (higher for children less than 6 years). The comorbidity index of a patient is another example, as it can influence the cost of some tests and, therefore, could be used to adjust the cost of the test. For a specific test, different patients may require additional tests, consequently with an increase in costs.

(b) In other circumstances, some tests might have been obtained previously and, logically, their original costs should not be considered again. Consequently, we adopt another approach for building trees where, for each new instance (patient), tests with known values are considered without cost and without "risk". For each new vector of distinct costs a new decision tree is built.

(c) Resources are not infinite neither are always available. A specific test may be conditional on the availability of a medical device in a specific time period. In these cases, it is possible to exclude that test and build a new tree. Optionally, it is possible to increase de factor of "risk" of that test, decreasing it probabilities of being selected. As the availability is not constant over time, this problem should be analyzed for each case (patient); this is a circumstantial cost and not a cost intrinsic to the patient.

5 Experimental Results

In this section we present some experimental results. We tested our cost-sensitive decision tree with several datasets, including the Pima Indians Diabetes. This dataset,

Table 1. Attribute costs for Pima Indians Diabetes

Test	Cost ($)	Group cost ($)
a. Number of times pregnant	1	
b. Glucose tolerance test	17.61	b + e = 38.29
c. Diastolic blood pressure	1	
d. Triceps skin fold thickness	1	
e. Serum insulin test	22.78	b + e = 38.29
f. Body mass index	1	
g. Diabetes pedigree function	1	
h. Age (years)	1	

donated by Vincent Sigillito, contains 768 instances and a class that assumes 2 values, "healthy" (500 instances) or "diabetes" (268 instances)[1]. We can see test costs (all attributes are numeric) in Table 1 and the cost matrix in Table 2.

Tests "glucose tolerance" and "serum insulin" are not immediate and have a distinct cost (other attributes have a symbolic cost of $1). Moreover they share a common cost, $2.1, from collecting blood. As these attributes are not immediate, when using the tree it is necessary to decide if both tests will be ordered together (in group) or not.

Table 2. Cost matrix for Pima Indians Diabetes

		Prediction	
		Healthy	**diabetes**
Reality	**healthy**	$ 0	FP cost
	diabetes	FN cost	$ 0

We ranged the cost scale factor from 0.0 to 1.0 and induced decision trees using 10-fold cross-validation. After that, we considered cost zero for cases classified correctly and varied misclassification costs, from $10 to $1,000, with equal costs for false negative (FN) and false positives (FP). In Table 3 we can see the results of the evaluation of five decision trees for that range of misclassification costs. Notice that for a cost scale factor equal to zero, the model is equal to the obtained by the traditional C4.5.

In Table 4, we considered different FN and FP costs, with a variable cost ratio. In this case, we can see that for low misclassification costs (less than or equal to $150) best results occurs with cost scale factor equal to 1,0. For high misclassification costs, the model obtained with cost scale factor equal to 0,0 (C4.5) has better results, but only with slight differences. These results are similar to the previous ones, obtained with equal FN and FP costs.

In this example, for low misclassification costs, the best results are obtained for a cost scale factor between 0.5 and 1.0. In these situations, average costs are much lower when compared with C4.5. For high misclassification costs, compared to test

---

[1] UCI Repository of Machine Learning Databases, ftp://ftp.ics.uci.edu/pub/machine-learning-databases/

Table 3. Average costs in the evaluation of 5 decision trees, for a range of misclassification costs (FN cost equal to FP cost) and cost scale factors; 95% confidence intervals included

Cost scale factor:	0.0	0.1	0.2	0.5	1.0
Accuracy (%):	73.8	73.7	72.3	68.5	68.2
$10	23.4 ± 0.53	21.9 ± 0.57	19.5 ± 0.66	6.4 ± 0.44	5.7 ± 0.35
$20	26.0 ± 0.79	24.5 ± 0.81	22.2 ± 0.89	9.5 ± 0.73	8.8 ± 0.68
$50	33.8 ± 1.67	32.4 ± 1.67	30.6 ± 1.75	19.0 ± 1.69	18.4 ± 1.66
$100	46.9 ± 3.20	45.5 ± 3.21	44.4 ± 3.30	34.7 ± 3.33	34.2 ± 3.31
$200	73.1 ± 6.31	71.8 ± 6.31	72.2 ± 6.45	66.2 ± 6.62	66.0 ± 6.61
$500	151.6 ± 15.7	150.8 ± 15.7	155.4 ± 16.0	160.8 ± 16.5	161.3 ± 16.5
$1,000	282.5 ± 31.2	282.3 ± 31.3	294.0 ± 31.8	318.3 ± 33.0	320.2 ± 33.0

Table 4. Average costs in the evaluation of 2 decision trees, for a range of misclassification costs (with FN cost different from FP cost) and cost scale factor (csf) 0.0 and 1.0.; 95% confidence intervals included

FN/FP ratio	FN	FP	csf = 0.0	csf = 1.0
1/10	$10	$100	34.3 ± 2.37	17.3 ± 2.35
10/1	$100	$10	36.0 ± 2.52	22.6 ± 2.75
3/1	$150	$50	47.9 ± 3.78	37.2 ± 4.12
1/3	$50	$150	46.0 ± 3.58	31.3 ± 3.59
1/2	$500	$1,000	212.2 ± 24.58	225.8 ± 25.20
2/1	$1,000	$500	221.9 ± 25.67	255.7 ± 28.09

costs, best results occur with cost scale factor equal to 0.0 (C4.5) and 0.1, but without substantial differences for the other models.

As misclassification costs rise, test costs tend to be negligible. That is, for sufficiently higher misclassification costs and when considering equal cost for false negatives and false positives, a higher accuracy rate corresponds to a lower average cost.

In this evaluation we also distinguished between individual and group costs, and realized that results considering group costs always had higher average costs. This happened because the serum insulin test, which shares group costs with the glucose tolerance test, only intermittently appeared in the decision trees and, therefore, there was an extra imputed cost that was nothing or little used.

We also tested the factor of "risk" with other datasets and obtained interesting results. Sometimes, with minimum increase in the total cost it is possible to obtain models that are more patient-friendly.

6   Conclusions

In this paper we present approaches to build cost-sensitive decision trees, considering different aspects of test costs, where are included economic and non-economical costs. Our framework integrates also a cost-sensitive meta-learner to consider the situations where misclassifications costs are different. Results show that it outperforms the traditional, non cost-sensitive, C4.5. We can clearly see that, in inductive learning, it is important to consider several types of costs as they can be substantially decreased.

As technologies became more expensive, it is even more rational to consider all the cost involved. A big challenge is to have better healthcare using less money. The factor of "risk" is an important item of the proposed framework as it can induce models patient-friendly. Decision trees represent a natural representation for classification problems (for diagnosis or prognosis) that includes costs.

In our future work we will continue to evaluate our strategies, with new experiments in other datasets, with real data and real costs. We will try to incorporate other costs associated with delayed tests, emergency situations, staff, facilities, and increased length of stay. The cost function will also be refined and tested for different scenarios. We will also study other methods for building and testing decision trees with the application of a cost-sensitive pruning strategy.

References

1. Cios, K.J. (ed.): Medical Data Mining and Knowledge Discovery. Physica-Verlag, New York (2001)
2. Fayyad, U.M., Piatetsky-Shapiro, G., Smyth, P., Uthurusamy, R. (eds.): Advances in Knowledge Discovery and Data Mining. AAAI/MIT Press (1996)
3. Witten, I.H., Frank, E.: Data mining: Practical Machine Learning Tools and Techniques, 2nd edn. Morgan Kaufmann, San Francisco (2005)
4. Coiera, E.: Guide to Health Informatics, 2nd edn. A Hodder Arnold Publication (2003)
5. Turney, P.: Types of Cost in Inductive Concept Learning. In: Proc. Workshop on Cost-Sensitive Learning, 17th Int. Conf. Machine Learning, pp. 15–21 (2000)
6. Fawcett, T., Provost, F.: Activity Monitoring: Noticing Interesting Changes in Behavior. In: Proc. 5th Int. Conf. Knowledge Discovery and Data Mining, pp. 53–62 (1999)
7. Breiman, L., Freidman, J.H., Olshen, R.A., Stone, C.J.: Classification and Regression Trees, Wadsworth, Belmont California (1984)
8. Elkan, C.: The Foundations of Cost-Sensitive Learning. In: Proc. 17th Int. Joint Conf. Artificial Intelligence, pp. 973–978 (2001)
9. Núñez, M.: The Use of Background Knowledge in Decision Tree Induction. Machine learning 6, 231–250 (1991)
10. Melville, P., Provost, F., Saar-Tsechansky, M., Mooney, R.: Economical Active Feature-Value Acquisition Through Expected Utility Estimation. In: Proc. 1st Int. Workshop on Utility-Based Data Mining, pp. 10–16 (2005)
11. Turney, P.: Cost-Sensitive Classification: Empirical Evaluation of a Hybrid Genetic Decision Tree Induction Algorithm. J. Artificial Intelligence Research 2, 369–409 (1995)
12. Zubek, V.B., Dietterich, T.: Pruning Improves Heuristic Search for Cost-Sensitive Learning. In: Proc. 19th Int. Conf. Machine Learning, pp. 27–35 (2002)
13. Greiner, R., Grove, A.J., Roth, D.: Learning Cost-Sensitive Active Classifiers. Artificial Intelligence 139(2), 137–174 (2002)
14. Arnt, A., Zilberstein, S.: Attribute Measurement Policies for Cost-effective Classification. In: Workshop Data Mining in Resource Constrained Environments, 4th Int. Conf. Data Mining (2004)
15. Chai, X., Deng, L., Yang, Q., Ling, C.X.: Test-Cost Sensitive Naive Bayes Classification. In: Proc. 4th Int. Conf. Data Mining (2004)

16. Ling, C.X., Yang, Q., Wang, J., Zhang, S.: Decision Trees with Minimal Costs. In: Proc. 21st Int. Conf. Machine Learning (2004)
17. Sheng, S., Ling, C.X., Yang, Q.: Simple Test Strategies for Cost-Sensitive Decision Trees. In: Proc. 16th European Conf. Machine Learning, pp. 365–376 (2005)
18. Sheng, S., Ling, C.X.: Hybrid Cost-sensitive Decision Tree. In: Jorge, A.M., Torgo, L., Brazdil, P.B., Camacho, R., Gama, J. (eds.) PKDD 2005. LNCS (LNAI), vol. 3721, Springer, Heidelberg (2005)
19. Zhang, S., Qin, Z., Ling, C.X., Sheng, S.: Missing Is Useful: Missing Values in Cost-Sensitive Decision Trees. IEEE Transactions on Knowledge and Data Engineering 17(12), 1689–1693 (2005)
20. Ling, C.X., Sheng, V.S., Yang, Q.: Test Strategies for Cost-Sensitive Decision Trees. IEEE Transactions on Knowledge and Data Engineering 18(8), 1055–1067 (2006)
21. Grobman, W.A., Stamilio, D.M.: Methods of Clinical Prediction. American Journal of Obstetrics and Gynecology 194(3), 888–894 (2006)
22. Quinlan, J.R.: C4.5: Programs for Machine Learning. Morgan Kaufmann, San Mateo (1993)
23. Drummond, C., Holte, R.C.: Exploiting the Cost (In)sensitivity of Decision Tree Splitting Criteria. In: Proc. 17th Int. Conf. Machine Learning, pp. 239–246 (2000)

Utilization of Global Ranking Information in Graph-Based Biomedical Literature Clustering

Xiaodan Zhang[1], Xiaohua Hu[1,2], Jiali Xia[2], Xiaohua Zhou[1],
and Palakorn Achananuparp[1]

[1] College of Information Science and Technology, Drexel University
3141 Chestnut street, Philadelphia, PA 19104, USA
{xzhang,thu}@ischool.drexel.edu,
{xiaohua.zhou,pkorn}@drexel.edu
[2] UFSoft School of Software, Jiangxi University of Finance and Economics
Nanchang, Jiangxi, China

Abstract. In this paper, we explore how global ranking method in conjunction with local density method help identify meaningful term clusters from ontology enriched graph representation of biomedical literature corpus. One big problem with document clustering is how to discount the effects of class-unspecific general terms and strengthen the effects of class-specific core terms. We claim that a well constructed term graph can help improve the global ranking of class-specific core terms. We first apply PageRank and HITS to a directed abstract-title term graph to target class specific core terms. Then k dense term clusters (graphs) are identified from these terms. Last, each document is assigned to its closest core term graph. A series of experiments are conducted on a document corpus collected from PubMed. Experimental results show that our approach is very effective to identify class-specific core terms and thus help document clustering.

Keywords: Document Clustering, Term Graph, Global ranking.

1 Introduction

It is shown that only a small portion of terms that has distinguishable power on documents clustering [9][10]. Steinbach et al. [9] argued that each document class has a "core" vocabulary of words and remaining "general" words may have similar distributions on different classes. Thus, two documents from different classes can share many general words (e.g. stop words) and will be treated similar in terms of vector cosine similarity. The ideal situation is that only distinguishable terms are used to cluster documents in a much lower dimensionality. However, to discover these distinguishable core terms is not trivial when we don't have knowledge about the document class beforehand.

HITS [6] and PageRank [8] based algorithms have been viewed as very effective approaches to calculate the global importance of a web document based on directed link information on world wide web. Moreover, LexRank [4] also showed its effectiveness on undirected graph for text summarization tasks. Therefore, in this

I.Y. Song, J. Eder, and T.M. Nguyen (Eds.): DaWaK 2007, LNCS 4654, pp. 313–322, 2007.

paper, we employ these two ranking methods on an undirected term co-occurrence graph of a given corpus to extract global important class specific core terms. However, when these algorithms are applied to a term co-occurrence graph, they face noise. The identified terms are very likely to be general terms that tend to co-occur with many "core" terms. If the noise of class-unspecific general terms is well discounted, we claim that the global ranking of class-specific core terms will be improved and only a small portion of top ranked terms will be good enough to form the initial clustering model.

We claim that this noise problem can be partially solved when the term graph is well constructed. We argue that different sections of the documents have different importance level on finding globally important class specific core terms. For example, title terms are usually more specific to the major topic of a document than that of abstract; the text of a document title usually contains much bigger percentage of topical terms than that of document abstract. Herein, a document abstract can be treated as an explanation of document title. In other words, abstract terms "cite" terms in the title. Based on this intuition, a directed abstract title term graph is constructed with abstract terms pointing to title terms. By this way, class specific core terms can get more in links from abstract terms than that of pure term co-occurrence graph.

Motivated from discussion above, a novel framework is presented to cluster a collection of documents utilizing term's global and local importance information. A collection of documents is first represented as directed title abstract term graph. PageRank and HITS based algorithms are then used to rank the terms in the graph. The top ranked terms are later clustered into k clusters. Last, a document is assigned to its closest term cluster.

Experiments are conducted on a selected PubMed document set. We make following main evaluations (1) terms' ranking on term co-occurrence graph and abstract title graph; (2) effects of different global ranking schemes; (3) quality of identified term cluster; (4) quality of identified document cluster.

Experimental results show that our approach is very effective on document clustering and can identify document clusters and core term clusters at the same time using only a small amount of distinguishable class specific core terms based on term's global and local importance information.

2 Related Work

Given the representation of documents or sentences as graph, there are some emerging works recently in text classification [5] [7] and text summarization [4].

[7] et al. represented a web document as a graph with consideration of semantic information and location of text and then extracted most frequent document Subgraphs. In the end, these document Subgraphs are used for document classification. However, in essence, this approach is equal to extract one-gram, two-gram, tri-gram, etc. from a document, and thus can not take advantage of the link information among documents and terms over the entire document set.

[5] developed an approach to cluster document by integrating term's PageRank score to documents' representation. PageRank is applied to term co-occurrence graph. Document vectors are then represented using (PageRank Score)*IDF. The author

shows it has a better performance than that of TF*IDF on clustering document by K-means. However, PageRank and IDF can be both treated as global ranking scores. Putting them together will cause information loss such as term frequency.

LexRank [4] is a PageRank based approach called power method to find globally important sentences in text summarization. It computes sentence importance based on the concept of eigenvector centrality in a graph representation of sentences. While they represent sentences as graph nodes, we take terms as graph nodes.

There are also some other works [1][2][3] that focused on how to use link information to enhance traditional content based text classification task. [2] [3] applied content based method to assign labels to part of the data and then used relaxation labeling techniques to estimate and re-estimate the class label using the hyper link information. In contrast, [1] combined content and connectivity information into a joint probabilistic model.

Although existing methods try to combine link information with content information, nonetheless, there is no approach for exploring how to identify class specific core terms using link information to facilitate initial document clustering.

3   Graph-Based Document Clustering

3.1   Framework of Our Approach

The proposed approach consists of the following five main steps: (1) document representation; (2) construction of abstract-title term graph and term co-occurrence graph; (3) ranking terms according to their global importance; (4) clustering the top ranked terms to k clusters from term co-occurrence graph utilizing local importance information; (5) assign each document to the closest term clusters. The whole clustering process is described in the figure below.

3.2   Document Representation

In biomedical domain, it is very common that a concept has more than one synonym and is composed of more than one word, which makes it very necessary to represent document using appropriate biomedical ontology.

MeSH Ontology. Medical Subject Headings (MeSH) [www.nlm.nih.gov./mesh] mainly consists of the controlled vocabulary and a MeSH Tree. The controlled vocabulary contains several different types of terms, such as Descriptor, Qualifiers, Publication Types, Geographics, and Entry terms. Among them, Descriptors and Entry terms are used in this research because they are the only terms that can be extracted from documents. Descriptor terms are main concepts or main headings. Entry terms are the synonyms or the related terms to descriptors. For example, "Neoplasms" as a descriptor has the following entry terms {"Cancer", "Cancers", "Neoplasm", "Tumors", "Tumor", "Benign Neoplasm", "Neoplasm, Benign"}. MeSH descriptors are organized in a MeSH Tree, which can be seen as the MeSH Concept Hierarchy. In the MeSH Tree there are 15 categories (e.g. category A for anatomic terms), and each category is further divided into subcategories. For each subcategory, corresponding descriptors are hierarchically arranged from most general to most specific.

Algorithm:

Input: an abstract title term collection graph G_AB-TI, a term co-occurrence collection graph G_CO, k (the desired number of clusters), p (initial # of vertices (terms) for term clustering), M (minimum number core terms in each cluster), Cluster quality ratio Q

Output: k document clusters

```
// 1: Calculating Salient Scores of vertices V
     Salience(vi) = GlobalRankingG_AB-TI(vi)(Eq (1), Eq(2))
  //sort V in the descending order of Salience(v)using abstract-title
graph G_AB-TI
  Sort(V,Salience(v),des)

  // 2: Detecting k core term clusters from Graph G_CO
  Do{
     For(i=1; i<=numOfNodesForClustering; i++){
        Get free cluster Ck from k free cluster pool
        Ck.add(Ti) //add term Ti to cluster Ck
        Check in_Cluster_Degree for all terms Ti_list that have
  edge with Ti //refer to Eq(3)
        Sort Ti_list descending
        For(j=size(Ti_list);j>=1,j--){
           //check cluster quality
           if (In_cluster_degree(Tj)/In_cluster_degree(Ti)>=Q){
              cutoff_point = j;
              break loop;
              }
           }
     }
     Add terms over cutoff point to cluster Ck
     If(numOfTerm(Ck)<M){
        remove all terms from cluster Ck
        put back Cluster Ck to free cluster Pool
        }
  }While reach the number of K Cluster
  // 3: Assign document to closest term cluster
  Assign remaining top ranked terms to K cluster by Max (Eq.(4))
  Match Document To Term Cluster by Max (Eq.(5))
```

Fig. 1. Clustering algorithm

Mesh Descriptor Term extraction. While processing an abstract, we map terms in each document to the Entry terms in MeSH and then maps the selected Entry terms into MeSH Descriptors to handle the synonyms. In this way, synonyms of a given MeSH descriptor are assigned a unique ID.

We create one stop term list for MeSH based on the analysis of PubMed documents from 1994-2004 using Zipf law [12]. Based on the stop term list, we exclude some MeSH terms that are too general (e.g. HUMAN, WOMEN or MEN) or too common in MEDLINE articles (e.g. ENGLISH ABSTRACT or DOUBLE-BLIND METHOD).

3.3 Global Ranking and Term Graph Construction

PageRank [8] is one of methods Google uses to determine a page's relevance or importance. The beauty of the method is that it integrates social reference knowledge into the page ranking procedure. Given a web page in a network, the PageRank score of this page is defined as follows:

$$PR(p_i) = \frac{1-d}{N} + d \cdot \sum_{p_j \in M(p_i)} \frac{PR(p_j)}{L(p_j)} \qquad (1)$$

where N is the number of pages under consideration, $M(p_i)$ is a set of pages that link to p_i, $L(p_j)$ is the number of outbound links on page p_j, and d is damping factor.

Hypertext Induced Topic Selection (HITS) [6] is a link analysis algorithm that rates Web pages by their authority and hub values. Authority value estimates the value of the content of the page; hub value estimates the value of its links to other pages. These values can be used to rank Web search results.

Let N be the set of nodes in the neighborhood graph. For every node n in N, let $H[n]$ be its hub score and $A[n]$ its authority score. Initialize $H[n]$ and $A[n]$ to 1 for all n in N. While the vectors H and A have not converged:

$$\text{For all n in N,} \quad A[n] := \sum_{(n',n) \in N} H[n'] \qquad (2)$$

$$\text{For all n in N,} \quad H[n] := \sum_{(n,n') \in N} A[n']$$

Normalize the H and A vectors.

Table 1. The PageRank score of title and abstract terms of a document

PageRank score of Title term		PageRank score of Abstract term	
		Gout	5.53E-06
		Association	3.93E-06
		Incidence	3.47E-06
		Epidemiology	2.52E-06
		Pain	1.68E-06
		Hyperuricemia	1.20E-06
Gout	5.53E-06	Arthritis, Gouty	1.18E-06
		Arthritis	4.10E-07
		Algorithms	2.14E-07
		Obesity	1.62E-07
		Patient Education	1.52E-07
		Diabetes Mellitus	1.37E-07
Arthritis, Gouty	1.18E-06	Hyperlipidemia	8.97E-08

The success of PageRank and HITS lies on calculating stationary distribution on a directed social link graph. However, for a term co-occurrence graph, it does not have this information, i.e., there is no reference relationship between terms. We can build symmetrical term co-occurrence graph where term co-occur with each other when they appear in the same document. However, ranking terms on pure term-concurrence graph can be very likely to assign general terms high ranking scores because they tend to co-occur with many other terms. Thus, how to let down the ranking of general terms will be crucial to let up the ranking of class-specific core terms. We claim that the different section of a document can provide the directed reference information like hyperlink environment. For example, let abstract terms link to title terms, so the link can be taken as a scenario in which abstract terms give a detailed explanation of title terms by referencing title terms. Since title terms usually contain much bigger percentage of topical terms than that of abstract, these topical terms can receive much higher score than within term co-occurrence graph as shown in table 1. In this way, a corpus level abstract-title term graph is built with abstract terms pointing to title terms.

3.4 Term Clustering by Local Density

We assume that the top ranked terms will form several dense areas within the co-occurrence term graph.

$$In\_Cluster\_Degree(t_i, C_k) = Num(Edge_{t_i, C_{kj}}) \tag{3}$$

In Fig. 1, our algorithm grows core term clusters from the top ranked terms by PageRank or HITS. For example, if the starting term is "Hepatitis B, Chronic", the algorithm will take all terms that connect to "Hepatitis B, Chronic" as a candidate cluster including itself. Then each cluster member's in cluster degree will be calculated and then sorted in descending order. The algorithm will start from the term with the least in cluster degree and calculating whether its ratio with the highest one is over threshold Q. If it is so, the algorithm will keep the terms over threshold as a core term cluster. We keep only a few higher ranked terms in the core cluster and remove lower ranked terms to the pool of reassignment. If this core term cluster has enough number of terms, a new core term cluster is formed. Otherwise, the algorithm will skip this term and grow core term cluster from the second top ranked term. If the second is already included in the cluster, then it will start from the third and so on. As discussed earlier, each class only contains a small number of class specific core terms. To guarantee the high quality of initial term cluster, we set the minimum number of core terms in each cluster to 3 and the quality ratio Q as 0.8.

The remaining top ranked terms are assigned to K term clusters according to its In_Cluster_Frequency(ICF) that is the number of edges connecting it to cluster members with edge weight counted (please refer to Fig.1for details):

$$In\_Cluster\_Freq(t_i, C_k) = \frac{\sum Weight(Edge_{t_i, C_{kj}})}{Num(Edge_{t_i, C_{kj}})} \tag{4}$$

where $Weight(Edge_{t_i, C_{kj}})$ is the edge weight between term i and term j in cluster C_k and $Num(Edge_{t_i, C_{kj}})$ is the number of edges. In this study, we use the frequency of term co-occurrence as edge weight. This is also extensible to other types of weight.

3.5 Document Clustering

After core term clusters are identified, each document is assigned to its closest cluster:

$$DocClusterCloseness(d_i, C_k) = \sum In\_Cluster\_Freq(d_{i,t_j}, C_k) \qquad (5)$$

where d_{i,t_j} is the term t_j in document d.

4 Experimental Results

4.1 Document Sets

For the extensive experiments, we collect document sets about various diseases from PubMed, which is the web interface to MEDLINE. We use "MajorTopic" tag along with the disease MeSH terms as queries to PubMed. In this way, 10 document classes are collected for the experiments (see table 2).

Table 2. The document sets and their sizes

Document Sets	#.of Docs
Gout	642
Chickenpox	1,083
Raynaud Disease	1,153
Jaundice	1,486
Hepatitis B	1,815
Hay Fever	2,632
Kidney Calculi	3,071
Age-related Macular Degeneration	3,277
Migraine	4,174
Otitis	5,233

4.2 Term's Global Ranking

Since class-unspecific general terms co-occur frequently with many other terms, how to reduce the effects of common terms will contribute to discover distinguishable class-specific core terms. We evaluate the impacts of term graph construction on term's global ranking. As shown in the table 3, the ATTG schemes has six class specific core terms, while the TCG scheme contains only two core terms. Obviously, PageRank algorithm assigns higher weight to class specific core terms on ATTG than on TCG, which indicates that this representation is very effective on discounting class-unspecific general terms.

4.3 Term Clustering Evaluation

We evaluate the quality of term cluster by whether it contains class name related core terms (Table 2). As shown in table 4, nine out of ten semantic related and graphical

Table 3. Top ten terms ranked by PageRank

Abstract-Title Term Graph (ATTG)	Term Co-occurrence Graph (TCG)
Otitis	Patients
Migraine Disorders	Therapeutics
Patients	Disease
Therapeutics	Child
Child	**Otitis**
Macular Degeneration	**Migraine Disorders**
Infection	Time
Chickenpox	Infection
Hepatitis B, Chronic	Serum
Kidney Calculi	Role

Table 4. Term cluster identified by our algorithm using PageRank

	Term cluster(corresponding class name)
1	Kidney Calculi, Shock, Lithotripsy (**Kidney Calculi**)
2	Macular Degeneration, Visual Acuity, Vision (**Macular Degeneration**)
3	Chickenpox, Viruses, Herpesvirus 3, Human(**Chickenpox**)
4	Migraine Disorders, Epilepsy, Women (**Migraine Disorders**)
5	Otitis Media with Effusion, Otitis, Observation (**Otitis**)
6	Hepatitis B, Chronic, Lamivudine, Hepatitis B virus, Hepatitis B, Antigens(**Hepatitis B**)
7	Kidney Calculi, Calcium Oxalate, Organization and Administration (**Kidney Calculi**)
8	Jaundice, "Jaundice, Neonatal", Bilirubin, Life (**Jaundice**)
9	Rhinitis, Pollen, Immunotherapy (**Hay Fever**)
10	Macular Edema, Cystoid, Visual Acuity, Edema (**Macular Degeneration**)

connected term graphs (clusters) (containing class name related terms) are identified through our core term cluster identification algorithm. This indicates our method can create initial cluster models with high quality.

4.4 Document Clustering Evaluation

Cluster quality is evaluated by three extrinsic measures, *purity*[13], *entropy*[13], *F-measure*[13] and *normalized mutual information (NMI)* [11].

To test the two schemes of term graph construction on document clustering, we run global ranking including PageRank and HITS on both term co-occurrence graph and abstract title term graph. The effects of the number of term nodes used for the detection of core term clusters. From table 5,6,7,8, we can see that (1) PageRank and HITS have very similar performances; and PageRank is slightly better; (2) abstract-title scheme is better than co-occurrence scheme when the terms used for clustering are very few. This is expected because abstract-title scheme give class-specific core terms higher ranking than term co-occurrence graph (Table 3); (3) when all the terms are used for clustering, there is not much difference between ATTG and TCG scheme because Lower ranked terms have the same chance to be grew into a cluster as top ranked terms; (4) the most promising result is with the ATTG scheme; only 100 terms are enough for the detection of core term clusters. It is worth noting that 200 instead of 100 are assigned for TCG scheme, because 100 terms are not enough to identify core term clusters. We also compare clustering results to that of spherical K-mean clustering using TF*IDF as document representation. Our clustering performance is slightly worse than K-mean (Entropy: 0.40, F-measure:0.754, Purity: 0.889 and NMI: 0.755). However, our algorithm (including term ranking, core term cluster identification and document matching) performs more efficiently. On a laptop PC with Duo core 1.83GHZ, and 1GB memory setting, our algorithm usually finishes within 20 seconds, while spherical K-means costs more than 30 seconds. Spherical K-means needs to iteratively re-estimate distance between each document and cluster center, while our algorithm can determine documents' class labels by one time calculation once the k core term clusters are discovered. Our contribution is beyond the document clustering itself since class-specific term clusters can serve as the interpretation of clustering results.

Table 5. PageRank on Abstract-Title term graph

Terms for clustering	Entropy	F-measure	Purity	NMI
100	0.500	0.811	0.885	0.661
200	0.620	0.772	0.850	0.640
All	0.633	0.763	0.840	0.654

Table 6. PageRank on term co-occurrence graph

Terms for clustering	Entropy	F-measure	Purity	NMI
200	0.780	0.737	0.812	0.581
300	0.731	0.758	0.825	0.610
All	0.646	0.764	0.839	0.650

Table 7. HITS on Abstract-Title term graph

Terms for clustering	Entropy	F-measure	Purity	NMI
100	0.522	0.799	0.785	0.641
200	0.656	0.742	0.760	0.623
All	0.641	0.763	0.775	0.642

Table 8. HITS on term co-occurrence graph

Terms for clustering	Entropy	F-measure	Purity	NMI
200	0.801	0.699	0.724	0.563
300	0.742	0.706	0.782	0.602
All	0.677	0.752	0.801	0.631

5   Conclusions and Future Work

In this paper, we present a framework of graph-based document clustering which utilizes both global and local term importance information on not only clustering documents but also identifying class-specific core term clusters. We mainly discuss:

(1) two schemes of graph construction: term co-occurrence graph and abstract-title; (2) the identification of class-specific core term clusters from these two schemes by considering terms' global and local importance; (3) clustering document based on the identified term clusters. Our main findings include: (1) the identified core dense term clusters are both important for document clustering and clustering results interpretation; (2) abstract-title graph scheme improves the global ranking of class-specific core terms and thus identifies more class-specific core term clusters than term Co-occurrence graph scheme (3) while performing more efficiently, our algorithm is comparable to top clustering algorithm such as spherical K-means. For our future work, we would like to evaluate our approach on other applications such as text classification and summarization; also we will extend our research to general domain.

Acknowledgments. This work is supported in part by NSF Career Grant IIS 0448023, NSF CCF 0514679, PA Dept of Health Tobacco Settlement Formula Grant (No. 240205 and No. 240196), and PA Dept of Health Grant (No. 239667).

References

[1] Angelova, R., Weikum, G.: Graph-based text classification: learn from your neighbors. In: SIGIR 2006, pp. 485–492 (2006)

[2] Charkrabarti, S., Dom, B.E., Indyk, P.: Enhanced hypertext categorization using hyperlinks. In: SIGMOD'98, pp. 307–318 (1998)

[3] Cohen, W.W., Hofmann, T.: The missing link—a probabilistic model of document conent and hypertext connectivity. In: NIPS 13 (2001)

[4] Erkan, G., Radev, D.R.: LexRank: Graph-based Lexical Centrality as Salience in Text Summarization. J. Artif. Intell. Res. (JAIR) 22, 457–479 (2004)

[5] Hassan, S., Banea, C.: Random-Walk TermWeighting for Improved Text Classification. In: Workshop on TextGraphs, at HLT-NAACL 2006, pp. 53–60 (2006)

[6] Kleinberg, J.: Authoritative sources in a hyperlinked environment. In: Proc. Ninth Ann. ACM-SIAM Symp. Discrete Algorithms, pp. 668–677. ACM Press, New York (1998)

[7] Marklv, A., Last, M., Kandel, A.: Model-based classification of web documents represented by Graphs. In: proceedings of WebKDD 2006 workshop on knowledge discovery (2006)

[8] Page, L., Brin, S., Motwani, R., Winograd, T.: The PageRank citation ranking:Bringing order to theWeb. Technical report, Stanford Digital Library Technologies Project (1998)

[9] Steinbach, M., Karypis, G., Kumar, V.: A Comparison of Document Clustering Techniques. Technical Report #00-034. Department of Computer Science and Engineering, University of Minnesota (2000)

[10] Wang, B.B., McKay, R.I., Abbass, H.A., Barlow, M.: Learning Text Classifier using the Domain Concept Hierarchy. In: Proceedings of International Conference on Communications, Circuits and Systems 2002, China (2002)

[11] Zhong, S., Ghosh, J.: A comparative study of generative models for document clustering. In: Proceedings of the workshop on Clustering High Dimensional Data and Its Applications in SIAM Data Mining Conference (2003)

[12] Zipf, G.K.: Human Behaviour and the Principle of Least-Effort. Addison-Wesley, Cambridge MA (1949)

[13] Zhao, Y., Karypis, G.: Criterion functions for document clustering: experiments and analysis, Technical Report, Department of Computer Science, University of Minnesota (2001)

Ontology-Based Information Extraction and Information Retrieval in Health Care Domain

Tran Quoc Dung and Wataru Kameyama

Graduate school of Global Information and Tele-communication study, Waseda University
dungtq@akane.waseda.jp, wataru@waseda.jp

Abstract. Ontology–based information extraction is considered as an effective method to improve the performance of information extraction (IE) systems. In order to build a better IE system using ontology-based technique, the two challenges should be most taken into account are semantic elements extraction, and ontology enhancement. Focusing on the two key points above, we propose an ontology-based information extraction system and its application information retrieval system in the health care domain. In ontology-based information extraction system, we propose a non part-of-speech based method for the semantic elements extraction and the ontology enhancement to overcome the above challenges. Experiments and Results show that all functions work well in the whole system. Many kinds of semantic elements can be extracted by the proposed extraction functions regardless parts-of-speech of words. High ratio of new semantic elements is detected for the ontology and database enrichment that yield very good results in information retrieval. The system is implemented in Vietnamese.

Keywords: Ontology, Ontology Enhancement, Information Extraction, Information Retrieval.

1 Introduction

The ontology allows us to define concretely objects and relations between them. Therefore, it is considered as a data model that is powerful in reasoning over data. With those features, ontology-based techniques have been used widely in many fields of computer science as well as information technology especially in knowledge management systems or intelligent systems [3], [4]. Ontology-based Information Extraction (IE) techniques are of the cases taking more and more researches on, where the ontology has been proved to be very helpful in improving accuracy of information extraction. However, these techniques are also facing to two major challenges: instances extraction using the ontology and the ontology enhancement.

The first challenge relates to identification of instances from the ontology in the text. Normally, there are five different types of information extraction from a text in a natural language: Entities, Mentions, Descriptions of entities, Relation between entities, and Events [3]. In general, the above information is extracted by Part-of-speech (POS)–technique which is mostly used in IE system [3], [4], [7]. This method

I.Y. Song, J. Eder, and T.M. Nguyen (Eds.): DaWaK 2007, LNCS 4654, pp. 323–333, 2007.

is not usable in applications that need to extract various semantic elements regardless of their part-of-speech. For example, at the same time, we need to extract named entities, descriptions of entities and events that happened to those entities. It can be argued that an extra gazetteer can be used in this case, however with such method; the whole system will not be synchronized because gazetteer is separated from the ontology and the knowledge as the result will not be well managed. Another week point of the part-of-speech based method is that it might not be effective with some languages which are facing to difficulties in natural language processing issues such as segmentation or part-of-speech.

The second challenging task in the ontology-based information extraction is the ontology enhancement [3], [6], [7]. The ontology enhancement is a process used for upgrading the ontology with new instances to cover the knowledge better in a domain. This is a very important issue in the ontology-based information extraction system because in order to extract exactly semantics elements, ontology should be well-defined to cover the knowledge as much as possible in a domain. This task, however, is rather difficult and takes time depending on the range of a domain such as how the ontology should be big enough for extraction in a domain, and how to construct such a large ontology at the beginning are really challenging tasks. So that, in the ontology-based IE system, there should be a strategy and mechanism to enrich the ontology for the better performance.

And again, in the circumstances of having diversity of part-of-speech of semantic elements, the ontology enhancement issue is getting more complex. At the same time, all new semantic elements ignoring their part-of-speech have to be detected and updated in the ontology for the better performance in the extraction process.

Taking these two challenging issues into account, we propose an XML-based schema that is used for extracting many types of semantic elements as well as detecting new elements for the ontology enhancement regardless of their part-of-speech to build an ontology-based information extraction and information retrieval system in health care domain. In our system, we extract health care information from web pages dedicated for only healthcare information. The following information is defined as semantic elements to be extracted:

- **Concept** (C): named entities about every parts of human body such as heart, lung, kidney...
- **Name of Disease** (N): words or phrases of disease names
- **Description** (D): any words or phrases that describe Concepts. "Description" refers to any kind of words or phrases that relates semantically to Concepts.
- **Pair** of Concept and Description (P): all possible combinations of Concepts and Descriptions. Combinations contain full meaning of relationships between C and D.

With these semantic elements as defined above, our system has to extract diversity of words regardless of their part of speech. Because normally, Concept or Name of disease elements are nouns and Description elements are adverbs or adjectives. Besides that, in Vietnamese, Description elements can be verbs. Moreover, our system has to be able to detect new semantic elements from various types of words in the web documents to upgrade ontology.

Organization of this paper is as follows: after this introduction part, we take a look of some related works to figure out the general picture of the IE issues in section 2. Section 3 is dedicated for the proposed system introduction. In this part, the system architecture, the proposed methods for semantic elements extraction and ontology enhancement which play as a core of system are presented. Experiments and results are shown in section 4. And base on the system performance, we present an application in health care information retrieval in section 5. We conclude the paper in the last section 6 together with future works.

2 Related Works

In this part, we introduce some representative researches on the ontology-based information extraction and information retrieval systems in which extraction, ontology enrichment and information retrieval methods are most focused on.

In many existing IE systems, extraction is mostly based on the part-of-speech mechanism to detect semantic elements. M. Abulaish in [7] has introduced a method to extract the relationship between entities in biological documents. In this method, words that are tagged as verb part-of-speech are processed to extract the fuzzy relationship to upgrade the ontology. With the same technique, L.Dey et al in [3] have proposed a method to extract imprecise description of wine entities. Wines have attributes such as color, taste but each attribute can be expressed more detail with additional descriptive words or phrases: light red and red for color or strong, light and very strong for taste. The author has designed an extension of ontology structure to include the imprecise information.

The ontology enhancement and ontology generation are also taken many efforts in researches recently [1],[2],[7]. Dealing with ontology generation, Shamsfard and Barforoush[2] have suggested an ontology building approach in which the system starts from a small ontology kernel and constructs the ontology through text understanding automatically. The kernel contains the primitive concepts, relations and operators to build the ontology. Michael Dittenbach, Helmut and Dieter Merk [6] have provided a connectionist approach to visualize semantic relation inherent in free form text documents related to a specific domain. They exploit word co-occurrence to catch the relationship of the word's context. And after that, they use the well-known neural network which is called self-organizing map with unsupervised learning function to map the high dimensional data onto a two-dimensional representation for convenient browsing.

Muller, Kenny, and Strenber [5] have developed an ontology-based text mining system TEXTPRESSO for scientific literature that splits documents into sentences, and sentences into words or phrases. Each word or phrase is then labeled according to the concepts defined in the Gene Ontology (GO). The labeling is based on identification of a set of terms associated to a concept. These terms and the set of entities to be looked for are handcrafted in consultation with biological databases. An index on all sentences with respect to labels and words is created to allow a rapid search for sentences that have a desired label and/or keyword. Ontology enhancement is not mentioned in TEXTPRESSO.

Another well-known platform for semantic annotation, indexing, and retrieval system called KIM has been introduced by Borislav Popov et.al [8]. It allows semi-automatic annotation and ontology population for the Semantic Web, using Information Extraction (IE) technology. KIM is based on two major platforms; GATE and Sesame/OMM. KIM has an upper-level ontology named KIMO having about 200 classes covering the most important entity types in a semantically sound fashion and providing ground for expansion to include more complex knowledge like relations, scenarios, events, domain or task-specific knowledge and integration with third party/customer information systems. Based on the ontology, large-scale and instance-base entity descriptions are maintained.

With approaches in the survey above, the extraction process mostly bases on part-of-speech method that has some weaknesses as described in the introduction part. The ontology enhancement, on the other hand does not get much efforts so that the issue has not yet been solved effectively. In addition to that, ontology enhancement and IE are dealt separately in systems. That is, ontology is constructed and enriched in the ontology engineering system and the completed ontology will be used as a knowledge base for information extraction.

3   Proposed System Architecture

In this part, we present our ontology-based information extraction system. The main functions of the system are to extract all semantics elements Cs, Ds, Ns and Ps in the health care domain from web pages and to detect new semantic elements for ontology enhancement during extraction process. The Fig.1 shows our proposed architecture:

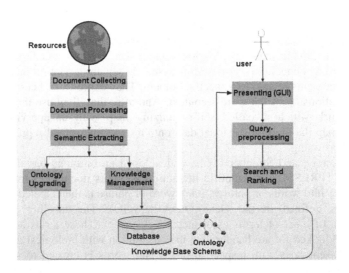

Fig. 1. Architecture of Proposed Information Extraction System and Information Retrieval

3.1 Knowledge Based Schema

The knowledge base includes the domain ontology and database. In the domain ontology, we construct a health care domain ontology which consists of sets of vocabularies about parts of body which are defined as Concept elements (Cs), names of diseases (Ns) and description of concepts (Ds) using protégé tool. All parts of body are divided into categories and sub-categories with respect to human organ systems such as circulatory system, immune system, respiratory system, heart, stomach. Each Concept may have a set of descriptions which are considered as attributes of entities. Descriptions are words or phrases that have descriptive relation with Concepts. Name of diseases is a list of diseases. Additionally, in order to provide a knowledge base for future use in other applications, we create a rational database to index web documents with respect to semantic elements. Database is also used to store all possible Pairs (Ps) which are combinations of Concepts and Descriptions.

3.2 Document Collecting Module and Document Processing Module

The document collecting module is designed to collect web pages for the extraction. A web spider is developed to craw and download new web pages form the list of specific web sites that are mainly dedicated for health care information in Vietnam.
The collected web documents are being processed to facilitate the semantic extraction in document processing module. All documents are filtered to remove other information except text. After that, all obtained text are reconstructed in a specific file format where the first line of a text is the address of the document and the second line is title of the document followed by the text body.

3.3 Semantic Elements Extracting Module

In order to extract various semantic elements at the same time, we implement a non part-of-speech method which is based on our proposed algorithm named "semantic element extracting algorithm" to build a structured XML file that represents content of web document along with extracted semantic elements.

```
<!ELEMENT vnhies (address, title, content)>
<!ELEMENT address (#PCDATA)>
<!ELEMENT title (#PCDATA)>
<!ELEMENT content (paragraph+)>
<!ELEMENT paragraph (sentence+)>
<!ELEMENT sentence (CDATA, concept*, description*,  disease*)>
<!ELEMENT concept (#PCDATA)>
<!ELEMENT description (#PCDATA)>
<!ELEMENT disease (#PCDATA)>
]>
```

Fig. 2. Structure of XML File for Semantic Extraction

Some of the key behavioral features of the algorithm are: (i) to covert a text file into XML file with structure defined by the DTD as in the Fig.2, (ii) to split document into paragraphs and to split paragraphs into sentences. Each sentence is scanned to extract Concept, Description of concept, and Name of diseases with a reference to the ontology. All extracted semantic elements are to be added to the XML with an element named Concept, Description and Diseases.

3.4 Ontology Upgrading Module

We designed a "semantic elements learning algorithm" so as to try to detect unknown semantic elements from resources for updating the knowledge base for the better performance in IE. The basic idea of the algorithm is using fuzzy relationships between Concepts and Descriptions in health care domain. An XML file created in the extraction phase is processed as follows. We traverse all sentences elements of the XML document and take a focus on which has some child elements "concept", "description" or "disease". And new semantic elements are guessed by the rules below:

- *If there are many Concept elements(Description elements) but just a few Description elements (Concept elements) in a sentence then there might be some unrecognized Description elements,* or
- *If there are some Concept elements and some Description elements in a sentence but no possible pair from database is found then there might be some unrecognized pairs of Concept-Description.*

The sentence that matches one of the above rules is being sent to a suggestion list and is used latter by the domain experts to extract real semantic elements. With this approach, new Concepts, Descriptions, and Pairs of Concept-Description might be detected regardless parts-of-speech of words. The algorithm itself is not dedicated for extracting new Names of disease elements. But with this open approach, some Names of disease elements might be found in the sentence that is added in the suggestion list.

3.5 Knowledge Base Management Module

In any web document, the system not only extracts semantic elements but also tries to catch more knowledge about the page. This can be done by calculating the summary of the page with respect to semantic elements extracted from the page. The knowledge management module is dedicated for this function. All the documents after being converted to XML file are summarized according to the extracted words. Then the document with summary information is indexed and stored in the database for later use by other applications.

4 Experiments and Results

We take some big web sites about the health care information currently published online on the Internet. For the experiment, we develop a crawler to download around 30000 pages from 6 indicated sites. The knowledge base comprises ontology and a database. The ontology has been constructed with 205 entities of parts of body, 30 Names of disease and 60 Descriptions. In the database, there are 10 records of Pairs of Concept-Description. In the first state, knowledge base has been purposely constructed not too big in order to test the performance of the knowledge base enhancement function.

At first, we take a test on the knowledge base enhancement function in order to get more semantic elements for the information extraction test. The Fig.3 shows the number of increased new semantic elements after processing 1500 web pages, semantic elements in knowledge base after processing 1500 pages comprising of 438 Concepts, 515 Descriptions, 971 Pairs and 120 Names of diseases.

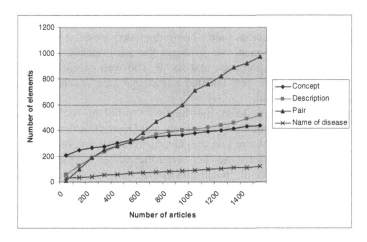

Fig. 3. Number of Semantic Elements Changes

To count how many percentages of new semantic elements are learned by the algorithm, in the first 100 web pages, we calculate manually the number of unrecognized elements and compare it with the number of new extracted elements by the algorithm. Table 1 below shows how many percentages of new semantic elements have been extracted in first 100 web pages using the original knowledge base.

Table 1. New Semantic Elements Detected in 100 Web Pages

Type of Extracted Information	Unrecognized Elements	Extracted Elements	Extraction Percentage <%>
Concept	316	201	63.61
Description	538	356	66.17
Pair	478	388	81.17
Name of Disease	67	23	34.33

Because the 100 articles are extracted using the same original ontology, there are some duplicated elements in the unrecognized elements set as well as the extracted ones from the suggestion set. Regardless of the duplicated elements, we just consider how our method can recognize new elements. The result shows that we extract the high number of new elements where Pairs can be detected most due to the very limited source in the database.

Table 2. Precision and Recall Ratios after Extracting 100 Articles with New Updated Ontology

Semantic Elements (SE)	SE in Sources	Correct Extracted SE	Incorrect Extract SE	P	R
Concept	1639	1498	63	96	91.4
Description	2064	1918	79	96	92.9
Pair	1896	1709	20	98.8	90.1
Name of Disease	402	351	10	97.2	87.3

The ontology and the database after being upgraded are used for the information extraction. For this experiment, we evaluate the performance of the information extraction function by the precision and recall. 100 web pages are used to extract semantic elements using the new updated knowledge base. Table 2 summarizes the results of precision and recall.

Wrong extraction comes from the linguistic problem such as antonyms or synonyms, and the context ambiguity. Concept and Description semantic elements are more affected from the above reason that causes the precision values lower than the precision of Pair or Name of disease. The linguistics issues and the context ambiguity also have the influence on recall. In addition to that, there are still number of semantic elements that are not recognized due to the limitation of the ontology and the database in the knowledge base. These make the recall result lower than the precision but this can be improved by upgrading the knowledge base process.

5 Discussion and Application

We have developed an ontology-based information extraction system for the health care domain. Our test result shows that our system can extract high rate of semantic elements in sources. And the most important thing is that we can upgrade the ontology during extracting phase. It makes ontology be enriched gradually and be useful so that it covers almost semantic elements in the health care domain. We also provide a ready used knowledge base with indexed documents according to semantic elements appearing in the document. With these advantages above, we develop an application that allows users to query some health care information.

5.1 User Interface and Query Processing

It provides a user with a GUI to create queries as well as present searching result. Before searching, the user query should be processed to facilitate the searching process. The functions of this module are to filter non-Vietnamese query, to normalize the query string and to try to catch the meaning of the query by extracted all possible semantic elements in the query. The query string is being scanned to extract any C, D, P or N by referring to ontology and database. The query is reformed as a set of semantic elements and being sent to search and ranking module to find most matched documents. Any query that does not contain any C or N is considered as a trivial query, and it needs not to be sent to the search and ranking phase.

5.2 Search and Ranking

Semantic elements which are extracted in the pre-processing are considered as keywords to search because those elements carry meaning of the query and imply what user wants to search about. In order to find the most matched document to a query in semantic point of view, the search and ranking procedures should satisfy the following constraints:

- Width priority: it bases on the number of semantic elements the document covers. The document that covers all semantics that are included in the query is considered to have a higher ranking than less covering documents.
- Depth priority: retrieval documents also can be ranked by the number of C, D, N and P in the query appearing in the document

Taking these constraints into the algorithm designing process, we propose the "Search and Ranking Algorithm". Assume that an arbitrary query has m Concepts (C1, C2,..,Cm), n Descriptions (D1, D2,..,Dn), q Names of diseases (N1,N2,..,Nq), and s Pairs of Concept and Description (P1,P2,..,Ps).

First, we find all web documents in the database that contain any of Concept and Description semantic elements in the query. Let X is the set of documents that has been found. Then, for each document in X, let's says: S is the total number of semantic elements C, D, N, P in the document that exactly match with those in the query.

Next step, the weight of each document in X is calculated by (1):

$$w = S * \left(\alpha 1 * \sum_{i=1}^{n} TD_i + \alpha 2 * \sum_{j=1}^{m} TC_j + \alpha 3 * \sum_{k=1}^{s} TP_k + \alpha 4 * \sum_{l=1}^{q} TN_l \right) \tag{1}$$

where

- $\alpha 1$, $\alpha 2$, $\alpha 3$, $\alpha 4$ are meaning coefficients of Description, Concept, Pair and Name of disease respectively and
- TD_i, TC_j, TP_k, TN_l are number of appearances of Descriptions D_i, Concepts C_j, Pairs P_k, and Names of diseases N_l respectively.

The value w of each document represents how much the document relates semantically to the user query. These values are being used to compare among documents to get the most matched one to the query. W = {w} which is set of weights for documents in X is sorted in descending order for document ranking.

5.3 Information Retrieval Part

The evaluation method is as follows. We use randomly five queries to search health care information in both keyword-based search engine and our system. Both the ontology-based and the keyword-based methods use the same data sources to implement five queries. There are three specific sites chosen for the experiments www.vnexpress.net, www.vietnamnet.net and www.bacsigiadinh.com that are called Dataset 1 (DS1), Dataset 2 (DS2), Dataset 3(DS3) hereafter respectively. For each query, taking the first 10 retrieval documents in retrieval result of both methods, we

Table 3. Precision and Recall Result in IR test

Query No.	DS1		DS2		DS3	
	P(%)	R(%)	P(%)	R(%)	P(%)	R(%)
Q1 - ONT	75.3	67.9	70.8	68.54	63.2	56.9
Q1 - KWD	34.6	31.2	33.2	23.4	19.8	18.8
Q2 – ONT	65.7	60.4	72.4	70.1	56.7	55.1
Q2 – KWD	45.6	24.5	38.5	30.4	23.6	22.0
Q3 – ONT	78.7	77.9	59.8	57.3	65.1	64
Q3 – KWD	23.5	20.6	19.7	17.8	33.1	29.8
Q4 – ONT	51.2	45.8	48.5	45.3	44.3	40.5
Q4 – KWD	35.6	34.5	8.3	4.9	9.4	7.3
Q5 – ONT	60.4	54.2	48.3	47.5	58.8	50.4
Q5 – KWD	18.4	16.5	21.1	13.9	9.6	8.3

calculate the precision and recall values. The measurement of recall and precision requires human assessment of the relevancy. However, to facilitate the measurement, relevant elements are calculated by exactly matched semantic elements in queries and in documents. Table 3 shows the precision and recall results. From the semantic elements point of view, it is clear that, the ontology-based method (ONT) is much better than the keyword-based method (KWD). However, some more evaluations should be done to assess the level of satisfaction to user requirements

6 Conclusion and Future Works

In this paper, we present a proposal of the information extraction and information retrieval system in health care domain. The proposed system can extract many kinds of semantic elements of the health care information regardless of parts-of-speech of words. The extracted information from Web documents is then summarized, indexed and stored in the database for the information retrieval. The system also provides the strategy and method to enrich the ontology and the database with new extracted semantic elements so that the knowledge base is upgraded gradually and supports better for information extraction. After all, the information retrieval which is one of application has been presented. The experiments and results show that our approach works well through all system. In the next, we will focus on taking more advantages of the ontology-based to improve IR function.

References

1. Cao, T.H., et al.: Automatic Fuzzy Ontology Generation for Semantic Web. IEEE Transactions on Data and Knowledge Engineering 18(5), 842–856 (2006)
2. Shamsfard, M., Barforoush, A.A.: Learning ontologies from natural language texts. International Journal of Human-Computer Studies 60(1), 17–63 (2004)
3. Davies, J., Studer, R., Warren, P.: Semantic Web Technology: trends and research in ontology-based systems, pp. 29–49. John Wiley & Sons Ltd, England (2006)

4. Taniar, D., et al.: Web Semantic Ontology, pp. 189–226. IDEA group publishing (2006)
5. Muller, H.M., Kenny: Textpresso: An ontology-based information retrieval and extraction system for biological literature. PloS Biol. 2(11), e309
6. Dittenbach, M., Berger, H., Merkl, D.: Improving Domain Ontologies by Mining Semantics from Text. In: Conference on Conceptual Modeling, Dunedin, New Zealand (2004)
7. Muhammad, A., et al.: Biological Ontology enhancement with Fuzzy Relation: A Text Mining Framework. In: International Conference on Web Intelligence WI'05 (2005)
8. Borislav, P., et al.: KIM – Semantic Annotation Platform. In: Fensel, D., Sycara, K.P., Mylopoulos, J. (eds.) ISWC 2003. LNCS, vol. 2870, pp. 20–23. Springer, Heidelberg (2003)

Fuzzy Classifier Based Feature Reduction for Better Gene Selection

Mohammad Khabbaz[1], Kievan Kianmher[1], Mohammad Alshalalfa[1], and Reda Alhajj[1,2]

[1] Computer Science Dept, University of Calgary, Calgary, Alberta, Canada
[2] Department of Computer Science, Global University, Beirut, Lebanon

Abstract. This paper presents a novel approach for identifying relevant genes by employing a fuzzy classifier. First a fuzzy classifier rule set is derived such that each rule involves a compact set of genes. Then, a correlation matrix is produced by considering the correlations between the genes in each rule. Apriori is applied on the correlation matrix to find the maximal sets of correlated genes after tuning the minimum support value. Experiments conducted on the Leukemia dataset demonstrate the effectiveness of the proposed approach in producing relevant genes.

1   Introduction

Feature selection in biological domain is to identify a small subset of genes which are informative and relevant to classification problems. This leads to dimensionality reduction and hence computational cost reduction. It prevents irrelevant genes from overshadowing the contribution of relevant genes. Further, identifying informative genes may be beneficial in revealing genes of particular importance in a biological process. The basic feature selection methods may be categorized into three broad groups: filter [3,4], wrapper [5,6] and embedded [7,8].

Many supervised machine learning algorithms such as neural networks, Bayesian networks and support vector machine (SVM), combined with feature selection techniques, have been previously applied to microarray gene expression. For instance, in the work described in [2], we achieved outstanding accuracy in gene expression classification by combining neural networks learning and SVM-based feature selection into an integrated approach. However, most of the existing gene selection approaches employed in the integrated solutions use a common technique which is based on removing or adding features from the actual training set to a smaller training set and measuring the quality of features one by one; they basically do not consider the correlation existing among genes while removing them. Further, some of the genes selected by these methods are not relevant to the disease being investigated, even though they result in building an accurate classifier model. Some other techniques that take the correlations among the genes into account suffer from unreliability, i.e., each run of the algorithm results in different genes extracted from the dataset; this degrades the reliability of the approach. To biologists, gene expression data can be a valuable source for

I.Y. Song, J. Eder, and T.M. Nguyen (Eds.): DaWaK 2007, LNCS 4654, pp. 334–344, 2007.
© Springer-Verlag Berlin Heidelberg 2007

understanding the biological associations between genes. Relevant associations among different genes may be discovered by using rule mining techniques.

In this study, we take advantage of using fuzzy classifier rules to capture the correlations between genes. The motivation is that a fuzzy classifier rule is essentially an "if-then" rule that contains linguistic terms to represent feature values. This way, the correlations among the genes are very simple to understand and interpret by domain experts. In our proposed gene selection procedure, instead of measuring the effectiveness of every single gene for building the classifier model, we incorporate the impotence of a gene correlation with other existing genes in the selection process. That is, we reject a gene if it is not in a significant correlation with other genes in the dataset. To improve reliability, we repeat the process several times in the experiments, and report genes that are selected in most experiments.

To demonstrate the effectiveness of the proposed method, we compare it with SVM-RFE [9] based gene selection approach, namely r-GeneSelect [10]. SVM-RFE [9] has been recognized as a powerful wrapper approach to conduct relevant gene selection for cancer classification. We obtained the source code of r-GeneSelect from [11] as developed by the authors of [10]. After conducting very precise biological analysis on several gene expression datasets, we noticed that some of the genes selected by this method are not biologically significant. Even though the selected genes resulted in building a highly accurate classifier, they are irrelevant to cancer. Using our gene selection method, we are able to build accurate classifier mostly with fewer number of genes compared to r-GeneSelect. Also, we argue that our genes are more related to cancer. We demonstrate this by performing a comprehensive biological analysis on the functionality of the genes selected from the ALL/AML leukemia dataset.

The rest of the paper is organized as follows. Our approach for discovering the correlation among the genes is presented in Section 2. Section 3 describes how relevant genes are extracted by employing a frequent pattern mining technique. Section 4 presents the selected evaluation model, the conducted experiments and the achieved results. Section 5 is summary and conclusions.

2   Constructing the Correlation Set

The first step in the proposed method is creating from the training set a classifier fuzzy rule set that highlights important existing correlations between informative genes. This is a mapping between the training set and a correlation set containing correlations between features. Since our main purpose here is not building a fuzzy rule based classifier, we propose an algorithm for generating fuzzy rules that rely on data samples from the training set. The target rule set is generated through a number of iterations, while changing the minimum expected value for standard deviation. The steps involved in each iteration of the process are described in the rest of this section.

Filtering Genes by Standard Deviation: Useful genes from every dataset are supposed to be able to discriminate patterns from different classes. To

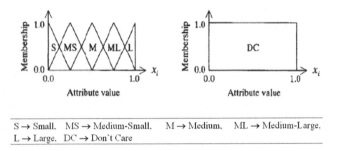

S → Small,   MS → Medium-Small,   M → Medium,   ML → Medium-Large.
L → Large,   DC → Don't Care

Fig. 1. Fuzzy membership functions

improve the performance of our fuzzy rule set classifier, we have added a simple filtering step before generating rules from data samples in the training set. This filtering procedure works as follows. Given a dataset and a threshold t, gene values for different classes are separated into different sets. Then, the mean value for each set is calculated, and the standard deviation of the mean values is computed. If the standard deviation is below t, the gene is filtered out without getting to the next step. The only concern here is selecting the value of t, which is variable in our final algorithm, i.e., we generate rule sets starting from low values of t and incrementally increase t in each subsequent step of the iterative process. We keep adding rules to the final rule set until t is so high that no more features can pass the initial filter.

Constructing Fuzzy Classifier Rule: A fuzzy classifier rule in our system is composed of a sequence of fuzzy linguistic values that indicate a class label. The employed fuzzy membership functions are shown in Figure 1. The length of a fuzzy classifier rule is the same as the number of genes. A fuzzy rule set consists of several fuzzy classifier rules.

The construction of a fuzzy rule set from the training set has already received considerable attention, e.g., [12,13]. Our proposed system generates the classifier rules by using existing patterns in the training set. For each feature (gene), we first check if it is in the list of features that passed the filter. If found, the attribute in the rule is set to the symbol of the membership function that gives the maximum value for the corresponding feature value. If not, the attribute in the rule is set to "*don't care*".

Assigning Class Label: In general, not all generated rules are useful. To assign class labels to the generated rules, we consider all data samples present in the training set. The degree of compatibility between a fuzzy classifier and a data sample is defined as the minimum of all membership values for the corresponding feature (gene) values in the fuzzy rule and the data sample; it is computed as:

$$\mu_j(x_p) = Min\{\mu_{j1}(x_{p1}), ..., \mu_{jn}(x_{pn})\} \ p = 1, 2, ... \tag{1}$$

where $\mu_j(x_p)$ is the degree of compatibility between rule j and sample x_p, and μ_{ji} is the membership function for the i-th attribute of the rule; p changes over the whole dataset for checking the compatibility of the rule with all samples in the set. After calculating the degree of compatibility between each rule and each sample, we determine the rule class using the sum of compatibility degrees for all classes as shown in Equation 2:

$$\beta_h(R_j) = \sum_{x_p \in Class h} \mu_j(x_p) \tag{2}$$

Here, $\beta_h(R_j)$ is the sum of all compatibility degrees of the j-th rule with samples of class h. Among all the classes, the one which achieves the maximum β value is the class selected for the newly created rule.

Evaluating Rule Effectiveness: Since our system works with individual rules created from data samples, we need a measure to evaluate each rule separately. The compatibility degree of each rule with each data sample is either 0 or a positive value. We compare the class labels of the rule and the data sample; if both class labels are the same, the sample is classified by the rule; otherwise the sample is misclassified by the rule. Associated with each rule, we have 2 variables in our system: one is the number of samples classified by the rule and the other is the number of samples misclassified by the rule. The former indicates the strength of the rule in correctly classifying samples and the latter indicates the risk of future misclassifications by the rule. These two factors are used to evaluate the effectiveness of each rule.

The Algorithm for Generating Rules: The process of generating the rule set used in constructing the correlation set invokes the following functions. *CandidateSamples* refers to the list of data samples which are candidate patterns for rule generation. If a rule created from a candidate sample is not able to satisfy the initial criteria, the sample is removed from the list of candidate samples. *GetRandomSample* picks a data sample randomly and returns it as a candidate for rule generation. The selection is done randomly because there is no guarantee that creating the rule set starting from the beginning of the list gives the best order. Due to this effect, *GetRules* does not return the same rule set each time. However, since there are some distinct genes which are important for the classification, important genes are identified by the algorithm regardless of the order of rule generation. To neutralize this randomness effect and to improve the reliability, we repeated the process several times in the experiments, and genes selected in most of the experiments are reported in the result. *DetermineRuleClass* determines the class label for the rule. *EvaluateRule* counts from the training set the number of data samples classified and misclassified by the current rule. We use these values to make an initial decision whether the rule is a good candidate for gene selection or not. No misclassification is allowed for the initial rule since we have enough candidate samples to generate rules and hence do not want to take any risk in rule selection. Before adding a generated rule to the rule set, it is reduced using the following function, *Reduce*, which is basically the feature reduction unit.

Given the list of features other than "don't care", we select a random feature from the rule and changes it to *"don't care"*. If this does not affect the effectiveness in terms of the number of samples classified and misclassified by the rule, the change is committed, otherwise the feature is identified as playing an important role in the classification power of the rule, and hence we rollback the change. Here, we still have the random selection for the order in which we remove the features because there is no guarantee that starting from the head of the list is the best order. This randomness effect for feature selection is natural. Other techniques which were attempted for gene selection also suffer from this randomness. For instance, SVM-RFE [9] extremely suffers from this randomness and each run of the algorithm returns different features. In [10], the authors tried to remove this randomness by reliability analysis using 10-fold cross validation. Although the most important genes reported from their approach are the same in all the experiments, it still suffers from randomness because some genes change in the final result of each run. To resolve this problem, we repeated the experiment and reported genes that appear as reliable in most experiments.

Eventually, the algorithm that generates the final rule set is simply a loop over the *GetRules* function. Before the loop, we identify the maximum standard deviation among all genes. Then, we define the step size for the threshold according to the number of times we want to repeat the algorithm with different thresholds for the initial filtering. We generate the sub-rule sets starting from threshold 0, and then change the threshold by the step size and add the sub-rule sets to the final rule set until the threshold value is so high that no gene can pass the filter. Using this approach in changing the value of the threshold, we consider the standard deviation as a factor for the effectiveness of individual genes and pay more attention to them, while at the same time we do not ignore the fact that some features that do not have high standard deviation values may also be important due to their correlation with other genes. These genes are given the chance to show their importance in the first iterations of the algorithm, and as the algorithm proceeds to the next iterations more chance is given to genes with higher values of standard deviation because discovering correlations between extremely important genes is a priority.

Generating the Correlation Set: The correlation set consists of correlations between genes where each row of the set contains few genes which are highly correlated to each other in the classification process. We create the correlation set using the final rule set. After evaluating each rule, it is associated with the samples it classifies. Here, it is worth noting that the number of misclassified samples for each rule is zero because we don't let any misclassification to happen by enforcing feature reduction when we generated the rule. The sequence of active features in the rule forms a correlation between them. The number of times a correlation appears in the correlation set is the same as the number of data samples classified by the rule.

3   Mining Frequent Patterns

We used the Apriori algorithm to discover frequent patterns in the correlation set. Using Apriori requires setting two initial parameters, Support and

Confidence. The confidence parameter is set to 0 for all experiments because our purpose is discovering frequent patterns and we don't want to generate association rules. Assuming the value ε as the support for the frequent itemsets, we extract all frequent itemsets that satisfy the minimum support. Assuming we have N frequent itemsets, the final list of genes is obtained using Equation 3.

$$FeatureSet = \bigcup_{i=1}^{N} \{MaximalSet_i | Size(MaximalSet_i) > 1\} \tag{3}$$

We exclude all singleton frequent itemsets from the final feature set. This is because we believe that frequency of individual features in rules should not be the only important factor, particularly for the case of gene selection. The only concern is adjusting the value of support. Since we already reduced each rule to the shortest rule containing the minimal set of features that can represent its characteristics, we do not expect very high frequency for itemsets. Also, because we exclude singleton frequent itemsets, we need to select a small value for support that allows enough frequent itemsets to be selected. For most experiments, a value around 2% demonstrated to be a good selection.

We select the value of support according to the ratio between the number of features in the selected frequent itemsets and the number of all singleton frequent itemsets. Since we are only interested in non-singleton frequent itemsets, the value of support must be small enough to find useful existing correlations and in order not to miss some important correlations due to a very large value; and on the other hand it should be large enough in order not to accept infrequent itemsets. We observed that this balance is achieved when the number of features in the union of the selected frequent itemsets is equal to the number of singleton frequent itemsets. Since equal number of features in the two mentioned groups is not possible in all cases, we select the value of support that leads to the best balanced in terms of the number of features in the two groups. Eventually, the lower and upper bounds of the support could be derived using the statistics of the correlation set. For instance, lower and upper bounds of the support for the ALL/AML Leukemia dataset are 0.3% and 2.8%, respectively.

4   Experimental Results

To demonstrate the applicability and effectiveness of the approach described in this paper, we conducted some experiments using the acute lymphoblastic leukemia (ALL)- acute myeloid leukemia (AML) dataset taken from [14]; it contains 7129 genes; the training set size is 38 (27/11) and the test is 34. First, we used the provided training set to extract useful genes and then we used the test set for the final evaluation. In order to achieve better reliability, we repeated the process 5 times.

After extracting the correlation set, the final set of genes is selected using Apriori. Table 1 summarizes the statistics obtained after the first phase of the algorithm. These statistics highlight the usefulness of the rule generation phase. Next, we elaborate more on the different parameters reported in Table 1. Number of lines is the number of feature sequences added to the correlation set.

Table 1. Statistics of the Leukemia correlation dataset

Parameter	R1	R2	R3	R4	R5	Average
Number of lines(P1)	279	279	279	279	279	279
Number of distinct lines(P2)	210	216	211	220	225	216.4
Total number of genes appeared(P3)	374	390	375	372	392	380.6
Average line length(P4)	3.16	3	3.04	2.91	2.96	3.01
Number of binary correlations(P5)	1401	1203	1328	1138	1239	1261.8
Maximum repetition of a binary correlation(P6)	8	10	7	7	9	8.2
Number of correlations appeared only once(P7)	423	409	386	397	471	417.2

Because more important sequences are repeated in the set, number of lines is generally more than the number of distinct lines in the correlation set. And the difference between these two numbers indicates the generalization that the final rule set achieves. Total number of genes shows the feature reduction power of the first phase. We started from 7129 genes, and after the first phase most of the redundant and unnecessary genes are removed; the number of genes is reduced to 380 genes on average. Average line length is the average reduction power of the *Reduce* function. The sequence length is 3.01 on average, which means each rule can maintain its classifier power and characteristics with this number of genes, and other genes are unnecessary. Number of binary correlations is the number of times a pair of features appeared together. Compared to the total number of possible combinations from the remaining genes ($C(374,2) = 69751$), the reported value of 1401 is rather small, i.e., the correlation matrix derived from the correlation set is sparse, contains only strong correlations between genes.

Maximum repetition of binary correlations is the maximum number of times that two possible genes have appeared together in correlations. This number is 8.2 on average, and can be used as a guideline to adjust the value of support. Simply maximum support for the correlation set derived from the Leukemia dataset can be the maximum appearance over the number of lines, while the minimum support value is one over the total number of lines. Number of correlations that appeared only once may be an indicator of the randomness attribute of the algorithm since a correlation that has appeared only once in the dataset is not definitely of great importance.

Table 2. Results of running Apriori and feature extraction on the correlation set

Experiment	Features
R1	**6201-5688-1882-4342-5290-2592**-4377-5552-**758**-4569-6224-1054- 2150-4237-4825-5163-5211
R2	**5688-758-5290-6201-4342-1882-4680**-5599-**6277-2592**-2830
R3	**5688-5290-6201-758-1882-4342**-6200
R4	**5688-5290-4680-758-6201**-5808-**4342-2592-6277**-3646-7001-5367-5712
R5	**5688-5290**-2642-**6201**-6606-2349-**4342-4680-6277**-5808-630-**2592**-5449-671-4409-4110-6806

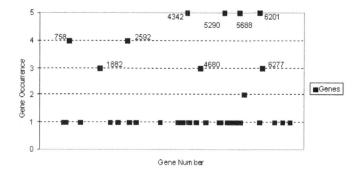

Fig. 2. Most frequent genes that appear in half or more of the results

The final step is to extract from the correlation set genes which are repeated in most experiments. First, the lower and upper bounds of support were calculated using the statistics, and then genes were selected. Table 2 and Figure 2 demonstrate extracted features and the process of picking features that appear in most results as important features. Numbers that appear at the head of each sequence are gene numbers that had existed in more frequent itemsets. Genes highlighted in bold font are frequent genes that have appeared in the results from at least half of the experiments. Finally, to measure its classification power, our approach reported 9 genes with 97.05% accuracy, while r-GeneSelect reported 5 genes and 88.23% accuracy; on the other hand, genes numbered 1882 and 6201 are in common. Biologists might find these features that are reported by both approaches and also other reported genes from our approach of great importance as highlighted in the next section.

4.1 Biological Significance of the Extracted Genes

Our approach discovered nine genes, namely L33930, D88270, Y00787, X59871, L00058, U02020, X82240, M30703 and M27891, which have strong correlation with Lymphoma cancer. Gene L33930 is CD24 signal transducer; it codes for sialoglycoprotein which is expressed in B cells; CD24 regulates E-cadherin, which plays major role in cell to cell interaction and TGF-beta3 expression. Also, CD24 expression is associated with breast and ovarian cancer. Gene D88270 is Immunoglobin lambda gene; it encodes a protein with similarity to Human topisomerase (DNA) III beta, which is thought to relax supercoiled DNA upon replication and cell division. Moreover, this gene transduces signals for cell proliferation, differentiation from the pro-B cell to pre-B cell stage. Gene U02020 has a pre-B-cell colony enhancing factor 1. It is upregulated in neutrophils by IL-1beta and functions as a novel inhibitor of apoptosis in response to a variety of inflammatory stimuli. It is more expressed in ALL patients. Gene X82240 is T-cell leukemia/lymphoma 1 A (TCL1A). Increased TCL1 expression correlates with PBC-ALL progression. TCL1 expression is important for the maturation of precursor lymphocytes. Also, it is a proto-oncogene in some cancers. These four genes are more expressed in ALL patients. However, the rest of our genes are more expressed in AML patients.

Gene Y00787 encodes Interleukin-8, which is one of the mediators of the inflammatory response. Secretion levels of IL-8 and TNF-alpha are elevated in activated T-cells in large granual lymphocytic leukemia associated with autoimmune disorders. IL-8 gene transcription is induced by p38 and NF-kB, which plays role in different cancers through degradation of IkB. Gene X59871 is a transcription factor 7 (T-cell factor-1) involved in WNT signaling pathway. It does not only regulate T-cell development, but also peripheral T-cell differentiation. Gene L00058 encodes a multi-functional protein that plays role in cell cycle progression, apoptosis. Mutations, over-expression, rearrangement and translocation of this gene have been associated with a variety of hematopietic tumors, leukemia and lumphoma. Gene M30703 is amphiregulin (AREG); it is a member of epidermal growth factor family. AREG gene expressed by purified primary myeloma cells. AREG plays an important role in biology of multiple myeloma. Finally, gene M27891 encodes a protein, which belongs to the cystatin superfamily; some members of those proteins are cysteine protease inhibitors.

As a result, it can be easily seen that almost all of our genes have related functions and play role in B and T cells proliferation and lymphocyte cells differentiation, besides their role in cancer and cell cycle regulation. The genes discovered by the other method are M19507, M27891, Y00787, M96326, and L20688. The authors claimed that those genes are the most significant genes to distinguish between AML and ALL. However, the function of the most significant gene they discovered (M19507) is Myeloperoxidase (MPO), which is a heme protein synthesized during myeloid differentiation. Although this protein is working in cell differentiation, there is no relation between MPO and cell cycle regulation. Both M27891, Y00787 are in common with our genes. Although the former gene is kind of unrelated to cancer, the second one plays role in inflammatory response.

M96326 and L20688 are Azurocidn and GDP, respectively. The former gene encodes an azurophil granule antibiotic protein, and the later has weak relation to cancer. Their method has considered FTH1 (L20941) as significant gene. This gene encodes the heavy subunit of ferritin, the major intracellular iron storage protein in prokaryotes and eukaryotes. Proteoglycan 1 gene is not related to cell cycle regulation. They also have discovered several genes like Enolase alpha, EIF4E. Those genes showed to play role in cell cycle, but no specific study showed their role in Lymphocyte differentiation. In conclusion, we realized that the genes we have discovered have strong relation to pre-B to Pro-B stage development. Some of the genes discovered by others are unrelated to cancer such Azurocidn and Cystatin. Some of them are related to cancer but not related to Leukemia or lymphocyte development. Another point is that they gave the same weight for ENO1 , EIF4E and FTH1. The formers have high expression in AML patients, but the later has weak ALL/AML expression; we argue that their method has to distinguish between such genes.

5   Summary and Conclusions

In this study, we have demonstrated the power of generating fuzzy classifier rules and using them for identifying relevant genes. We first applied feature reduction to get a compact set of representative features. The process involves creating fuzzy classifier rule set and then deriving a correlation matrix between the genes appearing in the rules. The correlation matrix is the input to the apriori algorithm to find frequent itemsets, which lead to the relevant genes. The proposed approach reported high accuracy on the ALL/AML Leukemia dataset. Also, the genes identified by the proposed approach are highly significant from biological perspective; the same may not be said regarding our competitors. Several studies stressed the role of our genes in Leukemia in particular and cancer in general. Currently, we are conducting experiments on other gene expression datasets and we are working on a comprehensive self-adaptive fuzzy classifier.

References

1. Zeng, F., Yap, C.H.R., Wong, L.: Using Feature Generation and Feature Selection for Accurate Prediction of Translation Initiation Sites. Genome Informatics 13, 192–200 (2002)
2. Kianmehr, K., Zhang, H., Nikolov, K., Özyer, T., Alhajj, R.: Combining Neural Network and Support Vector Machine into Integrated Approach for Biodata Mining. In: Proc. of ICEIS, pp. 182–187 (2005)
3. Dash, M., Choi, K., Scheuermann, P., Liu, H.: Feature Selection for Clustering-a Filter Solution. In: Proc. of IEEE ICDM, pp. 115–122 (2002)
4. Hall, M.A.: Correlation-Based Feature Selection for Discrete and Numeric Class Machine Learning. In: Proc. of ICML, pp. 359–366 (2000)
5. Caruana, R., Freitag, D.: Greedy Attribute Selection. In: Proc. of ICML, pp. 28–36 (1994)
6. Dy, J.G., Brodley, C.E.: Feature Subset Selection and Order Identification for Unsupervised Learning. In: Proc. of ICML, pp. 247–254 (2000)
7. Das, S.: Filters, Wrappers and a Boosting-Based Hybrid for Feature Selection. In: Proc. of ICML, pp. 74–81 (2001)
8. Ng, A.Y.: On Feature Selection: Learning with Exponentially Many Irrelevant Features as Training Examples. In: Proc. of ICML, pp. 404–412 (1998)
9. Guyon, I., Weston, J., Barnhill, S., Vapnik, V.: Gene selection for cancer classification using support vector machines. Machine Learning 46(1-3), 389–422 (2002)
10. Fu, L.M., Youn, E.S.: Improving reliability of gene selection from microarray Functional Genomics data. IEEE Transactions on Information Technology in Biomedicine 7(3) (2003)
11. Fu, L.M.: Cancer Subtype Classification Based on Gene Expression Signatures (Last Accessed on 17/4/2007), URL: http://www.cise.ufl.edu/ fu/NSF/ cancer_classify_GES.html

12. Ishibuchi, H., Nakashima, T., Muratam, T.: Performance Evaluation of Fuzzy Classifier Systems for Multi-dimensional Pattern Classification Problems. IEEE Trans. on Systems, Man, and Cybernetics (October 1999)
13. Abe, S., Lan, M.-s.: A Method for Fuzzy Rules Extraction Directly from Numerical Data and Its Application to Pattern Classification. IEEE Trans. on Fuzzy Systems 3(1), 18–28 (1995)
14. URL: `http://sdmc.lit.org.sg/GEDatasets/Datasets.html` (Last Accessed on 17/4/2007)

Two Way Focused Classification

Manoranjan Dash[1] and Vivekanand Gopalkrishnan[2]

[1] (Nanyang Technological University), 50 Nanyang Avenue, Singapore
asmdash@ntu.edu.sg,
[2] asvivek@ntu.edu.sg

Abstract. In this paper we propose TwoWayFocused classification that performs feature selection and tuple selection over the data before performing classification. Although feature selection and tuple selection have been studied earlier in various research areas such as machine learning, data mining, and so on, they have rarely been studied together. The contribution of this paper is that we propose a novel distance measure to select the most representative features and tuples. Our experiments are conducted over some microarray gene expression datasets, UCI machine learning and KDD datasets. Results show that the proposed method outperforms the existing methods quite significantly.

1 Introduction

Data are increasing in both ways: dimensions or features and instances or examples or tuples; not all the data are relevant though. So, it behooves for a data mining expert to remove the noisy, irrelevant and redundant data before proceeding with classification because many traditional algorithms fail in the presence of such noisy and irrelevant data [4]. In addition to this advantage, removing such noisy features and tuples improves the understanding of the user by bringing focus.

Thus, as a solution, in this paper we study how to remove the noisy and irrelevant features and tuples while improving or at least not significantly reducing the prediction accuracy. This task is called "focusing" in this paper, i.e, select the relevant features and tuples in order to improve the overall learning. In the literature many feature selection and tuple selection methods have been proposed [5,14]. Feature selection algorithms can be broadly grouped as wrapper and filter [4,11,14,8]. In the wrapper method the classifier is used as an evaluation function to evaluate and compare the candidate feature subsets. In filter method evaluation of feature subsets is usually independent of the final classifier. In tuple selection, a broad category of methods let the learning algorithm to select the relevant tuples [4,15,7].

Contribution of this paper is that we propose a distance measure to select representative features and tuples. Experimental results over several microarray datasets and UCI machine learning repository datasets and UCI KDD archive [1,3] show that the proposed TwoWayFocused algorithm performs better than the existing feature selection methods.

I.Y. Song, J. Eder, and T.M. Nguyen (Eds.): DaWaK 2007, LNCS 4654, pp. 345–354, 2007.

Notations. Let the data D have n number of tuples and m number of features. I be the set of tuples where I_i is the ith tuple for $i = 1, ...n$. F be the set of features where f_j is the jth feature for $j = 1, ...m$. C is the set of class labels. We will introduce other notations as and when they become necessary.

Histogram based Distance Measure. The proposed TwoWayFocused method tries to find a small subset having 1-itemset frequencies that are close to those in the entire database. The *distance* or *discrepancy* of any subset S_0 from its superset S with respect to 1-itemset frequencies is computed by the L_2-norm (Euclidean distance) as follows:

$$dist(S_0, S) = \sum (freq(A; S_0) - freq(A; S))^2 \qquad (1)$$

where A is an 1-itemset, and each element in subset S and S_0 is a set of items. Using this distance function we can compare two subsets S_1 and S_2 for their representativeness of the superset S: if $dist(S_1, S) < dist(S_2, S)$, then arguably S_1 is a better representative of the whole set than S_2. In place of L_2-norm, we can also use L_1-norm (Manhattan distance) and L_∞-norm.

2   Feature Selection

The proposed feature selection method is a two-phase algorithm where in the first phase we rank the features based on their relevance to the class label, and in the second phase we select from the top-ranking features in order to remove the redundant features.

Feature Ranking. In this section we show how the proposed distance function can be used to rank features based on their relevance. Relevance of a feature is defined in [10] as features whose removal would affect the original class distribution. In the previous section we showed that it is possible to compare two subsets for their representativeness of a superset. We extend this idea to compare individual features. Given that F is the total set of features, we can compare two features f_1 and f_2 as follows:

if $dist(F - \{f_1\}, F) < dist(F - \{f_2\}, F)$ then
f_2 is arguably better than f_1.

The idea is that if by removing feature f_1 distance of the resultant set $\{F - f_2\}$ from the whole set F is greater than the distance of $\{F - f_1\}$ (after removing f_1) from the whole set F, then it indicates that f_2 is arguably a better representative of F than f_1. We can easily extend this idea to rank all the features. For each feature f_j, $j = 1, ..., m$, compute $dist(F - \{f_1\}, F)$. Features having larger distances are given higher ranks. In Equation 1 the frequencies are determined for 1-itemsets. To compute in a similar fashion data must be binary valued (i.e., 0 or 1). Later in this section, we describe how to deal with other types of data such as nominal, discrete, and continuous. As we are dealing with classification task, data has class labels. We can make use of class information simply by first separating the data into groups based on the class labels. Then we apply the ranking algorithms for each group of tuples. Next we add the ranks of each feature for all groups to find out the overall rankings. Ties are broken arbitrarily. As our experiments show, this is a simple yet powerful way of utilizing the class information.

Feature Selection to remove Redundancy. After ranking the features we need to select a subset of features. The main goal here is to remove redundancy because while ranking the features we did not consider the fact that there may be redundant features.

In phase 2, we use a measure called *inconsistency measure* [6] to remove redundancy. Inconsistency of a feature subset is measured by finding out how inconsistent it is vis-a-vis the class label. Although inconsistency measure can be used directly to remove the redundant features, in order to search through the feature space we need a more efficient mechanism than a simple exhaustive method. We propose to use forward selection algorithm. A forward selection algorithm works as follows. In the beginning the highest ranked feature (according to phase 1) is selected. Note that the phase 2 is invoked after phase 1 where features are ranked by the distance measure. Then if the task is to select $|S'|$ number of features we take $2 * |S'|$ top-ranked features. Then we find out the best feature to add to the highest ranked feature by sequentially going through $2 * |S'| - 1$ features. We compare the subsets by measuring their redundancy and relevance. Redundancy is measured by inconsistency rate, and relevance is measured by determining the correlation between the subset of features. and the class. We use mutual information to estimate the correlation. It is described as follows.

Mutual information: Let there X and Y be two random variables with joint distribution $p(x, y$ and marginal distribution $p(x)$ and $p(y)$. The mutual information $M(X; Y)$ is given as:

$$M(X; Y) = \sum_x \sum_y p(x, y) \log \frac{p(x, y)}{p(x)p(y)} \tag{2}$$

It can be seen that mutual information is conceptually a kind of "distance" between the joint distribution $p(x, y)$ and the product distribution $p(x)p(y)$. When X and Y are independent, i.e., $p(x, y) = p(x)p(y)$, the "distance" becomes zero. The larger value of M indicates variables are more dependent. Thus, mutual information is appropriate to measure the correlation between a feature and the class.

We combine these two measures, i.e., mutual information M, and inconsistency measure I_R as follows. As discussed initially S' is the highest ranked feature f_r, i.e., $S' = \{f_r\}$. Suppose that at a certain stage we are considering feature f_j and we denote $S' = S' \cup f_r$, then the next feature to be added to S' is:

$$f^* = argmax_{f_j}(w_1 * M(f_j, C) + \frac{w_2}{I_R(S')}) \tag{3}$$

Our empirical experiments suggest that w_1 is 0.9 and w_2 is 0.2. Note that rank 1 or the highest rank corresponds to the highest distance, and so on.

Tuple Selection. Each tuple in the set of tuples I is first of all ranked using the same method described for ranking of features. Distance is computed by removing each tuple I_i from I in turn for $i = 1, ..., n$, and finally sorting the distances in descending order in order to produce ranks of the tuples. The difference between feature selection and tuple selection is the manner in which they utilize the class information. Like feature selection, here we first separate out the tuples based on their class labels. In an experimental set up there is a training set and a testing set. So, the tuples in the training set are separated out based on the class labels whereas the test set is untouched. Next, like feature selection, we compute the distance of each tuple under each class label. But notice that, unlike feature selection where each class had all the features, in tuple selection that is not the case. The total number n of tuples are divided among $|C|$ number of class labels. So, here we select a proportionate number of top-ranking tuples from each group and the combination of such top-ranking tuples are output as the selected tuples. The idea is that the final sample should have proportionate number of tuples from each class label. Ties in rankings are arbitrarily broken. Finally we propose a new algorithm called *TwoWayFocused* which denotes first applying feature selection followed by tuple selection. In order to compare we also propose *OneWayFocused* which denotes only feature selection without performing tuple selection.

Handling Different Data Types. In this section we describe how to apply the proposed focused classification method for different data types such as binary, nominal, discrete and continuous. In this section, for ease of understanding we consider the data to be a matrix of rows and columns where the task is to select a set of rows. This holds true (a) for feature selection where features are considered rows and tuples are considered as columns, and (b) for tuple selection where tuples are considered rows and features are considered columns. Note that the proposed distance function works for binary data only. So, for binary type of data there is no need to modify it. For nominal data type, it is converted to binary using the following technique. First of all, count the number of different nominal values that appear in a column l. If column l takes p different values, then we convert it to p binary columns $l_1, l_2, ..., l_p$.

Discrete and Continuous Data Types. If a column has continuous data type, it is discretized first of all. In the literature there are many discretization methods which can be broadly grouped as supervised and unsupervized depending on whether they use the class information. A good survey is available in [13]. In this work we use a simple and popular discretization method called '0-mean 1-sigma' where mean is the statistical mean μ of the values in a column f_j and sigma is the standard deviation σ of column f_j [19]. This discretization method assumes that μ is at 0. Each value in the column takes a new discretized value from the set 1, 2, 3 based on the following conditions. Any numerical value in the range $[\mu - \sigma, \mu + \sigma]$ is assigned the discrete value 2, numerical values in the range $[-\infty, \mu - \sigma]$ is assigned the discrete value 1, and numerical values in the range $[\mu + \sigma, +\infty]$ is assigned the discrete value 3. After discretization each column is converted to 3 columns. An original value v will be converted to a vector of 3 binary values which can be {1 0 0}, or {0 1 0}, or {0 0 1}.

If a column has discrete data type, it is treated as continuous data and is discretized using the same method described here.

3 Empirical Study

We select three methods: ReliefF, t-test, and RBF [12,9,19]. We mainly used two groups of datasets – group 1: UCI machine learning repository and UCI KDD archive, and group 2: gene expression microarray datasets [1,3]. Although both groups of data are high-dimensional, the difference is that in group 1 (UCI repository) the number of tuples usually far exceed the number of features, whereas for microarray data the number of features or genes are in thousands but the tuples or samples are in 10s. The two groups have a mixture of continuous and nominal data types, and different number of classes. In Table 1 we give a summary of the datasets used in the empirical study.

Table 1. Summary of Datasets

Title	Features	Tuples	Classes	Title	Features	Tuples	Classes
UCI ML/KDD Repository				Gene Expression Microarray Data			
Multi Features	649	2000	10	Prostate Cancer	12600	136	2
Lung Cancer	56	32	3	Colon Cancer	2000	62	2
Arrhy-thmia	279	452	16	Lung Cancer	12533	181	2
Coil2000	85	5822	2	Leukemia	7129	72	2
Promoters	57	106	2				
Splice	60	3190	3				
Musk2	166	6598	2				
Isolet	617	1560	26				

Framework for Comparison. Comparison is done based on prediction accuracy and time of a classifier. We chose C4.5 [16] to obtain the accuracy, and we use system time. We used the WEKA implementation of C4.5 [18]. Each data is divided into a training set and a testing set. We use 10-fold cross-validation for each experiment, i.e, 9 folds are used for training and one fold is used for testing. Note that feature selection, tuple selection, and training of C4.5 decision tree are all done using the 9 folds, and finally the prediction accuracy is tested using the last fold. That means, reported accuracy and time are average of 10

Table 2. P-values of the t-test comparing TwoWayFocused with existing algorithms

	Feature Ranking					Feature Selection
	ReliefF					**RBF**
P-value	0.025-0.05	0.005-0.01	<0.0005	<0.0005	<0.0005	0.0005-0.005
	t-test					
P-value	0.0005-0.005	<0.0005	<0.0005	<0.0005	<0.0005	

(for 10 folds) runs. All experiments are conducted in a system with configuration Pentium 4, 1.6 GHz and 512 MB RAM.

The proposed algorithm *TwoWayFocused* performs both feature selection (FS) and tuple selection (TS). We implemented two versions of the algorithm: (a) without TS (OneWayFocused) and (b) with TS (TwoWayFocused). Comparison is done with the three existing algorithms which are ReliefF, *t*-test, and RBF. The reason for choosing these algorithms are: (a) Relief (and its extension ReliefF) is a proven robust feature ranking method; also Relief is based on tuple selection, (b) *t*-test is a commonly used feature ranking method that usually works well, and (c) RBF is a recently proposed and very successful feature selection method [19]. Among these three, ReliefF and *t*-test are feature ranking methods whereas RBF is a feature selection method. We compare TwoWayFocused and OneWayFocused with these three methods in two ways: (a) ReliefF and *t*-test rank the features; so we compare the four algorithms for various number of top-ranked features; the range of the number of top-ranked features vary across datasets because total number of features vary, and (b) RBF outputs a single subset S'; so we take $|S'|$ top-ranked features for TwoWayFocused and OneWayFocused.

Both RBF and the proposed algorithms TwoWayFocused and OneWayFocused first discretize the data for feature selection. Then the original (numerical) data is used for training and testing the C4.5 classifier using the selected features.

Results and Analysis. In Tables 3 and 4 we give the comparison results of feature ranking using four algorithms: TwoWayFocused, OneWayFocused, ReliefF and *t*-test. In Table 5 we give the comparison results of feature using three algorithms: TwoWayFocused, OneWayFocused and RBF. The best prediction accuracy for each data is high-lighted.

In **ALL** experiments over 12 datasets, for both feature ranking and feature selection, the highest prediction accuracy is obtained by either TwoWayFocused or OneWayFocused. In feature ranking (see Tables 3 and 4), TwoWayFocused outperforms the other three algorithms including OneWayFocused. Only in 8 out of 60 experiments (12 datasets times 5 tests) for feature ranking, OneWayFocused is better than TwoWayFocused. Similarly, in feature selection (see Table 5), TwoWayFocused is better than RBF and OneWayFocused for 11 out of 12 datasets, and for the remaining one time, OneWayFocused is better than RBF and TwoWayFocused. The prediction accuracy results of

Table 3. Comparison results for feature ranking for UCI ML/KDD repository datasets

Datasets	Accuracy					Time (sec)
	UCI ML/KDD Repository Datasets					
	#Features	30	40	50	60	70	
	ReliefF	93.9	92.3	90.1	88.2	88.4	6.37
	t-test	94	93.4	89.5	86.7	88.9	0.4
CoiL2000	OneWayFocused	94.7	93.1	93.1	91.3	90	1.2
	TwoWayFocused	**95**	**94.6**	**94.1**	**93.2**	**92**	2.13
	#Features	10	20	30	40	50	
	ReliefF	75	51.9	66.4	66.2	70.9	0.09
	t-test	77.6	74.3	76	75.1	74.8	0.06
Promoters	OneWayFocused	**80.2**	80.2	**81.3**	79	79.2	0.02
	TwoWayFocused	79.3	**80.2**	80.1	**80.5**	**82.4**	0.02
	#Features	20	30	40	50	60	
	ReliefF	85.4	86.7	84.2	87.7	89.1	2.68
	t-test	86.1	88.9	85.2	89.2	87.4	3.25
Splice	OneWayFocused	87.2	90	88.3	90.3	91	0.71
	TwoWayFocused	**89.8**	**91.5**	**90.4**	**92.1**	**92.6**	0.98
	#Features	60	70	80	90	100	
	ReliefF	80.5	78.6	78.4	75.2	77.4	18.6
	t-test	80.6	79.5	77.2	76.3	79.7	34.69
Musk2	OneWayFocused	80.2	83.1	83.5	81.2	82.1	2.2
	TwoWayFocused	**82**	**85.2**	**84.9**	**84.9**	**85.8**	3.45
	#Features	65	85	105	125	150	
	ReliefF	65.4	67.4	71.8	69.1	74	20.8
	t-test	77.6	79.7	75.3	70.9	74.6	28.52
Isolet	OneWayFocused	81.5	81.9	**83.7**	80.8	**83**	3.03
	TwoWayFocused	**82.5**	**83.5**	82.4	**82.9**	81.9	4.14
	#Features	50	100	150	200	250	
	ReliefF	85.5	86.7	84.2	83.6	82.1	22.48
	t-test	92	92.1	90.3	89.7	87.5	18.92
Multi-Features	OneWayFocused	87.2	91	92.4	89.3	91.1	1.78
	TwoWayFocused	**93.1**	**93.1**	**92.9**	**91.7**	**93.8**	2.98
	#Features	20	25	30	35	40	
	ReliefF	83.1	82.8	81.7	80.9	80.2	0.06
	t-test	80.2	79.3	79.3	78	77.6	0.03
Lung Cancer	OneWayFocused	**83.2**	84.6	**85.9**	84	83.7	0.01
	TwoWayFocused	81.7	**85.3**	85.2	**85.8**	**84.5**	0.01
	#Features	30	50	70	90	110	
	ReliefF	66.2	65	65.6	64.1	62	2.76
	t-test	60.8	59.5	58.3	57.9	58.2	1.86
Arrhythmia	OneWayFocused	63.1	64.4	63.2	66.9	63.1	0.03
	TwoWayFocused	**67.9**	**65.8**	**65.7**	**68.8**	**66.3**	0.03

TwoWayFocused and OneWayFocused are more stable than the other algorithms. The possible reason is that the proposed algorithms are able to select important set of features that is much close to the optimal set.

Table 4. Comparison results for feature ranking for microarray data

Datasets	Accuracy						Time (sec)
	#Features	31	62	124	248	496	
	ReliefF	80.5	78.8	74.1	75.7	75	3.5
	t-test	81.6	75.5	80.4	80.5	74.1	0.03
Colon Cancer	OneWayFocused	81.6	80.4	80.2	82.7	77	0.05
	TwoWayFocused	**82.9**	**84.8**	**83.1**	**82.7**	**79.1**	0.06
	#Features	56	112	224	448	896	
	ReliefF	83.2	85.5	84.6	85.6	81.7	13.2
	t-test	78.1	83.7	83.1	86.3	80.3	0.13
Leukemia	OneWayFocused	84.5	87.7	93.7	92.9	84.1	0.3
	TwoWayFocused	**86.5**	**89.6**	**94.4**	**93.2**	**87.4**	0.42
	#Features	98	196	392	784	1568	
	ReliefF	91.1	87.1	85.3	84.2	82.9	71.5
	t-test	73.6	76.3	76.9	75.3	78.5	1.31
Prostate Cancer	OneWayFocused	**91.4**	88.9	90.1	89.4	87.6	0.95
	TwoWayFocused	89.8	**90.1**	**90.1**	**90.3**	**87.7**	1.14
	#Features	19	38	76	152	304	
	ReliefF	95.6	96.2	96.1	95.1	95.6	124.7
	t-test	81	83.9	82.8	89.2	91.7	1.52
Lung Cancer (μArray)	OneWayFocused	**95.6**	96.2	97.1	94.3	95.8	2.12
	TwoWayFocused	93.1	**97.6**	**97.4**	**95.3**	**96.4**	2.89

Table 5. Comparison results for feature selection

Datasets	#Selected Features	RBF		OneWayFocused		TwoWayFocused	
		Accuracy	Time	Accuracy	Time	Accuracy	Time
UCI ML/KDD Repository Datasets							
CoIL2000	5	93	1.59	93.5	1.23	**94.3**	0.78
Promoters	6	72.6	0.11	83.6	0.03	**84.1**	0.02
Splice	12	91.7	1.67	94.8	1.61	**95.2**	0.8
Musk2	2	83.8	4.99	85.5	3.23	**86.8**	1.9
Isolet	25	83.5	9.05	84	8.2	**85.7**	9.3
Multi-Features	47	93.7	29.56	93.6	23.57	**95.9**	24.23
Lung Cancer	4	84.5	0.01	**84.9**	0.01	82.8	0.01
Arrhythmia	5	70.4	0.04	71.6	0.03	**72.2**	0.02
Microarray Datasets							
Colon Cancer	4	85.3	0.23	91.2	0.06	**92.6**	0.87
Leukemia	4	88.9	1.17	94.4	0.23	**94.4**	0.31
Prostate Cancer	78	86.7	14.34	92.2	1.23	**93.3**	2.41
Lung Cancer (μarray)	1	97.2	17.48	97.2	2.98	**97.2**	3.76

Similarly for time, for both ranking and selection, the lowest execution time is reported by either TwoWayFocused or OneWayFocused except for 3 datasets. For these 3 datasets, t-test reported the lowest execution time.

In Table 2 we showed the p-values for feature ranking (5 different number of top ranking features) and feature selection. We use one-tailed (upper tail)

paired t-test between the accuracy of TwoWayFocused and the accuracy of an existing algorithm (ReliefF, t-test, and RBF). For level of significance 5% all values are significant, i.e., the null hypothesis of equality is rejected in all tests. The p-values are quite significant in most cases (e.g., < 0.0005).

An interesting outcome of this empirical study is that although the number of tuples or sample tests for gene expression microarray data is quite small (e.g., Leukemia dataset has only 72 tuples), still tuple selection improves the prediction accuracy even after reducing the number of tuples by half. Until now researchers have not considered the idea that some tuples may be unnecessary or even detrimental for prediction accuracy for classification of microarray data. They had only focused on how to reduce the thousands of genes or features (e.g., Leukemia dataset has 7129 genes) to 10s of genes. This new found information is going to assist the biologist in their study of gene functionalities.

4   Conclusion

In this paper we proposed focused classification using feature selection and tuple selection. The novelty of the work is that a distance based ranking method based on frequency histogram is used both for feature selection and tuple selection. The proposed algorithm TwoWayFocused consistently outperformed the three chosen existing algorithms Relief, t-test, and RBF. Also, results over a range of selected features showed that TwoWayFocused and OneWayFocused produce very good prediction accuracy consistently.

References

1. Kent ridge biomedical dataset repository, http://sdmc.lit.org.sg/GEDatasets/Datasets.html
2. Alon, U., Barkai, N., et al.: Broad patterns of gene expression revealed by clustering analysis of tumor and normal colon tissues probed by oligonucleotide arrays. Proceedings of National Academy of Science, Cell Biology 96, 6745–6750 (1999)
3. Blake, C., Merz, C.: UCI repository of machine learning databases (1998), http://www.ics.uci.edu/ mlearn/MLRepository.html
4. Blum, A.L., Langley, P.: Selection of relevant features and examples in machine learning. Artificial Intelligence 97, 245–271 (1997)
5. Dash, M., Liu, H.: Feature selection for classification. International Journal of Intelligent Data Analysis 1(3) (1997)
6. Dash, M., Liu, H.: Consistency-based search in feature selection. Artificial Intelligence 151(1-2), 155–176 (2003)
7. Fragoudis, D., Meretakis, D., Likothanassis, S.: Integrating feature and instance selection for text classification. In: Proceedings of ACM SIGKDD international conference on Knowledge discovery and data mining, pp. 501–506. ACM Press, New York (2002)
8. Inza, I., Larranaga, P., Blanco, R., Cerrolaza, A.J.: Filter versus wrapper gene selection approaches in dna microarray domains. Artificial Intelligence in Medicine 31(2), 91–103 (2004)

9. Jaeger, J., Sengupta, R., Ruzzo, W.L.: Improved gene selection for classification of microarrays. In: Proceedings of Pacific Symposium on Biocomputing, pp. 53–64 (2003)
10. John, G.H., Kohavi, R., Pfleger, K.: Irrelevant features and the subset selection problem. In: Proceedings of International Conference on Machine Learning, pp. 121–129 (1994)
11. Kohavi, R., John, G.H.: Wrappers for feature subset selection. Artificial Intelligence 97, 273–324 (1997)
12. Kononenko, I.: Estimating attributes: Analysis and extension of RELIEF. In: Proceedings of European Conference on Machine Learning, pp. 171–182 (1994)
13. Liu, H., Hussain, F., Tan, C.L., Dash, M.: Discretization: An enabling technique. Journal of Data Mining and Knowledge Discovery 6(4), 393–423 (2002)
14. Liu, H., Motoda, H. (eds.): Feature Extraction, Construction and Selection: A Data Mining Perspective. Kluwer Academic Publishers, Boston (1998)
15. Liu, H., Motoda, H.: Instance Selection and Construction for Data Mining. Kluwer Academic Publishers, Dordrecht (2001)
16. Quinlan, J.R.: C4.5: Programs for Machine Learning. Morgan Kaufmann, San Mateo, California (1993)
17. Quinlan, R.: Machine Learning - An Artificial Intelligence Approach. In: Learning efficient classification procedures and their application to chess end games, pp. 463–482 (1983)
18. Witten, I.H., Frank, E.: Data Mining: Practical machine learning tools and techniques, 2nd edn. Morgan Kaufmann, San Francisco (2005)
19. Yu, L., Liu, H.: Redundancy based feature selection for microarry data. In: Proceedings of ACM SIGKDD International Conference on Knowledge and Discovery, pp. 22–25. ACM Press, New York (2004)
20. Clarke, F., Ekeland, I.: Nonlinear oscillations and boundary-value problems for Hamiltonian systems. Arch. Rat. Mech. Anal. 78, 315–333 (1982)
21. Clarke, F., Ekeland, I.: Solutions périodiques, du période donnée, des équations hamiltoniennes. Note CRAS Paris 287, 1013–1015 (1978)
22. Michalek, R., Tarantello, G.: Subharmonic solutions with prescribed minimal period for nonautonomous Hamiltonian systems. J. Diff. Eq. 72, 28–55 (1988)
23. Tarantello, G.: Subharmonic solutions for Hamiltonian systems via a $\mathbb{Z}_p$ pseudoindex theory. Annali di Matematica Pura (to appear)
24. Rabinowitz, P.: On subharmonic solutions of a Hamiltonian system. Comm. Pure Appl. Math. 33, 609–633 (1980)

A Markov Blanket Based Strategy to Optimize the Induction of Bayesian Classifiers When Using Conditional Independence Learning Algorithms

Sebastian D.C. de O. Galvão and Estevam R. Hruschka Jr.

DC/UFSCar – Universidade Federal de São Carlos, Depto. de Computação,
São Carlos, Brazil
{sebastian_galvao,estevam}@dc.ufscar.br

Abstract. A Bayesian Network (BN) is a multivariate joint probability distribution graphical representation that can be induced from data. The induction of a BN is a NP problem. Two main approaches can be used for inducing a BN from data, namely, Conditional Independence (CI) and the Heuristic Search (HS) based algorithms. When a BN is induced for classification purposes (Bayesian Classifier - BC), it is possible to impose some specific constraints aiming at an increase in computational efficiency. In this paper a new CI based algorithm (MarkovPC) to induce BCs from data is proposed. MarkovPC uses the Markov Blanket concept in order to impose some constraints and optimize the traditional PC algorithm. Experiments performed with ALARM BN, as well as other UCI and artificial domains revealed that MarkovPC tends to execute fewer comparisons than the traditional PC. The experiments also show that the MarkovPC produces competitive classification rates when compared with both, PC and Naïve Bayes.

Keywords: Bayesian Networks, Bayesian Classifiers, Markov Blanket, Conditional Independence.

1 Introduction

In the last years, Bayesian Networks (BNs) have been applied in many supervised and unsupervised learning applications and presented good results. Therefore, many new BNs learning algorithms have been proposed [9]. The search space for learning a BN from data, however, has an exponential dimension, thus, this is a difficult problem.

There are two main approaches, described in the literature, to the BN learning problem. The first one is based on dependency analysis and is called Conditional Independence (CI), while the second approach searches for good network structures according to a heuristic (or metric) and is called Heuristic Search (HS). The Markov Blanket concept has been applied in conjunction with both CI [3] and HS [8] learning methods as a feature selection strategy. In this sense, after inducing a BN from data, the relevant variables to a specific query can be identified, thus, the model can be pruned in order to reduce its complexity.

I.Y. Song, J. Eder, and T.M. Nguyen (Eds.): DaWaK 2007, LNCS 4654, pp. 355–364, 2007.

In this work, the Markov Blanket is also applied trying to reduce the model complexity. Instead of using the MB concept after the complete BN induction (as done in [3] and [8]), however, our approach, named MarkovPC, explores the MB of the class variable during the BN induction process. Therefore, it is possible to simplify the model structure (BN) while building it, and consequently, to reduce the learning algorithm complexity. To do so, our method requires a supervised learning. In other words, MarkovPC is a BN learning algorithm designed to classification problems. Thus, it is possible to define that MarkovPC is a Bayesian Classifier (BC) learning algorithm based on the CI approach.

The remainder of this work is structured as following, the next section gives an overview of Bayesian Networks and Bayesian Classifiers foundations. Section 3 presents the classic PC learning algorithm. Section 4 introduces the proposed MarkovPC algorithm which can be seen as an extension of the PC algorithm. Section 5 shows the performed experiments and discusses the obtained results. Finally, Section 6 describes the conclusions and points out some future works.

2 Bayesian Networks and Bayesian Classifiers

A Bayesian Network (BN) [10] G has a directed acyclic graph (DAG) structure. Each node in the graph corresponds to a discrete random variable in the domain. An edge, $Y \rightarrow X$ in the graph describes a parent–child relation, where Y is the parent and X is the child. All parents of X constitute the parent set of X, which is denoted by π_X. Each node of the BN structure is associated to a conditional probability table (CPT) specifying the probability of each possible state of the node, given each possible combination of states of its parents. If a node has no parents, its CPT gives the marginal probabilities of the variable it represents.

In a Bayesian network where λ_{x_i} is the set of children of x_i, the subset of nodes containing π_{x_i}, λ_{x_i}, and the parents of λ_{x_i} is called Markov Blanket of x_i (Fig. 1). As shown in [10], the only nodes that have influence on the conditional distribution of a given node x_i (given the state of all remaining nodes) are the nodes that form the Markov Blanket (MB) of x_i. Thus, after constructing the network structure from data, the Markov Blanket of the class attribute can be used as a criterion for selecting a subset of relevant attributes for classification purposes. As it will be detailed in the sequel, this concept plays an important role in our proposed method. Meanwhile, let us address how Bayesian networks can be constructed.

BNs can be induced directly from domain knowledge or they can be automatically learned from data. It is also possible to combine both strategies. Learning BNs from data became an effervescent research topic in the last decade [9], and there are two main classes of methods to perform this task: methods based on heuristic search and methods based on conditional independence tests.

Instead of encoding a joint probability distribution over a set of random variables, as done by a BN, a Bayesian Classifier (BC) aims at correctly predicting the value of a discrete class variable, given the value of a vector of attribute variables (predictors). Bayesian Network learning methods may be used to induce a BC and this is done in this work. The BN learning algorithm used in the experiments described in Section 4 is based on the PC algorithm [12], which constructs a BN from data based on Conditional Independence tests as addressed in the next section.

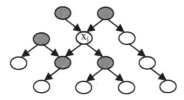

Fig. 1. The shadowed nodes represent the Markov Blanket of x_i

3   The PC Algorithm

The PC algorithm [12] is based upon statistical conditional independence tests. It works looking for a Bayesian Network that represents the independence relationship among variables in a dataset. This is done based on the conditional independence

```
PC Algorithm
Problem: Given a set IND of d-separations, determine the DAG pattern
faithful to IND if there is one.
Inputs: a set V of nodes and a set IND of d-separations among subsets of
the nodes.
Outputs: If IND admits a faithful DAG representation, the DAG pattern gp
containing the d-separations in this set.
1  begin
2  A.) form the complete undirected graph gp over V;
3  B.)
4  i = 0;
5  steps = 0;     //helps to measure the algorithm effort
6  repeat
7  for (each X ∈ V) {
8      for (each Y ∈ ADJX) {
9          determine if there is a set S ⊆ ADJX-{Y} such
10         that |S| = i and I({X},{Y}|S) ∈ IND; //|S|returns
11         steps = steps + 1;               //the size of set S
12         if such a set S is found {
13             S_XY = S;
14             remove the edge X - Y from gp;
15         }
16      }
17 }
18 i = i + 1;
19 until (|ADJX| < i for all X ∈ V);
20 C.)for each triple of vertices X, Y, Z such that the pair
21     X, Y and the pair Y, Z are each adjacent in C but the pair
22     X, Z are not adjacent in C, orient X-Y-Z as X->Y<-Z if and
23     only if Y is not in S_XY
24 D.) repeat
25         If A -> B, B and C are adjacent, A and C are not
26         adjacent, and there is no arrowhead at B, then
27         orient B-C as B->C.
28         If there is a directed path from A to B, and edge
29         between A and B, then orient A-B as A->B.
30     until no more edges can be oriented.
31 end.
```

Fig. 2. PC algorithm (adapted from [9]and [12])

criteria $I(X_i,X_j|A)$ defined in [10] where A is a subset of variables, X_i and X_j are variables. If $I(X_i,X_j|A)$ is true, variable X_i is conditionally independent of X_j given A (d-separation criterion).

To verify whether X and Y are conditionally independent given A, we compute the cross entropy $CE(X,Y|A)$ where the probabilities are their maximum likelihood estimators extracted from the data (i.e. relative frequencies). Other measures can also be used [4][12]. The main steps of the PC algorithm can be summarized as in Figure 2.

Having as input a list with all the independencies ($I(X_i,X_j|A)$) and adjacencies of each node (ADJ_{X_i}), PC first finds the graph skeleton (undirected graph) that best represents the d-separations expressed by $I(X_i,X_j|A)$. Afterwards, it starts establishing the orientation of the edges.

As stated in [11], "if the population, from which the sample input was drawn perfectly fits a DAG C all of whose variables have been measured, and the population distribution P contains no conditional independence except those entailed by the factorization of P according to C, then in the large sample limit the PC algorithm produces the true pattern" (present in the data).

4 MarkovPC

MarkovPC can be seen as an extension of the traditional PC algorithm. It is designed to explore the Class' Markov Blanket (CMB) trying to build more accurate and simpler classifiers. The main idea is excluding from possible structures those having attributes out of the CMB.

The use of the CMB allows the reduction of the computational effort needed to build the classifier structure (DAG), as well as, the simplification of its structure. Thus, MarkovPC can minimize the effort needed to both, build the classifier and to use it as a class prediction tool. In addition, having a simpler classifier makes easier the data visualization and understanding, reduce the measurement and storage requirements, reduce training and utilization times, and defy the curse of dimensionality to improve prediction performance. Fig. 3 summarizes MarkovPC in an algorithmic fashion based on the PC algorithm description given by [9] and [12].

Observing figures 2 and 3 it is possible to verify that the main difference between PC and MarkovPC are in lines 8 and 9 of Figure 3. These lines define the CMB test. In other words, the UNDIRECTED_MB(CLASS) procedure represents a set of nodes that may be contained in the CMB. It is important to say that the CMB created by the MarkovPC algorithm is an approximation of the CMB identified by the PC algorithm. It happens because at this point (lines 8 and 9 of the algorithm) the graph is not oriented, so the exact CMB can not be defined yet. In this sense, UNDIRECTED_MB(CLASS) is regardless of graph orientation and it means that this set contains all the nodes that can be reached from the CLASS in at most 2 edges. These nodes are eligible candidates to be part of the CMB when the graph is directed (steps C and D of Figure 3). Following this strategy, all nodes identified by the PC algorithm as part of the CMB will also be present in the CMB defined by MarkovPC. Some nodes present in the CMB defined by MarkovPC, however, may not be present in the CMB defined by the PC algorithm.

```
MarkovPC Algorithm
Problem:  Given a set IND of d-separations, determine the DAG
          pattern, optimized as a classifier and faithful to
          IND if there is one.
Inputs:   a set V of nodes, a set IND of d-separations among
          subsets of the nodes and a CLASS class-node ∈ V.
Outputs:  If IND admits a faithful DAG representation, the DAG
          pattern gp containing the d-separations in this set
          optimized as a Bayesian classifier.
```

```
1  begin
2  A.) form the complete undirected graph gp over V;
3  B.)
4  i = 0;
5  steps = 0;        //helps to measure the algorithm effort
6  repeat
7  for (each X ∈ V) {       //first X should be the CLASS node
8      if UNDIRECTED_MB(CLASS) ⊄ X {
9          remove X from gp;
10     }
11     else {
12         for (each Y ∈ ADJ_x) {
13             determine if there is a set S ⊆ ADJX-{Y} such
14             that |S| = i and I({X},{Y}|S) ∈ IND;
15             steps = steps + 1;
16             if such a set S is found {
17                 S_xy = S;
18                 remove the edge X - Y from gp;
19             }
20         }
21     }
22     steps = steps + 1;
23  }
24  i = i + 1;
25  until (|ADJX| < i for all X ∈ V OR i<=|S|);
26  C.)for each triple of vertices X, Y, Z such that the pair
27      X, Y and the pair Y, Z are each adjacent in C but the pair
28      X, Z are not adjacent in C, orient X - Y - Z as X -> Y <- Z
29      if and only if Y is not in S_xy
30  D.)repeat
31      If A -> B, B and C are adjacent, A and C are not
32      adjacent, and there is no arrowhead at B, then
33      orient B-C as B->C.
34      If there is a directed path from A to B, and edge
35      between A and B, then orient A-B as A->B.
36  until no more edges can be oriented.
37  end.
```

Fig. 3. MarkovPC algorithm

The resulting CMB induced by means of the strategy applied by the MarkovPC algorithm can be summarized as follows: considering a consistent list of independences (IND), the CMB induced by the MarkovPC algorithm is not the minimal CBM, but it contains at least all the relevant nodes (present in the CMB induced by the traditional PC) to the class variable.

When concerning the induced classifier complexity, as the MarkovPC algorithm selects the most relevant variables, it tends to induce classifiers containing a reduced number of variables. It is important to notice that the variable selection is based on a approximate CMB, thus the number of selected variables depends upon the CMB size. Considering the characteristics described in the MarkovPC algorithm, on one hand is possible to say that, when working with domains having a high number of features and a small CMB, the MarkovPC effort tends to be smaller than the one done by the traditional PC. On the other hand, in a situation where a domain has a CMB containing a high number of variables, MarkovPC tends to lead to no variable elimination, and, thus, its effort can be higher than the one needed by the traditional PC.

Considering that testing whether a variable is present in the CMB has the same complexity present in line 10 of PC algorithm (Figure 2), it is possible to analyze the above situation in an extreme case, in which a domain with N variables (all present in the CMB) and a list of indecencies sets (IND) having cardinalities from 0 to M (see [12] for a more detailed description of independency cardinality) are given. In such a situation, the MarkovPC algorithm will need $(N * (M+1))$ tests more that the Traditional PC algorithm. It happens because, for each independency cardinality set, the MarkovPC will test whether each variable is present in the CMB or not. In most of the domains, however, this extreme situation does not happen, thus, the MarkovPC tends to need less effort than the traditional PC.

Another difference between Figure 2 and Figure 3 is present in the "until" clause (line 19 in Figure 2 and line 25 in Figure 3). This clause in MarkovPC has an "or" operator which is not present in the original PC algorithm. The motivation for inserting this "or" operator is to eliminate unnecessary tests. This modification is not necessary to the Markov Blanket strategy proposed in MarkovPC, but it is implemented trying to reduce the computational complexity of the algorithm. It is important to observe that this modification (inserting the "or" operator) do not chance the algorithm behavior in terms of the classifier to be induced.

Taking into account the Average Correct Classification Rates (ACCRs), the MarkovPC algorithm should produce results consistent to the ones produced by the traditional PC. Next section describes the conducted experiments and analyzes the obtained results.

5 Experiments and Results

Trying to verify the soundness of the proposed MarkovPC algorithm, when compared to the traditional PC algorithm, a number of empirical classification experiments were conducted. The main aspects to be considered when concerning the MarkovPC behavior are twofold: the ACCRs and the classifier (structure) complexity. The remaining of this section initially describes the knowledge domains used in the experiments as well as the experimental methodology adopted. The results from the experiments are then presented and analyzed.

Ten domains were used in our simulations. A well-known Bayesian Network domain, namely ALARM [1]; six benchmark problems from the U. C. Irvine repository [2], namely, Car, kr-vs-kp, Lung Cancer (Lung), Postoperative-Patient-Data (Patient), Solar-flare 1 (Flare 1) and Solar-flare2 (Flare 2); and finally, three synthetic domains namely, Synth 1, Synth 2 and Synth 3. Table 1 summarizes datasets characteristics.

Table 1. Datasets Description with dataset name (Data), number of attributes plus class (AT), number of instances (IN) and number of classes (Cl)

	ALARM	Car	Kr-vs-kp	Lung	Patient	Flare 1	Flare 2	Synth 1	Synth 2	Synth 3
AT	38	7	37	57	9	13	13	32	32	32
IN	100000	1728	3196	32	90	323	1066	100000	100000	100000
Cl	See Table 2	4	2	3	3	7	7	2	2	2

Table 2.

Variable	Anaphylaxis	Intubation	KinkedTube	Disconnect	Hypovolemia	InsuffAnesth	LVFailure	PulmEmboulus
IDomainI	2	3	2	2	2	2	2	2

The ALARM network has 8 prediction variables and, in this work, all these variables are used as classes. In this sense, 8 different experiments were conducted with the ALARM domain; in each one, a different class is assumed. Table 2 shows the prediction variables (classes) names and their domain size.

The description of each UCI domain can be downloaded from the UCI Repository site[1]. The artificial domains (Synth 1, Synth 2 and Synth 3) have 32 variables and were created using a sample strategy applied to the Bayesian Classifiers structures described in Figure 4. Synth 1 describes a domain in which only one variable directly influences the class variable (the CMB has a single variable forming it). Synth 2, however, describes a domain in which 14 variables directly influence the class variable. Finally, Synth 3 represents a domain in which all the 32 variables are contained in the CMB.

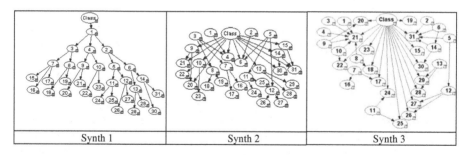

| Synth 1 | Synth 2 | Synth 3 |

Fig. 4. Bayesian Networks representing Synth 1, Synth 2 and Synth 3 domains. The graphical representations were created using GeNie Software [5].

Intending to perform a more robust comparative analysis, besides presenting classification results (ACCRs) obtained using the proposed MarkovPC and the PC algorithms, this section also shows the performance of the traditional Naïve Bayes Classifier [6] when applied to all the 10 described domains. In all the performed

---

[1] UCI Repository of Machine Learning Databases, [http://www.ics.uci.edu]. Irvine, CA, Univ. of California, Dept. Information and Computer Science.

Table 3. Number of steps and the Average Correct Classification Rates obtained with each domain using PC, MarkovPC and Naive Bayes algorithms

Domain	Effort (Number of steps)		ACCR (%)			|MB|
	PC	MarkovPC	PC	MarkovPC	NB	
ALARM Hypovolemia	2758	1121	97.98	98.38	96.55	3
ALARM LVFailure	2758	1135	98.95	99.03	96.41	4
ALARM Anaphylaxis	2758	1052	98.98	98.98	97.26	1
ALARM Insuff. Anesthesia	2758	1442	81.85	84.77	63.34	4
ALARM PulmEmboulus	2758	1160	99.21	99.42	97.30	3
ALARM Intubation	2758	1458	97.55	98.46	84.96	8
ALARM KinkedTube	2758	1327	98.87	98.91	85.30	4
ALARM Disconnect	2758	1145	96.16	98.74	92.95	2
Synth 1	1663	797	89.16	89.16	83.42	1
Synth 2	823	599	93.20	93.20	93.20	14
Synth 3	908	610	85.30	86.74	83.16	31
Car	76	34	89.81	93.57	85.64	N/A
Kr-vs-kp	6129	2127	96.08	94.08	87.76	N/A
Lung-Cancer	5881	2151	75.00	75.00	46.87	N/A
Patient	36	54	71.11	71.11	68.88	N/A
Solar-flare 1	418	176	72.75	73.06	66.25	N/A
Solar-flare 2	342	238	75.23	74.85	73.92	N/A
Average	2255.2	977.94	89.25	89.86	82.54	-
Standard Dev.	1787.1	643.45	10.33	10.38	14.28	-

classification a 10-fold cross validation strategy was applied, and the same training and test files were used by all algorithms. Table 3 presents the average obtained results (ACCRs).

Table 3 reveals that, considering the aforementioned domains, the computational effort (number of tests) performed by MarkovPC is (on average) only 43.36% of the total effort performed by the PC algorithm. Only with the Patient domain, the MarkovPC executed more comparative tests than the traditional PC. It happened, because MarkovPC did not eliminate any variable form the model and considered all them forming the CMB. This is a very interesting result to illustrate the extreme situation described in Section 4. As the Patient domain has 9 attributes and its independency list (IND) has two cardinalities sets (cardinality 0 and cardinality 1), the MarkovPC performed (9 * 2) tests more than the traditional PC.

Trying to get more robust conclusions (when analyzing the ACCRs) the t-test [7] was applied to compare PC and MarkovPC, as well as MarkovPC and Naïve-Bayes ACCRs. The t-test assesses whether the means of two groups are statistically different from each other. A "rule of thumb" was used to set the alpha level at .05. The obtained results allow concluding that the difference between the means for the two groups (in both cases) is significantly different (even given the variability). Therefore, it is possible to consider that, besides needing less computational effort, MarkovPC performed better than the traditional PC when the ACCRs are concerned. Still concerning the ACCRs, MarkovPC performed sensibly better (on average) than the NB.

When concerning the classifier complexity, it is possible to state that PC and Naïve Bayes generate classifiers containing all variables present in the dataset domains. The MarkovPC, however, tended to reduce the number of variables present in the induced classifiers. Considering the 17 simulations reported in table 3, on average, the PC and Naïve Bayes classifiers produced models having 31.52 variables. The MarkovPC, on the other hand, produced classifiers having 7.52 variables on average. Thus, the classifiers induced by the MarkovPC have only 23.85% of the variables present in the classifiers induced by the PC algorithm and the Naïve Bayes approach.

6   Conclusions and Future Work

This paper proposes and discusses a new CI based BC learning algorithm, named MarkovPC, to induce BCs from data. Instead of using the Markov Blanket concept to select variables after the BN induction, as done in other works described in the literature, MarkovPC uses the Markov Blanket concept in order to impose some constraints and optimize the traditional PC algorithm while inducing the BC. It is important to state that MarkovPC is designed specifically to classification tasks.

Experiments performed with a number of domains revealed that MarkovPC tends to be more accurate (in terms of ACCRs) than the traditional PC and Naïve-Bayes. In addition, MarkovPC produced simpler classifiers demanding less comparisons test during the learning procedure than the PC algorithm. The approximate CMB found by MarkovPC is consistent with the PC CMB. Therefore, MarkovPC can be considered consistent and promising.

Authors intend to continue along this line of investigation and plan to introduce the main ideas present in MarkovPC in other CI learning algorithms. Another interesting future work is to use the MarkovPC Markov Blanket strategy when performing CI statistical tests.

Acknowledgements

Authors acknowledge the Brazilian research agency FAPESP for its financial support.

References

[1] Beinlich, I., Suermondt, H.J., Chavez, R.M., Cooper, G.F.: The ALARM monitoring system: A case study with two probabilistic inference techniques for belief networks. In: Proc. of the Second European Conf. on Artificial Intelligence in Medicine, London, UK, vol. 38, pp. 247–256 (1989)

[2] Blake, C.L., Merz, C.J.: UCI Repository of Machine Learning Databases, University of California, Department of Information and Computer Science, Irvine, CA (1998)

[3] Cheng, J., Greiner, R., Kelly, J., Bell, D., Liu, W.: Learning Bayesian networks from data: an information-theory based approach. Artificial Intelligence 137(1), 43–90 (2002)

[4] Chickering, D.M., Meek, C.: On the incompatibility of faithfulness and monotone DAG faithfulness. Artificial Intelligence 170, 653–666 (2006)

[5] Druzdzel, M.J.: SMILE: Structural Modeling, Inference, and Learning Engine and GeNIe: A development environment for graphical decision-theoretic models (Intelligent Systems Demonstration). In: Proceedings of the Sixteenth National Conference on Artificial Intelligence (AAAI-99), Menlo Park, CA, pp. 902–903. AAAI Press/The MIT Press (1999)

[6] Duda, R.O., Hart, P.E.: Pattern Classification and Scene Analysis. Wiley, New York (1973)

[7] Hays, W.: Statistics. 5th edn. Wadsworth Publishing (1994)

[8] Hruschka Jr., E.R., Hruschka, E.R., Ebecken, N.F.F.: Feature Selection by Bayesian Networks. In: Tawfik, A.Y., Goodwin, S.D. (eds.) Canadian AI 2004. LNCS (LNAI), vol. 3060, pp. 370–379. Springer, Heidelberg (2004)

[9] Neapolitan, R.E.: Learning Bayesian networks. Prentice-Hall, Upper Saddle River, NJ (2003)

[10] Pearl, J.: Probabilistic Reasoning in Intelligent Systems: Networks of Plausible Inference. Morgan Kaufmann, San Mateo, CA (1988)

[11] Spirtes, P., Meek, C.: Learning Bayesian networks with discrete variables from data. In: KDD95, pp. 294–299 (1995)

[12] Spirtes, P., Glymour, C., Scheines, R.: Causation, Prediction, and Search. In: Adaptive Computation and Machine Learning, MIT Press, Cambridge, Massachusetts (2001)

Learning of Semantic Sibling Group Hierarchies - K-Means vs. Bi-secting-K-Means

Marko Brunzel[1,2]

[1] DFKI GmbH - German Research Center for Artificial Intelligence
[2] Otto-von-Guericke Universität Magdeburg, Germany
marko.brunzel@dfki.de

Abstract. The discovery of semantically associated groups of terms is important for many applications of text understanding, including document vectorization for text mining, semi-automated ontology extension from documents and ontology engineering with help of domain-specific texts. In [3], we have proposed a method for the discovery of such terms and shown that its performance is superior to other methods for the same task. However, we have observed that (a) the approach is sensitive to the term clustering method and (b) the performance improves with the size of the results'list, thus incurring higher human overhead in the postprocessing phase. In this study, we address these issues by proposing the delivery of a hierarchically organized output, computed with Bisecting K-Means. We compared the results of the new algorithm with those delivered by the original method, which used K-Means using two ontologies as gold standards.

1 Introduction

The discovery of semantic relations is important for the explication of shared domain knowledge. Prominent examples of such relations are those describing "co-hyponyms", i.e. subterms (hyponyms) of the same, possibly unknown or never articulated term, and "co-meronyms", i.e. parts (meronyms) of the same whole entity/term. In our earlier work [1,3,2], we have proposed methods for the extraction of sibling groups of terms from Web documents, exploiting the mark-up similarities of documents.

A disadvantage of methods that discover arbitrary groups of sibling terms is the potentially large number of output groups. One might expect that such groups are themselves correlated, e.g. when they share some terms. The organization of semantic sibling groups into a hierarchy would allow for a more concise and informative representation of the groups. First, a hierarchy can be rolled up to a high level of abstraction or drilled down, if the ontology engineer is interested in a particular sibling group of terms. Second, the hierarchical structure highlights vertical and horizontal relationships among groups, which remain hidden in a flat representation. In this study, we modify our original approach of [3] to deliver a hierarchy of sibling groups. For this purpose, we use the Bisecting

I.Y. Song, J. Eder, and T.M. Nguyen (Eds.): DaWaK 2007, LNCS 4654, pp. 365–374, 2007.

K-Means algorithm – a hierarchical variation of K-Means, which has been shown to be more robust than K-Means towards outliers [16].

The paper is organized as follows: In the next section, we discuss related work on the discovery of semantic sibling groups and on the performance of different clustering algorithms for text mining. In Section 3, we present the underpinnings of the original algorithm XTREEM-SG that discovers a flat list of sibling groups of and modify it to deliver a semantic hierarchy of groups. Our evaluation methodology is discussed in Section 4. Our experiments with two ontologies from the domain "Tourism" are described in Section 5. The last section summarizes our study and elaborates on open issues for future work.

2   Related Work

The discovery of semantic relations among terms belongs to the domain of ontology learning. A comprehensive overview on this subject has appeared in [4]. There are methods in the ontology learning field which are used to obtain a concept hierarchy, some using *Formal Concept Analysis* [6], others using clustering [5]. In [12], prototype ontologies are learned using hierarchical Latent Semantic Indexing. A prototype ontology is similar to a hierarchy of sibling groups, but sibling relations are not a primary characteristic of prototype ontologies.

The approaches above are focusing on ontology learning from unstructured texts. In contrast, we are performing ontology learning from semi-structured documents. The importance of this research topic is stressed among else in [11]. Kruschwitz [10] uses markup sections of Web documents to learn a domain model. Similarly to our approach, Kruschwitz exploits the markup for the representation of similar concepts inside Web documents, but does not incorporate the tree structure of HTML/XHTML documents. The work of [15] also exploits HTML structure to acquire hyponymy relations, but they focus on list itemizations only.

So-called *Hearst patterns* [9] are used to find relations among terms in text collections and are also appropriate for the discovery of co-hyponymy relations. This is done in [7,8], where Web documents are analyzed. However, Hearst patterns are rare, and the coverage is low even on big document collections.

For the discovery of sibling groups of terms from texts, we need clustering algorithms that are appropriate for the particularities of document collections, among else for large feature spaces. In [16], Steinbach, Karypis and Kumar compare document clustering techniques and point out (among else) that Bi-Secting-K-Means is superior to K-Means for the traditional bag-of-words representation of documents (a document being a vector in the feature space of terms). Cimiano, Hotho and Staab compare clustering methods, as well, whereby they model a term as a vector of documents [5]. Since we opt for a vectorization of documents on the feature space of terms, we orient our work towards the findings of [16], albeit we use a more elaborate vectorization, as presented in [3].

Hierarchical Clustering is frequently used for obtaining semantical structures [7,8]. In hierarchical clustering, we distinguish between *agglomerative* and *divisive* methods. The time complexity of *agglomerative hierarchical clustering* is n^3 or $log(n) \cdot n^2$, where n is the number of documents. For large document collections, this complexity is a serious caveat. In our approach, we use Web collections of possibly millions of documents, whereby each document contributes not one but several vectors. Thus, we need a method that combines the advantages of hierarchical clustering but scales better. This motivates the selection of Bisecting K-means.

The underpinnings of our method have been presented in [1,3,2]: The core idea is to exploit the Xhtml structure as the basis of similarity among document vectors: Similar vectors are clustered and terms characterizing the same cluster form sibling groups. In this work, we use the same core idea, but we perform hierarchical clustering and thus build a hierarchy of sibling groups.

3   Building a Sibling Group Hierarchy

We present first the principle of XTREEM (Xhtml TREE Mining). Both the method XTREEM-SG for sibling group discovery [3] and the new hierarchical method XTREEM-SGH for the establishment of a semantic hierarchy are based on this principle. Then, we present and juxtapose XTREEM-SG and XTREEM-SGH. A complete description of XTREEM-SG can be found in [3].

3.1   Foundations of XTREEM

Xhtml TREE Mining is based on markup conventions that can be found in almost all Web documents: Different authors use different nested tags to structure pieces of information in Web documents, but tend to adhere to similar structures. We find terms that adhere to the same syntactic structure within an Xhtml document and apply data mining to reduce the potential large amounts of candidate sets to find semantically related sibling terms. Such semantically related "pieces of text" are not necessarily physically "co-located" in the same narrow context window.

We use the expression *tagpath* for a sequence of tags and the expression *textspan* for the plain text, at which a tagpath ends. We first group textspans that share the same tagpath ("Group-By-Path" operation), eliminate rare textspans and use the retained ones as a feature space. Then, a tagpath becomes a vector in this feature space and similar vectors are clustered. Features/textspans that appear in the same cluster constitute sibling groups [3].

3.2   The XTREEM-SG and XTREEM-SGH Procedures

XTREEM takes as input a collection of documents, observing each document as collection of textspan sets. The collection is not a conventional corpus but is acquired by issuing domain-specific queries towards the whole Web.

Step 1 - Querying & Retrieving: The XTREEM procedure operates on a *Web Document Collection*. Such a Web document collection is obtained by querying a *Archive+Index Facility* on a *Query* with a Web document collection as result. The Query constitutes the domain of interest whereupon semantics should be discovered. It should therefore encircle the documents which are supposed to entail domain relevant content, e.g. "touris*".

The Web document collection should be big enough to contain manifold occurrences of the desired concepts. The Web document collection is not supposed to be a small manually handcrafted document collection; bigger amounts of web content which have an appropriate coverage of the domain are more desirable. Here, recall is more important than precision.

Step 2 - Group-By-Path: For each document the Group-By-Path algorithm is applied. It groups textspans that share a common syntactic structure, i.e. the tagpaths of Xhtml tags leading to them. As result we obtain the collection of textspan sets. Textspans that appear rarely (with respect to a threshold) are eliminated: They correspond to rare terms, for which we cannot expect to find sibling groups of statistical significance. Their elimination speeds up the whole process.

Step 3 - Filtering: This step is performed only if there exists a domain-specific vocabulary. Textspans that are not appearing in the vocabulary are eliminated as irrelevant to the domain. In this study, we do use a vocabulary of terms from the existing ontology, so that sibling groups among those terms (and only those) are discovered.

Step 4 - Vectorization: The feature space is the set of textspans retained after Group-By-Path, optionally reduced with respect to the domain-specific vocabulary (Step 3). The textspans or vocabulary terms constitute the feature space. Each document contributes several vectors – one per tagpath in it. The cells in the vectors are values in $[0,1]$ according to the classic *TF-IDF* [13] weighting scheme.

Step 5 - Bi-Secting-K-Means Clustering: While XTREEM-SG [3] invokes K-Means to group vectors into clusters, in XTREEM-SGH we invoke Bi-Secting-K-Means with cosine distance function. As with K-Means, the target number of clusters is specified in advance. But Bi-Secting-K-Means builds a tree of clusters in a top-down way by splitting the least homogeneous cluster into two more homogeneous ones. We retain this hierarchy of clusters, assigning labels to each of them, as described in the next step.

Step 6 - Cluster Labeling: We define a "cluster label" as the set of frequent features in the cluster. A frequent feature is a feature which has a *in-cluster-support* over a threshold τ. The *in-cluster-support* of a feature is the relative share of vectors (instances) which have this feature.

Here has to be mentioned that we use the resulting cluster hierarchy in a way different to the way cluster dendrograms are conventionally used. In traditional hierarchical clustering, a level of the dendrogram is selected. For Bi-Secting-K-Means, the K leaf nodes may be selected. In XTREEM-SGH, we retain all clusters in all levels of the dendrogram: This is permissible for our approach, since we are not interested in assigning instances to clusters but rather in understanding the cluster labels - the sibling groups. These groups are acquired by traversing the dendrogram in a breadth-first manner.

4   Evaluation Methodology

There is an ongoing discussion on how evaluation of ontology learning can be performed. Despite that gold standard evaluation can be criticized; we will compare the automatic obtained results against reference semantics. The measured quality is not easily comparable (over different references), but it can help to show tendencies.

Our evaluation objective is: "How good is K-Means clustering compared to Bi-Secting-K-Means clustering on structuring a given vocabulary into sibling groups. We evaluate against gold standards, i.e. ontologies that deliver both the vocabulary (the terms) and the sibling relations among them. Our goal is to find those relations. The evaluation of sibling relations is performed in [8] with the average sibling overlap measure. We will compare our results on this measure.

Different numbers of clusters can be obtained. We will show the resulting quality of sibling overlap in dependence to the generated/evaluated number of clusters.

We will also show the distribution of the best matching clusters, revealing if a hierarchy where different tree levels are allowed can be helpful.

4.1   Description of Experimental Influences

Evaluation Reference. The Evaluation is performed on two gold standard ontologies (GSO), from the tourism domain. The concepts of these ontologies are also terms, thus in the following the expressions "concepts and "terms are used interchangeably. The "Tourism GSO[1] (GSO1) contains 293 concepts grouped into 45 sibling sets; the "Getess annotation GSO[2]) (GSO2) contains 693 concepts grouped into 90 sibling sets.

There are three Inputs to the XTREEM procedure described in the following:

Input(1)- Archive+Index Facility: We have performed a topic focused web crawl on the "tourism related documents. The overall size of the document collection is about 9.5 million Web documents. The Web documents have been converted to XHTML. With an n-gram based language recognizer non-English documents have been filtered out. The Documents are indexed, so that for a given Query a Web Document Collection can be retrieved.

Input(2)-Query: For our experiments we consider a document collections which result from querying the Archive+Index Facility on the Query "touris*.

---

[1] http://www.aifb.uni-karlsruhe.de/WBS/pci/TourismGoldStandard.isa
[2] http://www.aifb.uni-karlsruhe.de/WBS/pci/getess_tourism_annotation.daml

Input(3)- Vocabulary: The GSO's described before, are lexical ontologies. Each concept is represented by a term. These terms constitute the vocabulary and the feature space.

The XTREEM-SGH (XTREEM-SG) approach was performed with the previously listed inputs. From 1,468,279 documents adhering to the query "touris*, 222,037 instances with at least two terms of the GSO1 vocabulary and 318,009 instances for GSO2 vocabulary have been obtained.

On the processing we vary the following:

Processing(1) - Clustering Method: K-Means vs. Bi-Secting-K-Means. We apply flat K-Means clustering and Bi-Secting-K-Means clustering. For Bi-Secting-K-Means there are various strategies for selecting a cluster to split next are manifold (1) size (cardinality), (2) homogeneity - variance minimization or more sophisticated quality measures [14] or (3) splitting all clusters. We use the strategy (3); the secting iterations are conducted to up to a depth of 15 levels (splitting is also only possible on clusters with two instances). From such a full hierarchy, different subsets are also obtained and evaluated against the reference.

Processing(2) - Number of Clusters: The number of clusters varies, for K-Means it has to be given in advance. For K we used values of 50, 100, 150, 200, 350, 500, 750, 1000, 2000 and 3000. These numbers include the range of numbers of clusters which are appropriate to be shown to a human ontology engineer. For the Bi-Secting-K-Means clustering different numbers of clusters result from different strategies of generating an accessing the cluster hierarchy. A hierarchy with up to 15 splits (tree depth is 15 levels) is produced. The cluster labelling threshold was set to 20 percent.

4.2   Evaluation Criteria

Each of the gold standard ontologies delivers a number of reference sets of terms in sibling relation. Each run of XTREEM-SG and XTREEM-SGH delivers a number of term clusters that are suggested as potential sibling groups. Intuitively, one would compare each of the suggested clusters against each of the reference sets, select the best match and then count the number of matches; clusters without match and reference sets for which no match was found would be observed as false positives or false negatives. However, the identification of a "best match is not straightforward, nor is the selection of a "single best match the most appropriate evaluation strategy.

The F-Measure on average sibling overlap (FMASO) [8] can be used to compare two collections of term sets against each other to obtain a statement about how the two collections overlap. This evaluation measure is described in detail in [3].

5   Experiments

5.1   Experiment 1: K-Means vs. Bi-secting-K-Means

Our setting - and algorithm is in so far different, that the Bi-Secting-K-Means is not used to obtain K Clusters. The criteria for selecting a cluster to split are manifold size (cardinality), homogeneity - variance minimization or more sophisticated quality measures [14]. The secting iterations are conducted to up to a depth of 15 levels (splitting is also only possible on clusters with two instances). From such a full hierarchy, different subsets are also obtained and evaluated against the reference.

Figure 1(a) shows that the quality of the results obtained by the K-Means Clustering algorithm are in generally as good as or better than those obtained via Bi-Secting-K-Means Clustering. For the second reference, shown in figure 1(b), K-Means shows even better results, Bi-Secting-K-Means results are with exception of some outliers generally worse.

Conclusion: For our setting, we have to change the conclusion of Steinbach, Karypis and Kumar [16] that "The bisecting K-means technique is better than the standard K-means approach [16]. Our conclusion is that K-Means is as good as or better than Bi-Secting-K-Means clustering; for our setting, our goal, and our data. But since the difference is rather small and a hierarchy has advantages on its own, Bi-Secting-K-Means clustering is nevertheless worth considering. We will investigate which settings cause a Bi-Secting-K-Means clustering which is as good as possible.

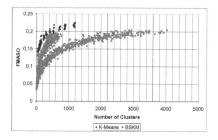 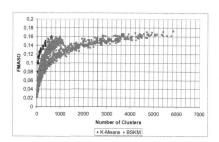

Fig. 1. FMASO on different K for K-Means Clustering and Bi-Secting-K-Means Clustering towards (a) GSO1 and (b) GSO2

5.2   Experiment 2: Accessing the Hierarchy

In this experiment we will differentiate the way the cluster hierarchy is accessed. We have performed experiments for the following strategies:

Access Strategy 1 - Separate Levels: Only the clusters which are obtained at a certain hierarchy level are evaluated as one automatically obtained result which is evaluated against the gold standard.

Access Strategy 2 - Up To Level X: All clusters up to the hierarchy Level X are evaluated together. Different hierarchy levels are mixed together.

Access Strategy 3 - Merged Levels: This is the special case of Up-ToLevelX, but here X is the deepest level which was calculated.

Figure 2 shows that the different ways of accessing the cluster hierarchies lead to different evaluated quality of the results. The best results are obtained for only using the clusters from a certain hierarchy level. This is comparable to creating only a flat clustering. But here has to be mentioned that if the hierarchy is used more deeply, also a lot of similar clusters is evaluated, but the evaluation criteria allows only for one best match. In a real world setting, where a human would inspect the cluster hierarchy, he could stop unfolding the cluster hierarchy, if the clusters do not contribute anymore.

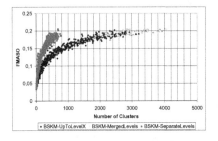 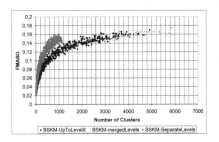

Fig. 2. FMASO for Bi-Secting-K-Means Clustering towards (a) GSO1 and (b) GSO2. Different Strategies of accessing the Hierarchy are differentiated.

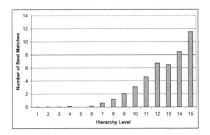 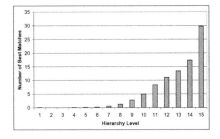

Fig. 3. Best Matching Hierarchy Level towards (a) GSO1 and (b) GSO2. Number of Best Matches is averaged over 60 clusterings.

5.3   Experiment 3: The Hierarchy Level of Best Sibling Group Overlaps

In the following experiment we will investigate on which hierarchy levels the best overlap with the reference sibling sets occurred. In figure 3 the distribution of *best matching hierarchy level* is shown for GSO1 and GSO2. The results are averaged over 60 Bi-Secting-K-Means clusterings. The hierarchy is accessed in *breath first traversal* order. This corresponds to the way a human would access the hierarchy, as a human user would rather start from near the root, than looking at a potentially very depth tree. As can be seen, the best matching hierarchy levels are distributed over certain levels. From this we conclude that it is appropriate to present the hierarchy to the human ontology engineer. If the access to the hierarchy would be limited to only the levels near the root, many sibling groups which are evaluated as 'good' would be missed.

6   Conclusions and Future Work

On a data set which is different to traditional text document vectorization, we investigated the quality obtained by using a flat K-Means clustering compared to using a hierarchical Bi-Secting-K-Means clustering. We can not state that the results obtained by Bi-Secting-K-Means clustering are as good as those obtained by a K-Means clustering, regarding a gold standard evaluation. But the results are not much worse. If the added value of the hierarchy can support the human ontology engineer on semi-automatic ontology learning, the minor worse results obtained by Bi-Secting-K-Means clustering are acceptable. The best matching clusters of the generated sibling set hierarchy can be found on various hierarchy levels, giving raise to the recommendation that the hierarchy should be presented to the user for subsequent human inspection. Future work circumference a comparison of Bi-Secting-K-Means clustering to the computational expensive agglomerative hierarchical clustering. Since this work is intended to be used in a semi-automatic setting we also want to investigate improvements on presenting the cluster hierarchy to the user.

Acknowledgements. Parts of this work have been supported by European Union IST fund (Grant FP6-027705, project Nepomuk).

References

1. Brunzel, M., Spiliopoulou, M.: Discovering multi terms and co-hyponymy from xhtml documents with XTREEM. In: Nayak, R., Zaki, M.J. (eds.) KDXD 2006. LNCS, vol. 3915, pp. 22–32. Springer, Heidelberg (2006)
2. Brunzel, M., Spiliopoulou, M.: Discovering semantic sibling associations from web documents with XTREEM-SP. In: Tjoa, A.M., Trujillo, J. (eds.) DaWaK 2006. LNCS, vol. 4081, pp. 469–480. Springer, Heidelberg (2006)
3. Brunzel, M., Spiliopoulou, M.: Discovering semantic sibling groups from web documents with XTREEM-SG. In: Staab, S., Svátek, V. (eds.) EKAW 2006. LNCS (LNAI), vol. 4248, pp. 141–157. Springer, Heidelberg (2006)
4. Buitelaar, P., Cimiano, P., Magnini, B.: Ontology Learning from Text: Methods, Evaluation and Applications. IOS Press, Amsterdam (2005)
5. Cimiano, P., Hotho, A., Staab, S.: Comparing conceptual, divise and agglomerative clustering for learning taxonomies from text. In: de Mántaras, R.L., Saitta, L. (eds.) ECAI, pp. 435–439. IOS Press, Amsterdam (2004)
6. Cimiano, P., Hotho, A., Staab, S.: Learning concept hierarchies from text corpora using formal concept analysis. Technical report, Insittue AIFB, University of Karlsruhe (November 2004)
7. Cimiano, P., Staab, S.: Learning by googling. SIGKDD Explorations 6(2), 24–33 (2004)
8. Cimiano, P., Staab, S.: Learning concept hierarchies from text with a guided agglomerative clustering algorithm. In: Biemann, C., Paas, G. (eds.) Proceedings of the ICML 2005 Workshop on Learning and Extending Lexical Ontologies with Machine Learning Methods, August 2005, Bonn, Germany (2005)
9. Hearst, M.A.: Automatic acquisition of hyponyms from large text corpora. In: Proceedings of the 14th conference on Computational linguistics, Morristown, NJ, USA, 1992, pp. 539–545. Association for Computational Linguistics (1992)

10. Kruschwitz, U.: Exploiting structure for intelligent web search. In: HICSS '01: Proceedings of the 34th Annual Hawaii International Conference on System Sciences (HICSS-34), Washington, DC, USA, 2001, vol. 4, p. 4010. IEEE Computer Society Press, Los Alamitos (2001)
11. Nayak, R., Zaki, M.J.: Knowledge discovery from xml documents. In: Ng, W.-K., Kitsuregawa, M., Li, J., Chang, K. (eds.) PAKDD 2006. LNCS (LNAI), vol. 3918, Springer, Heidelberg (2006)
12. Paaß, G., Kindermann, J., Leopold, E.: Learning prototype ontologies by hierachical latent semantic analysis. In: Abecker, A., Bickel, S., Brefeld, U., Drost, I., Henze, N., Herden, O., Minor, M., Scheffer, T., Stojanovic, L., Weibelzahl, S. (eds.) LWA, pp. 193–205. Humbold-Universität, Berlin (2004)
13. Salton, G., Buckley, C.: Term weighting approaches in automatic text retrieval. Technical report, Ithaca, NY, USA (1987)
14. Schaal, M., Müller, R.M., Brunzel, M., Spiliopoulou, M.: Relfin - topic discovery for ontology enhancement and annotation. In: Gómez-Pérez, A., Euzenat, J. (eds.) ESWC 2005. LNCS, vol. 3532, pp. 608–622. Springer, Heidelberg (2005)
15. Shinzato, K., Torisawa, K.: Acquiring hyponymy relations from web documents. In: HLT-NAACL, pp. 73–80 (2004)
16. Steinbach, M., Karypis, G., Kumar, V.: A comparison of document clustering techniques. In: KDD Workshop on Text Mining (2000)

Mining Top-K Multidimensional Gradients

Ronnie Alves[*], Orlando Belo, and Joel Ribeiro

Department of Informatics, School of Engineering, University of Minho
Campus de Gualtar, 4710-057 Braga, Portugal
{ronnie,obelo}@di.uminho.pt

Abstract. Several business applications such as marketing basket analysis, clickstream analysis, fraud detection and churning migration analysis demand gradient data analysis. By employing gradient data analysis one is able to identify trends, outliers and answering "what-if" questions over large databases. Gradient queries were first introduced by Imielinski et al [1] as the *cubegrade* problem. The main idea is to detect interesting changes in a multidimensional space (MDS). Thus, changes in a set of measures (aggregates) are associated with changes in sector characteristics (dimensions). MDS contains a huge number of cells which poses great challenge for mining gradient cells on a useful time. Dong et al [2] have proposed *gradient constraints* to smooth the computational costs involved in such queries. Even by using such constraints on large databases, the number of interesting cases to evaluate is still large. In this work, we are interested to explore best cases (Top-K cells) of interesting multidimensional gradients. There several studies on Top-K queries, but preference queries with multidimensional selection were introduced quite recently by Dong et al [9]. Furthermore, traditional Top-K methods work well in presence of convex functions (gradients are non-convex ones). We have revisited iceberg cubing for complex measures, since it is the basis for mining gradient cells. We also propose a gradient-based cubing strategy to evaluate interesting *gradient regions* in MDS. Thus, the main challenge is to find maximum gradient regions (MGRs) that maximize the task of mining *Top-K gradient cells*. Our performance study indicates that our strategy is effective on finding the most interesting gradients in multidimensional space.

1 Introduction

Gradient queries were first introduced by Imielinski et al [1] as the *cubegrade* problem. The main idea is to detect interesting changes in a multidimensional space (MDS). Thus, changes in a set of measures (aggregates) are associated with changes in sector characteristics (dimensions). By employing gradient data analysis one is able to identify trends, outliers and answering "what-if" questions over large databases. MDS contains a huge number of cells which poses great challenge for mining gradient cells on a useful time. To reduce search space, Dong et al [2] propose several

---

[*] Supported by a Ph.D. Scholarship from FCT-Foundation of Science and Technology, Ministry of Science of Portugal.

I.Y. Song, J. Eder, and T.M. Nguyen (Eds.): DaWaK 2007, LNCS 4654, pp. 375–384, 2007.

constraints which are explored by a set-oriented processing. Even by using such constraints on large databases, the number of interesting cases to evaluate is still large. Furthermore, at a first experience for finding gradients in MDS, users often don't feel confident of what set of probe constraints to use. Thus, it should be the case to provide some Top-K facility over MDS.

As illustration of the motivation behind this problem, consider the following example of generating alarms for potential fraud situations on mobile telecommunications systems. Generally speaking, those alarms are generated when abnormal utilization of the system is detected, meaning that sensitive changes happened concerning the normal behavior. For example, one may want to see what are associated with significant changes of the average (w.r.t., the number of calls) in the Porto area on Monday compared against Tuesday, and the answer could be one in the form "the average of number of calls for callings during working time in Campaña went up 55 percent, while those callings during night time in Bonfim went down 15 percent". Expressions such as "callings during working time" correspond to cells in data cubes and describe sectors of the business modeled by the data cube. Given that the number of calls generated in a business day in Porto area is extremely large (hundred million calls); the fraud analyst would be interest to evaluate just the Top-10 higher changes in that scenario, specially those ones in Campaña area. Then the fraud analyst would be able to "*drillthrough*" most interesting customers.

The problem of mining Top-K cells with multidimensional selection conditions and raking gradients is significantly more expressive than the classical Top-K and cubegrade query [1]. In this work, we are interested to mine best cases (Top-K cells) of multidimensional gradients in a MDS.

2 Problem Formulation and Assumptions

A *cuboid* is a multidimensional summarization by means of aggregating functions over a subset of dimensions and contains a set of cells, called cuboid cells. A data cube can be viewed as a *lattice of cuboids*, which also integrates a set of cuboid cells. Data cubes are built from *relational base tables* (fact tables) from large data warehouses.

Definition 1 (*K-Dimensional Cuboid Cell*). In a n-dimension data cube, a cell $c = (i_1, i_2, \ldots, i_n : m)$ (where m is a measure) is called a k-dimensional cuboid cell (i.e., a cell in a k-dimensional cuboid), if and only if there are exactly k ($k \leq n$) values among $\{i_1, i_2, \ldots, i_n\}$ which are not * (i.e., all). If k = n, then c is a *base cell*. A base cell does not have any descendant. A cell c is an *aggregated cell* only when it is an ancestor of some base cell (i.e., where k < n). We further denote $M(c) = m$ and $V(c) = (i_1, i_2, \ldots, i_n)$.

Definition 2 (*Iceberg Cell*). A cuboid cell is called *iceberg cell* if it satisfies a threshold constraint on the measure. For example, an iceberg constraint on measure count is $M(c) \geq min\_supp$ (where *min_supp* is an user-given threshold).

Definition 3 (*Closed and Maximal Cells*). Given two cells $c = (i_1, i_2, \ldots, i_n : m)$ and $c' = (i_1', i_2', \ldots, i_n' : m')$, we denote $V(c) \leq V(c')$ if for each i_j (j = 1,\ldots,n) which is not *, $i_j' = i_j$. A cell c is said to be covered by another cell c' if for each c'' such that $V(c) \leq V(c'') \leq V(c')$, $M(c'') = M(c')$. A cell is called a *closed cuboid cell* if it is not covered by any other cells. A cell is called a *maximal cuboid cell* if it is closed and has no other cell c which is superset of it.

Definition 4 (*Matchable Cells*). A cell c is said to match a cell c' when they differ from each other in one, and only one, cube modification at time. These modifications could be: Cube *generalization/specialization* iff c is an *ancestor* of c' and c' a *descendant* of c; or *mutation* iff c is a sibling of c', and vice versa.

Definition 5 (*Probe Cells*). Probe cells (pc) are particular cases of iceberg cells which are significant according to some selection condition. This selection condition is a query constraint on base table's dimensions.

Definition 6 (*Gradient Cells*). A cell c_g is said to be gradient cell of a probe cell c_p, when they are matchable cells and their *delta change*, given by $\Delta g(c_g, c_p) \equiv (g(c_g, c_p) \geq \psi)$ is *true*, where ψ is a constant value and g is a *gradient function*.

In this work we confine our discussion with average gradients $(M(c_g)/M(c_p))$. Average gradients are effective functions for detecting interesting changes in multidimensional space, but they also pose great challenge to the cubing computation model [2] [10].

Problem Definition. Given a base table R, iceberg condition IC, probe condition PC, the mining of *Top-k Multidimensional Gradients* from R is: Find the most interesting (Top_K) gradient-probe pairs (c_g, c_p) such that $\Delta g(c_g, c_p)$ is true.

3 Mining Top-K Gradient Cells

Before start the discussion about mining Top-K cells in MDS, lets first look to Figure 1. This figure shows how aggregating values are distributed along cube's dimensions. The base table used for cubing was generated according to a uniform distribution. It has 100 tuples, 3 dimensions (x, y, z) with cardinalities varying from 1 to 10, and a measure (m) varying from 1 to 100.

From Figure 1 we can see that different aggregating functions will provide different density regions in data cubes. When searching for gradients, one may want to start this task by evaluating a region in which presents higher variance. In Figure 1(b) the central region corresponding to the rectangle *R1 {[4, 7] : [6, 5]}* covers all five *bins* $\{b_0=\{0\text{-}20\}, b_1=\{20\text{-}40\}, b_2=\{40\text{-}60\}, b_3=\{60\text{-}80\}, b_4=\{80\text{-}100\}\}$. If it is possible to identify such region, chances are that the most interesting gradients will take place there. We denote those regions as *gradient regions(GR)*. The problem is that average is an algebraic function and it has a *spreading factor(SF)* with respect to the distribution of aggregating values over the cube quite unpredictable. Thus, there will be sets of possible gradient regions to looking for in the cube before selecting the most Top-K gradient cells.

Another interesting insight from Figure 1(b) is that gradients are maximized when walking from bins b_0 to b_4. Therefore, even if a region doesn't cover all bins but if at least has the lowest and highest ones; it should be a good candidate GR for mining Top-K cells. We expect that GRs with largest aggregating values will provide higher gradient cells. This observation motivates us to evaluate gradients cells by using a *gradient ascent* approach.

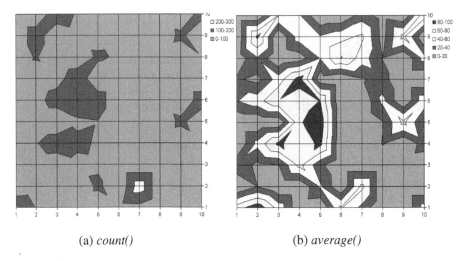

(a) *count()*   (b) *average()*

Fig. 1. It presents the distribution of aggregating values from a 2-D(x, y) cube

Gradient ascent is based on the observation that if the real-valued function $f(x)$ is defined and differentiable in a neighborhood of a point Y, $f(x)$ then increases fastest if one goes from Y in the direction of the gradient of f at Y, $\Delta f(Y)$. It follows that, if

$$Z = Y + \gamma \Delta f(Y) \tag{1}$$

for, $\gamma > 0$ a small enough number, then $f(Y) \leq f(Z)$. With this observation, one could start with x_0 for a local maximum of f, and considers the sequence $x_0, x_1, x_2, \ldots$ such that

$$x_{n+1} = x_n + \gamma_n \Delta f(x_n), n \geq 0 \tag{2}$$

We have $f(x0) \leq f(x1) \leq f(x2) \leq \ldots$, so we expect the sequence (x_n) converge to the local maximum. Therefore, when evaluating a *GR* we first search for the *maximal probe cells*, i.e. the highest aggregating values (non-closed and non-maximal cells) on it and then calculates its gradients from all possible *matchable cells*.

To allow this gradient ascent traversal through cube's lattice, one needs to incorporate the main observations mentioned above into the cubing process.

3.1 Spreading Factor

The spreading factor is measured by the variance. This statistical measure indicates how the values are spread around the expected value. From 1-D cuboid cell we want to capture how large the differences are in a list of measures values $M_L=(M_1,\ldots,M_n)$ from the base relation R. Thus, dimensions are projected onto cube's lattice according to this variation. This variation follows

$$\frac{\sum (X - \mu)^2}{N} \tag{3}$$

where, X is a particular value from M_L, μ is the mean from M_L and N is the number of elements in M_L. For large datasets one could use a variation of Equation 3, using a sample rather than the population. In this sense one can replace N with $N-1$.

3.2 Cubing Gradient Regions

Our cubing method follows *Frag-Cubing* approach [11]. We start by building *inverted indices* and *value-list indices* from the base relation R, and then assembling high-dimensional cubes from low-dimensional ones. For example, to compute the cuboid cell $\{x_1, y_3, *\}$ from Table 1, we intersect the *tids* (see Table 2) of x_1 $\{1, 4\}$ and y_3 $\{4\}$ to get the new list of $\{4\}$.

Table 1. Example of a base table R

tid	X	Y	Z	M
1	x1	y2	z3	1
2	x2	y1	z2	2
3	x3	y1	Z2	3
4	x1	y3	z1	4

The order in which we compute all cuboids is given by a processing tree GR_{tree} following a DFS traversal order. Before creating GR_{tree} we can make use of the first heuristics for pruning (p_1, **pruning non-valid tuples**). To smooth the evaluation of iceberg-cells, it is interesting to build such tree only with tuples satisfying the iceberg condition.

After applying p_1, we are able to calculate the spreading factor (*sf*, see Section 3.1) of all individual dimensions X, Y and Z.

Table 2. Inverted indices and spreading factor for all individual dimensions of Table 1

Attr. Value	Tid.Lst	Tid.Sz	sf
x1	1, 4	2	2.25
x2	2	1	0
x3	3	1	0
y1	2, 3	2	0.25
y2	1	1	0
y3	4	1	0
z1	4	1	0
z2	2, 3	2	0.25
z3	1	1	0

From Table 2 we can say that we have a total of nine GRs to grow.

The GR_{tree} will follow the order of X>>Y>>Z. Given that we want to find large gradients, chances are that higher gradients will take place on projecting GRs having higher spreading factors. Here comes the second heuristic for pruning (p_2, **pruning**

non-valid regions). For example, a threshold constraint on a GR is $sf(GR) \geq min\_sf$ (where min_sf is a user-given threshold). Let's say that we are interested to search for gradients on GRs having a sf ≥ 0.25. From this constraint we only need to evaluate 3 GRs rather than 9 ones. Cubing is carried out through projecting x_1; y_1 and finally z_2 (see Figure 2).

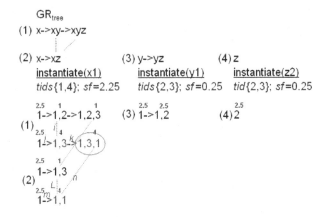

Fig. 2. The lattice formed by projecting GRs $\{x_1, y_1, z_2\}$. Aggregating values are placed "upper" all cuboid cells. Possible probe cells are denoted with shadow circles. The letters "matchable cells" shows valid gradient pairs.

Since we have cubing those three regions in Figure 2, it is possible now to mine the Top-K gradient cells from them. After this cubing, we also augment each region with its respective *bin* [agg$_{min}$; agg$_{max}$] according to the minimum and maximum aggregating value on it. For example $b_{x1}=[1; -4]$, $b_{y1}=[2,5; -2,5]$ and $b_{z2}=[2,5; -2,5]$. With all those information we can make use of the third heuristic for pruning (p_3, pruning **non-TOPk regions**). Given that we have an iceberg condition on average such as $M_{avg}(c) \geq 2.7$, we can confine our search for Top-K cells only for GR=x_1.

Even by using such constraint, the number of gradient pairs to evaluate is still large in x_1. So, we must define our set of probe cells pc{} to mine. Remember the discussion about gradient ascent; by taking the local maximum (i.e., a cuboid cell having the highest aggregating value) from a particular region, all matchable cells containing this local maximum will increase gradient factors. Thus, our probe cells are given by the set of the Top-K aggregating values in that region TopK$_{pc}$. For a cuboid cell to be eligible as a probe cell, it cannot be a *closed and maximal cell*. For example, in Figure 2 the cuboid cell =$\{1,3,1\}$ cannot be selected as a probe cell. Next, for each probe cell in TopK$_{pc}$ we calculate its gradients. Finally we select the Top-K gradient cells TopK$_{gr}$. For example, with we are looking for the Top-3 cell, the TopK$_{pc}$ is formed by $\{\{1,3\};\{1,1\};\{1\}\}$ and TopK$_{gr}$ is formed by letters $\{i, L, j\}$.

Usually we will have more valid-ToPk regions rather than in the previous example. Thus, the final solution must take into account all local TopK$_{gr}$ before raking TOP-K cells. Besides, after having calculated all local TopK$_{gr}$, we continue searching for other valid gradient cells resulting from matchable cells (i.e., projecting probe cells from a GR_i over cuboid cells in GR_j).

Pseudo Algorithm 3.1 TopK$_{gr}$-Cube

Input: a table relation t_{rel}; an iceberg condition on average min_{avg}; an iceberg condition on GR min_sf; the number of Top-K cells K
Output   : the set of gradient-cell pairs $TopK_{gr}$.
Method :

1. Let ft_{rel} be the relation fact table
2. Build the *processing tree* GR_{tree} // evaluating heuristic p$_1$
3. Call TopK$_{gr}$-Cube ().

```
Procedure TopK_gr-Cube (ft_rel, GR_tree, min_avg, min_sf, K)

1: Get 1-D cuboids from GR_tree
2: For each 1-D cuboids do {
3:   Set GR ← {subTree_GR(V(c))} //build gradient regions
4:   For each GR having sf ≥ min_sf do //apply p2
       Aggregate valid GR // do projections, cubing
5:   For each valid GR having a bin[min,max] satisfying min_avg //
apply p3
6:   Set TopK_pc ← {topKpc_GR(GR,K)} //select probe cells
7:   For each valid TopK GR
       Set TopK_gr ←{topKgr_GR(TopK_pc)} // calculate its gradients
8:   Rank Top-K cells (TopK_gr,K) //rank gradient cells,
                                DESC order of ψ
```

4 Evaluation Study

All the experiments were performed on a 3GHz Pentium IV with 1Gb of RAM memory, running Windows XP Professional. TopK$_{gr}$-Cube was coded with Java 1.5, and it was performed over synthetic datasets (Table 3). These datasets were generated using the data generator described in [2, 10].

Table 3. The overall information of each dataset

Dset	Tuples	Dims	Card[1]	M[min,max]
D1	5000	7	100	100,1000
D2	10000	10	100	100,1000

When running the next performance figures, we confine our tests with TopK$_{gr}$-Cube to the following scenarios: *one*, (Figure 3) we want to see the effects of using variance as a criteria for mining Top-K cells: *two*, (Figure 4) we want to see the effects of iceberg condition while selecting interesting regions; and *three*, (Figure 5) the pruning effects by using the heuristics used by our Top-K cubing method.

---

[1] Those cardinalities provide very sparse data cubes, thus, poses more challenge to the Top-K cubing computation model.

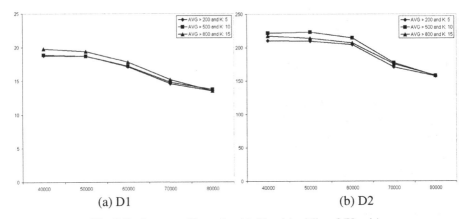

Fig. 3. Performance (Runn.time(s), Y-axis) x Min_sf (X-axis)

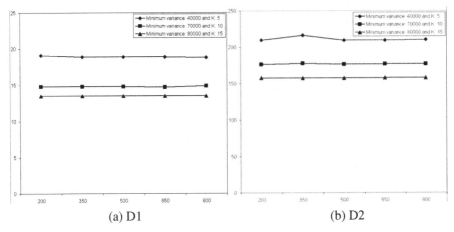

Fig. 4. Performance (Runn.time(s), Y-axis) x Min_AVG (X-axis)

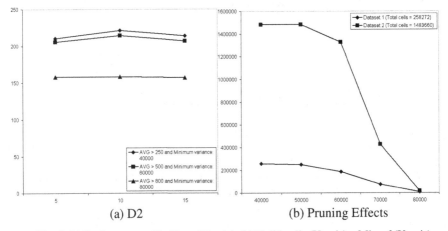

Fig. 5. (a) Performance x K effects (X-axis). (b) Valid-cells (Y-axis) x Min_sf (X-axis)

5   Final Discussion

5.1   Related Work

The problem of mining changes of sophisticated measures in a multidimensional space was first introduced by Imielinski et al. [1] as a cubegrade problem. The main idea is to explore how changes (*delta changes*) in a set of measures (*aggregates*) of interest are associated with changes in the underlying characteristics of sectors (*dimensions*). In [2] was proposed a method called *LiveSet-Driven* leading to a more efficient solution for mining gradient cells. This is achieved by group processing of live probe cells and pruning of search space by pushing several constraints deeply. There are also other studies [3, 4, 5] by Sarawagi for mining interesting cells on data cubes. The idea of interestingness in these works is quite different from that explored by gradient-based ones. Instead of using a specified gradient threshold in relevance to the cells' ancestor, descendants, and siblings, it relies on the *statistical analysis* of neighborhood values of a cell to determine its interestingness (or also *outlyingness*).

Those previous methods employ the idea of interestingness supported by either statistical or ratio-based approach. Such approach still provides a large number of interesting cases to evaluate, and on a real application scenario, one could be interested to explore just such a small number (best cases of *Top-K cells*) of gradient cells. There are several research papers on answering Top-queries [6, 7, 8] on large databases, which could be used as baseline for mining Top-K gradient cells. However, the range of complex delta functions provided by the cube gradient model complicates the direct application of those traditional Top-K query methods. To the best of our knowledge, the problem of mining Top-K gradient cells in large databases is not well addressed yet. Even the idea of Top-K queries with multidimensional selection was introduced quite recently by Dong et al. [9]. Furthermore, the model still relies on the computation of convex functions.

5.2   Conclusions

In this paper, we have studied issues and mechanisms on effective mining of Top-K multidimensional gradients. Gradient are interesting changes in a set of measures (aggregates) associated with the changes in the core characteristics of cube cells. We have also proposed a gradient-based cubing strategy (TopK$_{gr}$-Cube) to evaluate interesting gradient regions in MDS. This strategy relies on cubing gradient regions which show high variance. Thus, the main challenge is to find maximum gradient regions (MGRs) that maximize the task of mining Top-K gradient cells through a set of Top-K probe cells. To do so, we use a gradient ascent approach with a set of pruning heuristics guided by a specific GR$_{tree}$. Our performance study indicates that our strategy is effective on mining the most interesting gradients in multidimensional space. To end up our discussion we set the follow topics as interesting issues for future research:

– *Top-K average pruning into GRs*. Although, we make pruning of GRs according to an iceberg condition. One can take advantage of Top-K average [10] for cubing only GRs satisfying this condition.

- *Looking ahead for Top-K gradients.* Given that $GR_1 >> GR_2 >> GR_3$, it should be the case that by looking ahead for a gradient cell in GR_2 will not generate gradients higher than that in GR_1.
- *Mining high-dimensional Top-K cells.* The idea is to select small-fragments [11] (with some measure of interest) and then explore on-line query computation to mine high-dimensional Top-K cells.

Acknowledgements

Ronnie thanks the valuable discussion with Prof. Jiawei Han, Hector Gonzalez Xiaolei Li and Dong Xin during his visiting research to Data Mining Research Group (at DAIS Lab, University of Illinois at Urbana-Champaign).

References

1. Imielinski, T., Khachiyan, L., Abdulghani, A.: Cubegrades: Generalizing Association Rules. Data Mining and Knowledge Discovery (2002)
2. Dong, G., Han, J., Lam, J.M.W., Pei, J., Wang, K., Zou, W.: Mining Constrained Gradients in Large Databases. IEEE Transactions on Knowledge Discovery and Data Engineering (2004)
3. Sarawagi, S., Agrawal, R., Megiddo, N.: Discovery-Driven Exploration of OLAP Data Cubes. In: Proc. Int. Conference on Extending Database Technology (EDBT) (1998)
4. Sarawagi, S., Sathe, G.: i3: Intelligent, Interactive Investigaton of OLAP data cubes. In: Proc. Int. Conference on Management of Data (SIGMOD) (2000)
5. Sathe, G., Sarawagi, S.: Intelligent Rollups in Multidimensional OLAP Data. In: Proc. Int. Conference on Very Large Databases (VLDB) (2001)
6. Chang, Y., Bergman, L., Castelli, V., Li, M.L.C., Smith, J.: Onion technique: Indexing for linear optimization queries. In: Proc. Int. Conference on Management of Data (SIGMOD) (2000)
7. Hristidis, V., Koudas, N., Papakonstantinou, Y.: Prefer: A system for the efficient execution of multi-parametric ranked queries. In: Proc. Int. Conference on Management of Data (SIGMOD) (2001)
8. Bruno, N., Chaudhuri, S., Gravano, L.: Top-k selection queries over relational databases: Mapping strategies and performance evaluation. ACM Transactions on Database Systems (2002)
9. Dong, X., Han, J., Cheng, H., Xiaolei, L.: Answering Top-k Queries with Multi-Dimensional Selections: The Ranking Cube Approach. In: Proc. Int. Conference on Very Large Databases (VLDB) (2006)
10. Han, J., Pei, J., Dong, G., Wank, K.: Efficient Computation of Iceberg Cubes with Complex Measures. In: Proc. Int. Conference on Management of Data (SIGMOD), (2001)
11. Li, X., Han, J., Gonzalez, H.: High-dimensional OLAP: A Minimal Cubing Approach. In: Proc. Int. Conference on Very Large Databases (VLDB) (2004)

A Novel Similarity-Based Modularity Function for Graph Partitioning

Zhidan Feng[1], Xiaowei Xu[1], Nurcan Yuruk[1],
and Thomas A.J. Schweiger[2]

[1] The Department of Information Science, UALR
{zxfeng,xwxu,nxyuruk}@ualr.edu
[2] Acxiom Corporation
Tom.Schweiger@acxiom.com

Abstract. Graph partitioning, or network clustering, is an essential research problem in many areas. Current approaches, however, have difficulty splitting two clusters that are densely connected by one or more "hub" vertices. Further, traditional methods are less able to deal with very confused structures. In this paper we propose a novel similarity-based definition of the quality of a partitioning of a graph. Through theoretical analysis and experimental results we demonstrate that the proposed definition largely overcomes the "hub" problem and outperforms existing approaches on complicated graphs. In addition, we show that this definition can be used with fast agglomerative algorithms to find communities in very large networks.

1 Introduction

Many problems can be represented as networks or graphs, and identifying cluster in the network or partitioning the graph is a fundamental problem for computer networks analysis, VLSI designing, biologic network analysis, social networks analysis [18], business networks analysis, and community detection [7]. In the literature, graph partitioning has many names, and is sometimes called network analysis, network clustering, detecting communities in networks, etc. This area of research has seen a lot of efforts in this problem over decades, and many algorithms have been proposed, studied, and used.

The problem is defined thus: Given a graph $G = \{V, E\}$, where V is a set of vertices and E is a set of weighted edges between vertices, optimally divide G into k disjoint sub-graphs $G_i = \{V_i, E_i\}$, in which $V_i \bigcap V_j = \Phi$ for any $i \neq j$, and $V = \sum_{i=1}^{k} V_i$. The number of sub-graphs, k, may or may not be prior known.

In this paper, we focus on the partitioning problem of un-weighted graphs, that is, graphs for which the weight of all edges is 1.

Two questions need to be answered, one that is mathematical and one that is algorithmic. First, what is the definition of *optimal* when partitioning a graph, and second, how does one find the optimal partition efficiently.

I.Y. Song, J. Eder, and T.M. Nguyen (Eds.): DaWaK 2007, LNCS 4654, pp. 385–396, 2007.

Regarding to the first question, there is no consensus in the literature. Different approaches use different criteria for different applications. Examples are the min-max cut [1, 2, 3], modularity [4, 5, 6] and betweenness [7, 8]. To the second question, since finding the optimal partition in a graph is normally a NP-complete problem, most current approaches apply heuristics to shorten the search and find only near-optimal partitions.

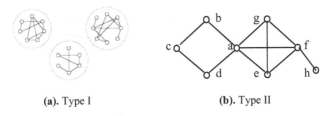

(a). Type I   (b). Type II

Fig. 1. Two types of graphs

These approaches, though they work well for some applications, performance deteriorates as graphs become more confused. Graphs may present three kinds of complications. Figure 1 illustrates two types of unweighed graphs. The first and main source of confusion is "random" interconnections between clusters, illustrated by the Type I graph. These have dense clusters that are sparsely inter-connected. As the number of interconnections increase, discerning the underlying structure becomes more challenging. Such structures have been the focus of past study.

The second and third sources of confusions are what we call hubs and outliers, and these have been little discussed in the literature. The Type II graph in Figure 1 has clusters that are connected by a "hub" vertex (i.e. vertex 'a') that is difficult to place in one cluster or another. It also has an outlier vertex that may be best placed in a cluster to itself (i.e. vertex 'h'). Past approaches do not deal with the "hub" and "outlier" problems very well. That is, they cannot split two clusters very well that are densely connected by one or more "hub" vertices, and they often fail to detect and isolate outliers.

In this paper, we propose a similarity-based graph partitioning definition that globally measures if one partitioning is better than another. Through theoretical analysis and experimental results, we demonstrate that the proposed definition outperforms existing approaches on complicated graphs and handles with agility hubs and outliers. Further, we show that this novel definition can be used with the fast agglomerative algorithm [4, 5] to find communities in very large networks.

The rest of this paper is organized as follows: in section 2 we briefly review related works; in section 3 we propose our novel similarity-based modularity measurement; in section 4 and 5 we discuss two techniques for finding optimal graph partitions by maximizing the modularity; in section 6 we apply them to network with known structures and present experimental results; and finally in section 7 we summarize our conclusions.

2   Related Work

The most traditional definition of graph partitioning is probably the min-max cut [1, 2, 3], which seeks to partition a graph G={V, E} into two sub graphs A and B. The principle of min-max clustering is minimizing the number of connections between clusters A and B and maximizing the number of connections within each. For a weighted graph the number of connection is the sum of weights of edges involved. Thus the connection between A and B is the *cut*:

$$cut(A,B) = W(A,B) = \sum_{u \in A, v \in B} w(u,v) \qquad (2.1)$$

A bi-partition of the graph is defined as minimizing the following objective function:

$$M_{cut} = \frac{cut(A,B)}{W(A,A)} + \frac{cut(A,B)}{W(B,B)}. \qquad (2.2)$$

The above function is called the min-max cut function. It minimizes the *cut* between two clusters while maximizing the connections within a cluster. However, a pitfall of this definition is that if we only cut out one node from a graph, M_{cut} will probably get the minimum value. So, in practice, M_{cut} must be accompanied with some constraints, such as A and B should be of equal or similar size, or $|A| \approx |B|$. These constraints are not always applicable for all applications, e.g., in clustering problems where some communities are much larger than the others.

To amend the issue above, a *normalized cut* (N_{cut}) was proposed [2]:

$$N_{cut} = \frac{cut(A,B)}{assoc(A,V)} + \frac{cut(A,B)}{assoc(B,V)}, \qquad (2.3)$$

where $assoc(A,V) = \sum_{u \in A, v \in V} w(u,v)$ is the total connection from vertices in A to all vertices in the graph. Using the N_{cut}, cutting out one vertex or some small part of the graph will no longer always yield a small N_{cut} value.

Using the M_{cut} or N_{cut} one can partition a graph into two sub-graphs. As the natural consequence, to divide a graph into k sub-graphs, one has to adopt a top-down approach, i.e., splitting the graph into two sub-graphs, then further splitting these sub-graphs into next level sub-graphs; repeat this process until k sub-graphs have been detected. The disadvantage [9] of this method is that, like all other bisection algorithms, it only partitions a graph into two sub-graphs. Though one can repeat this procedure on the sub-graphs, there is no guarantee of the optimality of partitioning, and one has no clue on when to stop the repeated the bisection procedure or how many sub-graphs should be produced in a graph.

To answer the question "what is the best graph partition", other researchers proposed global measurements, such as M. J. Newman's modularity [4, 5, and 10].

$$Q_N = \sum_{s=1}^{NC} \left[\frac{l_s}{L} - \left(\frac{d_s}{L} \right)^2 \right], \qquad (2.4)$$

where NC is the number of clusters, L is the number of edges in the network, l_s is the number of edges between vertices within cluster s, and d_s is the sum of the degrees of the vertices in cluster s. The optimal clustering is achieved by maximizing the modularity Q_N. Modularity is defined such that Q_N is 0 at the extremes of all vertices clustered into a single cluster and of vertices randomly clustered. As this modularity definition require no constraints, it is better than the min-max cut definition. Modularity is a quality measure of graph partitioning, while normalized cut is not. Note that with this definition both top-down (divide and conquer) and bottom-up (agglomerative) algorithms can be used for graph partitioning.

Finding the best Q_N is NP-complete. Instead of performing an exhaustive search, Newman used a bottom-up approach [4,5] which begins by merging two vertices into clusters that increase Q_N, then likewise merging pairs of clusters, until all have merged to form a single cluster. At each stage the value of Q_N is recorded. The partition with the highest Q_N is the solution of this algorithm. This is a typical hill-climbing greedy search algorithm, which generally has high speed but easily falls into a local optimum. In [10], the authors use a simulated annealing algorithm to find the optimal partition with the highest Q_N. This is a random search based algorithm, which usually can achieve higher quality than the greedy search (like the authors claimed in [10]) but has much lower speed.

According to our study [11], while Newman's modularity works well for type I graphs (Figure 1 (a)); it fails to deal well with type II graphs due to the "hub" and "outliner" vertices. To reiterate, a "hub" vertex is a vertex that densely connects two or more groups in a graph. For example, in Figure 1 (b), the central vertex 'a' connects two sub-groups. Note also that this same vertex has the largest degree. Using Newman's modularity definition, this graph in Figure 1 (b) will be clustered into two groups, $\{a, b, c, d\}$ and $\{e, f, g, h\}$. However, the more reasonable clustering schedule could be $\{b, c, d\}$, $\{a, e, f, g\}$, and $\{h\}$, where 'h' is in a cluster by itself. Newman's modularity definition will not identify outliers; rather it will assign them membership to some larger cluster.

Generally, when there are clear sub-graph structures existed in a graph, i.e., the connections between sub-graphs are sparse while the connections within a sub-graph are much dense, both min-max cut or its variations and Newman's modularity works very well. But when the sub-graph structure becomes more confused, i.e., the between connections become denser, the performance of these methods deteriorates rapidly.

3 A Novel Similarity-Based Modularity

Based on our observation, it is not very accurate to use connections within clusters and between clusters as the criteria of partitioning for all types of graphs, especially for graphs with complicated cluster structures. Instead, we propose a more general concept, similarity, to measure the graph partitioning quality. A good partition **is** one for which the similarity of vertices within a cluster is higher than the similarity of vertices between clusters.

If two connected vertices also share a lot of neighbors, we say they are similar. However, merely counting the number of shared neighbors doesn't tell us what proportion of the vertices' neighbors is shared, and therefore is insufficient for distinguishing a hub from a normal vertex. To handle this problem, we adopt the normalized similarity [12]. Let Γ_i be the neighborhood of vertex i in a network, the cosine normalization similarity is defined as:

$$S(u, v) = \frac{\left|\Gamma_u \cap \Gamma_v\right|}{\sqrt{\left|\Gamma_u\right|\left|\Gamma_v\right|}} , \tag{3.1}$$

Then we define the Similarity-based Modularity (Q_S) function as the following:

$$Q_S = \sum_{i=1}^{NC} \left(\frac{IS_i}{TS} - \left(\frac{DS_i}{TS} \right)^2 \right)$$

(3.2)

$$IS_i = \sum_{u,v \in V_i} S(u,v), \quad DS_i = \sum_{u \in V_i, v \in V} S(u,v) \quad \text{and} \quad TS = \sum_{u,v \in V} S(u,v),$$

where NC is the number of clusters, IS_i is the total similarity of vertices within cluster i; DS_i is the total similarity between vertices in cluster i and any vertices in the graph; TS is the total similarity between any two vertices in the graph; $S(u,v)$ is defined in (3.1); V is the vertex set of the graph and V_i is the vertex set of cluster i.

If we define

$$\delta(u,i) = \begin{cases} 0 & \text{if } u \text{ is not in cluster } i, \\ 1 & \text{if } u \text{ is in cluster } i, \end{cases}$$

then (3.2) is written as

$$Q_S = \sum_{i=1}^{NC} \left(\frac{\sum_u \sum_v S(u,v)\delta(u,i)\delta(v,i)}{\sum_u \sum_v S(u,v)} - \left(\frac{\sum_u \sum_v S(u,v)\delta(u,i)}{\sum_u \sum_v S(u,v)} \right)^2 \right)$$

(3.3)

In other words, Q_S is a measure of the similarity of a partitioning of the actual network versus the similarity of the same partitioning if the network's vertices were actually randomly connected. As with Newman's modularity, if we put all vertices either into one cluster or place them in clusters randomly, Q_S will be 0.

Referring again to the type II graph in Figure 1 (b), the similarity between the "hub" vertex 'a' and its neighbors is quite low, despite the hub's high degree. Likewise, the similarity between the "outlier" vertex 'h' and its lone neighbor is low. Table 1 summarizes Q_N and Q_S respectively for several possible partitionings of the network in Fig 1. (b).

Table 1. Q_N and Q_S for different clustering structures

Clustering Schedule	Q_N	Q_S
All vertices in one cluster	0.0	0.0
$\{a\}, \{b, c, d\}, \{e, f, g\}, \{h\}$	0.119835	0.343714
$\{a, b, c, d\}, \{e, f, g\}, \{h\}$	0.177686	0.312469
$\{a, b, c, d\}, \{e, f, g, h\}$	**0.227273**	0.289307
$\{b, c, d\}, \{a, e, f, g\}, \{h\}$	0.185950	**0.368656**
$\{b, c, d\}, \{a, e, f, g, h\}$	0.214876	0.321978

We acknowledge that preferring one graph partitioning over another may be subjective; however, we think the original modularity definition classifies the hub node_a into the wrong cluster $\{a, b, c, d\}$ and fails to segregate the outliner node $\{h\}$. The proposed Similarity-based Modularity successfully detects the outlier node $\{h\}$ and classifies the hub node $\{a\}$ into the more reasonable cluster $\{a, e, f, g\}$, which was our aim.

4   A Genetic Graph Partitioning Algorithm

Recall that finding the optimum Similarity-based Modularity in a graph is NP-hard, and heuristic approaches all suffer from the local optimum drawback. To evaluate the capability of the proposed modularity function, we would like to use a less heuristic approach that avoids the local optimum traps that greedy search algorithms suffer [13-17], namely a Genetic Algorithm (GA),

Encoding – For a graph with n vertices, a partition can be represented as an array of integers, for which each index corresponds to a vertex and the value to its cluster ID of this vertex. For example, the string "2 3 2 3 5 1 5 4 5" indicates that there are 5 clusters in the graph, the 1st and the 3rd vertices belong to cluster 2, the 2nd and the 4th vertices belong to cluster 3, and so on. Since each array element can have a value from 1 to n, the total search space is n^n, which is potentially very large.

Initial Population – the initial population can be generated randomly or using some task related strategies. Here we use a random initialization.

Fitness Measure – the function to be optimized. For the graph partitioning problem we use formulae (3.1) and (3.2).

Selection – some individuals are selected according to their fitness or ranks and they mate to produce offspring. We use a probabilistic selection:

$$\Pr(h_i) = \frac{Fitness(h_i)}{\sum_{j=1}^{k} Fitness(h_i)} \tag{4.1}$$

where k is the number of clusters.

Crossover – this operator determines how to produce offspring from parents. Either single point or multiple point crossovers can be used. An important parameter is the cross rate, which determines how many genetic elements are exchanged. We use single point crossover. The exchanged point is chosen at random and the length of the array block that is exchanged is set by the Crossover Rate parameter. There are two cases in single point crossover: with or without roll back.

   a). Single point crossover (Crossover Rate = 50%, no roll back):

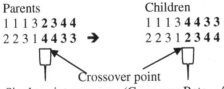

   b). Single point crossover (Crossover Rate = 50%, with roll back):

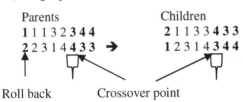

Mutation – mutation adds random variation and diversity to the population. Usually the mutation rate should be kept very small.

$$2\ 2\ 3\ 1\ 2\ 3\ 4\ 4 \qquad \rightarrow \qquad 2\ 2\ 1\ 1\ 2\ 3\ 4\ 4$$

Replacement – new offspring are inserted into the original population, replacing individuals with lower fitness. Usually, the replacement population size is constant.

The above procedure is repeated until a predefined number of generations have elapsed or the fitness of the population stops increasing. The string with highest fitness is the solution.

5 A Fast Agglomerative Algorithm

Although GA has the potential to jump out of local optimums, it has the drawback of being slow and thus may be unsuitable for large graphs. Here we extend Newman's fast hierarchical agglomerative clustering (FHAC) algorithm [4, 5] to the similarity-based modularity.

Formula (3.2) can be re-written as:

$$Q_S = \sum_{i=1}^{k} Q_{S_i} \ \text{With} \ Q_{S_i} = \frac{IS_i}{TS} - \left(\frac{DS_i}{TS}\right)^2 \tag{5.1}$$

The algorithm is as follows:

1. Initialize by placing every vertex v_i into its own cluster c_i, and calculates Q_{S_i} for each cluster.
2. Considering merging every possible pair of cluster c_g and c_h, $(g \neq h)$ and note the change to Q_S, which is given by:

$$\Delta Q_{S_{g+h}} = Q_{S_{g+h}} - Q_{S_g} - Q_{S_h} \tag{5.2}$$

1. Choose the merge producing the largest $\Delta Q_{S_{g+h}}$ and update and record Q_S.
2. Update all affected values of $\Delta Q_{S_{g+i}}$ and $\Delta Q_{S_{h+j}}$ $(i \neq g; j \neq h)$ using formula (5.2).
3. Repeat step 3 and 4 until all vertices are grouped into one cluster.
4. Retrieve the partitioning with the highest Q_S value as the best result.

In step 4, only clusters with edges connecting them to the merged pair need to be updated. The number of clusters to update that are connected to cluster g is $|i|$ and the number connected to cluster h is $|j|$. There are at most $|i|+|j|$ updates, each update taking $O(\log k) < O(\log n)$ time to access the data structure, where n is the number of vertices in the Graph. Thus each iteration takes $O((|i|+|j|)\log n)$ time [5].

The overall running time is the sum of all joining steps, which, as Newman indicated is at most $O(md \log n)$ [4], where m is number of edges and d is the depth of the dendrogram. More details of this algorithm may be found in [4, 5].

6 Evaluation Results

In this section, we present experimental result from analyzing synthetic graphs, detecting community structure in social networks, and clustering entities from a customer database. The size of the graphs varies from tiny to very large. Each example has a known structure that we are seeking to reproduce. We introduce a measure of "error" to gauge the performance of different clustering algorithms: if a node is classified into a cluster other than the pre-defined cluster, we count it as an error.

6.1 Tests on Synthetic Graphs

Synthetic graphs are generated based on predefined cluster structure and properties, so as to facilitate systematic study of the performance of our similarity-based modularity definition and graph partitioning algorithms. Here we use the same construction as used in [4, 7 and 10]. The synthetic graph consists of 128 vertices that are divided into 4 equal-size clusters. Each vertex connects to vertices within its cluster with a probability P_i, and connects to vertices outside its cluster with a probability $P_o < P_i$. On average, each vertex is connected to $K_{out} = 96 P_o$ outside vertices and to $K_{in} = 31 P_i$ inside vertices. We generate several random graphs using different K_{out} and K_{in}, and then test the performance of our two algorithms. Figure 2 illustrates examples of these synthetic random graphs with different levels of interconnectivity [10]. Notice that these graphs belong to type I defined in Figure 1.

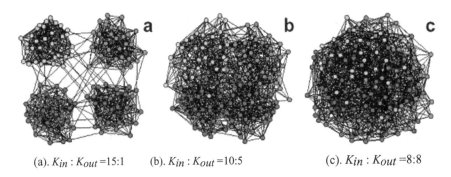

(a). $K_{in} : K_{out} = 15:1$ (b). $K_{in} : K_{out} = 10:5$ (c). $K_{in} : K_{out} = 8:8$

Fig. 2. Graphs with different $K_{in} : K_{out}$ Ratio

Table 2 summarizes the results of the Genetic algorithm using Newman's modularity (Q_N) and the similarity-based modularity (Q_S). Since the Genetic algorithm falls into local minimums, we run the algorithm several times for each parameter setting and report the best result. At $K_{in}:K_{out} >= 11:5$, the structures are very clear, both modularity definitions perfectly identify the structure. At $K_{in}:K_{out} = 10:6$, Q_N has 2 errors, but Q_S yields only one error. Since then, along with the confusion increases, Qs begins slightly lagging from Q_N. This is due to the synthetic graphs are generated based on the ratio of $K_{in}:K_{out}$, which exactly matches with the definition of Q_N. The accuracy comparison is plotted in Figure 3. We can conclude that the two modularity measures yield a comparable accuracy for the synthetic graphs.

Table 2. GA Performances

Kin : Kout	Best Q_N	Errors	Best Q_S	Errors
12:4	0.50	0	0.66	0
11:5	0.44	1	0.61	0
10:6	0.38	2	0.56	1
9:7	0.32	4	0.48	6
8:8	0.26	15	0.45	20

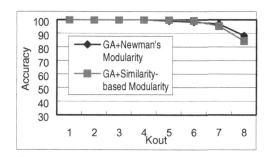

Fig. 3. Accuracy curves of GA using Q_N and Q_S

Likewise, we tested the fast hierarchical agglomerative clustering algorithm (FHAC) to detect structure in the synthetic graphs. The results are summarized in Table 3, and we plot accuracy curves in Figure 4.

Comparing Figure 3 and 4, one can see that the FHAC keeps accuracy with GA when $K_{out}<=6$ ($K_{in}=16-K_{out}$), then accuracy drops significantly when $K_{out}>6$, when

Table 3. Results of FHAC optimizing Q_S

Kin : Kout	Best Q_S	# of clusters	Error
11:5	0.61	4	0
10:6	0.57	4	4
9:7	0.48	5	18
8:8	0.46	8	44

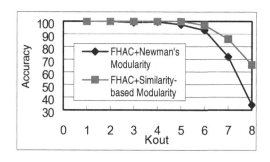

Fig. 4. Accuracy curves of FHAC using Q_N and Q_S

the cluster structure becomes confused. However, one may notice that the FHAC algorithm works much better when combining with the proposed Similarity-based Modularity. It demonstrates that the proposed measurement cooperates with the greedy search algorithm FHAC much better even when the pre-defined cluster structure is not in favor of it.

While the synthetic network is a Type I network that does suffers from no hubs or outliers, it is a significant problem to study to demonstrate the stability that similarity adds. In that regard, hubs and outliers are elements that add confusion.

6. 2 Real Applications

The first real application we report in our experiment is a social network – detecting communities (or conferences) of American college football teams [4, 7, 8, 10]. The NCAA divides 115 college football teams into 13 conferences. The question is how to find out the conferences from a network that represents the schedule of games played by all teams. We presume that because teams in the same conference are more likely to play each, that the conference system can be mapped as a structure despite the significant amount of inter-conference play. We analyze the graph by using both GA and FHAC. The results are reported in Table 4. From which, FHAC with Q_N partitions the graph into 6 conferences with 45 misclassifications [4], while FHAC with Q_S partitions to 12 conferences with only 14 errors. Again, Q_S significantly outperforms Q_N when combining with FHAC.

Table 4. Detecting conferences in college football teams by using Q_N and Q_S respectively

Alg.	Best Q_N	# of clusters	Errors	Best Q_S	# of clusters	Errors
GA	0.60	12	14	0.82	12	14
FHAC	0.58	6	45	0.82	12	14

The second application is detecting the individuals from customer records coming from Acxiom Corporation, where data errors blur the boundaries between individuals with similar information (see Figure 5). This example represents a Type II graph, with

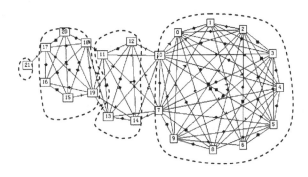

Fig. 5. Customer record networks

hubs and outliers. In this dataset, there are 3 groups of customers, a number of hubs (vertices 7, 10 , 11, and 19) and a single outlier (vertex 21).

We test both GA and FHAC by using Q_N and Q_S respectively. The results are summarized in Table 5. Both algorithms make 3 errors by using Q_N , they misclassify hub node vertex 7 and vertex 10 into wrong cluster and fail in detecting the outlier (vertex 21). However, by using the proposed Q_S , both algorithms perfectly classify this graph, which means the Q_S has better ability to deal with hub and outlier vertices.

The final experiment was on a very large dataset – the DBLP authors' collaboration networks of 2005 [19], which consist of 52,909 vertices (authors) and 245,300 edges (co-author relationship). The purpose of this experiment is testing the speed of FHAC on a very large network. The running time on an Intel PC with P3 2G CPU and 2GB memory are reported in Table 6. One can see that optimizing Q_S it runs marginally slower than optimizing Q_N, which means Q_S cooperates with the FHAC algorithm very well.

Table 5. Performance comparison on Acxiom dataset using Q_N and Q_S respectively

Alg.	Exp. # clusters	Exp. Q_N	Best Q_N	# of clusters	Err.	Exp. Q_S	Best Q_S	# of clusters	Err.
GA	4	0.38	0.40	3	3	0.47	0.47	4	0
FHAC	4	0.38	0.38	3	3	0.47	0.47	4	0

Note: "Exp." is the abbreviation for "Expected" and "Err." for "Error".

Table 6. FHAC running time for very large network

Vertices	Edges	FHAC+ Q_N	FHAC+ Q_S
52909	245300	591 Sec.	658 Sec.

7   Conclusion

In this paper, we propose a novel similarity-based modularity (Q_S) to measure the quality of a graph partitioning. Through theoretical analysis and extensive experiments, we can conclude that the propose measure is significantly more accurate and robust than Newman's connection-based modularity with respect to the result of clustering. Furthermore, it has a better ability to deal with hubs and outliners. The proposed similarity-based modularity in combination with the Genetic clustering algorithm (GA) and the greedy search algorithm FHAC yields an improved accuracy for even dense, confused graphs. The FHAC often converges to the global optimal for real applications using the proposed modularity. However, in some very tough cases, such as in very confused synthetic graphs, FHAC significantly lags the global optimal obtained by GA. This suggests us to further study more powerful fast clustering algorithm in the future to exert the potential of the proposed modularity definition.

References

[1] Ding, C.H.Q., He, X., et al.: A min-max cut algorithm for graph partitioning and data clustering. In: Proc. of ICDM 2001 (2001)

[2] Shi, J., Malik, J.: Normalized cuts and image segmentation. IEEE Trans. On Pattern Analysis and Machine Intelligence 22(8) (2000)

[3] Hegan, L., Kahng, A.B.: New spectral methods for ratio cut partitioning and clustering. IEEE Trans. On Computed Aided Design 11, 1074–1085 (1992)

[4] Newman, M.: Fast algorithm for detecting community structure in networks. Phys. Rev. E 69, art. No. 066133 (2004)

[5] Clauset, A., Newman, M., Moore, C.: Finding community in very large networks (2004)

[6] Newman, M.: Scientific collaboration networks: II. Shortest paths, weighted networks, and centrality. Phys. Rev. E 64, 15132 (2001)

[7] Girvan, M., Newman, M.: Community structure in social and biological networks. Proc. Natl. Acad. Sci. USA 99, 7821–7826 (2002)

[8] Newman, M.: Detecting community structure in networks. Eur. Phys. J. B 38, 321–330 (2004)

[9] Freeman, L.: A set of measures of centrality based upon betweeness. Sociometry 40, 35–41 (1977)

[10] Guimera, R., Amaral, L.A.N.: Functional cartography of complex metabolic networks. Letters to nature (February 2005)

[11] Feng, Z., Xu, X., Schweiger, T.: Genetic Clustering Algorithm for Graph Partitioning, technique report (2006)

[12] Leicht, E.A., Holme, P., Newman, M.E.J.: Vertex similarity in networks. Phys. Rev. E 73, 26120 (2006)

[13] Dias, C.R., Ochi, L.S.: Efficient Evolutionary Algorithms for the Clustering Problem in Directed Graphs. In: The Congress on Evolutionary Computation, CEC '03 (2003)

[14] Wang, J., Xu, L., Zhang, B.: A Genetic Annealing Hybrid Algorithm based Clustering Strategy in Mobile Ad hoc Network. Proc. on Communications, Circuits and Systems (2005)

[15] Sheng, W., Swift, S., Zhang, L., Liu, X.: A Weighted Sum Validity Function for Clustering With a Hybrid Niching Genetic Algorithm. IEEE Trans. On Sys., Man and Cybernetics, part B:Cybernetics 35(6) (December 2005)

[16] Hernadez, G., Bobadilla, L., Sanchez, Q.: A Genetic Word Clustering Algorithm. In: The Congress on Evolutionary Computation, CEC '05 (2005)

[17] Zhang, J., Chung, H., Hu, B.: Adaptive Probabilities of Crossover and Mutation in Genetic Algorithms Based on Clustering Technique. In: The Congress on Evolutionary Computation, CEC'04 (2004)

[18] Wasserman, S., Faust, K.: Social Network Analysis. Cambridge University Press, Cambridge (1994)

[19] http://www. informatik. uni-trier. de/ley/db/

Dual Dimensionality Reduction for Efficient Video Similarity Search

Zi Huang, Heng Tao Shen, Xiaofang Zhou, and Jie Shao

School of ITEE, The University of Queensland, Australia
{huang,shenht,zxf,jshao}@itee.uq.edu.au

Abstract. With ever more advanced development in video devices and their extensive usages, searching videos of user interests from large scale repositories, such as web video databases, is gaining its importance. However, the huge complexity of video data, caused by high dimensionality of frames (or feature dimensionality) and large number of frames (or sequence dimensionality), prevents existing content-based search engines from using large video databases. Hence, dimensionality reduction on the data turns out to be most promising. In this paper, we propose a novel video reduction method called Optimal Dual Dimensionality Reduction ($ODDR$) to dramatically reduce the video data complexity for accurate and quick search, by reducing the dimensionality of both feature vector and sequence. For a video sequence, $ODDR$ first maps each high dimensional frame into a single dimensional value, followed by further reducing the sequence into a low dimensional space. As a result, $ODDR$ approximates each long and high dimensional video sequence into a low dimensional vector. A new similarity function is also proposed to effectively measure the relevance between two video sequences in the reduced space. Our experiments demonstrate the effectiveness of $ODDR$ and its gain on efficiency by several orders of magnitude.

1 Introduction

With the rapid advances in multimedia devices and processing technologies, more and more applications with quickly enlarging video databases are emerging, such as WWW video search, advertising, news video broadcasting, TV and movie edition, video copyright detection, medical video identification, and personal video archives, etc. Content-based retrieval, such as Query by Video Clip, is generating tremendous and growing interests [4,2,8,9]. A video is typically defined as a sequence of frames, each of which is a high-dimensional feature vector [2,8]. Video similarity search is to find similar videos with respect to a user-specified query from a video collection underlying a defined similarity measure. However, as the video database becomes very large, this problem is very difficult due to extremely high complexity of video data mainly due to high dimensionality of feature vectors (or feature dimensionality) and large number of frames (or sequence dimensionality) in video sequences. Existing data organization methods fail to indexing a very large and continuously growing video database for fast

I.Y. Song, J. Eder, and T.M. Nguyen (Eds.): DaWaK 2007, LNCS 4654, pp. 397–406, 2007.

search. Towards an effective and efficient solution of video similarity search, the following challenges have to be addressed: (a) effective and compact video representations, (b) effective similarity measurement, and (c) efficient indexing on the compact representations with a fast search strategy. It is indispensable to reduce the complexity of original video data before any indexing structure can be deployed.

In this paper, we address the first two issues and propose a novel video reduction method called Dual Dimensionality Reduction (DDR) to dramatically reduce the video data complexity. DDR is the first to explore the potential to reduce the dimensionality of both feature vector and sequence. It consists of two steps. In the first step, DDR maps each high dimensional frame into a single dimensional value via distance-based transformation. A video sequence is then represented as a sequence of one dimensional values, which can be viewed as a time series. Given that the number of frames in a video sequence is typically very large, in the second step, DDR further reduces the long one dimensional sequence into a low dimensional vector via an efficient reduction approach called Segment Approximation (SA). SA partitions a sequence into a few segments, each of which is represented by a couple of its mean and length. As a result, DDR approximates each long and high dimensional video sequence into a low dimensional vector, where each dimension is an approximation of a segment represented by its mean and length. An optimal method based on DDR, called *Optimal Dual Dimensionality Reduction* (ODDR), is then proposed for further improvement on sequence segmentation. The method of dynamic programming is applied in ODDR for optimal sequence segmentation. A new similarity function is proposed to effectively measure the relevance of two video sequences in the reduced dimensional space. Our theoretical analysis and experiments on large real life video dataset confirm the effectiveness of DDR/ODDR and its gain on search efficiency by several orders of magnitude over the traditional method.

The rest of paper is organized as follows. We formulate the problem and review some related work in Section 2. Dual Dimensionality Reduction method is presented in Section 3. The results of performance study are reported in Section 4, and finally, we conclude our paper in Section 5.

2   Video Similarity Search

A video can be considered as a sequence of frames, each of which is represented as a high dimensional image feature vector. Formally, we denote a video sequence as $V = \{v_1, ..., v_{|V|}\}$, where $v_i = (v_i^1, ..., v_i^d)$ is a d-dimensional feature vector and $|V|$ is the number of frames in V. The problem of video similarity search can be defined as follows:

Definition 1 (Video Similarity Search). *Given a query video $Q = \{q_1, ...,$ $q_{|Q|}\}$ and a video database $\mathbb{V} = \{V_1, ..., V_{|\mathbb{V}|}\}$, where $|Q|$ and $|\mathbb{V}|$ denote the number of frames in Q and the number of videos in $\mathbb{V}$, respectively. Video similarity search is to find videos from $\mathbb{V}$, whose similarity to Q are not less than a predefined threshold underlying a similarity function.*

Very often, the value of similarity threshold is nontrivial to be set. The above definition can be extended to also find the top-k most similar videos to Q.

To measure the similarity of two video sequences, many approaches have been proposed. One widely used video similarity measure is the percentage of similar frames shared by two sequences [2,8]. Warping distance [7] considers the temporal differences between two sequences. Hausdorff distance [6] is used to measure the maximal dissimilarity between two shots. However, all these measures need to compare most, if not all, frames pairwise. The time complexity is at least linear in the number of frames multiplied by the dimensionality of frame. For very large video databases, such expensive computations are strongly undesirable.

To reduce the computational cost, one promising approach is to summarize the video sequences into compact representations, hence the video similarity can be approximated based on the compact representations. Existing works focus on reducing the sequence dimensionality by clustering similar frames into a single representative. In [8], similar frames are summarized into a single cluster represented by a video triplet including cluster center, cluster radius, and cluster density. The original videos are then represented by a smaller number of high dimensional feature vectors. In [2], a randomized summarization method which randomly selects a number of seed frames and assigns a small collection of closest frames (called video signature) to each seed is introduced. Finally, a multidimensional indexing structure is applied to facilitate the search. However, high feature dimensionality still resists on their limited performance.

3   Dual Dimensionality Reduction

In this section, we present DDR, a novel dimensionality reduction method to reduce both feature and sequence dimensionality dramatically, as a significant step for effective and efficient video search in very large video databases.

3.1   Dimensionality Reduction

The high computational complexity of video search is mainly caused by high dimensionality of feature vectors and large number of frames. To overcome this bottleneck for efficient search, we propose to reduce the both feature dimensionality and sequence dimensionality. Fig. 1 outlines our Dual Dimensionality Reduction (DDR) method, which consists of two major steps.

Feature Dimensionality Reduction (lines 1-2). Given a video sequence $V = \{v_1, ..., v_{|V|}\}$ where $v_i = (v_i^1, ..., v_i^d)$ is a d-dimensional feature vector, in the first step, DDR employs a distance-based transformation method to map a d-dimensional feature vector into a single distance value. To do so, a simple way is the compute the distance between v_i and a reference point $D(v_i, O)$ as proposed in [3], where O is a reference vector/point selected by users. Hence, a d-dimensional feature vector v_i can be represented by its distance to the reference point O. In this paper, we choose the origin point $[0]^d$ as the reference point and

DDR Algorithm.
Input: V - a video sequence
Output: SA - segment approximation of V
//*Step 1: feature dimensionality reduction*
1. **For** i=1 to $|V|$
2. $V'[i] \leftarrow D(v_i, O)$;
//*Step 2: sequence dimensionality reduction - SA*
3. min = max = sum = $V'[1]$; $l = 1$;
4. $L = 0$;
5. $V'[|V| + 1] = V'[|V|] + 2\delta$
6. **For** i = 2 to $|V|+1$
7.   **if** $|V'[i]\text{-min}| > \delta$ or $|V'[i]\text{-max}| > \delta$
8. $SA[++L] = < sum/l, l >$;
9.   min = max = sum = $V'[i]$; $l = 1$;
10.   **else**
11.   min$\leftarrow$ Min(min,$V'[i]$);
12.   max$\leftarrow$ Max(max,$V'[i]$);
13.   sum = sum + $V'[i]$;
14. $l++$;
15. return SA;

Fig. 1. DDR Algorithm

Euclidean Distance function is applied. Obviously, such a reduction from d to 1 dimensionality would cause lots of information loss. One most important feature for video data is the sequence order along temporal line. Our inspiration comes from the intuition that *information lost during feature dimensionality reduction might be compensated by taking sequence order into consideration.* As we will see later in the experiments, integrating sequence order to compute the similarity indeed has compensation action for similarity search.

After feature dimensionality reduction, a sequence of d-dimensional feature vectors V is reduced to be a sequence of 1-dimensional values V', which can be regarded as a time series, as shown from Fig. 2a to Fig. 2b. Obviously, $|V| = |V'|$.

Sequence Dimensionality Reduction (lines 3-14). Since V' is 1-dimensional sequence, motivated by dimensionality reduction approaches, such as APCA [5], from the literature of signal/time series indexing, we propose an efficient segmentation method called Segment Approximation (SA) to divide the whole V' into several segments/pieces, each of which is represented by its mean and length. The key idea of SA is to partition the signal into segments whose amplitudes are less than a threshold - δ.

As shown in the second step of Fig. 1, the parameters for a current segment, where $(max - min)$ suggests its amplitude and l is its length, are first initialized (line 3). The number of segments L is also initialized (line 4). If a valid segment is detected (line 7), the segment is approximated as a couple of (μ, l), where $\mu = sum/l$ (line 8), and the parameters are re-initialized (line 9). Otherwise, the statistics of current segment are updated correspondingly (lines 10-14).

(a) A sequence of d-dimensional feature vectors

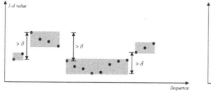

(b) A sequence of 1-dimensional values

(c) A sequence of segment approximations

Fig. 2. Graphical illustration of DDR algorithm

After sequence dimensionality reduction, V' is further reduced to be a sequence of segments called SA, and each segment is represented by a couple of its mean and length, as shown from Fig. 2b to Fig. 2c. Finally, SA is returned by DDR as the compact representation of V (line 15). Clearly, SA can also be viewed as a low dimensional vector, where each dimension is a couple of mean and length values.

3.2 Optimal Dual Dimensionality Reduction

The sequence dimensionality reduction in the above DDR algorithm involves a parameter δ to determine the number of segments. Manual adjustment of this parameter is ad-hoc and impractical in real world applications. Therefore, in this section, we propose a method based on dynamic programming for parameter optimization to automatically determine the optimal video sequence segmentation without any parameter. Consequently, an Optimal Dual Dimensionality Reduction (ODDR) algorithm is show in Fig. 3, where the output $err(|V|, n)$ is the error function which indicates the minimal approximation error caused by dividing the sequence into a number of n segments. Based on ODDR, we can get the optimal segmentation SA to achieve the minimal approximation error under any specific n ($n \leq |V|, |SA| = n$). Therefore, under the same number of segments n, ODDR will give the better segmentation for DDR. Obviously, it is a tradeoff between n and $err(|V|, n)$. The bigger n, the smaller $err(|V|, n)$. However, the variance of $err(|V|, n)$ will get smooth when n goes beyond a critical value. We

ODDR Algorithm.
Input: V - a video sequence
Output: err - error function
//*Step 1: feature dimensionality reduction*
1. **For** i=1 to $|V|$
2. $V'[i] \leftarrow D(v_i, O);$
//*Step 2: sequence dimensionality reduction*
3. **For** i = 1 to $|V|$
4.   err[i][i] = 0;
5.   err[i][1] $\leftarrow \Delta(1,i)$
6. **For** i = 3 to $|V|$
7.   **For** j = 2 to i-1
8.   err[i][j] $\leftarrow \min\limits_{k=j-1..i-1}(err[k][j-1]+\Delta(k+1,i))$

9. return err;

Fig. 3. ODDR Algorithm

consider this value as an optimal n to balance both the number of segments and the approximation error. Note that,

$$mean(i,j) = (\sum_{k=i}^{j} V'[k])/(j-i+1) \qquad (1)$$

$$\Delta(i,j) = \sum_{k=i}^{j} |V'[k] - mean(i,j)| \qquad (2)$$

3.3   Complexity Comparison

Given a video sequence V, its size is $d*|V|$, where $|V|$ is its number of frames. For its corresponding SA, the size is $2*|SA|$, where $|SA|$ is the number of segments generated from V. Hence the size of SA is just $\frac{2*|SA|}{d*|V|}$ of that of V. For example, given a V in 64-dimensional feature space with 500 frames and its $|SA| = 10$, the size of SA is just $\frac{1}{1600}$ of original video size. The threshold δ affects $|SA|$. Later in the experiments, we will see how it affects the search accuracy and efficiency.

3.4   Similarity Measure

The similarity of two videos can be approximated by the similarity of their SAs. Given two segment approximations SA_V and SA_Q generated from a database video V and a query video Q, they may have different lengths, i.e., $|SA_V| \neq |SA_Q|$. To compute their similarity, we first reduce the longer SA to be the same length of the shorter one, by iteratively combining two consecutive segments which have closest μ values. This process is done by dynamic programming. The reason we force two SAs to have the same length is to preserve their sequence property for fast similarity computation. After two SAs become equally long, the dissimilarity of their corresponding videos is approximated as follows:

$$Dis(V, Q) = \sqrt{\sum_{i=1}^{|SA_Q|} (SA_V[i].\mu * SA_V[i].l - SA_Q[i].\mu * SA_Q[i].l)^2}$$

The above similarity measure is an extended Euclidean distance function by taking local segment length into consideration and its complexity is linear in the length of SA. However, most of existing sequence similarity measures preserving temporal order are quadratic to the sequence length, since they involve pairwise frame distance comparisons [7,6].

4 Performance Study

4.1 Experiments Setup

Our test database consists of about 8,500 video clips, including TV commercials, movie trailers, news clips, etc. They are recorded by VirtualDub [1] at PAL frame rate of 25fps in AVI format at the resolution of 384 x 288 pixels. The video lengths vary from seconds to minutes (more than 500 hours in total), and the average number of frames is about 380. Each video frame is processed and two popular feature spaces are used: RGB color space and HSV color space. Four feature datasets in 8-, 16-, 32- and 64-dimensionality for each color space were extracted for evaluation purpose, and the value of each dimension was normalized by dividing the total number of pixels. We will present the experimental results performed on the 64-dimensional RGB dataset in the following subsections.

All the experiments were performed on Window XP platform with Intel Pentium 4 Processor (2.8GHz CPU) and 1GB RAM. All results are averaged on 20 query video clips for top-50 search. The indicators *precision* and *search time* are used to measure the effectiveness and efficiency of our method.

4.2 Effectiveness

In this experiment, we tested the effectiveness of DDR and ODDR. The ground truth results are computed on the original 64-dimensional video data using the same similarity measure defined above. Since we use top-k search, precision is sufficient for evaluation. δ value affects the degrees of approximation in DDR. Table 1 shows the average sequence dimensionality remained (i.e., the average number of segments) for different δ values. Note that the value of δ is often small, as the original histogram feature vectors have already been normalized. The larger the parameter δ, the smaller the number of segments generated (i.e., the smaller size of SA). We can observe that when δ reaches the value of 0.01, the number of segments is reduced by 77% comparing to the original sequence dimensionality (i.e., $\delta = 0$).

Table 1. $|SA|$ *vs.* δ for DDR

δ	0	0.002	0.004	0.006	0.008	0.01	1
$|SA|$	376	237	167	129	104	88	1

Next, we see the effect of δ on search precision. From Fig. 4, we have the following observations. First, when $\delta=0$, i.e., only feature dimensionality is reduced, precision is above 80% which is quite acceptable. Second, the precision decreases slowly as δ increases. This is reasonable since a larger δ causes more information loss during sequence dimensionality reduction. The precision drops dramatically when δ is increased to above 0.008, suggesting a threshold beyond which the original data cannot be well preserved any more.

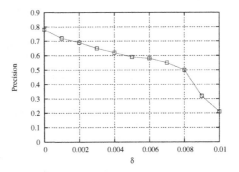

Fig. 4. Precision vs. δ

Fig. 5 reveals the relationship between the approximation error and the number of segments n in ODDR. Comparing Fig. 6 and Fig. 4, it can be observed that ODDR is better than DDR method in term of search precision at the same number of n.

The results indicate that *although reducing the original 64-dimensional feature vectors into 1-dimensional values causes significant information loss, the preservation of sequence property compensates the loss to an acceptable extent.*

Traditionally, a video can be represented as a set of key frames. We test different sampling rates to select key frames. The disadvantage of this approach is high information loss. Its precision is not comparable to that of our proposed methods. Due to the page limit, we will not present the precision curve in this paper.

4.3   Efficiency

In this experiment, we test the efficiency of DDR by reporting the search time on reduced video representations - SAs. As we analyzed in Section 3.3, the size of SA is smaller than the original video size by many orders of magnitude. The

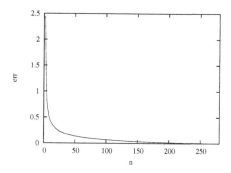

Fig. 5. Error function vs. n

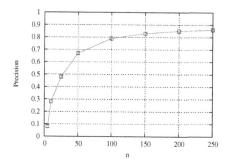

Fig. 6. Precision vs. n for ODDR

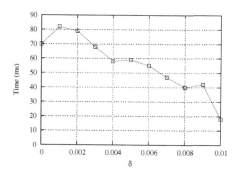

Fig. 7. Search time vs. δ

search time on the original data is clearly incomparable to that on SAs and cannot fit into the figure. Hence, we just show the search time on SAs, without applying any multidimensional indexing structure for this experiment. As shown in Fig. 7, search time is in the unit of millisecond for our large video database. It decreases as δ increases since a larger δ leads to smaller sized approximations of original video data. By taking both the effectiveness and efficiency of DDR into account, it seems an optimum value of δ would be in the range of (0.006, 0.008). The search time of ODDR is similar to that of DDR since they share the same complexity of video representation under a specific number of segments (i.e. $|SA|$) and the same similarity measure.

5 Conclusion

In this paper, we propose a video reduction method called Dual Dimensionality Reduction (DDR) and its optimal version ODDR to dramatically reduce the video data complexity for efficient similarity search, by reducing both feature dimensionality and sequence dimensionality. DDR and ODDR can approximate each long and high dimensional video sequence into a low dimensional vector. Such a reduced video representation is extremely small in size comparing with original data. A new similarity function is proposed to effectively measure the relevance between two video sequences in the reduced space. Our theoretical analysis and experiments show the superiority our method.

References

1. http://www.virtualdub.org
2. Cheung, S.-C.S., Zakhor, A.: Efficient video similarity measurement with video signature. IEEE Trans. Circuits Syst. Video Techn. 13(1), 59–74 (2003)
3. Jagadish, H.V., Ooi, B.C., Tan, K.-L., Yu, C., Zhang, R.: idistance: An adaptive b$^+$-tree based indexing method for nearest neighbor search. ACM Trans. Database Syst. 30(2), 364–397 (2005)
4. Jain, A.K., Vailaya, A., Xiong, W.: Query by video clip. Multimedia Syst. 7(5), 369–384 (1999)
5. Keogh, E.J., Chakrabarti, K., Mehrotra, S., Pazzani, M.J.: Locally adaptive dimensionality reduction for indexing large time series databases. In: SIGMOD Conference, pp. 151–162 (2001)
6. Kim, S.H., Park, R.-H.: An efficient algorithm for video sequence matching using the modified hausdorff distance and the directed divergence. IEEE Trans. Circuits Syst. Video Techn. 12(7), 592–596 (2002)
7. Naphade, M.R., Wang, R., Huang, T.S.: Multimodal pattern matching for audiovisual query and retrieval. In: Storage and Retrieval for Image and Video Databases (SPIE), pp. 188–195 (2001)
8. Shen, H.T., Ooi, B.C., Zhou, X., Huang, Z.: Towards effective indexing for very large video sequence database. In: SIGMOD Conference, pp. 730–741 (2005)
9. Yuan, J., Duan, L.-Y., Tian, Q., Ranganath, S., Xu, C.: Fast and robust short video clip search for copy detection. In: PCM, pp. 479–488 (2004)

Privacy-Preserving Genetic Algorithms for Rule Discovery*

Shuguo Han and Wee Keong Ng

Center for Advanced Information Systems, School of Computer Engineering,
Nanyang Technological University, Singapore. 639798
{hans0004,awkng}@ntu.edu.sg

Abstract. Decision tree induction algorithms generally adopt a greedy approach to select attributes in order to optimize some criteria at each iteration of the tree induction process. When a decision tree has been constructed, a set of decision rules may be correspondingly derived. Univariate decision tree induction algorithms generally yield the same tree regardless of how many times it is induced from the same training data set. Genetic algorithms have been shown to discover a better set of rules, albeit at the expense of efficiency. In this paper, we propose a protocol for secure genetic algorithms for the following scenario: Two parties, each holding an arbitrarily partitioned data set, seek to perform genetic algorithms to discover a better set of rules without disclosing their own private data. The challenge for privacy-preserving genetic algorithms is to allow the two parties to securely and jointly evaluate the fitness value of each chromosome using each party's private data but without compromising their data privacy. We propose a new protocol to address this challenge that is correct and secure. The proposed protocol is not only privacy-preserving at each iteration of the genetic algorithm, the intermediate results generated at each iteration do not compromise the data privacy of the participating parties.

1 Introduction

In the modern world of business competition, collaboration between industries or companies is one form of alliance to maintain overall competitiveness. Two industries or companies may find that it is beneficial to collaborate in order to discover more useful and interesting patterns, rules or knowledge from their joint data collection, which they would not be able to derive otherwise. Due to privacy concerns, it is impossible for each party to share its own private data with one another if the data mining algorithms are not secure. Privacy-preserving data mining (PPDM) has been proposed to resolve the data privacy concerns [1,8]. Conventional PPDM makes use of Secure Multi-party Computation [15] or randomization techniques to allow the participating parties to preserve their data privacy during the mining process.

---

* This work is supported in part by grant P0520095 from the Agency for Science, Technology and Research (A*STAR), Singapore.

I.Y. Song, J. Eder, and T.M. Nguyen (Eds.): DaWaK 2007, LNCS 4654, pp. 407–417, 2007.

In recent years, PPDM has emerged as an active area of research in the data mining community. Several traditional data mining algorithms have been adapted to be become privacy-preserving: decision trees, association rule mining, k-means clustering, SVM, Naïve Bayes. These algorithms generally assume that the original data set has been horizontally and/or vertically partitioned, with each partition privately held by a party. Jagannathan and Wright introduced the concept of arbitrarily partitioned data [6], which is a generalization of horizontally and vertically partitioned data. In arbitrarily partitioned data, different disjoint portions are held by different parties.

Decision tree induction algorithms iteratively perform greedy selection on attributes when constructing the decision tree. At each iteration, the best attribute that optimize some criteria, such as entropy, information gain, or the gini index, is chosen to split the current tree node. Univariate decision tree induction is known to produce the same decision tree regardless of the number of times it is induced from the same data set. As identified by Freitas [4], genetic algorithms are able to discover a better set of rules than decision trees, albeit at the expense of efficiency. Hence, the genetic algorithm approach is a better alternative for generating and exploring a more diverse set of decision rules than decision tree induction.

In this paper, we propose a protocol for two parties each holding a private data partition to jointly and securely apply genetic algorithms to discover a set of decision rules for their private data partitions without compromising individual data privacy. As genetic algorithms are iterative, it is a challenge to not only preserve the privacy of data at each iteration, it is also a challenge to ensure that the intermediate results produced at each iteration do not compromise the data privacy of the participating parties. We shall show that our proposed protocol satisfy these two requirements for data privacy.

This paper is organized as follows. In the following section, we review various privacy-preserving data mining algorithms. In Section 3, we propose a protocol for performing privacy-preserving genetic algorithms for arbitrarily partitioned data involving two parities. Section 4 analyzes the correctness, security, and complexity of the protocol. We conclude our work in the final section.

2   Related Work

In this section, we review current work on privacy-preserving data mining algorithms that are based on Secure Multi-party Computation [15].

Lindell and Pinkas [8] proposed a privacy-preserving ID3 algorithm based on cryptographic techniques for horizontally partitioned data involving two parties. Vaidya and Clifton [10] presented privacy-preserving association rule mining for vertically partitioned data based on the secure scalar product protocol involving two parties. The secure scalar product protocol makes use of linear algebra techniques to mask private vectors with random numbers. Solutions based on linear algebra techniques are believed to scale better and perform faster than those based on cryptographic techniques.

Vaidya and Clifton [11] presented a method to address privacy-preserving k-means clustering for vertically partitioned data involving multiple parties. Given a sample input which are held partially by different parties, determining which cluster the sample is closest must be done jointly and securely by all the parties involved. This is accomplished by the secure permutation algorithm [3] and the secure comparison algorithm based on circuit evaluation protocol [15]. Jagannathan and Wright [6] proposed a new concept of arbitrarily partitioned data which is a generalization of horizontally and vertically partitioned data. They provided an efficient privacy preserving protocol for k-means clustering in an arbitrarily partitioned data setting. To compute the closest cluster for a given point securely, the protocol also makes use of secure scalar product protocols.

Yu et $al.$ [17] proposed a privacy-preserving SVM classification algorithm for vertically partitioned data. To achieve complete security, the generic circuit evaluation technique developed for secure multiparty computation is applied. In another paper, Yu et $al.$ [16] securely constructed the global SVM classification model using nonlinear kernels for horizontally partitioned data based on the secure set intersection cardinality protocol [12]. Laur et $al.$ [7] proposed secure protocols to implement the Kernel Adaption and Kernel Perception learning algorithms based on cryptographic techniques without revealing the kernel and Gramm matrix of the data.

To the best of our knowledge, there is no work to date on a privacy-preserving version of genetic algorithms. In this paper, we propose a protocol for two parties each holding a private data partition to jointly and securely apply genetic algorithms to discover a set of decision rules from their private data partitions. We propose protocols to securely compute a fitness function for rule discovery and ensure that the intermediate results of the protocols do not compromise data privacy.

3   Genetic Algorithms for Rule Discovery

In this section, we present a protocol for privacy-preserving genetic algorithms for rule discovery. When applying genetic algorithms to a problem domain, the solutions in the problem domain are mapped to chromosomes (individuals) which are crossovered and mutated to evolve new and better chromosomes as the genetic process iterates. When using genetic algorithms for rule discovery, each rule—representing a possible solution—is mapped to a chromosome (individual). An example of a rule is $(X = x_i) \longrightarrow (C = c_i)$, where x_i is an attribute value of attribute X and c_i is a class label. As rules evolve in the genetic iteration process, the antecedent of rules may include more attributes, thereby becoming more complex and accurate in predicting the class labels of the data set.

To evaluate the goodness of each rule, a fitness function is defined. In a two-party setting, the fitness function is jointly applied by the two parties to evaluate each rule with respect to the private data partitions held by each party.

Protocol 1. PPGA Protocol

1: Two parties jointly initialize population randomly.
2: Evaluate the fitness of each individual in the population based on the Secure Fitness Evaluation Protocol.
3: **repeat**
4:    Two parties jointly select the best-ranking individuals for reproduction.
5:    Breed new generation through crossover and mutation (genetic operations) and give birth to offspring by two parties together.
6:    Evaluate the individual fitness of the offspring based on the Secure Fitness Evaluation Protocol.
7:    Replace the worst ranked part of population with offspring by two parties together.
8: **until** ⟨terminating condition⟩.

		Actual Class	
		c_i	not c_i
Predicted	c_i	N_{TP}	N_{FP}
Class	not c_i	N_{FN}	N_{TN}

Fig. 1. Confusion matrix for the rule $(X = x_i) \longrightarrow (C = c_i)$

The challenge here is to perform this joint evaluation without compromising each party's data privacy. In Section 3.1, we describe a general protocol for privacy-preserving genetic algorithms. We propose protocols to securely compute the fitness function in Section 3.2.

3.1   General Protocol for Privacy-Preserving Genetic Algorithms

Party A and Party B each holds an arbitrarily partitioned data portion respectively. They wish to jointly and securely use genetic algorithms to discover a set of decision rules from their private data partitions so that these rules can be used to predict the class label of future data samples. The general protocol for Privacy-Preserving Genetic Algorithms (PPGA) in the setting of arbitrarily partitioned data involving two parties is shown in Protocol 1.

In this paper, the steps that we use for population initialization and genetic operations (i.e., crossover and mutation) in genetic algorithms for rule discovery are the same as those by Freitas [4].

3.2   Secure Fitness Function Evaluation for Rule Discovery

In this section, we describe how a fitness function for rule discovery can be securely computed. The performance of a classification rule with respect to its predictive accuracy can be summarized by a confusion matrix [14]. For example, for the rule $(X = x_i) \longrightarrow (C = c_i)$, its confusion matrix is shown in Fig. 1. Each quadrant of the matrix refers to the following:

- N_{TP} (True Positives): Number of tuples belonging to class c_i whose class labels are correctly predicted by the rule; i.e., these tuples satisfy the TP condition $(X = x_i) \wedge (C = c_i)$.
- N_{FP} (False Positives): Number of tuples not belonging to class c_i whose class labels are incorrectly predicted by the rule; i.e., these tuples satisfy the FP condition $(X = x_i) \wedge (C \neq c_i)$.
- N_{FN} (False Negatives): Number of tuples belonging to class c_i whose class labels are incorrectly predicted by the rule; i.e., these tuples satisfy the FN condition $(X \neq x_i) \wedge (C = c_i)$.
- N_{TN} (True Negatives): Number of tuples not belonging to class c_i whose class lables are correctly predicted by the rule; i.e., these tuples satisfy the TN condition $(X \neq x_i) \wedge (C \neq c_i)$.

Note that there is a confusion matrix for each classification rule. The goodness (fitness) of each rule can be evaluated using its confusion matrix:

$$\text{Fitness} = \text{True\_Positive\_Rate} \times \text{True\_Negative\_Rate}. \qquad (1)$$

where True_Positive_Rate $= N_{TP}/(N_{TP} + N_{FN})$ and True_Negative_Rate $= N_{TN}/(N_{TN} + N_{FP})$. Note that other forms of fitness functions may also be defined using combinations of Precision, True_Positive_Rate, True_Negative_Rate, and/or Accuracy_rate, with Precision $= N_{TP}/(N_{TP}+N_{FP})$ and Accuracy_Rate$= (N_{TP} + N_{TN})/(N_{TP} + N_{FP} + N_{FN} + N_{TN})$.

Note that the fitness function as defined in Eq. 1 has been used for classification rule discovery using genetic algorithms [2]. It has also been used as a rule quality measure in an ant colony algorithm for classification rule discovery [9].

It is clear that the fitness function as defined in Eq. 1 is defined in terms of the four values N_{TP}, N_{FP}, N_{FN}, and N_{TN}. If these values can be computed in a secure manner, the overall fitness function can also be securely computed. In the remaining part of this section, we show how each of the four values can be securely computed as the scalar product of two binary vectors.

Binary Vector Notation: Define a binary vector $\mathbf{V}_\beta^\alpha$, where $\alpha \in \{a, b\}$ ($a \equiv$ 'Party A' and $b \equiv$ 'Party B') and $\beta \in \{\text{TP, FP, FN, TN}\}$. For a given classification rule r, vector $\mathbf{V}_\beta^\alpha$ encodes its TP, FP, FN, or TN condition with respect to the private data partition of a party α. More precisely, for a particular rule $r: (X = x_i) \longrightarrow (C = c_i)$, α (either Party A or B) generates binary vector $\mathbf{V}_\beta^\alpha = [V_1, V_2, \dots V_m]^T$ (m is the number of tuples in data set $\mathbf{S}$) where

- $V_i = 1$ if (i) attribute X's value in the i-th tuple of $\mathbf{S}$ is held by another party but not α or (ii) X's value in the i-th tuple of $\mathbf{S}$ is held by α and the i-th tuple satisfies β's condition.
- $V_i = 0$ otherwise.

Vector $\mathbf{V}_\beta^a$ contains partial information of how partial tuples in α's partition satisfy β. For instance, $\mathbf{V}_{\mathrm{TP}}^a$ captures partial information of those tuples in Party A's data partition satisfying the TP condition with respect to the rule r.

In this way, the value N_β ($\beta \in \{\mathrm{TP}, \mathrm{FP}, \mathrm{FN}, \mathrm{TN}\}$) for rule r can be computed by applying secure scalar product protocols on vector $\mathbf{V}_\beta^a$ from Party A and vector $\mathbf{V}_\beta^b$ from Party B:

$$N_\beta = \mathbf{V}_\beta^a \bullet \mathbf{V}_\beta^b. \tag{2}$$

Therefore, the four values N_{TP}, N_{FP}, N_{FN} and N_{TN} may be securely computed using secure scalar products. When there are more than one attributes in the rule antecedent, we use a logical conjunction of their corresponding binary vectors to obtain a single binary vector.

Although the fitness value of a rule may now be securely computed, there is one weakness with the use of secure scalar product protocols. Vaidya and Clifton [12] have pointed out that existing secure scalar product protocols are subject to *probing attacks*. Party A may probe Party B's binary vector by generating a vector $[0,0,1,0,0]$ containing only one element whose value is 1. Party A uses this probing vector to probe Party B's third element value when computing the scalar product. Party A could likewise generate all such probing vectors to probe Party B's vector. To prevent probing attacks, we could find a secure way to express the scalar product of two binary vectors as the sum of two component values; each component value is held privately by each party. In this way, each party does not know the other party's component value, but they are still able to jointly and securely come together to determine the scalar product. In the following paragraph, we describe a protocol to compute the fitness function (Eq. 1) in the manner just described.

Secure Fitness Evaluation Protocol: There exist known protocols on how to express the scalar product of two vectors as the sum of two component values [3,5]. Party A and B may use any of these protocols to derive component values when computing N_{TP}, N_{FP}, N_{FN}, and N_{TN}. At the end of these protocols, each party only knows their component values, but not those of the other party. This prevents probing attacks. The protocol to securely compute the fitness value of a rule is shown above. Steps 1 to 4 of the protocol yield the component values corresponding to N_{TP}, N_{FP}, N_{FN}, and N_{TN}.

With the component values, Party A and B may now jointly and securely compute the fitness value (Eq. 1) of rule r in the genetic iteration process as shown:

$$
\begin{aligned}
\text{Fitness} &\Leftrightarrow \frac{\mathbf{V}_{\mathrm{TP}}^a \bullet \mathbf{V}_{\mathrm{TP}}^b}{\mathbf{V}_{\mathrm{TP}}^a \bullet \mathbf{V}_{\mathrm{TP}}^b + \mathbf{V}_{\mathrm{FN}}^a \bullet \mathbf{V}_{\mathrm{FN}}^b} \times \frac{\mathbf{V}_{\mathrm{TN}}^a \bullet \mathbf{V}_{\mathrm{TN}}^b}{\mathbf{V}_{\mathrm{FP}}^a \bullet \mathbf{V}_{\mathrm{FP}}^b + \mathbf{V}_{\mathrm{TN}}^a \bullet \mathbf{V}_{\mathrm{TN}}^b} \\
&\Leftrightarrow \frac{v_{\mathrm{TP}}^a + v_{\mathrm{TP}}^b}{(v_{\mathrm{TP}}^a + v_{\mathrm{TP}}^b) + (v_{\mathrm{FN}}^a + v_{\mathrm{FN}}^b)} \times \frac{v_{\mathrm{TN}}^a + v_{\mathrm{TN}}^b}{(v_{\mathrm{FP}}^a + v_{\mathrm{FP}}^b) + (v_{\mathrm{TN}}^a + v_{\mathrm{TN}}^b)} \\
&\Leftrightarrow \frac{(v_{\mathrm{TP}}^a + v_{\mathrm{TP}}^b) \times (v_{\mathrm{TN}}^a + v_{\mathrm{TN}}^b)}{[(v_{\mathrm{TP}}^a + v_{\mathrm{FN}}^a) + (v_{\mathrm{TP}}^b + v_{\mathrm{FN}}^b)] \times [(v_{\mathrm{FP}}^a + v_{\mathrm{TN}}^a) + (v_{\mathrm{FP}}^b + v_{\mathrm{TN}}^b)]}
\end{aligned}
$$

Protocol 2. Secure Fitness Evaluation Protocol

Input: Party A has four input vectors $\mathbf{V}_{\mathrm{TP}}^a$, $\mathbf{V}_{\mathrm{FP}}^a$, $\mathbf{V}_{\mathrm{FN}}^a$ and $\mathbf{V}_{\mathrm{TN}}^a$ with respect to rule r. Party B also has four input vectors $\mathbf{V}_{\mathrm{TP}}^b$, $\mathbf{V}_{\mathrm{FP}}^b$, $\mathbf{V}_{\mathrm{FN}}^b$ and $\mathbf{V}_{\mathrm{TN}}^b$ for rule r.
Output: Fitness = True_Positive_Rate × True_Negative_Rate is the fitness value of rule r.

1: Party A and Party B compute $\mathbf{V}_{\mathrm{TP}}^a \bullet \mathbf{V}_{\mathrm{TP}}^b$ using existing secure scalar product protocols. At the end of the protocol, Party A and B each holds a private component values v_{TP}^a and v_{TP}^b respectively, where $v_{\mathrm{TP}}^a + v_{\mathrm{TP}}^b = \mathbf{V}_{\mathrm{TP}}^a \bullet \mathbf{V}_{\mathrm{TP}}^b$.
2: Similarly, Party A and Party B compute $\mathbf{V}_{\mathrm{FP}}^a \bullet \mathbf{V}_{\mathrm{FP}}^b$ and at the end, each holds a private component values v_{FP}^a and v_{FP}^b respectively, where $v_{\mathrm{FP}}^a + v_{\mathrm{FP}}^b = \mathbf{V}_{\mathrm{FP}}^a \bullet \mathbf{V}_{\mathrm{FP}}^b$.
3: Similarly, Party A and Party B compute $\mathbf{V}_{\mathrm{FN}}^a \bullet \mathbf{V}_{\mathrm{FN}}^b$ and at the end, each holds a private component values v_{FN}^a and v_{FN}^b respectively, where $v_{\mathrm{FN}}^a + v_{\mathrm{FN}}^b = \mathbf{V}_{\mathrm{FN}}^a \bullet \mathbf{V}_{\mathrm{FN}}^b$.
4: Similarly, Party A and Party B compute $\mathbf{V}_{\mathrm{TN}}^a \bullet \mathbf{V}_{\mathrm{TN}}^a$ and at the end, each holds a private component values v_{TN}^a and v_{TN}^b respectively, where $v_{\mathrm{TN}}^a + v_{\mathrm{TN}}^b = \mathbf{V}_{\mathrm{TN}}^a \bullet \mathbf{V}_{\mathrm{TN}}^b$.
5: Party A uses v_{TP}^a and v_{TN}^a and Party B uses v_{TP}^b and v_{TN}^b as inputs to execute Protocol 3. The output is $u_1 = (v_{\mathrm{TP}}^a + v_{\mathrm{TP}}^b) \times (v_{\mathrm{TN}}^a + v_{\mathrm{TN}}^b)$.
6: Party A uses $v_{\mathrm{TP}}^a + v_{\mathrm{FN}}^a$ and $v_{\mathrm{FP}}^a + v_{\mathrm{TN}}^a$ and Party B uses $v_{\mathrm{TP}}^b + v_{\mathrm{FN}}^b$ and $v_{\mathrm{FP}}^b + v_{\mathrm{TN}}^b$ as inputs to execute Protocol 3. The output is $u_2 = [(v_{\mathrm{TP}}^a + v_{\mathrm{FN}}^a) + (v_{\mathrm{TP}}^b + v_{\mathrm{FN}}^b)] \times [(v_{\mathrm{FP}}^a + v_{\mathrm{TN}}^a) + (v_{\mathrm{FP}}^b + v_{\mathrm{TN}}^b)]$.
7: Fitness = u_1/u_2.

Protocol 3

Input: Party A has input values w_1^a and w_2^a. Party B has input values w_1^b and w_2^b.
Output: $o = (w_1^a + w_1^b) \times (w_2^a + w_2^b)$.

1: Party A and Party B securely compute the scalar product of vectors $\mathbf{W}^a = [w_1^a, \ w_2^a]^T$ and $\mathbf{W}^b = [w_2^b, \ w_1^b]^T$. At the end of this protocol, Party A and B each holds a private component values w^a and w^b respectively, where $w^a + w^b = \mathbf{W}^a \bullet \mathbf{W}^b$.
2: Party A sends the partial result $o^a = w_1^a \times w_2^a + w^a$ to Party B.
3: Party A computes the final result $o = o^a + o^b$ and sends it to Party B.

The above shows that the fitness value of a rule is expressed in terms of private component values corresponding to N_{TP}, N_{FP}, N_{FN}, and N_{TN} of the rule. Thus, what remains is to compute the numerator and denominator of the fraction. This is accomplished in Steps 5 and 6 of the protocol. As shown in Protocol 2, another protocol is required to perform Steps 5 and 6. This is shown in Protocol 3.

4   Protocol Analysis

In this section, we show that Protocols 2 and 3 are correct and privacy-preserving during the iteration process of genetic algorithms.

4.1   Correctness

Protocol 3: The correctness of the protocol is given:

$$o \Leftrightarrow o^a + o^b \Leftrightarrow \{w_1^a \times w_2^a + w^a\} + \{w_1^b \times w_2^b + w^b\}$$
$$\Leftrightarrow w_1^a \times w_2^a + \mathbf{W}^a \bullet \mathbf{W}^b + w_1^b \times w_2^b$$
$$\Leftrightarrow w_1^a \times w_2^a + [w_1^a, w_2^a]^T \bullet [w_2^b, w_1^b]^T + w_1^b \times w_2^b$$
$$\Leftrightarrow (w_1^a + w_1^b) \times (w_2^a + w_2^b)$$

Protocol 2: Party A has four input vectors $\mathbf{V}_{\text{TP}}^a$, $\mathbf{V}_{\text{FP}}^a$, $\mathbf{V}_{\text{FN}}^a$ and $\mathbf{V}_{\text{TN}}^a$. Party B has four input vectors $\mathbf{V}_{\text{TP}}^b$, $\mathbf{V}_{\text{FP}}^b$, $\mathbf{V}_{\text{FN}}^b$ and $\mathbf{V}_{\text{TN}}^b$. The output is Fitness = True_Positive_Rate × True_Negative_Rate. It is clear that the protocol is correct by definition of the fitness function in the previous page.

Protocol 1: It is clear that the protocol is correct as it uses Protocol 2, which has been shown above to be correct.

4.2   Privacy Preservation

Protocol 3: We show that it is impossible for Party B (dishonest) to guess the values of Party A (honest) regardless of the number of times Protocol 3 is invoked. If Protocol 3 is invoked once, Party B only knows the intermediate values of o and o^a from Party A as follows:

$$o = (w_1^a + w_1^b) \times (w_2^a + w_2^b) = w_1^a \times w_2^a + w_1^a \times w_2^b + w_2^a \times w_1^b + w_2^a \times w_2^b$$
$$o^a = w_1^a \times w_2^a + w^a$$

There are three unknown values w^a, w_1^a and w_2^a for Party B in two nonlinear equations. Even if the equations are linear, with only three unknowns, there are infinitely possible solutions as the number of unknowns is more than the number of equations.

If Protocol 3 is invoked more than once, the values of w^a, w_1^a and w_2^a are different, as they are randomly generated and known only to Party A. The more number of times the protocol is executed, the more unknowns Party B has. Therefore, it is impossible for Party B to guess the values of w^a, w_1^a and w_2^a using the intermediate results from the two or more times the protocol is executed. Hence, Protocol 3 is secure.

Protocol 2: As shown above, value v_{TP}^a (which corresponds to w_1^a in Protocol 3) of Party A as used in Step 5 of Protocol 2 cannot be disclosed during Protocol 3. Without v_{TP}^a, Party B is not able to guess, even by probing, any element values of vector $\mathbf{V}_{\text{TP}}^a$ of Party A. Likewise, Party B will not know any element values of vectors $\mathbf{V}_{\text{FP}}^a$, $\mathbf{V}_{\text{FN}}^a$ and $\mathbf{V}_{\text{TN}}^a$ of Party A. Therefore, if Protocol 2 is invoked once, the data privacy of Party A is not compromised. However, the numerator u_1 and denominator u_2 in Protocol 2 are revealed. To achieve the complete security, the secure division protocol proposed by Vaidya *et al.* [13] that securely computes the division can be applied. According to the authors, the protocol would significantly increase the complexity.

If Protocol 2 is invoked more than once, the values of v_{TP}^a, v_{FP}^a, v_{FN}^a and v_{TN}^a are different, as they are randomly generated and known only to Party A. It is impossible for Party B to guess any element values of the private vectors of Party A using the intermediate results. Thus, Protocol 2 is secure.

On the whole, no matter how many times Protocol 2 is invoked by Protocol 1, the dishonest party is not able to guess any element values of the honest party's private vectors.

Protocol 1: The protocol is secure as it uses Protocol 2, which has been shown above to be secure.

4.3 Complexity Analysis

The computational complexity and communication cost of the secure scalar product protocol used in Protocol 2 and Protocol 3 are defined as $O(\phi(z))$ and $O(\phi'(z))$ respectively, where (1) z is the number of elements in vectors; and (2) $\phi(z)$ and $\phi'(z)$ are expressions for the computational complexity and communication cost respectively of z with respect to some chosen secure scalar product protocol.

As Protocol 3 invokes the secure scalar product protocol for two vectors of length 2 once (Step 1), the overall computational complexity of Protocol 3 is $O(\phi(2)) = O(1)$ and the communication cost is $O(\phi'(2)) = O(1)$.

As Protocol 2 invokes the secure scalar product protocol for two vectors of length m four times (Steps 1–4) and Protocol 3 twice (Steps 5 and 6), the computational complexity of Protocol 2 is $O(4\phi(m)) + O(2\phi(2)) = O(\phi(m))$, where m is the number of tuples in the data sets. The communication cost is $O(4\phi'(m)) + O(2\phi'(2)) = O(\phi'(m))$.

Since Protocol 1 uses Protocol 2 once (Step 6) at each iteration, the total computational complexity for Protocol 1 is $T \times O(\phi(m)) = O(T \times \phi(m))$ and the communication cost is $T \times O(\phi'(m)) = O(T \times \phi'(m))$ where T is the total number of iterations before the termination condition is satisfied.

5   Conclusions

Several traditional data mining algorithms have been adapted to be become privacy-preserving: decision trees, association rule mining, k-means clustering, SVM, Naïve Bayes, Bayesian network. These algorithms generally assume that the original data set has been horizontally and/or vertically partitioned, with each partition privately held by a party. In this paper, we proposed protocols for securely using genetic algorithms for rule discovery on private arbitrarily partitioned data that are held by two parties without compromising their data privacy. As genetic algorithms are iterative, it is a challenge to not only preserve the privacy of data at each iteration, it is also a challenge to ensure that the intermediate results produced at each iteration do not compromise the data privacy of the participating parties. We showed that the proposed protocols satisfy these two requirements for data privacy. As the protocol currently only works for two parties, extending to multiple parties is part of the future work.

References

1. Agrawal, R., Srikant, R.: Privacy-preserving data mining. In: Proceedings of the ACM international Conference on Management of Data, Dallas, Texas, United States, pp. 439–450 (2000)
2. Carvalho, D.R., Freitas, A.A.: A genetic algorithm with sequential niching for discovering small-disjunct rules. In: Proceedings of the Genetic and Evolutionary Computation Conference, 2002, Sanfrancisco, CA, USA, pp. 1035–1042. Morgan Kaufmann Publishers Inc., San Francisco (2002)
3. Du, W., Atallah, M.J.: Privacy-preserving cooperative statistical analysis. In: Proceedings of the 17th Annual Computer Security Applications Conference, New Orleans, Louisiana, USA, December 10–14, 2001, pp. 102–110 (2001)
4. Freitas, A.A.: Data Mining and Knowledge Discovery with Evolutionary Algorithms. Spinger, Berlin (2002)
5. Goethals, B., Laur, S., Lipmaa, H., Mielikainen, T.: On private scalar product computation for privacy-preserving data mining. In: Proceedings of the 7th Annual International Conference in Information Security and Cryptology, Seoul, Korea, December 2–3, 2004, pp. 104–120 (2004)
6. Jagannathan, G., Wright, R.N.: Privacy-preserving distributed k-means clustering over arbitrarily partitioned data. In: Proceedings of the 8th ACM International Conference on Knowledge Discovery in Data Mining, Chicago, Illinois, USA, pp. 593–599 (2005)
7. Laur, S., Lipmaa, H., Mielikainen, T.: Cryptographically private support vector machines. In: Proceedings of the 12th ACM International Conference on Knowledge Discovery and Data Mining, Philadelphia, PA, USA, pp. 618–624 (2006)
8. Lindell, Y., Pinkas, B.: Privacy preserving data mining. In: Bellare, M. (ed.) CRYPTO 2000. LNCS, vol. 1880, pp. 36–53. Springer, Heidelberg (2000)
9. Parpinelli, R., Lopes, H., Freitas, A.: An ant colony based system for data mining: Applications to medical data. In: Spector, L., G., E., et al. (eds.) Proceedings of the Genetic and Evolutionary Computation Conference, San Francisco, USA, July 2001, pp. 791–798. Morgan Kaufmann, San Francisco (2001)
10. Vaidya, J., Clifton, C.: Privacy preserving association rule mining in vertically partitioned data. In: Proceedings of the 8th ACM International Conference on Knowledge Discovery and Data Mining, Edmonton, Alberta, Canada, July 23-26, 2002, pp. 639–644 (2002)
11. Vaidya, J., Clifton, C.: Privacy-preserving k-means clustering over vertically partitioned data. In: Proceedings of the 9th ACM International Conference on Knowledge Discovery and Data Mining, Washington, DC, pp. 206–215 (2003)
12. Vaidya, J., Clifton, C.: Secure set intersection cardinality with application to association rule mining. Journal of Computer Security, 13(4) (2005)
13. Vaidya, J., Kantarcioglu, M., Clifton, C.: Privacy-preserving naive bayes classification. The International Journal on Very Large Data Bases
14. Weiss, S.M., Kulikowski, C.A.: Computer Systems That Learn: Classification and Prediction Methods from Statistics, Neural Nets, Machine Learning, and Expert Systems. Morgan Kaufmann Publishers Inc., San Francisco, CA, USA (1991)

15. Yao, A.C.: How to generate and exchange secrets. In: Proceedings of the Annual IEEE Symposium on Foundations of Computer Science, pp. 162–167 (1986)
16. Yu, H., Jiang, X., Vaidya, J.: Privacy-preserving svm using nonlinear kernels on horizontally partitioned data. In: Proceedings of the ACM Symposium on Applied Computing, Dijon, France, pp. 603–610 (2006)
17. Yu, H., Vaidya, J., Jiang, X.: Privacy-preserving svm classification on vertically partitioned data. In: Ng, W.-K., Kitsuregawa, M., Li, J., Chang, K. (eds.) PAKDD 2006. LNCS (LNAI), vol. 3918, pp. 9–12. Springer, Heidelberg (2006)

Fast Cryptographic Multi-party Protocols for Computing Boolean Scalar Products with Applications to Privacy-Preserving Association Rule Mining in Vertically Partitioned Data*

Dragoş Trincă and Sanguthevar Rajasekaran

Department of Computer Science and Engineering
University of Connecticut
Storrs, CT 06269, USA
{dtrinca,rajasek}@engr.uconn.edu

Abstract. Recently, the problem of privately mining association rules in vertically partitioned data has been reduced to the problem of privately computing boolean scalar products. In this paper, we propose two cryptographic multi-party protocols for privately computing boolean scalar products. The proposed protocols are shown to be secure and much faster than other protocols for the same problem.

1 Introduction

With the recent concerns about privacy of personal information, considerable work has been done in the relatively new area of privacy-preserving data mining [2,9]. The goal is not only to develop new data mining algorithms that incorporate privacy constraints, but also to augment the existing ones with privacy-preserving capabilities. Such algorithms should also be fast and as efficient as possible in terms of accuracy of the results. Most of these algorithms are increasingly important in homeland security and counterterrorism-related applications, where the data is usually sensitive.

In this paper, we focus on the problem of privately mining association rules in vertically partitioned data, which was recently addressed in [13], [12], [14], [7], [4], [11], and [6]. In [13] and [7], the authors have proposed two distinct algebraic protocols for the special case of the problem when there are only two parties involved in the computation of the scalar product. Both of these algebraic techniques are interesting, although the protocol given in [7] requires at least twice the bitwise communication cost of the protocol presented in [13]. Another protocol for the special case when there are only two parties involved in the computation of the scalar product has been proposed in [14], but their protocol is cryptographic in nature, since it is based on homomorphic encryption. The protocol proposed in [4] is a two-party cryptographic protocol for privately computing boolean scalar products, but was recently proven insecure in [6]. To the

---

* This research has been supported in part by the NSF Grant ITR-0326155.

I.Y. Song, J. Eder, and T.M. Nguyen (Eds.): DaWaK 2007, LNCS 4654, pp. 418–427, 2007.
© Springer-Verlag Berlin Heidelberg 2007

best of our knowledge, the only protocols proposed for the general case when multiple parties are involved in the computation of the scalar product are as follows.

1. In [11], the authors have extended the algebraic protocol given in [13] to the multi-party case. However, the complexity of the extended protocol is about $\mathcal{O}(n^k)$, where n is the number of transactions in the database and k is the number of parties. So, for reasonable values of n and k, the extended protocol is not even implementable. Moreover, this protocol was recently proven insecure in [6].
2. In [12], the author has proposed two distinct multi-party cryptographic protocols. In the first one, the author has reduced the problem to secure set intersection and then solved the latter using some cryptographic tools. However, this protocol was shown to produce leakage of information for all parties. The second cryptographic protocol produces no leakage of information, but it is not fast at all (according to the results reported in [12] and this paper).
3. In [6], the authors have proposed a multi-party cryptographic protocol. This protocol was proven to be secure, but it has two significant problems. First, it was proven in [6] to be collusion-resistant only if the number of colluding parties is strictly less then $\frac{k}{2}$, where k is the total number of parties. Second, it introduces a large overhead in the communication part of the problem.

In this paper, we propose two cryptographic protocols for privately computing boolean scalar products. Our protocols are shown to be secure and much faster than the protocol given in [6] and the second protocol proposed in [12]. The paper is organized as follows. In Section 2, we provide the reader with a concise introduction to the problem of privately mining association rules in vertically partitioned data. In Section 3, we propose two cryptographic multi-party protocols for privately computing boolean scalar products. The proposed protocols are shown to be secure. Results of our implementations are reported in Section 4. The protocol given in [6] and the second cryptographic protocol proposed in [12] are the only protocols that we compare our protocols against, since these are the only previously proposed multi-party protocols for privately computing boolean scalar products that were not proven insecure yet. Experimental results show that the proposed protocols are much faster than other cryptographic protocols for the same problem, and this is due to the fact that the proposed protocols do not make use of any modular exponentiations or other time-consuming primitives. Finally, in the last section, we provide some conclusions and future work directions.

2   Privacy-Preserving Association Rule Mining in Vertically Partitioned Data

The problem of mining association rules, as it was first posed in [3], can be formally stated as follows. Let $\mathcal{I} = \{I_1, I_2, \ldots, I_m\}$ be a set of attributes, usually called items, and let $\mathcal{D}$ be a set of transactions. Each transaction in $\mathcal{D}$ is a set $T \subseteq \mathcal{I}$ of items. An *association rule* is an implication of the form $X \implies Y$, where $X \subset \mathcal{I}$, $Y \subset \mathcal{I}$, and $X \cap Y = \emptyset$. The rule $X \implies Y$ has *support s* in $\mathcal{D}$ if $s\%$ of the transactions in $\mathcal{D}$ contain $X \cup Y$. On the other hand, we say that the rule $X \implies Y$ holds with *confidence c* in $\mathcal{D}$ if $c\%$ of the transactions in $\mathcal{D}$ that contain X also contain Y. The problem is to find all association rules with support and confidence above certain thresholds (usually referred to as *minsup* and *minconf*). The association rule mining problem can be decomposed into two distinct subproblems:

1. Generate all combinations of items that have support at least *minsup*.
2. For every frequent itemset $Y = \{I_{i_1}, I_{i_2}, \ldots, I_{i_k}\}$ found in the first step, consider all rules of the form $X \implies I_{i_j}$ ($1 \le j \le k$), where $X = Y - \{I_{i_j}\}$. Clearly, every such rule has support at least *minsup*. However, not all such rules may have confidence at least *minconf*. In order to see whether the rule $X \implies I_{i_j}$ is satisfied or not, just divide the support of Y by the support of X. If the ratio is at least *minconf*, then the rule is satisfied; otherwise, it is not.

Most of the work done so far has focused on the first subproblem, since generating the corresponding association rules from the frequent itemsets is a trivial task. The best known algorithm for mining frequent itemsets is the Apriori algorithm [1]. In [13], the authors have proposed a variant of the Apriori algorithm for the case when the database is vertically splitted among several parties. They showed that the problem of privately mining association rules in such a distributed setting can be reduced to the problem of privately computing boolean scalar products in the same distributed setting. In this paper, we propose two cryptographic multi-party protocols for privately computing boolean scalar products that are much faster than the only previously proposed protocols for the same problem that were not proven insecure yet, namely the protocol proposed in [6] and the second protocol proposed in [12].

3   Secure Cryptographic Multi-party Protocols for Computing Boolean Scalar Products Under Privacy Constraints

We have seen in Section 2 that the problem of privately mining association rules in vertically partitioned data can be reduced to the problem of privately computing boolean scalar products. The latter the can be formally stated as follows. Let $\mathcal{P}_1, \mathcal{P}_2, \ldots, \mathcal{P}_k$ be k parties, and let $\boldsymbol{X}_i$ be the boolean column vector corresponding to party $\mathcal{P}_i$ (all vectors have the same size n). The parties want to collaboratively compute the scalar product $\boldsymbol{X}_1 \cdot \boldsymbol{X}_2 \cdot \ldots \cdot \boldsymbol{X}_k$ without any party revealing its own vector to the other parties. In this section, we propose two cryptographic multi-party protocols for privately computing boolean scalar products, and then analyze their security.

3.1   SECProtocol-I

The aim of this section is to propose a new cryptographic protocol for privately computing boolean scalar products. We describe the new protocol by considering a running example involving three parties. The extension to more than three parties follows the same idea. Let $\mathcal{P}_1$, $\mathcal{P}_2$, $\mathcal{P}_3$ be three parties that want to collaboratively compute the scalar product $X_1 \cdot X_2 \cdot X_3$, where

$$X_1 = \begin{bmatrix} 1 \\ 1 \\ 0 \end{bmatrix}, X_2 = \begin{bmatrix} 0 \\ 1 \\ 1 \end{bmatrix}, X_3 = \begin{bmatrix} 0 \\ 1 \\ 0 \end{bmatrix}$$

are the vectors associated with $\mathcal{P}_1$, $\mathcal{P}_2$, $\mathcal{P}_3$, respectively. Each of the vectors has size $n = 3$. Let $\mathcal{P}_3$ be the initiator of the protocol. The protocol makes use of a randomly generated parameter, denoted by t, which is available only to the initiator of the protocol (in our case, $\mathcal{P}_3$), and is used to hide the input vectors within the matrices constructed during the protocol. In our running example, $t = 7$.

Step 1. First, $\mathcal{P}_3$ forms an $n \times 2^n$ matrix

$$M_3 = \begin{bmatrix} 0 1 1 1 0 0 0 1 \\ 0 1 1 0 1 0 1 0 \\ 1 0 1 0 1 0 0 1 \end{bmatrix},$$

where the t-th column of M_3 is X_3 and the rest of the entries are randomly generated in such a way that M_3 contains all possible boolean column vectors of size n within its columns. Then, $\mathcal{P}_3$ sends M_3 to $\mathcal{P}_2$.

Step 2. Upon receiving M_3 from $\mathcal{P}_3$, $\mathcal{P}_2$ forms an $n \times 2^n$ matrix

$$M_2 = \begin{bmatrix} 0 0 0 0 0 0 0 0 \\ 0 1 1 0 1 0 1 0 \\ 1 0 1 0 1 0 0 1 \end{bmatrix},$$

where $M_2[l, c] = X_2[l] M_3[l, c]$ for all l, c. Then, $\mathcal{P}_2$ splits the set of column indices, namely $C = \{1, 2 \ldots, 8\}$, into equivalence classes, in such a way that two indices i_1 and i_2 are in the same equivalence class if the i_1-th and the i_2-th column are the same. In the case of M_2, the equivalence classes are $C_1 = \{1, 8\}$, $C_2 = \{2, 7\}$, $C_3 = \{3, 5\}$, $C_4 = \{4, 6\}$. For each equivalence class $C_i = \{i_1, i_2, \ldots, i_x\}$ that has at least two indices, that is, $x \geq 2$, $\mathcal{P}_2$ randomly selects $x - 1$ indices from C_i, and for those $x - 1$ indices, it replaces the corresponding columns with vectors that are not already present in the matrix. Let us suppose that it replaces the 1-st, the 7-th, the 5-th, and the 6-th column with

$$\begin{bmatrix} 1 \\ 1 \\ 1 \end{bmatrix}, \begin{bmatrix} 1 \\ 0 \\ 1 \end{bmatrix}, \begin{bmatrix} 1 \\ 1 \\ 0 \end{bmatrix}, \text{ and } \begin{bmatrix} 1 \\ 0 \\ 0 \end{bmatrix},$$

respectively. After this replacement, the new matrix is

$$M_2^{full} = \begin{bmatrix} 1 0 0 0 1 1 1 0 \\ 1 1 1 0 1 0 0 0 \\ 1 0 1 0 0 0 1 1 \end{bmatrix},$$

which contains all possible boolean column vectors of size n within its columns. The correspondence between M_2 and M_2^{full} is given by the tuple

$$full_2 = (8, 2, 3, 4, 3, 4, 2, 8).$$

This tuple is interpreted as follows. If the i-th element in the tuple is x, $\mathcal{P}_2$ knows that the i-th column in M_2 moved on position x in M_2^{full}. The matrix M_2^{full} is further processed by applying a random permutation, say

$$perm_2 = (3, 1, 7, 5, 6, 4, 8, 2).$$

The new matrix is

$$M_2^{perm} = \begin{bmatrix} 0 0 1 1 0 1 0 1 \\ 1 0 1 0 0 1 1 0 \\ 0 1 1 0 0 0 1 1 \end{bmatrix}.$$

This permutation is interpreted as follows. If the i-th element in the permutation is x, $\mathcal{P}_2$ knows that the i-th column in M_2^{full} moved on position x in M_2^{perm}. Finally, $\mathcal{P}_2$ sends M_2^{perm} to $\mathcal{P}_1$, along with

$$product_2 = (2, 1, 7, 5, 7, 5, 1, 2),$$

where the i-th element in $product_2$ is x if the i-th element in $full_2$ is j and the j-th element in $perm_2$ is x. So, $product_2$ is the product between $full_2$ and $perm_2$.

Step 3. Upon receiving M_2^{perm} from $\mathcal{P}_2$, $\mathcal{P}_1$ forms an $n \times 2^n$ matrix

$$M_1 = \begin{bmatrix} 0 0 1 1 0 1 0 1 \\ 1 0 1 0 0 1 1 0 \\ 0 0 0 0 0 0 0 0 \end{bmatrix},$$

where $M_1[l, c] = \boldsymbol{X}_1[l] M_2^{perm}[l, c]$ for all l, c. As in the previous step, $\mathcal{P}_1$ splits the set of column indices into equivalence classes. In the case of M_1, the equivalence classes are $C_1 = \{1, 7\}, C_2 = \{2, 5\}, C_3 = \{3, 6\}, C_4 = \{4, 8\}$. Now, let us suppose that $\mathcal{P}_1$ replaces the 1-st, the 2-nd, the 4-th, and the 6-th column of M_1 with

$$\begin{bmatrix} 0 \\ 0 \\ 1 \end{bmatrix}, \begin{bmatrix} 1 \\ 0 \\ 1 \end{bmatrix}, \begin{bmatrix} 1 \\ 1 \\ 1 \end{bmatrix}, \text{ and } \begin{bmatrix} 0 \\ 1 \\ 1 \end{bmatrix},$$

respectively. After this replacement, the new matrix is

$$M_1^{full} = \begin{bmatrix} 0 1 1 1 0 0 0 1 \\ 0 0 1 1 0 1 1 0 \\ 1 1 0 1 0 1 0 0 \end{bmatrix},$$

which contains all possible boolean vectors of size n within its columns. The correspondence between M_1 and M_1^{full} is given by

$$full_1 = (7, 5, 3, 8, 5, 3, 7, 8).$$

M_1^{full} is further processed by applying a random permutation, say

$$perm_1 = (6, 4, 3, 8, 1, 5, 2, 7).$$

The new matrix is

$$M_1^{perm} = \begin{bmatrix} 0 & 0 & 1 & 1 & 0 & 0 & 1 & 1 \\ 0 & 1 & 1 & 0 & 1 & 0 & 0 & 1 \\ 0 & 0 & 0 & 1 & 1 & 1 & 0 & 1 \end{bmatrix}.$$

The product between $full_1$ and $perm_1$ is

$$product_1 = (2, 1, 3, 7, 1, 3, 2, 7),$$

and the product between $product_2$ and $product_1$ is

$$product = (1, 2, 2, 1, 2, 1, 2, 1).$$

Finally, $\mathcal{P}_1$ sends to $\mathcal{P}_3$ the tuple

$$sp = (0, 1, 1, 0, 1, 0, 1, 0),$$

where the i-th element of sp is x if the i-th element of $product$ is j and the number of 1's in the j-th column of M_1^{perm} is x.

Step 4. When $\mathcal{P}_3$ receives sp from $\mathcal{P}_1$, it knows that the i-th element of sp corresponds to the scalar product $X_1 \cdot X_2 \cdot M_3[-, i]$. Since $t = 7$, $\mathcal{P}_3$ concludes that $X_1 \cdot X_2 \cdot X_3 = 1$, since the 7-th element of sp is 1.

Denote this multi-party cryptographic protocol by SECProtocol-I. The computations involved in SECProtocol-I take $\mathcal{O}(kn2^n)$ time. The communication complexity is $\mathcal{O}(kn2^n)$ as well.

3.2   SECProtocol-II

In this section, we propose some changes to SECProtocol-I that will reduce its complexity significantly. Let $\mathcal{P}_1$, $\mathcal{P}_2$, $\mathcal{P}_3$ be three parties that want to collaboratively compute the scalar product $X_1 \cdot X_2 \cdot X_3$, where

$$X_1 = \begin{bmatrix} 1 \\ 1 \\ 0 \end{bmatrix}, X_2 = \begin{bmatrix} 0 \\ 1 \\ 1 \end{bmatrix}, X_3 = \begin{bmatrix} 0 \\ 1 \\ 0 \end{bmatrix}$$

are vectors of size $n = 3$ each. First, $\mathcal{P}_3$ (which is the initiator of the protocol) forms the matrices

$$M_3^1 = \begin{bmatrix} 0 & 1 \end{bmatrix}, M_3^2 = \begin{bmatrix} 1 & 0 \end{bmatrix}, M_3^3 = \begin{bmatrix} 1 & 0 \end{bmatrix}.$$

Also, let $c_1 = 1$, $c_2 = 1$, and $c_3 = 2$. It can be easily seen that the c_1-th element of M_3^1 is $\boldsymbol{X}_3[1]$, the c_2-th element of M_3^2 is $\boldsymbol{X}_3[2]$, and the c_3-th element of M_3^3 is $\boldsymbol{X}_3[3]$. Each of the matrices M_3^1, M_3^2, M_3^3 contains all possible boolean column vectors of size 1 within its columns.

Step 1. $\mathcal{P}_3$ runs SECProtocol-I in order to find the scalar product $\boldsymbol{X}_1[1] \cdot \boldsymbol{X}_2[1] \cdot \boldsymbol{X}_3[1]$. The parameters used are M_3^1 and c_1.

Step 2. $\mathcal{P}_3$ runs SECProtocol-I in order to find the scalar product $\boldsymbol{X}_1[2] \cdot \boldsymbol{X}_2[2] \cdot \boldsymbol{X}_3[2]$. The parameters used are M_3^2 and c_2.

Step 3. Finally, $\mathcal{P}_3$ runs SECProtocol-I in order to find the scalar product $\boldsymbol{X}_1[3] \cdot \boldsymbol{X}_2[3] \cdot \boldsymbol{X}_3[3]$. The parameters used are M_3^3 and c_3.

Step 4. Summing up the three scalar products, $\mathcal{P}_3$ finds the desired scalar product $\boldsymbol{X}_1 \cdot \boldsymbol{X}_2 \cdot \boldsymbol{X}_3$.

Remark 1. The protocol given above can be generalized, in the sense that we do not have to restrict ourselves to n splits, each of size 1. The idea is as follows. Each party $\mathcal{P}_i$ splits its own vector $\boldsymbol{X}_i$ into x subvectors of size $\frac{n}{x}$ each. The subvectors associated with party $\mathcal{P}_i$ are denoted by $\boldsymbol{X}_i^1$, $\boldsymbol{X}_i^2$, ..., $\boldsymbol{X}_i^x$. Additionally, $\mathcal{P}_k$ randomly generates $c_1, c_2, \ldots, c_x$, where $c_i \in \{1, 2, 3, \ldots, 2^{\frac{n}{x}}\}$ for all i, and then forms x boolean matrices M_k^1, M_k^2, ..., M_k^x (each of size $\frac{n}{x} \times 2^{\frac{n}{x}}$), in such a way that the c_i-th column of M_k^i is $\boldsymbol{X}_k^i$, for all i. The matrices M_k^1, M_k^2, ..., M_k^x are also generated in such a way that each of them contains all possible boolean vectors of size $\frac{n}{x}$ within its columns. Then, for each $i \in \{1, 2, \ldots, x\}$, $\mathcal{P}_k$ initiates SECProtocol-I in order to find the scalar product $\boldsymbol{X}_1^i \cdot \boldsymbol{X}_2^i \cdot \ldots \cdot \boldsymbol{X}_k^i$. The parameters used for finding the scalar product $\boldsymbol{X}_1^i \cdot \boldsymbol{X}_2^i \cdot \ldots \cdot \boldsymbol{X}_k^i$ are M_k^i and c_i. Summing up these x scalar products, $\mathcal{P}_k$ finds the desired scalar product, namely $\boldsymbol{X}_1 \cdot \boldsymbol{X}_2 \cdot \ldots \cdot \boldsymbol{X}_k$.

Denote this improved version of SECProtocol-I by SECProtocol-II. The computations involved in SECProtocol-II take $\mathcal{O}(xk\frac{n}{x}2^{\frac{n}{x}})$ time. The communication complexity is $\mathcal{O}(xk\frac{n}{x}2^{\frac{n}{x}})$ as well.

3.3 Security

In this section, we analyze the security offered by SEC-Protocol-I and SECProtocol-II. To simplify the exposition, we consider the following definition.

Definition 1. *Let M be an $n \times q$ boolean matrix. If each of the 2^n possible boolean vectors of size n is present within the columns of M, then M is called full with respect to the columns.*

Given the nature of the problem, there is some inherent leakage of information that exists in every protocol for this problem, no matter how secure it is. For example, the initiator of the protocol can work with an input vector which has a 1 on some position p and 0's on the other positions. In such a case, if the scalar product is 1, the initiator will know that each of the participating parties has a 1 on position p. SECProtocol-I and SECProtocol-II make no exception from this leakage of information. However, except for this inherent leakage of information, SECProtocol-I leaks no other information. SECProtocol-II leaks some extra information, namely the x intermediate scalar products. However, the intermediate scalar products cannot be used in any way to find the actual values of the vectors, which means that in practice this small leakage of information is useless for an adversary.

It can be easily seen that SECProtocol-I is secure. The security stems from the fact that each of the matrices exchanged during the protocol is full with respect to the columns. SECProtocol-II is basically SECProtocol-I in which the input vectors are splitted into several parts. So, in conclusion, SECProtocol-II is also secure.

4 Comparisons Between SECProtocol-II and Other Protocols

Since the complexity of SECProtocol-I is $\mathcal{O}(kn2^n)$, it is clear that SECProtocol-I is practical only for small values of n, say $n = 10$. Assuming $x = \frac{n}{2}$, the complexity of SECProtocol-II is $\mathcal{O}(kn)$, which is indeed fast. So, we will compare SECProtocol-II (under the assumption that $x = \frac{n}{2}$) against two protocols: the cryptographic protocol proposed in [6] and the second cryptographic protocol proposed in [12], which are the only previously proposed multi-party protocols for privately computing boolean scalar products that were not proven insecure yet (and which will be denoted by GLLM and OU, respectively).

4.1 Implementation

We have implemented all three protocols, which were tested on a GenuineIntel server with an Intel Pentium 4 CPU at 2.40GHz. As in [12], we have implemented OU using the Okamoto-Uchiyama public-key cryptosystem [10]. The results are reported in Table 1. As one can see, SECProtocol-II is much faster than GLLM and OU, mostly due to the fact that there are no modular exponentiations or other cryptographic functions involved in SECProtocol-II.

Table 1. Comparisons between SECProtocol-II and other protocols

k	n	SECProtocol-II	OU	GLLM
5	1,000	0.03s	9.27s	6.97s
	5,000	0.07s	25.10s	19.63s
	10,000	0.14s	52.65s	40.48s
	50,000	0.68s	6.67m	5.04m
	100,000	1.34s	17.41m	14.28m
10	1,000	0.03s	17.70s	14.35s
	5,000	0.14s	49.40s	37.19s
	10,000	0.27s	1.75m	1.32m
	50,000	1.33s	13.32m	10.71m
	100,000	2.70s	34.86m	26.82m

Definition 2. *The communication overhead of a multi-party scalar product protocol P is defined to be equal to $\frac{C(P)}{n}$, where $C(P)$ is the number of bits communicated in the protocol P.*

The results reported in Table 1 include only the computations involved in the protocols. The communication part is not included. However, if the communication part is also to be included, then SECProtocol-II is expected to be even faster, since the communication overhead of SECProtocol-II is $4k$ (assuming $x = \frac{n}{2}$), whereas the communication overhead of GLLM is $k \log n$ and the communication overhead of OU is $kS_1 + \frac{kS_2}{n}$, where S_1 is the keysize and S_2 is the size of an encrypted message.

5   Conclusions and Future Work

In this paper, we have focused on the problem of privately mining association rules in vertically distributed boolean databases. This problem can be reduced to the problem of privately computing boolean scalar products. We proposed two cryptographic multi-party protocols for privately computing boolean scalar products. The two protocols are shown to be not only secure, but also much faster than similar protocols. As future work, it would be interesting to design and test parallel variants of the proposed multi-party protocols.

References

1. Agrawal, R., Srikant, R.: Fast Algorithms for Mining Association Rules. In: Proceedings of the 20th International Conference on Very Large Databases, pp. 487–499 (1994)
2. Agrawal, R., Srikant, R.: Privacy-Preserving Data Mining. In: Proceedings of the ACM SIGMOD International Conference on Management of Data, pp. 439–450. ACM Press, New York (2000)
3. Agrawal, R., Imielinski, T., Swami, A.: Mining Association Rules between Sets of Items in Large Databases. In: Proceedings of the 1993 ACM SIGMOD International Conference on Management of Data, pp. 207–216. ACM Press, New York (1993)
4. Du, W., Atallah, M.J.: Privacy-Preserving Cooperative Statistical Analysis. In: Proceedings of the 17th Annual Computer Security Applications Conference (ACSAC 2001), pp. 102–110 (2001)
5. Evfimievski, A.V., Srikant, R., Agrawal, R., Gehrke, J.: Privacy Preserving Mining of Association Rules. In: Proceedings of the 8th ACM SIGKDD International Conference on Knowledge Discovery and Data Mining, 2002, pp. 217–228 (2002)
6. Goethals, B., Laur, S., Lipmaa, H., Mielikäinen, T.: On Private Scalar Product Computation for Privacy-Preserving Data Mining. In: Park, C.-s., Chee, S. (eds.) ICISC 2004. LNCS, vol. 3506, pp. 104–120. Springer, Heidelberg (2005)
7. Ioannidis, I., Grama, A., Atallah, M.J.: A Secure Protocol for Computing Dot-Products in Clustered and Distributed Environments. In: Proceedings of the 31st International Conference on Parallel Processing, 2002, pp. 379–384 (2002)

8. Kantarcioglu, M., Clifton, C.: Privacy-Preserving Distributed Mining of Association Rules on Horizontally Partitioned Data. In: Proceedings of the 2002 ACM SIGMOD Workshop on Research Issues in Data Mining and Knowledge Discovery, ACM Press, New York (2002)

9. Lindell, Y., Pinkas, B.: Privacy Preserving Data Mining. Journal of Cryptology 15(3), 177–206 (2002)

10. Okamoto, T., Uchiyama, S.: A New Public-Key Cryptosystem as Secure as Factoring. In: Nyberg, K. (ed.) EUROCRYPT 1998. LNCS, vol. 1403, pp. 308–318. Springer, Heidelberg (1998)

11. Trincă, D., Rajasekaran, S.: Towards a Collusion-Resistant Algebraic Multi-Party Protocol for Privacy-Preserving Association Rule Mining in Vertically Partitioned Data. In: Proceedings of the 3rd IEEE International Workshop on Information Assurance, New Orleans, Louisiana, USA, 2007, pp. 402–409 (2007)

12. Vaidya, J.: Privacy Preserving Data Mining over Vertically Partitioned Data. Ph.D. Thesis, Purdue University (August 2004)

13. Vaidya, J., Clifton, C.: Privacy Preserving Association Rule Mining in Vertically Partitioned Data. In: Proceedings of the 8th ACM SIGKDD International Conference on Knowledge Discovery and Data Mining, pp. 639–644 (2002)

14. Zhan, J.Z., Matwin, S., Chang, L.: Private Mining of Association Rules. In: Proceedings of the 2005 IEEE International Conference on Intelligence and Security Informatics, pp. 72–80. IEEE Computer Society Press, Los Alamitos (2005)

Privacy-Preserving Self-Organizing Map*

Shuguo Han and Wee Keong Ng

Center for Advanced Information Systems, School of Computer Engineering,
Nanyang Technological University, Singapore. 639798
{hans0004,awkng}@ntu.edu.sg

Abstract. Privacy-preserving data mining seeks to allow the cooperative execution of data mining algorithms while preserving the data privacy of each party concerned. In recent years, many data mining algorithms have been enhanced with privacy-preserving feature: decision tree induction, frequent itemset counting, association analysis, k-means clustering, support vector machine, Naïve Bayes classifier, Bayesian networks, and so on. In this paper, we propose a protocol for privacy-preserving self-organizing map for vertically partitioned data involving two parties. Self-organizing map (SOM) is a widely used algorithm for transforming data sets to a lower dimensional space to facilitate visualization. The challenges in preserving data privacy in SOM are (1) to securely discover the winner neuron from data privately held by two parties; (2) to securely update weight vectors of neurons; and (3) to securely determine the termination status of SOM. We propose protocols to address the above challenges. We prove that these protocols are correct and privacy-preserving. Also, we prove that the intermediate results generated by these protocols do not violate the data privacy of the participating parties.

1 Introduction

Privacy-preserving data mining has become an active area of research in the data mining community since privacy issue gains significance and importance [1,10]. To date, various data mining algorithms have been enhanced with a privacy-preserving version for horizontally and/or vertically partitioned data. This includes decision trees, association rule mining, support vector machine (SVM), Naïve Bayes classifier, Bayesian network structure, and so on.

To the best of our knowledge, there is no work to date on a privacy-preserving version of Self-Organizing Map (SOM). Self-organizing map is a widely used algorithm for transforming data sets to a lower dimensional space to facilitate visualization. In this paper, we propose a protocol for two parties each holding a private vertically partitioned data to jointly and securely perform SOM. The challenges in preserving data privacy in SOM are (1) to securely discover the winner neuron from data privately held by two parties; (2) to securely update weight vectors of neurons; and (3) to securely determine the termination status of SOM.

---

* This work is supported in part by grant P0520095 from the Agency for Science, Technology and Research (A*STAR), Singapore.

I.Y. Song, J. Eder, and T.M. Nguyen (Eds.): DaWaK 2007, LNCS 4654, pp. 428–437, 2007.

We propose protocols to address the above challenges. We prove that these protocols are correct and privacy-preserving. Also, we show that the intermediate results produced by the proposed protocols during the execution of SOM do not compromise the data privacy of the participating parties.

The paper is organized as follows. In Section 2, we review various privacy-preserving data mining algorithms and briefly describe the general algorithm of self-organizing map. Section 3 propose a protocol with three sub-protocols for privacy-preserving SOM in the setting of vertically partitioned data involving two parties. Section 4 analyzes the correctness, complexity, and privacy of the proposed protocol and sub-protocols. We conclude the paper in Section 5.

2   Background

2.1   Related Work

In this section, we review current work on privacy-preserving data mining algorithms that are based on Secure Multi-party Computation [16].

Lindell and Pinkas [10] proposed a privacy-preserving ID3 algorithm based on cryptographic techniques for horizontally partitioned data involving two parties. Kantarcioglu and Clifton [8] proposed a method to securely mine association rules for horizontally partitioned data involving three or more parties. The method incorporates cryptographic techniques to reduce the information disclosed. Vaidya and Clifton [11] presented privacy-preserving association rule mining for vertically partitioned data based on the secure scalar product protocol involving two parties.

Kantarcioglu and Vaidya [7] presented a privacy-preserving Naïve Bayes classifier for horizontally partitioned data based on the secure sum—an instance the Secure Multi-party Computation [16]. Vaidya and Clifton [12] developed privacy-preserving Naïve Bayes classifier for vertically partitioned data. Secure scalar products protocols are used to give the appropriate probability for the data while preserving data privacy. Wright and Yang [15] presented a privacy-preserving Bayesian network computation for vertically distributed data involving two parties. They showed an efficient and privacy-preserving version of the $K2$ algorithm to construct the structure of a Bayesian network for the parties' joint data.

Han and Ng [5] presented a privacy-preserving genetic algorithms for rule discovery for arbitrarily partitioned data involving only two parties. To achieve data privacy of the participant parties, secure scalar product protocols was applied to securely evaluate the fitness value. Wan et al. [14] proposed a generic formulation of gradient descent methods for secure computation by defining the target function as a composition of two functions. They presented a secure two-party protocol for performing gradient descent.

To the best of our knowledge, there is no work to date on a privacy-preserving version of Self-Organizing Map (SOM). In this paper, we propose a protocol for two parties each holding a private, vertical data partition to jointly and securely perform SOM. We prove that these protocols are correct and privacy-preserving. Also, we prove that the intermediate results generated by these protocols do not violate the data privacy of the participating parties.

2.2 Self-Organizing Map (SOM)

SOM is a feed-forward neural network without any hidden layer [9]. The output layer is a grid map of neurons interconnected with weights to other neurons. The most common topology of the map is a 2-dimensional rectangular or hexagonal interconnection structure. The goal of SOM is to produce a low dimensional (usually 2-dimensional) projection of the data set while preserving the topological properties of the data set. This gives users a means to visualize low-dimensional views of high-dimensional data.

SOM executes a series of iterations each consisting of two phases: the competition and cooperation phases. At iteration t, during the competition phase, for each input data $\mathbf{X}(t) = [X_i(t), X_2(t), \ldots, X_d(t)]$ from the data set, the Euclidean distance between $\mathbf{X}(t)$ and each neuron's weight vector $\mathbf{W}_j(t) = [\mathbf{W}_{j,1}(t), \mathbf{W}_{j,2}(t), \ldots, \mathbf{W}_{j,d}(t)]$ $(1 \leqslant j \leqslant K$, where K is the total number of neurons in the grid) is computed to determine the neuron closest to $\mathbf{X}(t)$ as follows:

$$||\mathbf{X}(t) - \mathbf{W}_j(t)|| = \sqrt{\sum_{i=1}^{d}(X_i(t) - \mathbf{W}_{j,i}(t))^2} \qquad (1)$$

The neuron c with weight vector $\mathbf{W}_c(t)$ that has the minimum distance to the input data $\mathbf{X}(t)$ is called the winner neuron:

$$c = \underset{j}{\operatorname{argmin}} ||\mathbf{X}(t) - \mathbf{W}_j(t)||. \qquad (2)$$

During the cooperation phase, the weight vectors of the winner neuron and the neurons in the neighborhood $G(\mathbf{r}_c)$ of the winner neuron in the SOM grid are sheared towards the input data, where $\mathbf{r}_j$ is the physical position in the grid of the neuron j. The magnitude of the change decreases with time and is smaller for neurons that are physically far away from the the winner neuron. The function for change at iteration t can be defined as $Z(\mathbf{r}_j, \mathbf{r}_c, t)$.

The update expression for the winner neuron c and the neurons in neighborhood $G(\mathbf{r}_c)$ of the winner neuron is shown as follows:

$$\forall j \in G(\mathbf{r}_c), \ \mathbf{W}_j(t+1) = \mathbf{W}_j(t) + Z(\mathbf{r}_j, \mathbf{r}_c, t)[\mathbf{X}(t) - \mathbf{W}_j(t)]. \qquad (3)$$

3   Privacy-Preserving Self-Organizing Map

In this section, we introduce the protocol for privacy-preserving self-organizing map for vertically partitioned data involving two parties.

In vertically partitioned data, each input data point is partitioned into two portions. Each party holds one private portion. To preserve the data privacy of the participating parties, each weight vector is split into two private component vectors, each held by one party. The actual weight vector is the vector sum of the two component vectors.

Protocol 1. Privacy-Preserving Self-Organizing Map

Input: Input data $\mathbf{X} = (X_1, X_2, \ldots, X_d)$ from feature space.
Output: A grid with neurons $j = 1, 2, \ldots, K$.

1. Each neuron j is initialized with a weight vector $\mathbf{W}_j(0) = \mathbf{W}_j^A(0) + \mathbf{W}_j^B(0)$ for all $j = 1, 2, \ldots, K$ at step $t = 0$ where $\mathbf{W}_j^A(0)$ and $\mathbf{W}_j^B(0)$ are small component vectors that are securely and randomly generated by Party A and Party B respectively.
2. At each iteration t, randomly select an input data $\mathbf{X}(t)$.
3. Calculate the distance between input data and weight vector of neuron j $||\mathbf{X}(t) - \mathbf{W}_j(t)||$ for $1 \leqslant j \leqslant K$ and find the winning neuron c by running *Secure Computation of Closest Cluster Protocol*.
4. Update weight vectors of the winner neuron and all neurons in the neighborhood of the winner neuron based on *Secure Weight Vector Update Protocol*.
5. Two parties jointly update learning rate $h(t)$ and shrink neighborhood $\sigma(t)$.
6. Continue the training and repeat the steps 2-5 with iteration $t = t+1$. The training is executed until all $\mathbf{W}_j(t)$ change less than ϵ (a predefined small value). That is checked securely by *Secure Detection of Termination Protocol*.

To introduce the protocol formally, we describe the following notations for privacy-preserving self-organizing map, which are similar to those for k-means clustering [6].

In vertically partitioned data, for an input data $\mathbf{X}(t)$ at iteration t, Party A holds data portion $\mathbf{X}^A(t)$ for a subset of attributes, and Party B holds data portion $\mathbf{X}^B(t)$ for the remaining attributes. Without loss of generality, we assume that the values of the first s attributes of the input data $\mathbf{X}(t)$ belong to Party A; i.e., $\mathbf{X}^A(t) = [X_1^A(t), \ldots, X_s^A(t)]$). The values of the remaining $d - s$ attributes of the input data $\mathbf{X}(t)$ belong to Party B; i.e., $\mathbf{X}^B(t) = [X_{s+1}^B(t), \ldots, X_d^B(t)]$. We note that $\mathbf{X}^A(t) \cup \mathbf{X}^B(t) = \mathbf{X}(t)$ and $\mathbf{X}^A(t) \cap \mathbf{X}^B(t) = \emptyset$.

The weight vector $\mathbf{W}_j(t)$ of neuron j at iteration t is split into two private component vectors, each held by one party. Party A holds the component vector $\mathbf{W}_j^A(t)$ and Party B holds the component vector $\mathbf{W}_j^B(t)$. The weight vector $\mathbf{W}_j(t)$ for neuron j is the sum of two private component vectors: $\mathbf{W}_j(t) = \mathbf{W}_j^A(t) + \mathbf{W}_j^B(t)$.

The protocol for privacy-preserving self-organizing map is given in Protocol 1. As shown in Protocol 1, the difference between privacy-preserving self-organizing map and standard self-organizing map is that three subprotocols are required to perform some computations securely. The *Secure Computation of Closest Cluster Protocol* is used to compute the winner neuron without disclosing the private data of one party to the other party; the *Secure Weight Vector Update Protocol* is used to update new weight vectors securely; and the *Secure Detection of Termination Protocol* is used to securely determine the termination status of SOM. We note that although the sub-protocols are similar to those of [6], they are adapted here to implement privacy-preserving self-organizing map more efficiently and securely. These protocols are shown in Protocol 2, Protocol 3, and Protocol 4 respectively.

Protocol 2. Secure Computation of Closest Cluster Protocol

Input: The inputs of Party A are (i) data portion $\mathbf{X}^A(t)$ and (ii) weight component vector $\mathbf{W}_j^A(t)$. The inputs of party B (i) data portion $\mathbf{X}^B(t)$ and (ii) weight component vector $\mathbf{W}_j^B(t)$.
Output: The winner neuron $c = \text{argmin}_j \|\mathbf{X}(t) - \mathbf{W}_j(t)\|$.

1. **for** $j = 1$ to K **do**
2.   Party A and Party B jointly and securely compute the term $\sum_{m=1}^{s} \left((X_m^A(t) - \mathbf{W}_{j,m}^A(t)) \times \mathbf{W}_{j,m}^B(t) \right)$ by applying the secure scalar product protocols [2,4]. At the end of the protocol, Party A and Party B each holds a private component value a_j and b_j respectively, where $a_j + b_j = \sum_{m=1}^{s} \left((X_m^A(t) - \mathbf{W}_{j,m}^A(t)) \times \mathbf{W}_{j,m}^B(t) \right)$.
3.   Two parties jointly compute the term $\sum_{m=s+1}^{d} \left((X_m^B(t) - \mathbf{W}_{j,m}^B(t)) \times \mathbf{W}_{j,m}^A(t) \right)$ using the secure scalar product protocols [2,4]. At the end of the protocol, Party A and Party B each holds a private component value c_j and d_j respectively, where $c_j + d_j = \sum_{m=s+1}^{d} \left((X_m^B(t) - \mathbf{W}_{j,m}^B(t)) \times \mathbf{W}_{j,m}^A(t) \right)$.
4.   Party A separately computes $\alpha_j = P_A - 2 \times a_j - 2 \times c_j$ where $P_A = \sum_{m=1}^{s} (X_m^A(t) - \mathbf{W}_{j,m}^A(t))^2 + \sum_{m=s+1}^{d} (\mathbf{W}_{j,m}^A(t))^2$.
5.   Party B separately computes $\beta_j = P_B - 2 \times b_j - 2 \times d_j$ where $P_B = \sum_{m=s+1}^{d} (X_m^B(t) - \mathbf{W}_{j,m}^B(t))^2 + \sum_{m=1}^{s} (\mathbf{W}_{j,m}^B(t))^2$.
6. **end for**
7. Party A and Party B securely compute the index c such that $\alpha_c + \beta_c$ is the minimum among $\alpha_j + \beta_j$ for $1 \leqslant j \leqslant K$ based on secure comparison protocol.

In Protocol 2, two parties compute the index c of the winner neuron based on their private data portions $\mathbf{X}^A(t)$ and $\mathbf{X}^B(t)$, and their private component vectors $\mathbf{W}_j^A(t)$ and $\mathbf{W}_j^B(t)$ for $1 \leqslant j \leqslant K$. The square of distance between $\mathbf{X}(t)$ and $\mathbf{W}_j(t)$ is given:

$$\|\mathbf{X}(t) - \mathbf{W}_j(t)\|^2 \Leftrightarrow \|\mathbf{X}(t) - \mathbf{W}_j^A(t) - \mathbf{W}_j^B(t)\|^2$$

$$\Leftrightarrow \sum_{i=1}^{d} (X_i(t) - \mathbf{W}_{j,i}^A(t) - \mathbf{W}_{j,i}^B(t))^2 \Leftrightarrow \sum_{m=1}^{s} (X_m^A(t) - \mathbf{W}_{j,m}^A(t))^2$$

$$+ \sum_{m=s+1}^{d} (\mathbf{W}_{j,m}^A(t))^2 + \sum_{m=s+1}^{d} (X_m^B(t) - \mathbf{W}_{j,m}^B(t))^2 + \sum_{m=1}^{s} (\mathbf{W}_{j,m}^B(t))^2$$

$$-2 \times \sum_{m=1}^{s} \left((X_m^A(t) - \mathbf{W}_{j,m}^A(t)) \times \mathbf{W}_{j,m}^B(t) \right)$$

$$-2 \times \sum_{m=s+1}^{d} \left((X_m^B(t) - \mathbf{W}_{j,m}^B(t)) \times \mathbf{W}_{j,m}^A(t) \right) \tag{4}$$

The above shows that there are only two terms that are required to be jointly computed by two parties, which are $\sum_{m=1}^{s} \left((X_m^A(t) - \mathbf{W}_{j,m}^A(t)) \times \mathbf{W}_{j,m}^B(t) \right)$ and $\sum_{m=s+1}^{d} \left((X_m^B(t) - \mathbf{W}_{j,m}^B(t)) \times \mathbf{W}_{j,m}^A(t) \right)$. This can be computed using secure

---

Protocol 3. Secure Weight Vector Update Protocol

Input: The inputs of Party A are (i) data portion $\mathbf{X}^A(t)$ and (ii) weight component vector $\mathbf{W}_j^A(t)$. The inputs of party B (i) data portion $\mathbf{X}^B(t)$ and (ii) weight component vector $\mathbf{W}_j^B(t)$. The non-sensitive information $Z(r, r_c, t)$.
Output: The new weight component vectors $\mathbf{W}_j^A(t+1)$ and $\mathbf{W}_j^B(t+1)$ for Party A and Party B respectively.

1. **for** $j \in G(\mathbf{r}_c)$ **do**
2.   **for** $i = 1$ to d **do**
3.   **if** $\mathbf{X}_{j,i}(t)$ is held by Party A **then**
4. $\mathbf{W}_{j,i}^A(t+1) = \mathbf{W}_{j,i}^A(t) - Z(r, r_c, t) \times \left(\mathbf{X}_{j,i}(t) - \mathbf{W}_{j,i}^A(t)\right);$
                $\mathbf{W}_{j,i}^B(t+1) = \mathbf{W}_{j,i}^B(t) + Z(r, r_c, t) \times \mathbf{W}_{j,i}^B(t).$
5.   **else**
6. $\mathbf{W}_{j,i}^B(t+1) = \mathbf{W}_{j,i}^B(t) - Z(r, r_c, t) \times \left(\mathbf{X}_{j,i}(t) - \mathbf{W}_{j,i}^B(t)\right);$
                $\mathbf{W}_{j,i}^A(t+1) = \mathbf{W}_{j,i}^A(t) + Z(r, r_c, t) \times \mathbf{W}_{j,i}^A(t).$
7.   **end if**
8.   **end for**
9. **end for**

---

scalar product protocols [2,4] for two parties. At the end of the protocol, each party each holds one component value for each term without disclosing their inputs to the other. The square of distance can be securely computed as shown in Protocol 2.

In Protocol 3, each party updates the weight component vectors based on its own private data portions only. Hence, the update of the weight vectors separately by two parties is shown in Protocol 3.

In Protocol 4, two parties need to determine the termination status of SOM in Protocol 1 based on the changes to their private weight component vectors. If weight vector $\mathbf{W}_j(t)$ changes less than ϵ for $1 \leqslant j \leqslant K$, Protocol 1 will terminate. The difference between the new weight vector $\mathbf{W}_j(t+1)$ and the old weight vector $\mathbf{W}_j(t)$ is given as follows:

$$
\begin{aligned}
& ||\mathbf{W}_j(t+1) - \mathbf{W}_j(t)||^2 \\
\Leftrightarrow\ & || \left(\mathbf{W}_j^A(t+1) + \mathbf{W}_j^B(t+1)\right) - \left(\mathbf{W}_j^A(t) + \mathbf{W}_j^B(t)\right) ||^2 \\
\Leftrightarrow\ & \sum_{i=1}^{d} \left(\mathbf{W}_{j,i}^A(t+1) - \mathbf{W}_{j,i}^A(t)\right)^2 + \sum_{i=1}^{d} \left(\mathbf{W}_{j,i}^B(t+1) - \mathbf{W}_{j,i}^B(t)\right)^2 \\
& +2 \times \sum_{i=1}^{d} \left(\mathbf{W}_{j,i}^A(t+1) - \mathbf{W}_{j,i}^A(t)\right) \times \left(\mathbf{W}_{j,i}^B(t+1) - \mathbf{W}_{j,i}^B(t)\right) \quad (5)
\end{aligned}
$$

The only term $\sum_{i=1}^{d} \left(\mathbf{W}_{j,i}^A(t+1) - \mathbf{W}_{j,i}^A(t)\right) \times \left(\mathbf{W}_{j,i}^B(t+1) - \mathbf{W}_{j,i}^B(t)\right)$ that require to be securely computed by two parties can be done using existing secure scalar product protocols [2,4]. At the end of the protocol, two parties each holds a private component value. Protocol 4 shows how the two parties securely determine the termination status of SOM.

Protocol 4. Secure Detection of Termination Protocol

Input: (i) The old weight component vectors $\mathbf{W}_j^A(t)$ and $\mathbf{W}_j^B(t)$ and (ii) The new weight component vectors $\mathbf{W}_j^A(t+1)$ and $\mathbf{W}_j^B(t+1)$.

Output: True or False.

1. **for** $j = 1$ to K **do**
2.   Party A and Party B jointly compute the term $\sum_{i=1}^{d} \left(\mathbf{W}_{j,i}^A(t+1) - \mathbf{W}_{j,i}^A(t)\right) \times \left(\mathbf{W}_{j,i}^B(t+1) - \mathbf{W}_{j,i}^B(t)\right)$ by applying the secure scalar product protocols [2,4]. After execution of the protocol, Party A and Party B have a private component value e_j and f_j respectively where $e_j + f_j = \sum_{i=1}^{d} \left(\mathbf{W}_{j,i}^A(t+1) - \mathbf{W}_{j,i}^A(t)\right) \times \left(\mathbf{W}_{j,i}^B(t+1) - \mathbf{W}_{j,i}^B(t)\right)$.
3.   Party A separately computes $\mu_j = Q_A + 2 \times e_j$ where $Q_A = \sum_{i=1}^{d} \left(\mathbf{W}_{j,i}^A(t+1) - \mathbf{W}_{j,i}^A(t)\right)^2$.
4.   Party B separately computes $\nu_j = Q_B + 2 \times f_j$ where $Q_B = \sum_{i=1}^{d} \left(\mathbf{W}_{j,i}^B(t+1) - \mathbf{W}_{j,i}^B(t)\right)^2$.
5.   Party A and Party securely check the size of $\mu_j + \nu_j$ and ϵ (where ϵ is a specified value) based on secure comparison protocol [6].
6.   **if** $\mu_j + \nu_j \geq \epsilon$ **then return false**
7. **end for**
8. **return true**

---

4   Protocol Analysis

4.1   Correctness

Protocol 2: The square of distance between the input data $\mathbf{X}(t)$ and the weight vector $\mathbf{W}_j(t)$ is correctly computed by Protocol 2 as follows:

$$\alpha_j + \beta_j \Leftrightarrow (P_A - 2 \times a_j - 2 \times c_j) + (P_B - 2 \times b_j - 2 \times d_j)$$
$$\Leftrightarrow P_A + P_B - 2 \sum_{m=1}^{s} \left((X_m^A(t) - \mathbf{W}_{j,m}^A(t)) \times \mathbf{W}_{j,m}^B(t)\right)$$
$$-2 \sum_{m=s+1}^{d} \left((X_m^B(t) - \mathbf{W}_{j,m}^B(t)) \times \mathbf{W}_{j,m}^A(t)\right)$$

It is clear that the final expression above is equivalent to $||\mathbf{X}(t) - \mathbf{W}_j(t)||^2$ by definition of the square of distance between $\mathbf{X}(t)$ and $\mathbf{W}_j(t)$ in Eq. (4).

Protocol 3: The correctness of Protocol 3 is shown for the case where $\mathbf{X}_{j,i}(t)$ is held by Party A:

$$\left[\mathbf{W}_{j,i}^A(t) - Z(r, r_c, t) \times \left(\mathbf{X}_{j,i}(t) - \mathbf{W}_{j,i}^A(t)\right)\right] + \left[\mathbf{W}_{j,i}^B(t) - Z(r, r_c, t) \times \mathbf{W}_{j,i}^B(t)\right]$$
$$\Leftrightarrow \mathbf{W}_{j,i}^A(t) + \mathbf{W}_{j,i}^B(t) - Z(r, r_c, t) \times \left(\mathbf{X}_{j,i}(t) - \mathbf{W}_{j,i}^A(t) - \mathbf{W}_{j,i}^B(t)\right)$$
$$\Leftrightarrow \mathbf{W}_{j,i}(t) - Z(r, r_c, t) \times (\mathbf{X}_{j,i}(t) - \mathbf{W}_{j,i}(t)) \Leftrightarrow \mathbf{W}_{j,i}(t+1)$$

If $\mathbf{X}_{j,i}(t)$ is held by Party B, the correctness can be shown in a similar manner.

Protocol 4: The correctness of Protocol 4 is shown as follows:

$$\mu_j + \nu_j \Leftrightarrow Q_A + 2 \times e_j + Q_B + 2 \times f_j \Leftrightarrow Q_A + Q_B + 2 \times (e_j + f_j)$$

$$\Leftrightarrow \sum_{i=1}^{d} \left(\mathbf{W}_{j,i}^{A}(t+1) - \mathbf{W}_{j,i}^{A}(t) \right)^2 + \sum_{i=1}^{d} \left(\mathbf{W}_{j,i}^{B}(t+1) - \mathbf{W}_{j,i}^{B}(t) \right)^2$$

$$+2 \times \sum_{i=1}^{d} \left(\mathbf{W}_{j,i}^{A}(t+1) - \mathbf{W}_{j,i}^{A}(t) \right) \times \left(\mathbf{W}_{j,i}^{B}(t+1) - \mathbf{W}_{j,i}^{B}(t) \right)$$

It is clear that the final expression above is equivalent to $||(\mathbf{W}_j^A(t+1) + \mathbf{W}_j^B(t+1) - \mathbf{W}_j^A(t) + \mathbf{W}_j^B(t))||^2$ by definition of difference between $\mathbf{W}_j(t)$ and $\mathbf{W}_j(t+1)$ in Eq. (5).

Protocol 1: It is clear that the protocol is correct as it uses Protocol 2, Protocol 3, and Protocol 4, which have been shown above to be correct.

4.2   Complexity Analysis

We assume the computational complexity and communication cost of secure scalar product protocols used in Protocol 2 and Protocol 4 are defined $O(\phi(z))$ and $O(\phi'(z))$ respectively, where (1) z is the number of elements in vectors; and (2) $\phi(z)$ and $\phi'(z)$ are an expression for computational complexity and communication cost respectively of z with respect to some secure scalar product protocol applied.

Protocol 2: From Step 1 to Step 6, K iterations are performed. At each iteration, secure scalar product protocols are invoked twice for vectors of length d (Step 2) and vectors of length $d - s$ (Step 3). Hence, the overall computational complexity and communication cost of Step 1 to Step 6 are $K \times O(\phi(s)) + O(\phi(d-s)) = O(K \times \phi(d))$ and $K \times \left(O(\phi'(s)) + O(\phi'(d-s)) \right) = O(K \times \phi'(d))$ respectively. Step 7 invokes the secure comparison protocol K times to find the minimum values. The computational complexity and communication cost are both $O(K)$.

Overall, the computational complexity and communication cost of Protocol 2 are $O(K \times \phi(d))$ and $O(K \times \phi'(d))$ respectively.

Protocol 3: In Protocol 3, two parties update their weight component vectors based on the private attributes of input vector held by them. They do not need to communicate with each other. Hence, the computational complexity is $O(d \times K') = O(d \times K)$, where K' is the number of neurons in its neighborhood.

Protocol 4: From Step 1 to Step 9, K iterations are performed. At each iteration, secure scalar product protocols are invoked once (Step 2) between two vectors of length d. The overall computational complexity and communication cost are $K \times O(\phi(d))$ and $K \times O(\phi'(d))$ respectively. Step 5 invokes the circuit evaluation protocol once to compare the size of two values. It can be done efficiently as shown above with the computational complexity and communication cost are both $O(K)$.

Hence, the overall computational complexity and communication cost of Protocol 4 are $O(K \times \phi(d))$ and $O(K \times \phi'(d))$ respectively.

Protocol 1: Protocol 1 invokes Protocol 2, Protocol 3 and Protocol 4 once at each iteration. Hence, the overall computational complexity and communication cost of Protocol 1 are $O(T \times K \times \phi(d))$ and $O(T \times K \times \phi'(d))$ respectively where T is the total number of iterations before the termination condition is satisfied.

4.3   Privacy Preservation

Protocol 2: Protocol 2 securely computes the index of the winner neuron for a given input vector.

The distance between the given input vector to each weight vector is computed by secure scalar product protocols. At the end of the protocols, each parties holds a private component value and one cannot obtain extra information about vectors held by the other party. Finally, the winner neuron is computed securely by the secure comparison protocol. The only information revealed is the winner neuron index at the end of the protocol. We note that the only extra information obtained from the other party does not violate the data privacy of the other party. Hence, Protocol 2 is privacy-preserving.

Protocol 3: In protocol 3, two parties update their weight component values based only on their private attributes of a given input vector. One party obtains nothing from the other party without communicating with others. Hence, Protocol 3 is secure.

Protocol 4: The privacy is preserved as the secure scalar product protocol and secure comparison protocol used are privacy-preserving. The only information that one party obtains at the end of the protocol is the information of whether Protocol 1 can be terminated. Hence, Protocol 2 is privacy-preserving.

Protocol 1: The proposed protocol for privacy-preserving self-organizing map is an iterative protocol. During the execution of the entire Protocol 1, the only intermediate results that Party A and Party B learn are the indices of the winner neurons at different iterations. Party A and Party B each holds a weight component vector of weight vectors. So the dishonest party is not able to know the honest party's data using incomplete information about weight vectors. Hence, the proposed protocol for privacy-preserving self-organizing map does not leak any extra information about the honest party's data.

5   Conclusions

Privacy-preserving data mining seeks to allow the cooperative execution of data mining algorithms while preserving the data privacy of each party concerned. In this paper, we proposed a protocol for privacy-preserving self-organizing map for vertically partitioned data involving two parties. The challenges in preserving data privacy in SOM are (1) to securely discover the winner neuron from data privately held by two parties; (2) to securely update weight vectors of neurons; and (3) to securely determine the termination status of SOM. We proposed protocols to address the above challenges. We proved that these protocols are correct and privacy-preserving. Also, we proved that the intermediate results generated by these protocols do not violate the data privacy of the participating parties.

References

1. Agrawal, R., Srikant, R.: Privacy-preserving data mining. In: SIMGOD, Dallas, Texas, United States, pp. 439–450 (2000)
2. Du, W., Atallah, M.J.: Privacy-preserving cooperative statistical analysis. In: Proceedings of the 17th Annual Computer Security Applications Conference, New Orleans, Louisiana, USA, December 10–14, 2001, pp. 102–110 (2001)
3. Du, W., Zhan, Z.: Building decision tree classifier on private data. In: Proceedings of the IEEE International Conference on Privacy, Security and Data Mining, Maebashi City, Japan, 2002, pp. 1–8 (2002)
4. Goethals, B., Laur, S., Lipmaa, H., Mielikainen, T.: On private scalar product computation for privacy-preserving data mining. In: Proceedings of the 7th Annual International Conference in Information Security and Cryptology, Seoul, Korea, December 2–3, 2004, pp. 104–120 (2004)
5. Han, S., Ng, W.K.: Privacy-preserving genetic algorithms for rule discovery. In: SIGKDD, Regensburg, Germany (2007)
6. Jagannathan, G., Wright, R.N.: Privacy-preserving distributed k-means clustering over arbitrarily partitioned data. In: SIGKDD, pp. 593–599 (2005)
7. Kantarcioglu, M., Clifton, C.: Privacy preserving naive bayes classifier for horizontally partitioned data. In: IEEE ICDM Workshop on Privacy Preserving Data Mining, November 2003, Melbourne, FL (2003)
8. Kantarcioglu, M., Clifton, C.: Preserving data mining of association rules on horizontally partitioned data. TKDE 16(9), 1026–1037 (2004)
9. Kohonen, T.: Self-Organizing Maps. Springer, Heidelberg (1995)
10. Lindell, Y., Pinkas, B.: Privacy preserving data mining. In: Bellare, M. (ed.) CRYPTO 2000. LNCS, vol. 1880, pp. 36–53. Springer, Heidelberg (2000)
11. Vaidya, J., Clifton, C.: Privacy preserving association rule mining in vertically partitioned data. In: SIGKDD, July 23-26, 2002, pp. 639–644 (2002)
12. Vaidya, J., Clifton, C.: Privacy preserving naïve bayes classifier for vertically partitioned data. In: Proceedings of the SIAM International Conference on Data Mining, 2004, Lake Buena Vista, Florida, pp. 522–526 (2004)
13. Vaidya, J., Clifton, C.: Privacy-preserving decision trees over vertically partitioned data. In: Proceedings of the 19th Annual IFIP WG 11.3 Working Conference on Data and Applications Security, Storrs, Connecticut (2005)
14. Wan, L., Ng, W.K., Han, S., Lee, V.C.S.: Privacy-preservation for gradient descent methods. In: SIGKDD, San Jose, California, USA (2007)
15. Wright, R., Yang, Z.: Privacy-preserving bayesian network structure computation on distributed heterogeneous data. In: SIGKDD, 2004, pp. 713–718 (2004)
16. Yao, A.C.: How to generate and exchange secrets. In: Proceedings of the Annual IEEE Symposium on Foundations of Computer Science, pp. 162–167. IEEE Computer Society Press, Los Alamitos (1986)

DWFIST: Leveraging Calendar-Based Pattern Mining in Data Streams

Rodrigo Salvador Monteiro[1,2], Geraldo Zimbrão[1,3], Holger Schwarz[2],
Bernhard Mitschang[2], and Jano Moreira de Souza[1,3]

[1] Computer Science Department, Graduate School of Engineering, Federal University
of Rio de Janeiro, PO Box 68511, ZIP code: 21945-970, Rio de Janeiro, Brazil
{salvador,zimbrao,jano}@cos.ufrj.br
[2] Institute f. Parallel & Distributed Systems, University of Stuttgart,
Universitaetsstr. 38, 70569 Stuttgart, Germany
{Holger.Schwarz,Bernhard.Mitschang}@informatik.uni-stuttgart.de
[3] Computer Science Department, Institute of Mathematics, UFRJ, Rio de Janeiro, Brazil

Abstract. Calendar-based pattern mining aims at identifying patterns on specific calendar partitions. Potential calendar partitions are for example: every Monday, every first working day of each month, every holiday. Providing flexible mining capabilities for calendar-based partitions is especially challenging in a data stream scenario. The calendar partitions of interest are not known a priori and at each point in time only a subset of the detailed data is available. We show how a data warehouse approach can be applied to this problem. The data warehouse that keeps track of frequent itemsets holding on different partitions of the original stream has low storage requirements. Nevertheless, it allows to derive sets of patterns that are complete and precise. This work demonstrates the effectiveness of our approach by a series of experiments.

1 Introduction

Calendar-based schemas [9,2] were proposed as a semantically rich representation of time intervals and used to mine temporal association rules. An example of a calendar schema is (year, month, day, day_period), which defines a set of calendar patterns, such as *every morning of January of 1999* (1999, January, *, morning) or *every 16th of January of every year* (*, January, 16, *). In the research field of data mining, frequent itemsets derived from transactional data represent a particularly important pattern domain. Association rule mining is the most recognized application of frequent itemsets [1]. Other examples are generalized rule mining [12] and associative classification [10]. The combination of the rich semantics of calendar-based schemas with frequent itemset mining, namely calendar-based frequent itemset mining, corresponds to the first step of various calendar-based pattern mining tasks, e.g., calendar-based association rules. An example of calendar-based association rules provided in [9] is that eggs and coffee are frequently sold together in morning hours. Considering the transactions at the all-day granule would probably not reveal such a rule and its implicit knowledge.

Network traffic analysis, web click stream mining, power consumption measurement, sensor network data analysis, and dynamic tracing of stock fluctuation are

I.Y. Song, J. Eder, and T.M. Nguyen (Eds.): DaWaK 2007, LNCS 4654, pp. 438–448, 2007.

examples for applications that produce data streams. A data stream is continuous and potentially infinite. Mining calendar-based frequent itemsets in data streams is a difficult task that is described as follows:

Problem Statement: Let D be a transactional dataset provided by a data stream. Let X be a set of calendar-based constraints and T the subset of transactions from D satisfying X. The *frequency* of an itemset I over T is the number of transactions in T in which I occurs. The *support* of I is the *frequency* divided by the total number of transactions in T. Given a minimum support σ, the set of *calendar-based frequent itemsets* is defined by the itemsets with *support* $\geq \sigma$ over T.

Some examples of calendar-based constraints are: weekday in {Monday, Friday}; day_period = "Morning"; holiday = "yes". The main issues for mining calendar-based frequent itemsets are that the calendar partitions that reveal interesting temporal patterns are not known a priori and that at each point in time only a subset of the detailed data is available. Existing approaches cannot solve this problem because either they require all transactions to be available during the calendar-based mining task or they do not provide enough flexibility to consider a calendar-based subset of the data stream. In order to flexibly derive calendar-based patterns, we need some kind of summary for previous time windows. As the calendar partitions that will be interesting for analysis are not known in advance, it is not obvious how to build and store such a summary.

DWFIST is an approach that keeps track of frequent itemsets holding on disjoint sets of the original transactions [13,8]. It refers to each disjoint set as a *partition*. One partition may represent a period of one hour for example. A data warehouse stores all frequent itemsets per partition. A temporal dimension represents the calendar features, such as year, month, holiday, weekday, and many others. One may freely combine the partitions in order to retrieve the frequent itemsets holding on any set of partitions.

DWFIST does not comprise new algorithms for mining frequent itemsets on data streams. Instead, it builds on existing algorithms to perform regular frequent itemset mining on small partitions of stream data, and provides means to flexibly combine the frequent itemsets of different partitions. As a database-centered approach, DWFIST benefits from database index structures, query optimization, storage management facilities and so on. These are vital features because the data warehouse of frequent itemsets potentially comprises a large, but manageable, volume of data. In addition, a data warehouse of frequent itemsets can be implemented using commercial databases.

The main contributions of this paper are as follows: (1) show how to apply DWFIST in a data stream scenario, leveraging calendar-based pattern mining capabilities; (2) show how to cope with tight time constraints imposed by the data stream scenario; (3) show that the storage requirements of the data warehouse are kept at a manageable level; (4) discuss completeness and precision of retrieved sets of frequent itemsets; and (5) present experimental results related to these issues.

The remainder of the paper is organized as follows. Section 2 lists related work. The DWFIST approach is presented in Section 3. In Section 4, we discuss time constraints and storage requirements, whereas we address completeness and precision issues in Section 5. The experimental results are presented in Section 6. Finally, Section 7 concludes the paper and presents some future work.

2 Related Work

In order to discover temporal association rules, the work in [14] states very clearly the problem of omitting the time dimension. It is assumed that the transactions in the database are time stamped and that the user specifies a time interval to divide the data into disjoint segments, such as months, weeks, days, etc. Cyclic association rules are defined as association rules with a minimum confidence and support at specific regular time intervals. A disadvantage of the cyclic approach is that it does not deal with multiple granularities of time intervals. An example of a calendar pattern that cannot be represented by cycles is the simple concept of the first working day of every month. [2] introduces a calendar algebra, which basically defines a set of time intervals. Each time interval is a set of time units, e.g. days. A calendric rule is a rule that has the minimum support and confidence over every time unit in the calendar. Calendar schemas, as the example given in the introduction, are used in [9] to specify the search space for different possible calendars. The rules mined by such algorithms are presented together with their mined calendars. All these approaches perform a calendar-based frequent itemset mining step. However, they require all transactions to be available during the calendar-based mining task, which disables these approaches from being applied to data streams. Thus, they cannot be used to solve our problem.

The focus in data stream mining has been on stream data classification and stream clustering. Only recently, mining frequent counts in streams gained attention. An algorithm to find frequent items using a variant of the classic majority algorithm was developed simultaneously by [3] and [7]. Frameworks to compute frequent items and itemsets were provided in [11,5]. The presented algorithms for mining frequent patterns in data streams, called Lossy Counting and MFIS, assume that patterns are measured from the start of the stream up to the current moment. They always consider the whole stream and do not provide the flexibility to mine the frequent itemsets holding on a subset of the stream. The work presented in [4] is closer to our approach. It proposes FP-Stream, a new model for mining frequent patterns from data streams. This model is capable of answering user queries considering multiple time granularities. A fine granularity is important for recent changes whereas a coarse granularity is adequate for long-term changes. FP-Stream supports this kind of analysis by a tilted-time window, which keeps storage requirements very low but prevents calendar-based pattern analysis. An example of a tilted-time window (in minutes) is 15,15,30,60,120,240,480,etc. It is possible to answer queries about the last 15 minutes or the last 4 hours (15+15+30+60+120 minutes), but it is not possible to answer queries about *last Friday* or *every morning*, for example. The lack of a uniform partitioning, widely used in multimedia data mining, prevents calendar-based pattern analysis.

3 The DWFIST Approach

The acronym DWFIST stands for Data Warehouse of Frequent ItemSets Tactics. Figure 1 presents the components of the DWFIST approach [13,8].

The *pre-processing and loading step* is composed of three tasks: gather the transactions into disjoint sets (partitions); mine the frequent itemsets holding on a partition using a pre-defined mining minimum support; and load the mined frequent itemsets into the data warehouse. The *Data Warehouse of Frequent Itemsets* (referenced throughout the paper simply as DW) is the main component of the approach. Its task is to store and organize the frequent itemsets into partitions. The role of the *Basic Frequent Itemset Retrieval Capabilities* component is to retrieve a set of frequent itemsets holding on a user-specified set of DW partitions with a user-defined minimum support. The *Frequent Itemset Based Pattern Mining Engine* generates patterns that can be obtained from frequent itemsets. The *Advanced Analytical Tools* component comprises analysis and exploration tools built on top of the other components.

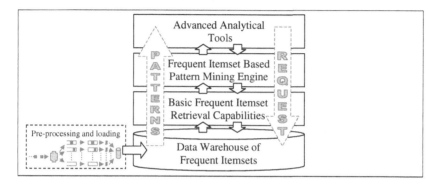

Fig. 1. Components of the DWFIST approach

One of the goals of the DWFIST approach is to provide flexible pattern retrieval capabilities without requiring access to the detailed original data. This is the key feature for leveraging calendar-based pattern mining on data streams. In this paper, we focus on this issue. Hence, we do no further discuss the Frequent Itemset Based Pattern Mining Engine and Advanced Analytical Tools components.

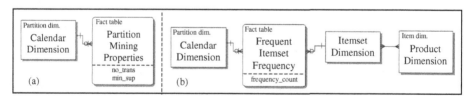

Fig. 2. (a) Partition Mining Properties and (b) Frequent Itemset Count Dimensional Schemas

In Figure 2, we present one sample instance of a DW of frequent itemsets. It has one partition dimension and one item dimension. Partition dimensions organize the space of the original transactions into disjoint sets, which we call DW partitions. In this example, we use a calendar dimension with a granularity of 1 hour. This means that each tuple in the calendar dimension represents a period of 1 hour, e.g., "02/15/2007 [08:00AM, 09:00AM)", "02/15/2007 [09:00AM, 10:00AM)", and so on. Of course, additional calendar information has to be stored, e.g., the period of the day (morning, afternoon, ...), weekday, holiday (yes, no) and any other calendar unit that represents an aggregation of the basic granularity. This basic granularity also sets the criteria for gathering the transactions in the pre-processing step. The item dimensions provide additional information about the items that appear in transactions. In our sample instance, we use a single product dimension. Examples of additional information are: description, department, category, etc.

A data warehouse of frequent itemsets is composed of two fact tables. The first one, presented in Figure 2(a), stores some partition properties and thus has relationships only with partition dimensions. It stores the number of transactions and the minimum support used for the frequent itemset mining task per calendar granule (1 hour). The second fact table, presented in Figure 2(b), stores the frequency count for each set of products mined as frequent in a specific calendar granule (1 hour). Figure 2(b) also presents an itemset dimension that works as a bridge table between the multivalued dimension and the fact table. This sample data warehouse of frequent itemsets tracks the frequent sets of products per hour. By such a minimum granularity, it is not possible to analyze time intervals shorter than one hour. On the other hand, one may analyze all possible combinations of one-hour periods.

4   Pre-processing and Loading in DWFIST

In this section, we discuss how the pre-processing and loading steps can cope with tight time constraints and how we keep storage requirements at a manageable level.

As far as we are able to pre-process and load the data of one DW partition before the next one is available, we cope with the time constraints that result from the data stream scenario. In our example, we defined the minimum granularity of analysis as 1 hour. Hence, each DW partition covers data for one hour and a one-hour window is available for pre-processing and load. The first step in pre-processing is to gather the transactions into DW partitions. This is straightforward because the partitioning is usually based on a timestamp. The second step is to mine frequent itemsets on the gathered partition using a pre-defined mining minimum support. Finally, we load the frequent itemsets into the DW. While we gather the transactions of the current partition, we can mine and load the previous partition in parallel.

Many frequent itemset mining algorithms have been proposed. In our mining pre-processing step, we may apply any efficient algorithm. As an example, [4] uses the FP-Growth algorithm [6] to mine frequent itemsets on batches of transactions from a data stream. The efficiency of this step depends mainly on the mining minimum support and the number of transaction. The load step mainly comprises the insertion of new frequent itemset frequency counts and the creation of new dimension tuples. An important property of the load step is that the time required is proportional to the number of frequent itemsets to be loaded instead of the number of transactions taking place in the minimum granularity of analysis. For a fixed minimum support, an increased number of transactions used in a frequent itemset mining task does not lead to a proportional increase in the number of frequent itemsets. In most cases, the number of frequent itemsets tends to stabilize or even decrease.

The definition of the minimum granularity of analysis plays a central role for coping with time constraints. First, it defines the time window available for processing. Secondly, an increased time window makes the task of the mining step harder and the task of the load step easier. The task of the mining step gets more expensive because a bigger window means a potentially higher number of transactions. Fortunately, current frequent itemset mining algorithms are able to process a considerable large number of transactions efficiently. Therefore, an increase in the time window is likely to bring more benefit for the load step than losses for the mining step. Our experimental results (see Fig. 4(a) in Section 6) show, that pre-processing is likely to be dominated by the load step. Anyway, in extreme scenarios, we can additionally apply the framework presented in [4], called FP-Stream, to perform the mining task. This framework is capable of answering queries such as "retrieve the frequent itemsets holding on the transactions of the last 15 minutes" or "of the last 4 hours" on a stream. Also, the required processing time is independent from the period being queried. This means that using this framework it is possible to increase the time window without increasing the time required for mining. The disadvantage of applying this framework is that it introduces an error in the frequency counts. We have to consider this additional error when computing the frequency upper bound (see Section 5) during the DW retrieval.

When it comes to storage requirement issues, an important assertion is that the size of DW increases in a lower rate than the size of the considered input stream. The reasoning that supports this assertion is two-fold. First, the frequent itemsets require less storage space than the original data stream transactions. Secondly, the reuse of information stored in the dimensions reduces the storage requirements.

Nevertheless, it is important to make a remark at this point. Dense correlated databases may produce a large number of frequent itemsets. An extensive amount of work has been done on condensed representations of frequent itemsets aiming to represent a set of frequent itemsets in a compact way. The DWFIST approach allows to use such condensed representations to describe the frequent itemsets holding on each partition. In this work, we do not apply any condensed representation because we understand that using regular frequent itemsets imposes a more difficult scenario for streams, as we have to deal with a larger amount of data. In this sense, coping with stream requirements using regular frequent itemsets suggests that we can achieve at least the same with condensed representations.

A similar discussion on adjusting the minimum granularity of analysis applies to storage requirements issues. In an analogous way, the storage requirements are proportional to the number of frequent itemsets being stored.

5 Retrieving Calendar-Based Frequent Itemsets from the DW

The Calendar Dimension provides attributes representing different calendar features. In our example, an attribute *day_period* in the calendar dimension may classify the periods of 1 hour into morning, afternoon, evening and dawn. Let us discuss how to perform the frequent itemsets retrieval task for the morning period. Considering our sample data warehouse, the Disjoint Partitions Property [8] tells us that we can sum the frequencies of a specific itemset over any set of partitions. This sum gives the frequency lower bound. The frequency upper bound for each itemset is computed using the Error Upper Bound Property [8]. The error upper bound is based on the set of partitions where the frequency of the itemset is unknown. In the relational paradigm, it is more efficient to compute the error upper bound related to the partitions where the itemset frequency is known and subtract it from the error upper bound related to the whole set of partitions being queried. Figure 3 presents a query to perform the

```
Select S.itemset_id, S.LB_Frequency, (S.LB_Frequency+G.Global_Error-S.Known_Part_Error) as UB_Frequency
From  ( Select FMF.itemset_id, sum(FMF.frequency_count) as LB_Frequency,
               sum(PMP.no_trans*PMP.min_sup) as Known_Part_Error,
        From  Calendar_Dimension CD, Frequent_Itemset_Frequency FMF, Partition_Mining_Properties PMP
        Where CD.CD_id = PMP.CD_id and CD.CD_id = FMF.CD_id and CD.day_period = 'Morning'
        Group by FMF.itemset_id) S,
      ( Select sum(PMP.no_trans) as Total_no_trans, sum(PMP.no_trans*PMP.min_sup) as Global_Error
        From  Calendar_Dimension CD, Partition_Mining_Properties PMP
        Where CD.CD_id = PMP.CD_id and CD.day_period = 'Morning') G
Where ( S.LB_Frequency+G.Global_Error-S.Known_Part_Error) >= (G.Total_no_trans*:query_minimum_support)
```

Fig. 3. Frequent itemset retrieval for the morning period

frequent itemset retrieval task for the morning period in our example using a query minimum support provided by the user.

As the retrieved itemset frequencies are represented by intervals, we can only discard an itemset when its frequency interval lies completely below the specified threshold. The completeness of the result can be guaranteed as far as the query minimum support (σ_q) is equal or greater than the mining minimum support (σ_m) applied in the pre-processing step [8]. Furthermore, the precision of the result is measured by the ratio between the error and the approximate frequency retrieved from the DW. We call this ratio frequency relative error. A worst-case upper bound on the frequency relative error can be computed as follows (see [8] for a proof):

$$\text{Worst\_case\_FRE} = \frac{\sigma_m}{2\left(\sigma_q - 0.5\sigma_m\right)}$$

Some important questions arise immediately: (1) Which range of query minimum support can be reasonably supported by the DW as it does not contain complete information about the original transactions, and (2) How close is the answer provided by the DW to the real set of frequent itemsets holding on the original transactions?. We address these issues in our experimental evaluation.

6 Experimental Results

We performed a series of experiments related to pre-processing and load time, storage requirements of DW, and the precision of the sets of frequent itemsets retrieved from DW. All experiments were performed on a PC Pentium IV 1.2 GHz with 768 MB of RAM. We used Oracle 10g to implement a data warehouse identical to our example.

The stream data was created using the IBM synthetic market-basket data generator [15]. Two data streams were generated both representing a period of one week. The first data stream (Stream1) has 60 million transactions, 1000 distinct items and an average of 7 items per transaction. The transactions of Stream1 were uniformly distributed over 168 one-hour partitions (1 week). This data stream does not present strong calendar pattern behavior. We created the second data stream (Stream2) in two steps. First, we built a dataset with 50 million transactions, 1000 distinct items and an average of 7 items per transaction. Secondly, we built another dataset with 10 million transactions, 1500 distinct items and an average of 7 items per transaction. The 50M-transactions dataset was uniformly distributed over 140 one-hour partitions and the 10M-transactions dataset over 28 one-hour partitions. Finally, the 28 one-hour partitions of the 10M-transactions dataset were associated to the morning period (08:00 to 12:00) over the weekdays as well as to the period from 08:00 to 17:00 excluding lunch (12:00 to 13:00) on the Saturday. The 140 one-hour partitions of the 50M-transactions were associated to the remaining slots completing a week. Hence, Stream2 presents strong calendar behavior.

Figure 4(a) presents the pre-processing and load times. The Y-axis represent the percentage of the one-hour window that was required for processing whereas the X-axis represent the one-hour partitions. We used a mining minimum support of 0.1% for all one-hour partitions. As this task took less than one minute for all partitions, it appears only as a thin layer at the bottom of the charts. Figure 4(a) shows that for most of the partitions pre-processing and loading is completed in less than 30% of the available time window. Even for the rare partitions with peek processing time, pre-processing is completed within the one-hour window.

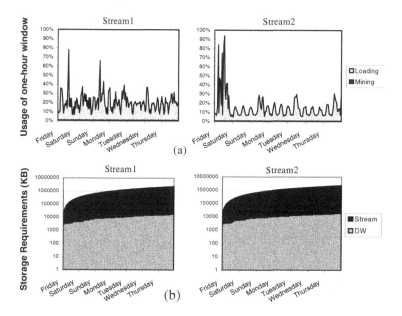

Fig. 4. (a) Pre-processing and load time (b) Storage Requirements

The detailed analysis of the loading step reveals that it is mainly influenced by three factors: the number of itemsets being loaded, the number of new itemsets being created, and the need for extending the physical storage area. The high processing peeks are related to periods where all three factors presented a negative influence. It is not likely that all three factors will negatively influence the load processing of several consecutive periods. For example, consecutive periods normally should not require an extension of the physical storage area. If they happen to do, it is most likely that the size of the extent is not properly set. Hence, it should be possible to compensate some processing peeks out of the window within a few consecutive periods. This would lead to a less conservative definition of the minimum granularity of analysis. Of course, we must also consider an increase in the storage requirements due to the need of a buffer to keep the itemsets with a delayed load.

Figure 4(b) presents the storage requirements comparing the sizes of the streams with the corresponding DW. It is clear that the storage requirements for the streams quickly reach unmanageable levels. For our examples, only one week already represents 2 GB. The size of the DW is orders of magnitude lower and increases at a lower rate than the size of the source data stream. For both examples, the DW size did not reach 20 MB including indexing structures.

Further experiments cover measurements related to the precision of the set of frequent itemsets retrieved from the DW. We retrieved frequent itemsets for three different calendars: (i) whole Saturday (00:00 to 24:00); (ii) Monday, Wednesday and Friday from 08:00 to 17:00; and (iii) morning periods of weekdays (08:00 to 12:00). These calendars are related to the strong calendar behavior introduced in Stream2 in different ways. "Saturday" comprises 24 partitions from which 8 come from the 10M-transactions dataset. "Monday, Wednesday and Friday from 08:00 to 17:00" presents 12 out of 27. "Morning periods of weekdays" is related only to partitions from the 10M-transactions dataset. They were defined in order to measure in Stream2 the effect of a strong calendar behavior on the precision of the result. We also executed the same queries on Stream1, which does not present a calendar behavior.

We report the precision of the retrieved set of frequent itemsets according to two aspects: the number of false positives and the frequency relative error. In order to obtain these measurements, we performed a regular frequent itemset mining task for each query directly on the corresponding stream transactions. The answer retrieved from the DW was compared with the exact answer.

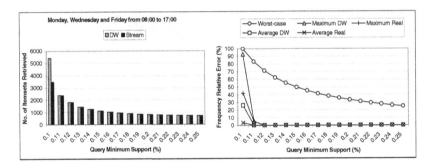

Fig. 5. Precision measurements on Stream1

As the results for the three queries on Stream1 presented the same behavior, we present only the results for one of them in Figure 5. The difference between the two bars on the left graph corresponds to the number of false positives. Note that with a query minimum support slightly greater than the mining minimum support (0.1%) it is already possible to retrieve the exact set of frequent itemsets. The graph on the right hand side of Figure 5 presents the frequency relative error measurements. The Worst-case corresponds to the value obtained applying equation (1). The Maximum DW and Average DW are computed calculating the error as the difference between the approximate frequency and the frequency upper bound thus corresponding to the tightness of the frequency interval. The Maximum Real and Average Real use the error as the difference between the approximate frequency and the exact real frequency retrieve directly from the stream. The error drops quickly to zero for query minimum support values slightly greater than the mining minimum support.

Figure 6 presents results for the three queries executed on Stream2. Again, for values of the query minimum support slightly greater than the mining minimum support the result presents a good precision. With a query minimum support of 0.12%, the Average DW frequency relative error already drops below 10%. It is especially important to observe the low values of the Average DW. As it does not require knowledge about the exact real frequency, it represents the average precision guarantee provided to a user. Furthermore, the query for the morning period of weekdays presents the same behavior as the queries executed on Stream1.

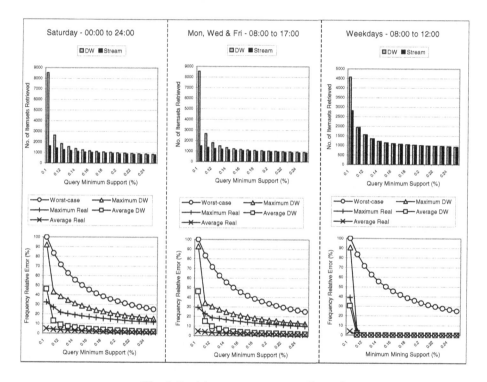

Fig. 6. Precision measurements on Stream2

7 Conclusions and Future Work

In this paper, we presented how to leverage calendar-based pattern mining in data streams using the DWFIST approach. Our experimental evaluation reveals that this approach copes with the tight time constraints imposed by data streams and keeps storage requirements at a manageable level. The minimum granularity of analysis, DW parameters (mining minimum support) and DBMS parameters (extent size) provide ways of tuning the performance and storage requirements. We can guarantee the completeness of the set of frequent itemsets retrieved from the DW and we provided a theoretical worst-case bound for the relative error of the itemset frequency retrieved from the DW. Furthermore, we measured the precision of different sets of frequent itemsets retrieved from the DW. The number of false positives decreases rapidly as the query minimum support moves away from the mining minimum support. Moreover, the tightness of the frequency interval presented an average behavior much better than the theoretical worst-case.

As future work, we intend to evaluate the use of condensed representations of frequent itemsets. Furthermore, the development of algorithms in the Frequent Itemset Based Pattern Mining Engine component of the DWFIST approach to derive other calendar-based patterns, such as calendar-based association rules, exploring the calendar features of the DW is an interesting issue.

References

1. Agrawal, R., Imielinski, T., Swami, A.: Mining Association Rules between Sets of Items in Large Databases. In: Proc. ACM SIGMOD, pp. 207–216. ACM Press, New York (1993)
2. Ramaswamy, S., Mahajan, S., Silberschatz, A.: On the discovery of interesting patterns in association rules. In: Proc. of the VLDB Conf, pp. 368–379 (1998)
3. Demaine, E.D., L´opez-Ortiz, A., Munro, J.I.: Frequency estimation of internet packet streams with limited space. In: Möhring, R.H., Raman, R. (eds.) ESA 2002. LNCS, vol. 2461, Springer, Heidelberg (2002)
4. Giannella, C., Han, J., Pei, J., Yan, X., Yu, P.S: Mining Frequent Patterns in Data Streams at Multiple Time Granularities. In: Kargupta, H., et al. (eds.) Data Mining: Next Generation Challenges and Future Directions, AAAI/MIT Press (2003)
5. Xie, Z.-j., Chen, H., Li, C.: MFIS-Mining Frequent Itemsets on Data Streams. In: Proceeding of the Advanced Data Mining and Applications, pp. 1085–1093 (2006)
6. Han, J., Pei, J., Yin, Y.: Mining frequent patterns without candidate generation. In: Proceeding of the 2000 SIGMOD Conference, Dallas, Texas, May 2000, pp. 1–12 (2000)
7. Karp, R.M., Papadimitriou, C.H., Shenker, S.: A simple algorithm for finding frequent elements in streams and bags. ACM Trans. Database Systems (2003)
8. Monteiro, R.S., Zimbrão, G., Schwarz, H., Mitschang, B., Souza, J.M: DWFIST: The Data Warehouse of Frequent Itemsets Tactics Approach. In: Darmont, J., Boussaid, O.(eds.) Processing and Managing Complex Data for Decision Support, pp. 185–214. Idea Group Publishing (2006)
9. Li, Y., Ning, P., Wang, X.S., Jajodia, S.: Discovering calendar-based temporal association rules. In: Proc. Int. Symp. Temp. Representation and Reasoning, pp. 111–118 (2001)
10. Liu, B., Hsu, W., Ma, Y.: Integrating classification and association rule mining. In: Proceedings KDD'98, pp. 80–86. AAAI Press, New York, USA (1998)
11. Manku, G., Motwani, R.: Approximate frequency counts over data streams. In: VLDB Conf., pp. 346–357 (2002)
12. Mannila, H., Toivonen, H.: Multiple Uses of Frequent Sets and Condensed Representations. In: Proceedings KDD'96, pp. 189–194. AAAI Press, Portland (1996)
13. Monteiro, R.S., Zimbrão, G., Schwarz, H., Mitschang, B., Souza, J.M.: Building the Data Warehouse of Frequent Itemsets in the DWFIST Approach. In: Proceedings of the 15th Int. Symp. on Methodologies for Intelligent Systems, May 2005, Saratoga Springs, NY, (2005)
14. Özden, B., Ramaswamy, S., Silberschatz, A.: Cyclic association rules. In: Proc. of the 14th Int'l Conf. on Data Engineering, pp. 412–421 (1998)
15. Agrawal, R., Arning, A., Bollinger, T., Mehta, M., Shafer, J., Srikant, R.: The quest data mining system. In: Proc. of the 2nd KDD, August 1996, Portland, Oregon (1996)

Expectation Propagation in GenSpace Graphs for Summarization

Liqiang Geng[1], Howard J. Hamilton[2], and Larry Korba[3]

[1] IIT, National Research Council Canada, Fredericton, Canada, E3B 9W4
[2] Department of Computer Science, University of Regina, Regina, Canada, S4S 0A2
[3] IIT, National Research Council Canada, Ottawa, Canada, K1A 0R6
liqiang.geng@nrc-cnrc.gc.ca, hamilton@cs.uregina.ca,
larry.korba@nrc-cnrc.gc.ca

Abstract. Summary mining aims to find interesting summaries for a data set and to use data mining techniques to improve the functionality of Online Analytical Processing (OLAP) systems. In this paper, we propose an interactive summary mining approach, called GenSpace summary mining, to find interesting summaries based on user expectations. In the mining process, to record the user's evolving knowledge, the system needs to update and propagate new expectations. In this paper, we propose a linear method for consistently and efficiently propagating user expectations in a GenSpace graph. For a GenSpace graph where uninteresting nodes can be marked by the user before the mining process, we propose a greedy algorithm to determine the propagation paths in a GenSpace subgraph that reduces the time cost subject to a fixed amount of space.

1 Introduction

Summarization at different concept levels is a crucial task addressed by online analytical processing (OLAP). For a multidimensional table, the number of OLAP data cubes is exponential. Exploring the cube space to find interesting summaries with basic rollup and drilldown operators is a tedious task for users. Many methods and tools have been proposed to facilitate the exploration process. The use of an iceberg cube has been proposed to selectively aggregate summaries with an aggregate value above minimum support thresholds [1]. This approach assumes that only the summaries with high aggregate values are interesting. Discovery-driven exploration has been proposed to guide the exploration process by providing users with interestingness measures based on statistical models [9]. The use of the diff operator has been proposed to automatically find the underlying reason (in the form of detailed data that accounts for the difference) for a surprising difference in a data cube identified by the user [8].

Nonetheless, these approaches for finding interesting summaries have three weaknesses. First, a huge number of summaries can be produced, and during exploration, the data analyst must examine summaries one by one to assess them before the tools can be used to assist in exploring the discovered relationships. Secondly, the capacity for incorporating existing knowledge is limited. Thirdly, the

I.Y. Song, J. Eder, and T.M. Nguyen (Eds.): DaWaK 2007, LNCS 4654, pp. 449–458, 2007.

ability to respond to dynamic changes in the user's knowledge during the knowledge discovery process is limited. For example, if the information that more Pay-TV shows are watched during the evening than the afternoon has already been presented to the user, it will subsequently be less interesting to the user to learn that more Pay-TV shows are watched starting at 8:00PM than shows starting at 4:00PM.

To overcome these limitations, we are developing a summarization method based on generalization space (GenSpace) graphs [3]. The GenSpace mining process has four steps [3]. First, a domain generalization graph (DGG) for each attribute is created by explicitly identifying the domains appropriate to the relevant levels of granularity and the mappings between the values in these domains. The expectations are specified by the user for some nodes in each DGG to form an ExGen graph and then propagated to all the other nodes in the ExGen graph. In this paper, the term *"expectations"* refers to the user's current beliefs about the expected probability distributions for a group of variables at certain conceptual levels. Second, the framework of the GenSpace graph is generated based on the ExGen graphs for individual attributes, and the potentially interesting nodes in this graph are materialized. Third, the given expectations are propagated throughout the GenSpace subgraph consisting of potentially interesting nodes, and the interestingness measures for these nodes are calculated [5]. Fourth, the highest ranked summaries are displayed. Expectations in the GenSpace graph are then adjusted and the steps are repeated as necessary. Having proposed the broad framework for mining interesting summaries in GenSpace graphs in [3], we now propose a *linear expectation propagation method* that can efficiently and consistently propagate the expectations in GenSpace graphs.

A large variety of previous work has been conducted on incorporating user's domain knowledge and applying subjective interestingness measures to find interesting patterns However, most of these methods only apply to discovered rules. In this paper, we incorporate the user's knowledge to find interesting summaries.

Expectation propagation in a GenSpace is similar to Bayesian belief updating in that they both propagate a user's beliefs or expectations from one domain to another. However, Bayesian belief updating deals with the conditional probability of different variables, and does not deal with multiple concept levels of a single variable. Therefore, it does not apply to the problem addressed in this paper.

The remainder of this paper is organized as follows. In Section 2, we give the theoretical basis for ExGen graphs and GenSpace graphs. In Section 3, we propose a linear expectation propagation method. In Section 4, we propose a strategy for selecting propagation paths in a GenSpace subgraph. In Section 5, we explain our experimental procedure and present the results obtained on Saskatchewan weather data. In Section 6, we present our conclusions.

2   ExGen Graphs and GenSpace Graphs

An ExGen graph is used to represent the user's knowledge relevant to generalization for an attribute, while a GenSpace graph is used for multiple attributes. First, we give some formal definitions.

Definition 1. Given a set $X = \{x_1, x_2, ..., x_n\}$ representing the domain of some attribute and a set $P = \{P_1, P_2, ..., P_m\}$ of partitions of the set X, we define a nonempty binary relation $\preceq$ (called a ***generalization relation***) on P, where we say $P_i \preceq P_j$ if for every section $S_a \in P_i$, there exists a section $S_b \in P_j$, such that $S_a \subseteq S_b$. For convenience, we often refer to the sections by labels. The graph determined by generalization relation is called ***domain generalization graph*** (***DGG***). Each arc corresponds to a generalization relation, which is a mapping from the values in the ***parent*** node to that of the ***child***. The ***bottom*** node of the graph corresponds to the original domain of values X and the ***top*** node T corresponds to the most general domain of values, which contains only the value *ANY*.

In the rest of the paper, we use "summary", "node", and "partition" interchangeably.

Example 1. Let *MMDDMAN* be a domain of morning, afternoon, and night of a specific non-leap year {*Morning of January* 1, *Afternoon of January* 1, *Night of January* 1, ..., *Night of December* 31}, and P a set of partitions {*MMDD, MM, Week, MAN*}, where *MMDD* = {*January* 1, *January* 2, ..., *December* 31}, *MM* = {*January, February, ..., December*}, *Week* = {*Sunday, Monday, ..., Saturday*}, and *MAN* = {*Morning, Afternoon, Night*}. Values of *MMDDMAN* are assigned to the values of the partitions in the obvious way, i.e., all *MMDDMAN* values that occur on *Sunday* are assigned to the *Sunday* value of *Week*, etc. Here *MMDD* $\preceq$ *MM* and *MMDD* $\preceq$ *Week*. Figure 1 shows the DGG obtained from the generalization relation.

Definition 2. An ***expected distribution domain generalization*** (or ***ExGen***) ***graph*** $\langle P , Arc , E \rangle$ is a DGG that has expectations associated with every node. Expectations represent the expected probability distribution of occurrence of the values in the domain corresponding to the node. For a node (i.e., partition) $P_j = \{S_1, ..., S_k\}$, we have $\forall S_i \in P_j, 0 \leq E(S_i) \leq 1$ and $\sum_{i=1}^{k} E(S_i) = 1$, where $E(S_i)$ denotes the expectation of occurrence of section S_i.

Example 2. Continuing Example 1, for the partition *MAN* = {*Morning, Afternoon, Night*}, we associate the expectations [0.2, 0.5, 0.3], i.e., $E(Morning) = 0.2$, $E(Afternoon) = 0.5$, and $E(Night) = 0.3$.

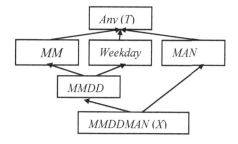

Fig. 1. An Example DGG

Definition 3. Assume node Q is a parent of node R in an ExGen graph, and therefore for each section $S_b \in R$, there exists a set of specific sections $\{ S_{a_1}, ..., S_{a_k} \} \subseteq Q$, such that $S_b = \bigcup_{i=1}^{k} S_{a_i}$. If for all $S_b \in R$, $E(S_b) = \sum_{i=1}^{k} E(S_{a_i})$, we say that nodes Q and R are **consistent**. An ExGen graph G is **consistent** if all pairs of adjacent nodes in G are consistent.

Theorem 1. An ExGen graph G is consistent iff every node in G is consistent with the bottom node.

Proof. A proof of Theorem 1 is given in [2].

An ExGen graph is based on one attribute. If the generalization space consists of more than one attribute, we can extend it to a **GenSpace graph** [2]. Each node in a GenSpace is the Cartesian product of nodes in all ExGen graphs. Theorem 1 for ExGen graphs can be adapted directly to GenSpace graphs.

3   Propagation of Expectations in GenSpace Graphs

Due to the combinational number of the nodes in a GenSpace graph, it is not practical to require that the user specify the expectations for all nodes. In exploratory data mining, the user may begin with very little knowledge about the domain, perhaps only vague assumptions about the (a priori) probabilities of the possible values at some level of granularity. After the user specifies these expectations, a data mining system should be able to create default preliminary distributions for the other nodes. We call the problem of propagating a user's expectations from one node or a group of nodes to all other nodes in a GenSpace graph the *expectation propagation problem*.

For each node in a GenSpace graph, the user can specify the expectations as an explicit probability distribution, or as a parameterized standard distribution, or leave the expectations unconstrained. If a standard distribution is specified, it is discretized into an explicit probability distribution. If expectations are not specified for a node, they are obtained by expectation propagation.

We propose the following three *expectation propagation principles*: (1) the results of propagations should be consistent, (2) if new expectations are being added, the new information should be preserved, while the old expectations should be changed as little as possible, and (3) if available information does not fully constrain the distribution at a node, the distribution should be made as even as possible.

To guarantee that propagation preserves consistency, we first propagate the new expectations directly to the bottom node, and then propagate the expectations up to the entire graph. Since bottom-up propagation makes all nodes consistent with the bottom node, the entire graph is consistent, according to Theorem 1. Bottom-up propagation will be elaborated in Section 4. Here we concentrate on the first step of our approach, *specific expectations determination*, i.e., how to propagate expectations from any non-bottom nodes specified by the user to the bottom node.

Based on expectation propagation principles, we treat the problem of specific expectations determination as an optimization problem. Given a set of nodes with new expectations, the expectations at the bottom node X are found by representing the constraint due to each node as an equation and then solving the set of equations. For a node i with j_i sections S_{ik}, where $1 \leq k \leq j_i$, assume that $E'(S_{ik})$ is the new expectation for each section S_{ik}, and s is an element of X, i.e., $s \in X$. We have

$$\sum_{s \subseteq S_{ik}} E'(s) = E'(S_{ik}).$$

Under these constraints, we minimize $\sum_{s \in X} (E'(s) - E(s))^2$, where $E'(s)$ and $E(s)$ are new and old expectations for element s in bottom node, to make the changes to the node X as small as possible. This minimization reflects the intuition that the old knowledge is preserved as much as possible. Since the number of variables equals to the number of the records in the bottom node, for large GenSpace graphs, the optimization process, like Lagrange multipliers, has a prohibitive time and space cost.

We propose the **linear propagation method** for finding an approximation to the solution to the optimization problem. We use the following linear equation to obtain the new expectations for the bottom node,

$$E'(s) = \frac{E'(S_{ik})}{E(S_{ik})} E(s),$$ where s is a section of X and $s \in S_{ik}$. According to this

equation, we accept the probability ratio among the sections and also retain the ratio among the elements of each section.

This method has three advantages. First, it is computationally efficient, because it involves only linear computation with time complexity of $O(|N||X|)$, where $|N|$ is the size of node N where the user changed his/her expectations. Secondly, it does not need to resolve conflicts among the user's new expectations, because we propagate new expectations in sequence rather than simultaneously as with the optimization method. Thirdly, we have identified an upper bound in terms of the changes of expectations specified by the user for the expectation changes between the old and new expectations for all the nodes in a GenSpace graph. In this sense, the user's old information is preserved.

Theorem 2. Let N be a node in a GenSpace graph G with m sections S_1, ..., S_m. Suppose the old expectations for node N are $\{E(S_1), ..., E(S_m)\}$, the new expectations for N are $\{E'(S_1), ..., E'(S_m)\}$ and $\max\limits_{i=1}^{m} \left(\frac{|E'(S_i) - E(S_i)|}{E(S_i)}\right) = \alpha$. After

propagating E' from N to the entire GenSpace graph using the linear propagation method, the relative variance v_P for any node P

satisfy $v_P = \sqrt{\dfrac{1}{m_P} \sum\limits_{i=1}^{m_P} \left(\dfrac{E'(S_{iP}) - E(S_{iP})}{E(S_{iP})}\right)^2} \leq \alpha$, where m_P denotes the number

of sections in node P.

Proof. A proof of Theorem 2 is given in [2].

Example 3. Theorem 2 tells us that if we change the expectations for each weekday by no more than 20%, after propagation, the expectation change for all other nodes (including nodes *Month*, *Day* and, *Morning-Afternoon-Night*) will not exceed 20%.

A drawback of the linear propagation method is that if expectations in two or more nodes are updated simultaneously, it cannot update expectations in other nodes simultaneously, as the optimization method does. However, we can obtain an approximation by applying the linear method multiple times, once for each updated node.

4   Propagation in GenSpace Subgraphs

If the user can identify some uninteresting nodes, we can prune them before propagation and only materialize the potentially interesting nodes. Pruning all uninteresting nodes saves the most storage, but it may increase propagation time.

Selecting views to materialize draws much attention from researchers [4, 6, 7]. Harinarayan et al. proposed a linear time cost model for aggregating a table [4]. They found that the time cost for producing a summarization is directly proportional to the size of the raw table. Therefore, in our case, when we propagate expectations from node A to node B, the time cost is directly proportional to the size of node A.

Example 4. In Figure 2, nodes are shown with a unique identifier and their size. The solid ovals denote interesting nodes and the blank ovals denote uninteresting nodes. If we prune node N_4, the propagation cost from node N_5 to N_1, N_2, and N_3 is size(N_5) * 3 = 3000. If we preserve N_4 as a hidden node, the propagation cost is size(N_5) + size(N_4) * 3 = 1600. Keeping hidden nodes reduces the propagation cost by nearly 50%.

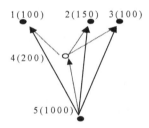

Fig. 2. Efficiency Improvement Due to Keeping Uninteresting Nodes

Sarawagi et al. encountered a similar problem when they calculated a collection of group-bys [10]. They converted the problem into a directed Steiner tree problem, which is an NP-complete problem. They used a heuristic method to solve their problem. Here, we consider our problem as a constrained directed Steiner tree problem, where there is a space limit for the uninteresting nodes. Furthermore, to use a Steiner tree model, we need to create additional arcs in the graph to connect all pairs of nodes that have a generalization relation, which causes extra storage space. We propose a greedy method to select hidden nodes and find an efficient propagation path for subgraphs.

Definition 4. (Optimal Tree) Given a GenSpace graph $G = <P, Arc, E>$ and a set of nodes $N \subseteq P$ such that the bottom node $X \in N$, a *optimal tree* OT of N in G is a tree that consists of nodes in N. The root of the tree is X. For every non-bottom node $n \in N$, its parent is an ancestor in G of minimum size that belongs to N.

An optimal tree for a set of nodes has the most efficient propagation paths for the subgraph consisting of these nodes, because every node obtains expectations from its smallest possible ancestor in the subgraph.

Algorithm *SelectHiddenNodes* for choosing hidden nodes and determining the propagation paths is given in Figure 3. We first create the optimal tree consisting of only the interesting nodes, then check the uninteresting nodes, and select the one that results in the greatest reduction in the **scanning cost**, which is defined as the sum of the sizes of the parent nodes in the optimal trees for all nodes. Then we modify the optimal tree to incorporate the selected nodes. This process continues until the space cost reaches the given threshold or no improvement can be obtained or no candidate uninteresting nodes are left. Selecting an uninteresting node will affect propagation efficiency in two ways. First it will cause an extra propagation cost for propagating expectations to this node; secondly, it may reduce the cost for its interesting descendents. Function Improvement calculates the efficiency improvement for an uninteresting node u. If it returns a positive value, it will reduce the scanning cost.

```
SelectHiddenNodes(GenSpace graph G, Interesting node set I, Uninteresting node set U) {
        Create initial optimal tree OT based on I;
        U' = ∅;
        While there is enough memory and U is not empty do {
                For each u ∈ U, do
                        Imp_u = Improvement(u, I, G, U', OT);
                Select u_max as the one with maximum improvement Imp_max;
                If Imp_max ≤ 0
                        break;
                U' = U' ∪ {u_max}; U = U − {u_max};
                For each node x ∈ I∪U' do {
                        Find its ancestor u_an with smallest size in I∪U';
                        Set u_an as x's parent;
                }
                OT = newly generated optimal tree;
        }
        Return selected node set U' and optimal tree OT for nodes I∪U';
}
Improvement(u, I, G, U', OT) {
        Improvement = 0;
        Find u's best ancestor u_an ∈ I∪U' in G;
        Find u's descendent set D ⊆ I∪U' in G;
        For each d in D, do {
                Find d's parent p in current OT;
                If size(p) > size (u)
                        Improvement += size(p) − size(u);
        }
        Improvement −= size(u_an);
        Return Improvement;
}
```

Fig. 3. Algorithm for Selecting Hidden Nodes and Creating an Optimal Tree

5   Experimental Results

We implemented the GenSpace summary mining software and applied it to three real dataset. Due to space limitations, we only present the results for the Saskatchewan weather dataset here. This dataset has 211,584 tuples and four attributes, *time*, *station*, *temperature*, and *precipitation*. The ExGen graphs for these attributes are available elsewhere [2].

First, we compared the efficiency of the linear propagation method and the optimization method on the ExGen graph in Figure 1. The experiments were conducted in Matlab 6.1 in a PC with 512M memory and 1.7 GHz CPU. We tested two cases. Table 1 compares the running time. We can see that for the optimization method, when the size of the bottom node (number of the variables) increases from 365 to 1095, the running time increases dramatically from 67 seconds to 1984 seconds. For the linear propagation method, in both cases, the running time is unperceivable.

Table 1. Comparison of Running Time between Optimization and Linear Methods

Cases	# of variables	# of constraints	Running time (Sec)	
			Optimization	Linear
Case 1	365	19	67	<1
Case 2	1095	3	1984	<1

To show the efficiency of the *SelectHiddenNodes* algorithm, we present two cases: (1) mark all nodes in the lowest 5 levels (out of 19 levels) as uninteresting and (2) mark all the nodes in the lowest 5 levels or with specific date values or specific temperature values as uninteresting. The results are shown in Table 2. The *Storage* column lists the storage cost in thousands of records. The *Scanning* column lists the number of the records scanned during propagation, in thousands of records. The *Time* column lists the propagation cost in seconds. We first compare the storage and the scanning costs between the subgraph obtained from *SelectHiddenNodes* and that from pruning all uninteresting nodes. In case 1, after selecting hidden nodes, the storage increased by 25%, while the scanning cost decreased by 60%. In case 2, the storage cost increased by 26%, and the scanning cost decreased by 82%. In both cases, the scanning cost decreased significantly while the storage cost increased by a smaller percentage. Then we compare the storage and time costs between the subgraph selected by *SelectHiddenNodes* and the entire GenSpace graph. For both cases, both storage and scanning costs are significantly less for the subgraph obtained from *SelectHiddenNodes*. Figure 4 shows the storage and scanning costs after selecting varying numbers of nodes for cases 1 and 2. The X-axis denotes the number of the nodes currently selected, and the Y-axis denotes the storage and scanning costs, in thousands of records. We can see that in both cases, the first few nodes contribute the most to the reduction of the scanning cost.

We also tested the scalability of the algorithm. Figure 5(a) compares the scanning costs of propagation in the full GenSpace graph, the subgraph composed of only interesting nodes, and the subgraph obtained from *SelectHiddenNodes*. Figure 5(b) compares the corresponding storage costs. The results show that the time savings are significant for *SelectHiddenNodes* regardless of the size of the bottom nodes, and that its storage cost falls in an acceptable range.

Table 2. Propagation Costs in GenSpace Subgraph

		Storage (K)	Scanning (K)	Time (Sec)
Entire GenSpace	Preserve all	15309	18669	2037
Case 1	Prune all	4128	21173	2149
	SelectHiddenNodes	5156	8391	1056
Case 2	Prune all	912	11703	983
	SelectHiddenNodes	1152	2155	152

The analysis for the detailed process and the quality of the results was presented in [2]. In this paper, we show the interestingness values (logarithm of the variances) for the top summaries and the top five summaries for twenty iterations (see Figure 6). As the process iterates, both measures are roughly decreasing functions of the number of iterations. This trend shows that the user's knowledge of the distribution, as measured by what the user has been told, gradually becomes closer to the real distribution.

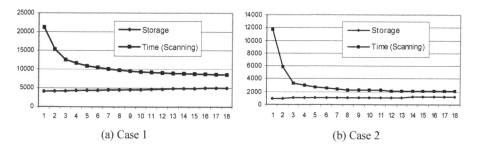

(a) Case 1   (b) Case 2

Fig. 4. Change Trend of Storage and Time When Selecting Hidden Nodes

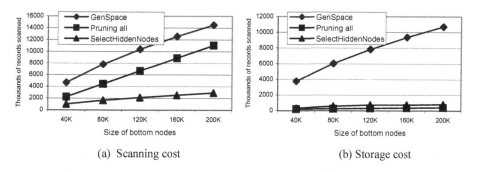

(a) Scanning cost   (b) Storage cost

Fig. 5. Scanning and Storage Costs Comparison

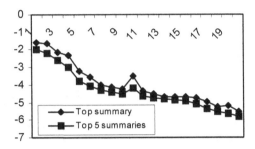

Fig. 6. Logarithm of Variances for the Top Summaries in the First 20 Iterations

6 Conclusions

We proposed an efficient method to propagate a user's expectations in a GenSpace graph, which describes a variety of ways of performing generalization. Based on the propagated expectations, the system finds interesting summaries in the GenSpace graph based on current expectations. We also proposed a strategy for selecting propagation paths in a GenSpace subgraph that reduces the time cost subject to a fixed amount of space. Experiments on real data sets show the effectiveness and efficiency of our approach. In the future, we will employ the known dependencies between attributes to facilitate the propagation process.

References

[1] Beyer, K., Ramakrishnan, R.: Bottom-up Computation of Sparse and Iceberg CUBEs. In: Proceedings of ACM SIGMOD'99, pp. 359–370 (1999)

[2] Geng, L.: Finding Interesting Summaries in GenSpace Graphs, Ph.D Thesis, Department of Computer Science, University of Regina, Regina, Canada (2005)

[3] Hamilton, H.J., Geng, L., Findlater, L., Randall, D.J.: Efficient Spatio-Temporal Data Mining with GenSpace Graphs. Journal of Applied Logic 4(2), 192–214 (2006)

[4] Harinarayan, V., Rajaraman, A., Ullman, J.D.: Implementing Data Cubes Efficiently. In: Proceedings of ACM SIGMOD '96, pp. 205–216 (1996)

[5] Hilderman, R.J., Hamilton, H.J.: Knowledge Discovery and Measures of Interest. Kluwer Acadamic, Dordrecht (2001)

[6] Lawrence, M., Rau-Chaplin, A.: Dynamic View Selection for OLAP. In: Proc. of DaWaK2006, September 2006, Krakow, Poland, pp. 33–44 (2006)

[7] Li, H., Huang, H., Liu, S.: PMC: Select Materialized Cells in Data Cubes. In: Tjoa, A.M., Trujillo, J. (eds.) DaWaK 2005. LNCS, vol. 3589, pp. 168–178. Springer, Heidelberg (2005)

[8] Sarawagi, S.: Explaining Differences in Multidimensional Aggregates. In: Proc. of the 25th Int. Conf. on Very Large Databases (VLDB), Edinburgh (September 1999)

[9] Sarawagi, S., Agrawal, R., Megiddo, N.: Discovery Driven Exploration of OLAP Data Cubes. In: Schek, H.-J., Saltor, F., Ramos, I., Alonso, G. (eds.) EDBT 1998. LNCS, vol. 1377, Springer, Heidelberg (1998)

[10] Sarawagi, S., Agrawal, R., Gupta, A.: On Computing the Data Cube. Research Report RJ10026, IBM Almaden Research Center (1996)

Mining First-Order Temporal Interval Patterns with Regular Expression Constraints[*]

Sandra de Amo[1], Arnaud Giacometti[2], and Waldecir Pereira Junior[1]

[1] Universidade Federal de Uberlândia, Faculdade de Computação,
Av. João Naves de Avila, 2121, Bloco B, Uberlândia-MG, Brazil
deamo@ufu.br, wpjunior@gmail.com
[2] Université de Tours, Laboratoire d'Informatique,
3, Place Jean Jaurès 45000 Blois, France
giaco@univ-tours.fr

Abstract. Most methods for temporal pattern mining assume that time is represented by points in a straight line starting at some initial instant. In this paper, we consider a new kind of first order temporal pattern, specified in Allen's Temporal Interval Logic, where time is explicitly represented by intervals. We present the algorithm MILPRIT for mining temporal interval patterns, which uses variants of the classical level-wise search algorithms. MILPRIT allows a broad spectrum of constraints over temporal patterns to be incorporated in the mining process. Some experimental results over synthetic and real data are presented.

Keywords: Temporal Data Mining, First-Order Temporal Interval Logic, Constraint-based Mining, Sequential Patterns.

1 Introduction

We present a new sequential pattern where time is mesured in terms of *intervals* instead of *points*. These patterns, which we call *temporal interval patterns*, aim at capturing how events taking place in time intervals relate to each other. For instance, (1) in a medical application, we could be interested in discovering if patients who take some medicine X during a certain period of time, and who presented the symptom Y *before* taking the medicine, will present the symptom Z *during or after* taking the medicine X, (2) in an agricultural application, we could be interested in discovering if the use of some organic fertilizer during a period of time has an effect on the way a plant grows *during* and *after* the fertilizer application.

In [6], Allen's Propositional Interval Logic [3] has been used for the first time, to treat the problem of discovering association rules over time series. However, to our knowledge the use of Interval Logic in temporal data mining has been restricted to *propositional temporal patterns*, that is, patterns not involving first order predicates. The need for more expressive kind of temporal patterns arises

---

[*] Research partially supported by CNPq-Brazil, project no. 473309/2004-1.

I.Y. Song, J. Eder, and T.M. Nguyen (Eds.): DaWaK 2007, LNCS 4654, pp. 459–469, 2007.
© Springer-Verlag Berlin Heidelberg 2007

for instance, when modelling Unix-users behaviour [7], as is pointed out in [8]. Our temporal pattern is defined as a set of atomic first order formulae where time is explicitly represented by an interval variable, together with a set of interval relationships (*before,during,starts,finishes,overlaps,meets*) described in terms of Allen's First Order Interval Logic [3]. An example of such temporal pattern is $\{Med(x,penicillin,i_1), Symp(x,dizziness,i_2), during(i_2,i_1)\}$, meaning that "*patients who take penicillin during the time interval i_1 will feel dizzy during this period, but eventually the dizziness will stop*". In a temporal data model using an interval time representation, the structure of the flow of time is far more complex than in a model using a punctual time representation: the latter is $(\mathbb{N}, <)$, where $<$ is the linear order of natural numbers and the former is $(\mathbb{N} \times \mathbb{N}, \mathcal{P}_T)$, where $\mathcal{P}_T = \{$ *before,during,starts,finishes,overlaps,meets* $\}$.

Some related work. The algorithm MILPRIT we introduce in this paper is designed to mine first order temporal patterns, where time is represented by *intervals*, and regular expression contraints are used to restrict the search space. A very preliminary version of the ideas underlying the specification of MIL-PRIT has been previously presented in [1]. In [8], the algorithm SeqLog, based on Inductive Logic Programming, was introduced for discovering first order sequential patterns where time is represented by *points* in a straight line (and not by intervals like in our case). Recently, many work on contrained-based temporal mining ([5], [10], [9]) have adopted regular expressions as a mechanism to specify user constraints. [5] is a pioneer work proposing to use regular expressions for specifying constraints on patterns. The authors introduced the family of algorithms SPIRIT for mining transaction sequential patterns (propositional patterns, where time is punctual) with constraints specified by regular expressions. In [10], the algorithm Spirit-Log, designed to mine first order temporal patterns was introduced. Spirit-Log extends SeqLog (where time is punctual), by pushing regular expression constraints in the mining process. We emphasize that both algorithms SeqLog and Spirit-Log deal with temporal patterns where time is punctual, differently from our approach where time is measured in terms of intervals.

2   Problem Formalization

Temporal i-Patterns. Let us suppose three disjoints sets of symbols $\mathcal{P}$ (predicates), $\mathcal{V}$ (data variables) and $\mathcal{C}$ (data constants). Let us also assume the set of temporal predicates $\mathcal{P}_T = \{$ *before, meets, overlaps, during, finishes, starts* $\}$ and two other disjoint sets of symbols $\mathcal{V}_t$ (temporal variables) and $\mathcal{C}_t$ (temporal constants). A *data atom* is an expression of the form $p(s_1, ..., s_n, e)$ or of the form $t_1 = t_2$, where each s_i is a data variable or a data constant, e is a temporal variable, t_1 is a data variable and t_2 is a data constant or a data variable. A *temporal atom* is an expression of the form $r(e, f)$ or $e = f$, where r is one of the temporal predicates in $\mathcal{P}_T$ and e, f are temporal variables. The predicates in $\mathcal{P}_T$, their dual forms *after, metby, overlappedby, contains, finishedby, startedby*

and the equality predicate = determine all the possible relationships which can
exist between two intervals (see [3] for details).

The following figure illustrates the semantics of the temporal predicates:

before(e,f)   meets(e,f)   overlaps(e,f)   during(e,f)   finishes(e,f)   starts(e,f)

Temporal variables are evaluated over *intervals* $[a, b]$, where a and b are dates
with $a < b$. A *valuation* of temporal variables is a mapping v associating an interval $v(e) = [a, b]$ to each temporal variable e. For instance, the formula *before(e,f)*
is true if the variables e and f are evaluated as $[10/12/2004, 15/12/2005]$ and
$[17/12/2004, 28/12/2005]$ respectively.

Definition 1 (Temporal i-pattern). A *temporal i-pattern* (temporal interval
pattern) is a triple $(K, \mathcal{D}, \mathcal{T})$ where (1) $\mathcal{D}$ is a set of data atoms (2) $\mathcal{T}$ is a set of
temporal atoms and (3) K is a special data atom $K(x_1, ..., x_n)$ whose variables
$x_1, ..., x_n$ appear in $\mathcal{D}$. The data atom K is called the *reference query*. It is
essential in the definition of an i-pattern, since it specifies *what* is counted. It
plays the same role in our setting as the *atom key* in the Datalog patterns of [4].

Example 1. Let us consider the temporal i-pattern $(Patient(x), D, T)$ where $\mathcal{D}$
$= \{Med(x,y,e), Symp(x,z,f)\}$ and $\mathcal{T} = \{ \ before(e,f) \ \}$. Intuitively, this temporal
pattern translates the fact that a patient x takes the medicine y during a period
of time e and during a further period of time f (after she has stopped taking the
medicine y) presents a disease symptom z. Note that we are always interested
in analyzing the behaviour of patients registered in the relation *Patient*, even
though relations *Med* and *Symp* may contain data related to other patients
(not only those registered in *Patient*). For this reason, *Patient(x)* is called *the
reference query.*

In what follows, we suppose that the set of temporal atoms $\mathcal{T}$ is *consistent*, i.e.,
the temporal variables can be instantiated in a consistent way. For instance,
the set of temporal atoms $\{before(e, f), \ before(f, g), \ before(g, e)\}$ is not consistent, because the first two atoms imply that e comes before g and so, there is
no way to instantiate the temporal variables e, f and g in order to make the
three temporal formulae simultaneously true. We also suppose that the set $\mathcal{T}$
is *complete*, i.e., for each pair of temporal variables e, f appearing in T it is
possible to infer a temporal relationship between e and f. For instance, the set
$\{before(e,f), meet(f,g)\}$ is complete because the relationship between e and g is
completely determined (even though it does not appear explicitly in the set): the
only possibility is *before(e,g)*. On the other hand, $\{during(e,f), \ overlaps(f,g)\}$ is
not complete since the relationship between e and g is not deterministically implied by the two relationships $during(e,f), \ overlaps(f,g)$: it could be *during(e,g)*,
before(e,g) or *overlaps(e,g)*. As we will see next, given a consistent and complete
set of temporal atoms, its temporal variables can be ordered in a natural way.
Due to space limitations, we will give an intuitive idea of this order relation,
omitting its rigorous formalization.

Definition 2 (Sequential representation of temporal i-patterns). Let $[a, b]$ and $[c, d]$ be two intervals in $\mathcal{I} = \{[i, j], i, j \in \mathbb{N}\}$. We say that $[a, b] \leq [c, d]$ if and only if one and only one of the following conditions is verified: (1) $a = c$ and $b = d$, (2) $b < d$ or (3) $b = d$ and $a < c$. This ordering over the interval structure $\mathcal{I}$ naturally induces an ordering over the set $\mathcal{T}$ of temporal variables appearing in an i-pattern. The following example illustrates the idea: let $\mathcal{T} = \{before(e, f), during(f, g)\}$. Then the induced ordering over the set of temporal variables $\{e, f, g\}$ is $e <_{\mathcal{T}} f <_{\mathcal{T}} g$. Indeed, for **all** instantiation v of the temporal variables, $v(e) = [a_e, b_e], v(f) = [a_f, b_f], v(g) = [a_g, b_g]$, such that the predicates in $\mathcal{T}$ are simultaneously true, the ending points b_e, b_f, b_g must satisfy $b_e < b_f < b_g$. That is, for **all** instantiation v, we have $v(e) \leq v(f) \leq v(g)$.

Theorem 1. If the set $\mathcal{T}$ of temporal predicates is *consistent and complete* then the relation $<_{\mathcal{T}}$ is an order relation.

The induced order $<_{\mathcal{T}}$ over the temporal variables appearing in a temporal i-pattern M allows us to view M as a *sequence* of data atoms. Indeed, let $M = (K, \mathcal{D}, \mathcal{T})$, where $\mathcal{D} = \{p_1(x_1^1, ..., x_n^1, e_1), ..., p_k(x_1^k, ..., x_n^k, e_k)\}$. Since we are assuming that $\mathcal{T}$ is complete and consistent, then its temporal variables can be ordered by $<_{\mathcal{T}}$. Assuming that $e_1 <_{\mathcal{T}} e_2 <_{\mathcal{T}} ... <_{\mathcal{T}} e_k$, then the set $\mathcal{D}$ can be viewed as the *sequence* $< p_1, ..., p_k >$.

The Dataset. Let $\mathbf{R} = \{K, R_1, ..., R_n\}$ be a temporal database schema, i.e., a database schema such that each relational schema R_i has arity $k_i \geq 2$ and sort *(data,...,data,time)* and K has arity m and sort *(data,...,data)*. The attributes appearing in the relation K are called *reference attributes*. A *temporal dataset* over $\mathbf{R}$ is a set of temporal relations $\{k, r_1, ..., r_n\}$ where each r_i is a set of tuples $(a_1, ..., a_{n_i}, e)$ over R_i and k is a set of tuples over K.

Definition 3 (Support). Let D be a temporal database over the schema $\mathbf{R} = \{K, R_1, ..., R_n\}$ and $M = (K(x_1, ..., x_m), \{D_1, ...D_s\}, \{T_1, ..., T_l \})$ be a temporal i-pattern over $\mathbf{R}$. The *support* of M with respect to D, denoted by $sup(M)$, is defined by $sup(M) = \frac{|\{u \in K \mid u \models Q_M\}|}{|K|}$ where Q_M is the *relational calculus* query $\exists y_1 \exists y_2 ... \exists y_k (D_1 \wedge D_2 \wedge ... \wedge D_s \wedge T_1 \wedge ... \wedge T_l)$, and $y_1, ..., y_k$ are the data variables not appearing in the reference query $K(x_1, ..., x_m)$. The expression $u \models Q_M$ means that u is an answer to the query Q_M when executed over D. Given $0 \leq \alpha \leq 1$, we say that the temporal i-pattern M is *frequent* with respect to D and α if $sup(M) \geq \alpha$.

Example 2. Let us consider the temporal database $\{Patient(NamePat),$ $Med(NamePat, NameMed, T), Symp(NamePat, NameSymp, T)\}$ and the temporal i-pattern $M = (Patient(x), \{Med(x, y, e), Symp(x, z, f)\}, \{before(e, f)\})$. Let D be the temporal database $Patient = \{(Paul), (Charles), (John)\}$, $Med = \{(Paul,penicillin,[1,3]), (Charles,tetracycline, [4,7]), (Mary,penicillin,[8,10])\}$ and $Symp = \{(Paul,dizziness,[4,6]), (Charles,dizziness, [5,6]), (John,fever,[8,10])\}$. The relational query Q_M can be specified in the relational algebra formalism by the expression $\Pi_{NamePat}(Patient \bowtie Med \bowtie Symp)$. The answer of Q_M is $\{(Paul)\}$. So, the support of M w.r.t. D is $\frac{1}{3}$.

3   Temporal i-Patterns with Restrictions

In this section, we introduce a formalism based on regular expressions to specify a broad spectrum of constraints over patterns in order to better satisfy user requirements as well as to reduce the search space of patterns.

Definition 4 (Mode). A *mode* over a temporal database $\mathbf{R} = \{K, R_1, ..., R_k\}$ is an expression of the form $R(u_1, ..., u_s, \#)$, where each u_i is a data variable or the (new) symbol $*$, $\#$ is a new symbol, and R is one of the predicates $R_i \in \mathbf{R}$. We say that a data atom A is *in accord with* the mode $R(u_1, ..., u_s, \#)$ if $A = R(y_1, ..., y_s, e)$ and for each $i = 1, ..., s$ we have: (1) If u_i is a variable then $y_i = u_i$; (2) If $u_i = *$, then y_i is a variable or a constant ; (3) e is a temporal variable. For instance, $Med(x, *, \#)$ is a mode and the atom $Med(x, penicillin, e)$ is in accord with $Med(x, *, \#)$. On the other hand, the atom $Med(y, penicillin, e)$ is not in accord with the mode $Med(x, z, \#)$.

Let Σ be a set of modes over $\mathbf{R} = \{K, R_1, ..., R_n\}$. In what follows, we will consider regular languages (sets of strings) over the alphabet Σ. These languages are specified by regular expressions. Given a regular expression E over the alphabet Σ, we denote by $W(E)$ the set of strings (words) verifying E. We denote by $P(E)$ the set of prefixes of strings in $W(E)$. We will need this notation later.

Definition 5 (The Search Space). Let $\mathbf{R} = \{K, R_1, ..., R_n\}$ be a database schema, Σ a set of modes over $\mathbf{R}$ and E a regular expression over Σ. The *search space* defined by E is the set of temporal i-patterns $(K(x_1, ..., x_m), \mathcal{D}, \mathcal{T})$ such that $\mathcal{D} = < p_1(t_1^1, ..., t_{l_1}^1, e_1), ..., p_n(t_1^n, ..., t_{l_n}^n, e_n) >$ satisfies the following properties: (1) There exists a string $w_1 w_2 ... w_n \in W(E)$, such that p_i is in accord with w_i for each $i = 1, ..., n$; (2) For all data atom $p_i(x_1^i, ..., x_{l_i}^i, e_i) \in \mathcal{D}$ let $w_i = p_i(u_1, ..., u_{l_i}, \#)$ be the associated mode; if $u_j = *$ and x_j is a variable, then x_j has only one occurrence in $\mathcal{D}$ (a data variable symbol x_j can be used to replace a symbol $*$ in a mode only once). We denote by $\mathcal{W}(E)$ the search space specified by E. We denote by $\mathcal{P}(E)$ the set of i-patterns defined in a similar way as $\mathcal{W}(E)$, by considering strings $w_1 w_2 ... w_n \in P(E)$ instead of $W(E)$ in condition (1). Patterns in $\mathcal{W}(E)$ are called *valid*. Patterns in $\mathcal{P}(E)$ are called *prefix valid* or *p-valid*.

Example 3. Let $E = Med(x, *, \#)\ Med(x, *, \#)^*\ Symp(x, *, \#)$. The i-pattern $M_1 = (Patient(x),\ \{Med(x, penicillin, e),\ Med(x, tetracyclin, f), Symp(x, diarrhea, g)\},\ \{before(e, f),\ overlaps(f, g)\})$ belongs to $\mathcal{W}(E)$. The pattern $M_2 = (Patient(x),\ \{Med(x, penicillin, e),\ Med(x, tetracyclin, f)\}, \{before(e, f)\})$ belongs to $\mathcal{P}(E)$. On the other hand, the pattern $M_3 = (Patient(x), \{Med(x, y, e), Symp(x, y, f)\}, \{before(e, f)\})$ does not belong to $\mathcal{P}(E)$, since property (2) of Definition 5 is not verified. Intuitively, the regular expression E captures the temporal i-patterns corresponding to a patient x taking certain medicines during successive periods of time and eventually presenting some symptom.

Our mining task is: Given a temporal database D, a minimum support threshold α, $0 \leq \alpha \leq 1$, and a regular expression E over Σ, find all temporal i-patterns in $\mathcal{W}(E)$ and which are frequent with respect to D and α.

4   The Algorithm MILPRIT

In this section, we present the Algorithm MILPRIT (Mining Interval Logic Patterns with Regular expressIons consTraints). In a high level, it follows the general Apriori strategy of Agrawal/Srikant [2], working in passes, and each pass producing patterns more *specific* than those produced in the previous pass.

The classical Apriori strategy uses a *Pruning Phase*, where all patterns more specific than a pattern p not belonging to the set of frequent patterns produced so far (in previous passes) are pruned. This pruning strategy relies on the Anti-monotone Property. In our setting, at the end of pass k, the algorithm discovers the frequent patterns F_k which satisfy a regular expression E. The constraint E is pushed into the generation phase at pass $k + 1$ in order to produce only patterns satisfying it. So, the number of generated patterns is in direct proportion to the restrictiveness of E. In the pruning phase, however, all patterns which are more specific than a pattern *satisfying* E and not belonging to F_k should be pruned. We notice that in this case, the size of the set of pruned patterns is *in inverse proportion* to the restrictiveness of the constraint E. Such tradeoffs are due to the fact that regular expression contraints are not anti-monotone, that is, if a pattern satisfies E it may have some subpattern not satisfying E. In order to find a suitable trade-off, we use a *relaxation* of the constraint E, that is, a less restrictive one, the prefix constraint associated to E, according which, only frequent p-valid i-patterns will be produced in the mining process. A post-processing phase will filter the valid patterns. The general structure of the algorithm MILPRIT is depicted below:

Procedure MILPRIT(D,α,E)
D is a temporal dataset over $\mathbf{R} = \{K, R_1, ..., R_n\}$, E is a regular expression over the alphabet of modes Σ over $\mathbf{R}$, α is a minimum support threshold. Let $\mathcal{M}_1$ be the set of data atoms $R_i(x_1, ..., x_{n_i}, e)$, where e is a temporal variable.
begin
1. $F := F_1 = \{M \in \mathcal{M}_1 : M$ is frequent and $M \in \mathcal{P}(E)\}$
2. $k := 2$
3. **repeat**
  4. $C_k := \text{Gen}(F_{k-1},E)$ (***Generation Phase***)
  5. $P := \{M \in C_k : \exists N \in \mathcal{P}(E)$,such that $N \notin F$ and M is more specific than $N\}$
  6. $C_k := C_k - P$ (***Pruning Phase***)
  7. $F_k := \{M \in C_k : \text{sup}(M) \geq \alpha\}$ (***Validation Phase***)
  8. $F := F \cup F_k$
  9. $k := k + 1$
10. **until** $F_{k-1} = \emptyset$
11. $F := \{M \in F : M \in \mathcal{W}(E)\}$ (***Post-Processing Phase***)

The Generation Phase. MILPRIT works in passes, each pass k producing patterns of "*level*" k. The "level" of a temporal i-pattern M is measured in terms of the *refinement vector* associated to M as defined below.

Definition 6. The *refinement vector* of $M = (K, \mathcal{D}, \mathcal{T})$ is $v(M) = (n, c)$ where n is the size of $\mathcal{D}$ and c is the number of constants appearing in $\mathcal{D}$. The *refinement level* of M is $l(M) = n + c$, where $v(M) = (n, c)$. For instance, let M be the temporal i-pattern $(Patient(x), \{Med(x, penicillin, e), Symp(x, z, f)\}, \{before(e, f)\})$. Then $v(M) = (2, 1)$ and $l(M) = 3$.

The procedure $\text{Gen}(F_{k-1}, E)$ is designed to generate temporal i-patterns whose refinement level is k, by *specializing* the frequent temporal i-patterns of F_{k-1} (whose refinement level is $k - 1$) in such a way that the patterns produced are prefixes of patterns satisfying E. This is accomplished by two *specialization operators* (ρ_E and ρ_I) defined below.

Definition 7 (Extension). Let E be a regular expression and $M \in \mathcal{P}(E)$, $M = (K(x_1, ..., x_n), \mathcal{D}, \mathcal{T})$, where $\mathcal{D} = < D_1, ..., D_m >$. The *extension operator* ρ_E executed over M is defined as follows: (a) if $v(M) = (n, c)$ with $c > 0$, then $\rho_E(M) = \emptyset$; (b) if $v(M) = (n, 0)$ then $\rho_E(M)$ is the set of temporal i-patterns $M' = (K, \mathcal{D}', \mathcal{T}') \in \mathcal{P}(E)$ such that $\mathcal{D}' = \mathcal{D} \cup \{p_{m+1}(x_1, \ldots, x_l, e_{m+1})\}$ where $p_{m+1} \in \mathbf{R}$, x_i are data variables, e_{m+1} is a temporal variable and $\mathcal{T}' = \mathcal{T} \cup \{r_1(e_1, e_{m+1}), \ldots, r_m(e_m, e_{m+1})\}$ where each r_i is a temporal predicate, and $\mathcal{T}'$ is complete and consistent.

Example 4. Let E be the regular expression $E = Med(x, *, \#)^* Symp(x, *, \#)$. The i-patterns $M = (Patient(x), \{Med(x, y_1, e_1)\}, \emptyset)$ and $M' = (Patient(x), \{Med(x, prozac, e_1)\}, \emptyset)$ belong to $\mathcal{P}(E)$. Let us consider the following i-patterns: $M_1 = (Patient(x), \{Med(x, y_1, e_1), Med(x, y_2, e_2)\}, \{overlaps(e_1, e_2)\})$, $M_2 = (Patient(x), \{Med(x, y_1, e_1), Med(x, y_2, e_2), Symp(x, z, e_3)\}, \{overlaps(e_1, e_2), overlaps(e_2, e_3)\})$ and $M_3 = (Patient(x), \{Med(x, prozac, e_1), Med(x, y, e_2)\}, \{before(e_1, e_2)\})$. We can verify that $M_1 \in \rho_E(M)$ (that is, M_1 is an *extension* of M), but $M_2 \notin \rho_E(M_1)$, because there is no explicit nor implicit relationship between the temporal variables e_2 and e_3. Moreover, $M_3 \notin \rho_E(M')$ because the refinement vector of M' is $(1,1)$, and so it cannot be extended, according to condition (a) of Definition 7.

Definition 8 (Instantiation). Let E be a regular expression and $M \in \mathcal{P}(E)$, $M = (K(x_1, ..., x_n), \mathcal{D}, \mathcal{T})$, where $\mathcal{D} = < D_1, ..., D_m >$. The *instantiation operator* executed over M produces the set $\rho_I(M)$ of temporal i-patterns $M' = (K(x_1, ..., x_n), \mathcal{D}', \mathcal{T})$, where $\mathcal{D}'$ is obtained by replacing some variable y_k ($y_k \neq x_l$ for $l \in [1, n]$) occurring in some $D_i = p(y_1, \ldots, y_p, e)$ by a constant c in such a way that if the string of modes corresponding to $\mathcal{D}$ is $w_1 w_2 ... w_m$ and $w_i = p(u_1, \ldots, u_p, \#)$, then: (a) $u_k = *$; (b) for every $l > k$, if $u_l = *$, then y_l is not a constant; (c) for every $j > i$, if $w_j = p'(v_1, \ldots, v_q, \#)$, then for every $l \in [1, q]$, if $v_l = *$ and $D_j = p'(z_1, \ldots, z_q, e)$, then z_l is not a constant.

Example 5. Let E be the regular expression $E = Med(x, *, \#)^*\ Symp(x, *, \#)^*$. Consider the following i-patterns in $\mathcal{P}(E)$:

$M_1 = (Patient(x), \{Med(x, w, e_1), Symp(x, y, e_2)\}, \{before(e_1, e_2)\})$
$M_2 = (Patient(x), \{Med(x, w, e_1), Symp(x, dizziness, e_2)\}, \{before(e_1, e_2)\})$
$M_3 = (Patient(x), \{Med(x, penicillin, e_1), Symp(x, dizziness, e_2)\}, \{before(e_1, e_2)\})$

According to Definition 8, M_2 is in $\rho_I(M_1)$, but $M_3 \notin \rho_I(M_2)$.

Theorem 2. The specialization operator ρ defined as $\rho(M) = \rho_I(M) \cup \rho_E(M)$ satisfies the following properties: (1) *completeness*: every i-pattern $M \in \mathcal{P}(E)$ can be obtained by successively applying ρ to some i-pattern in the refinement level 1. (2) *optimality*: suppose that the regular expression E verifies the following condition: *for all distinct modes m_1 and m_2 appearing in E, associated to the same predicate p, the positions containing the symbol * are the same in boths modes.* Under this condition, the instantiation operator ρ is optimal, in the sense that an i-pattern cannot be obtained by specializing successively two non-equivalent i-patterns. Two temporal i-patterns M and M' are said to be *equivalent* if M' is obtained from M by renaming its variables.

The support counting of the generated i-patterns is a rather technical procedure and its details are omitted here.

5 Experimental Results

MILPRIT has been evaluated through an extensive set of experiments using synthetic and real databases. Due to lack of space, only some of the results are presented in this paper. Our experiments have been executed on a Pentium 4 with 3GHz and 2GB of main memory, running on Linux OS.

5.1 Synthetic Data

We have developed a synthetic data generator which produces temporal databases according to some input parameters, shown in Table 1. For the tests

Table 1. Parameters used in the Synthetic Data Generator

Parameters		Default
P	Number of tables	2
M	Number of attributes per table	4
U	Number of tuples per table - in '000s	5
S	Size of the domain of reference attributes - in '000s	2
D	Size of the domain of non-reference attributes - in '000s	1
T	Size of the domain of temporal attributes - in '000s	1
Q	Minimum amount of i-patterns	20
E	Regular expression	a*b*

presented in this paper, the alphabet of modes considered was $\Sigma = \{a, b\}$, where $a = R(x, *, \#)$ and $b = S(x, *, \#)$.

The default number of attributes per table is 4. Two of these attributes are used for storing the time information, i.e., the beginning and the end of the intervals. We use the following notation for the input datasets: $Ux - Sy - Dz - Uu - Qv - Ew$, where x, y, z, u, v, w are values for the input parameters U, S, D, T, Q, E respectively. The remaining input parameters (P and M) are set as the default ones ($P = 2, M = 4$). We notice that parameter Q refers to the number of valid i-patterns which surely appear in the generated dataset.

For the performance analysis, the experiments have been carried out over databases containing 5000 tuples in each table. Figure 1(a) shows the performance of MILPRIT over dataset $U5 - S2 - D1 - T1 - Q20 - Ea^*b^*$ as the minimum support increases from 1% to 10%. As expected, as the minimum support threshold increases, the execution time of the algorithm decreases, since few candidates are potentially frequent for high values of minimum support. Figure 1(c) shows how the algorithm performs as the number of "(*)-blocks" in the regular expression increases. The datasets used in these tests were $U5 - S2 - D1 - T1 - Q20 - Ex$, where $x \in \{a^*, a^*b^*, a^*b^*a^*\}$. The minimum support has been set as 2.5%. As expected, as the number of different blocks in the regular expression increases, the number of sequential patterns that satisfies it increases as well. Thus, the execution time of the algorithm increases. It can be noticed that the increasing rate of the execution time is not linear.

For the scalability analysis, we set the regular expression as a^*b^* and the minimum support as 2.5%. Figure 1(b) shows how MILPRIT scales up as the number of tuples in each table increases from 5000 to 55000. We show the results for the datasets $Ux - Sy - Dz - Tw - Q20 - Ea^*b^*$, where, $x \in \{5, \ldots, 55\}, y = \frac{x}{2.5}$, $z = \frac{x}{5}$ and $w = \frac{x}{5}$. Clearly, the higher the number of tuples in the database tables, the longer the execution time of the algorithm, since the validation phase (the support evaluation) is the most computationnally costly. It can be noticed that MILPRIT scales almost linearly with respect to the number of tuples in the dataset.

The table depicted in Figure 1(d) shows how the execution time increases as the number of generated patterns increases. This experiment consisted in running the algorithm over datasets $U5 - S2 - D1 - T1 - Qx - Ea^*b^*$, for $x \in \{100, 500, 1000, 1500, 2000\}$ and minimum support 2.5%. We can see that MILPRIT scales up almost linearly with respect to the number of interesting patterns appearing in the dataset.

5.2 Real Data

The $AMDI$ database is part of a project in our laboratory, which aims at building a system supporting an indexed atlas of digital mammograms. This system intends to be used by radiologists, in order to help them in breast cancer diagnosis. The atlas stores, for each patient, a series of digital mammograms, taken during its lifetime. Besides image data, some important temporal data are stored, related to the

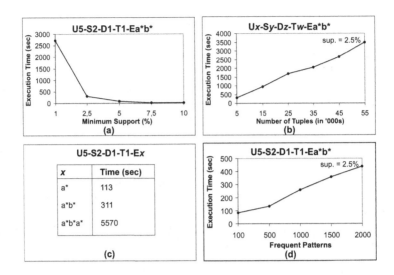

Fig. 1. Performance and scalability results

patient habits and life style: in which periods of time did she smoke, or take contraceptive drugs, for instance. For the time being, the database has 1365 records. Its schema is $\mathbf{R} = \{Patient(NamePat),\ Contraceptive(NamePat,Name,T),$ $Antidepressant(NamePat,Name,T),\ Hormone(NamePat,Name,T),\ Tobacco$ $(NamePat,Amount,T),\ Drink(NamePat,Type,T)\}$.

For the experiment carried out over this database, we used a minimum support threshold of 0.2%. The execution time of the algorithm was 2 seconds. Several interesting frequent i-patterns have been found. For instance, the i-pattern $(Patient(x),\ \{Antidepressant(x,\ fluoxetin,\ e_1),\ Hormone(x\ ,\ estrofen,\ e_2)\},\ \{finishes(e_1,e_2)\})$, meaning that the following behavior is frequent among patients: they start taking the antidepressant *fluoxetin* after starting taking the hormone *estrofen* (for menopause) and stop taking these two meds at the same time. The frequency of this i-pattern suggests that patients in menopause tend to suffer from depression. Another frequent interesting i-pattern which have been found is $(Patient(x),\ \{Contraceptive(x,diane,e_1),\ Antidepressant(x,fluoxetin,e_2),\ Tobacco(x,20,e_3)\},\ \{before(e_1,e_2),\ during(e_1,e_3),\ finishes(e_2,e_3)\})$. This i-pattern suggests the following behavior: patients who smoke 20 cigarettes per day, when starting taking contraceptive *diane* followed by the antidepressant *fluoxetin* stop smoking at the same time they stop taking the antidepressant.

6   Conclusion and Further Work

In this paper we have presented the algorithm MILPRIT to mine a new kind of temporal relational patterns where time is measured in terms of intervals instead of points as is the case of most classical methods for sequential pattern mining. Our ongoing research consists in generalizing the method in order to take into account both interval and punctual time representation as usually happens in temporal database applications, where temporal data contain attributes for valid (interval representation) and transaction time (punctual representation).

References

1. de Amo, S., Giacometti, A., Santana, M.S.: MILPRIT: Mining Interval Logic Patterns with Regular Expression Constraints. In: 1st Brazilian Workshop on Data Mining Algorithms, Uberlândia, Brazil (2004)
2. Agrawal, R., Srikant, R.: Mining Sequential Patterns: Generalizations and Performance Improvements. In: Proc. of the Fifth Int. Conference on Extending Database Technology, March 1996, Avignon, France (1996)
3. Allen, J.F., Ferguson, G.: Actions and Events in Interval Temporal Logic. Technical Report TR521 (1994)
4. Dehaspe, L., Toivonnen, H.: Discovery of Frequent Datalog Patterns. Data Mining and Knowledge Discovery 3, 7–36 (1999)
5. Garofalakis, M.N., Rastogi, R., Shim, K.: SPIRIT: Sequential Pattern Mining with Regular Expression Constraints. The VLDB Journal, 223–234 (1999)
6. Höppner, F.: Discovery of Temporal Patterns: Learning Rules about the Qualitative Behaviour of Time Series. In: Siebes, A., De Raedt, L. (eds.) PKDD 2001. LNCS (LNAI), vol. 2168, pp. 192–203. Springer, Heidelberg (2001)
7. Jacobs, N., Blockeel, H.: From Shell Logs to Shell Scripts. In: Rouveirol, C., Sebag, M. (eds.) ILP 2001. LNCS (LNAI), vol. 2157, pp. 80–90. Springer, Heidelberg (2001)
8. Lee, S.D., De Raedt, L.: Constraint Based Mining of First Order Sequences in Seqlog. In: Workshop on Knowledge Discovery in Inductive Databases (2002)
9. Albert-Lorincza, H., Boulicaut, J.-F.: Mining frequent sequential patterns under regular expressions: a highly adaptive strategy for pushing constraints. In: International Conference on Data Mining SDM'03, pp. 316–320 (2003)
10. Masson, C., Jacquenet, F.: Mining Frequent Logical Sequences with Spirit-Log. In: Matwin, S., Sammut, C. (eds.) ILP 2002. LNCS (LNAI), vol. 2583, pp. 166–181. Springer, Heidelberg (2003)

Mining Trajectory Patterns Using Hidden Markov Models

Hoyoung Jeung, Heng Tao Shen, and Xiaofang Zhou

National ICT Australia (NICTA), Brisbane, QLD, Australia
School of Information Technology and Electrical Engineering
The University of Queensland
{hoyoung,shenht,zxf}@itee.uq.edu.au

Abstract. Many studies of spatiotemporal pattern discovery partition data space into disjoint cells for effective processing. However, the discovery accuracy of the space-partitioning schemes highly depends on space granularity. Moreover, it cannot describe data statistics well when data spreads over not only one but many cells. In this study, we introduce a novel approach which takes advantages of the effectiveness of space-partitioning methods but overcomes those problems. Specifically, we uncover frequent regions where an object frequently visits from its trajectories. This process is unaffected by the space-partitioning problems. We then explain the relationships between the frequent regions and the partitioned cells using *trajectory pattern models* based on hidden Markov process. Under this approach, an object's movements are still described by the partitioned cells, however, its patterns are explained by the frequent regions which are more precise. Our experiments show the proposed method is more effective and accurate than existing space-partitioning methods.

1 Introduction

We are facing an unprecedented proliferation of mobile devices, many equipped with positional technologies such as GPS. These devices produce a huge amount of trajectory data which is described as geometry changes over time continuously. Since a large amount of trajectories can be accumulated for a short period of time, many applications need to summarize the data or extract valuable knowledge from it. As a part of the trend, discovery of trajectory patterns has been paid great attention due to many applications.

For the pattern discovery of spatiotemporal data, many techniques in the literature have partitioned data space into disjoint cells (e.g., fixed grid) [1,2,3,4]. The reasons are mainly two-folded. First, space-decomposition techniques bring efficiency of discovery process. Obviously, dealing with symbols identifying each cell is much simpler than handling real coordinates which should have bigger data size and give lower intuitions for data processing. Second, spatiotemporal data has a distinct characteristic from general data for mining studies (e.g., basket data). Assume Paul arrives at his work at 9 a.m. every weekday. Though

I.Y. Song, J. Eder, and T.M. Nguyen (Eds.): DaWaK 2007, LNCS 4654, pp. 470–480, 2007.
© Springer-Verlag Berlin Heidelberg 2007

his work is an identical place in semantic, the location may not be expressed by spatially the exact same coordinates because of the different vacancy of parking lots everyday. Therefore, pattern discovery methods need to regard slightly different locations as the same.

Despite the popularity of the *space-partitioning approaches*, it has two critical shortcomings. First, it cannot solve the *anwer loss* problem [5]. Suppose a problem finding dense regions in Figure 1. Data space is divided into nine cells from A to I and four objects (o_1, o_2, o_3, and, o_4) have moved in the space for two timestamps. If we define a dense region as a cell having more than two hitting points, r_1 cannot be a dense region though there are three close points since the points spread over three cells. Second, space-partitioning approaches have granularity problems. The precision of pattern discovery highly depends on how big or small the space divided. Especially, when there are many noises of movements (e.g., Paul unusually makes a trip to a far away for a few days), discovery process should manage a large size of data space, and thus the accuracy can decrease for efficient computation. For instance, r_2 and r_3 are distinct dense regions, however, both should be expressed by only one cell E due to the rough granularity.

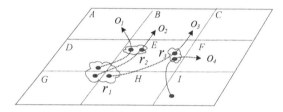

Fig. 1. Deficiencies of space-partitioning approaches

In order to overcome those problems, this study introduces a novel approach that takes both advantages of space-partitioning schemes and data-centric methods. Specifically, we reveal *frequent regions* that an object frequently visits by applying periodic data mining techniques [6] based on the data-centric approach. It is unaffected by space partitioning problems, hence, it should be more precise. However, it does not have the space-partitioning efficiency. For efficient data handling, we introduce *trajectory pattern model* (TPM) that explains the relationships between the regions and partitioned cells using hidden Markov models (HMMs). An HMM is a doubly embedded stochastic process with an underlying stochastic process that is not observable. We model partitioned cells to observable states and discovered frequent regions to hidden states. Therefore, the TPM let applications be able to deal with symbols of the cells (instead of using real coordinates) for effectiveness but have more precise discovery results than existing space-partition methods.

Building a TPM from historical trajectories can be useful for many applications. First, it computes the probability of a given observation sequence. It implies the TPM can explain how the current movements (a sequence of cell symbols) of an object is similar to its common movement patterns (a sequence of frequent regions). Second, given a cell symbol sequence, it can also compute the most likely sequence of frequent regions. Moreover, the model can be trained by newly added data. Hence, it can reflect not only historical movements of an object but also its current motion trends.

2   Related Work

In this Section, we study previous work based on two major discovery techniques, *Markov chain models* and *spatiotemporal data mining*, for extracting movement patterns of an object from historical trajectories.

Markov chain models have been widely used in order to estimate the probability of an object's movements from one region or state to another at next time period. Ishikawa et al. derive the Markov transition probabilities between cells from indexed trajectories [1]. In their further study [7], a special type of histogram, called mobility histogram, is used to describe mobility statistics based on the Markov chain model. They also represent the histogram as cube-like logical structures and support an OLAP-style analysis. Authors in [8] classify an object's mobility patterns into three states (stationary state, linear movement, and random movement) and apply Markov transition probabilities to explain a movement change one state to another. [9,10] consider the location tracking problem in PCS networks. Both studies are based on the same Markov process in order to describe users' movements from one or multiple PCS cells to another cell. However, they have different ways to model users' mobilities using Morkov models, thus, show distinct results to each other.

Spatiotemporal data mining methods have been also studied well for describing objects' patterns. [2] introduces mining algorithms that detect a user's moving patterns, and exploits the mined information to invent a data allocation method in a mobile computing environment. Another mining technique is shown in [11]. This study focuses on discovering spatio-temporal patterns in environmental data. In [6], authors do not only explore periodic patterns of objects but also present indexing and querying techniques of the discovered or non-discovered pattern information. [3,4] address *spatio-temporal association rules* of the form $(r_i, t_1, p) \longrightarrow (r_j, t_2)$ with an appearance probability p, where r_i and r_j are regions at time (interval) t_1 and t_2 respectively $(t_2 > t_1)$. It implies that an object in r_i at time t_1 is likely to appear in r_j at time t_2 with $p\%$ probability. Besides, [3] considers spatial semantic areas (i.e. sources, sinks, stationary regions, and thoroughfares) in each r_i as well as more comprehensive definitions and algorithms of spatio-temporal association rules.

All above studies except [6] are based on the space-partitioning schemes, thus, the discovery accuracy depends on how the system decides space granularity of data space. When they partition the data space into a large number of small size cells, the accuracy increases, however, managing such many cells in memory can be burden to the system. On the contrary, using large size cells for partitioning cause low precision of discovery. Moreover, they cannot avoid the answer-loss problem no matter how the spatial granularity is set to.

3   Problem Definition

We assume a database stores a large volume of accumulated trajectories and each location in each trajectory is sampled periodically with a discrete time interval. Let a point $p_i = (l_i, t_i)$ represent a d-dimensional location l_i at time t_i. A whole trajectory T of an object is denoted as $\{p_0, p_1, \cdots, p_{n-1}\}$, where n is the total number of points. The entire data space is partitioned into k fixed grid cells $C = c_1, c_2, \cdots, c_k$, each cell c_i of which has the same coverage size.

The goals of our study are two-folded. First, we aim to reveal spatial regions, namely *frequent regions*, where an object frequently and periodically appears. For example, given a trajectory having 10 days movements, we are interested in discovering areas where the object appears more than 5 days at the same time. Specifically, given an integer P, called *period*, we consider decomposing T into $\lfloor \frac{n}{P} \rfloor$ sub-trajectories and group points G_w having the same time offset w of P ($0 \leq w < P$) in each sub-trajectory. We formally define the frequent region as follows:

Definition 1. *A **frequent region** is a minimum bounding rectangle that consists of a set of points in G, each point of which contains at least MinPts number of neighborhood points in a radius Eps.*

Second, building hidden Markov models to explain relationships between frequent regions and partitioned cells is another goal in this study. A discrete Markov process is a stochastic process based on the Markov assumption, under which the probability of a certain observation only depends on the observation that directly preceded it. It has a finite number of states that make transitions according to a set of probabilities associated with each state at discrete times. A hidden Markov process is a generalization of a discrete Markov process where a given state may have several existing transitions, all corresponding to the same observation symbol [12]. In other words, the stochastic process is hidden and can only be observed through another set of stochastic processes that produce the sequence of observations.

There are many possible ways to model an object's mobility using hidden Markov models. Our approach is to construct a trajectory pattern model where each state corresponds to a frequent region. Observations are cells that change the current state with some probability to a new state (a new current frequent region). Formally,

Definition 2. *A **trajectory pattern model** describes an object's movement patterns based on hidden Markov process. It consists of a set of N frequent regions, each of which is associated with a set of M possible partitioned cells.*

4   Discovery of Trajectory Pattern Models

In this Section, we describe how to solve the two sub-problems of Section 3, which are discovery of frequent regions and building trajectory pattern models.

4.1   Extracting Frequent Regions

For the detection of frequent regions from an object's trajectories, we adopt a periodical pattern mining methods [6]. In these techniques, the whole trajectory is decomposed into $\lfloor \frac{n}{P} \rfloor$ sub-trajectories. Next, all locations from $\lfloor \frac{n}{P} \rfloor$ sub-trajectories which have the same time offset w of P will be grouped into one group G_w. P is data-dependent and has no definite value. For example, $P =$ 'a day' in a traffic control application since many vehicles have daily patterns, while animal's annual migration behaviors can be discovered by $P =$ 'a year'. A The clustering method DBSCAN [13] is then applied to find the dense clusters R_w in each G_w. Given two parameters *Eps* and *MinPts*, DBSCAN finds dense clusters, each point of which has to contain at least *MinPts* number of neighborhood points within a radius *Eps*.

In spite of many advantages of DBSCAN, it can not be applicable to our study directly. Our key methods of this research is to describe correlations between frequent regions, and between partitioned cells and frequent regions. It implies the size of a frequent region can affect explanation of the patterns considerably. Therefore, we decompose a cluster when it is 'large'. As each dataset may have different data distributions and map extents from others, a proper area size for cluster decompositions can not be easily detected. Due to this property, we use $(Eps * \kappa)^2$ to define a 'large' cluster, where κ is given by a user (named *area limiter*). κ limits a cluster's size merely with respect to *Eps*. For example, $\kappa = 2$ means that a cluster's size must be smaller than the square's area having double *Eps* length on a side. Thus, a user does not need to know about the data distributions. In our experiments, when $5 \leq \kappa \leq 10$ regardless of data characteristics, the cluster decomposition showed more precise discovery results.

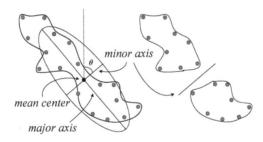

Fig. 2. Cluster decomposition using standard deviation ellipse

When partitioning is needed on a cluster we apply *standard deviation ellipse* for the decomposition process. The standard deviation ellipse captures the directional bias in a point distribution using three components: major (longer) axis, minor (shorter) axis, and θ between the north and the y axis rotated clockwise. Figure 2 illustrates how a cluster is partitioned into two co-clusters based on the standard deviation ellipse. This method first computes the minor axis of the cluster's standard deviation ellipse. Next, it classifies each point dividing the cluster into two co-clusters. This classification is performed by checking whether each point is left (above) of the minor axis or right (bellow). The shapes of co-clusters decomposed tend to be sphere-like, hence, the center position of each co-cluster can represent its cluster effectively. The partitioning method also distributes an almost equal number of points to each co-cluster. When the number of points in any co-cluster is smaller than $MinPts$, we cease the cluster partitioning. The reason for this is that a cluster having less than $MinPts$ points contradicts the definition of a frequent region and becomes less useful. The whole process of cluster decomposition is described in the Algorithm 1:

Algorithm 1. Cluster Decomposition

Input:
   given cluster C, area limiter κ, radius Eps, minimum number of points $MinPts$
Output:
   a set of clusters
Description:
1: **if** area of $C \geq (Eps \times \kappa)^2$ **then**
2:    get the minor axis $minx$ of C's standard deviation ellipse
3:    **for** each point p in C **do**
4:       **if** p is the left of $minx$ **then**
5:          add p to a co-cluster $c1$
6:       **else**
7:          add p to a co-cluster $c2$
8:    **if** $numPoints\ (c1) \geq MinPts$ and $numPoints\ (c2) \geq MinPts$ **then**
9:       return $c1$ and $c2$
10: **return** C

4.2   Building Trajectory Pattern Models

Let us recall the movements of objects in Figure 1. Assume an application needs to find an object which has the most similar movements to o_2. Though o_1 is obviously the closest answer, it is hard to be detected since there is only one common cell between $o_2 = GEB$ and $o_1 = DEA$, which has the same number of common cell as between o_2 and o_4. In our approach, *trajectory pattern model*, we illustrate real movements of an object as a sequence of frequent regions instead of a cell sequence. For example, $o_1 = r_1 r_2 *$, $o_2 = r_1 r_2 *$, $o_3 = r_1 r_3 *$, and $o_4 = *r_3 *$ ('*' for non-frequent regions). We then apply hidden Markov models to figure out relationships between a cell sequence (e.g., ABC) and a frequent region sequence (e.g., $r_1 r_2 *$). Under this approach, although an object's movements are denoted as a cell sequence, the problem of finding most similar movements to o_2 can be performed on the frequent region sequences, which are more precise.

In order to construct a trajectory pattern model, we must define the elements of an HMM, which are (i) the number of states N, (ii) the number of observation symbols per state M, (iii) an initial state π, (iv) a set of state transition probabilities A, and (v) a set of observation symbol probabilities B. This HMM parameter set can be denoted as $\lambda = (A, B, \pi)$. How to model this λ is critical to explain an object's mobilities since there are many possible ways to model.

Our approach is to construct λ where each state corresponds to a frequent region r_i when a cell c_i appears. Therefore, hidden states (frequent regions) are expressed by

$$S = \{r_i, r_i \atop 1 \ \ 2} \ r_i, \cdots, r_i \atop 3 \ \ \ \ \ N\},$$

and the number of states N is the number of r_i when c_i is found. We also define an observation sequence O as

$$O = \{c_i, c_i \atop 1 \ \ 2} \ c_i, \cdots, c_i \atop 3 \ \ \ \ \ M\}$$

where M corresponds to the total number of partitioned cells. The state transition probabilities can be formed as a matrix $A = \{a_{ij}\}$ where

$$a_{ij} = P[r_{t+1} = S_j | r_t = S_i], 1 \leq i, j \leq N,$$

and r_t denotes the actual state at time t in the sequence S. In the same manner, the observation transition probabilities B is expressed by following matrix

$$b_{ij} = P[c_t | r_t = S_i], 1 \leq i \leq N, 1 \leq j \leq M$$

with elements b_{ij} denoting the probability of observing symbol $j \in [1, M]$ that the system is in state $i \in [1, N]$. The initial state is defined as $\pi = \{\pi_i\}$ $(1 \leq i \leq N)$ which π_i describes the distribution over the initial frequent region set.

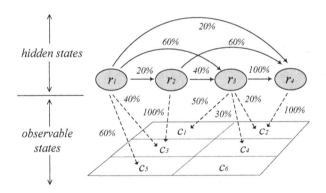

Fig. 3. An example of a trajectory pattern model

Figure 3 represents an example of a TPM modelling based on our scheme. Circles stand for frequent regions as hidden states and the grid does partitioned cells as observation symbols. In the example, the TPM parameters can be denoted as follows; hidden states $S = \{r_1, r_2, r_3, r_4\}$, thus the number of states $N = 4$, and observable symbols $O = \{c_1, c_2, c_3, c_4, c_5, c_6\}$ $(M = 6)$. Transition probabilities of states and observation symbol probabilities are illustrated as follows:

$$A = \{a_{ij}\} = \begin{bmatrix} 0 & 20 & 60 & 20 \\ 0 & 0 & 40 & 60 \\ 0 & 0 & 0 & 100 \\ 0 & 0 & 0 & 0 \end{bmatrix}, \qquad B = \{b_{ij}\} = \begin{bmatrix} 0 & 0 & 40 & 0 & 60 & 0 \\ 0 & 0 & 100 & 0 & 0 & 0 \\ 50 & 20 & 0 & 30 & 0 & 0 \\ 0 & 100 & 0 & 0 & 0 & 0 \end{bmatrix}$$

In order to compute A, we investigate whether each point p_w of a sub-trajectory is geometrically within any of r_i for all p_i. If so, we report the frequent region id otherwise a wild card mark '*'. Therefore, a sub-trajectory can be denoted by a state sequence as $r_1 r_2 * r_4$. We then compute the probabilities of each state transition from the sequences. Likewise, elements of π are calculated by the same way. The computation of B is done by checking if p_i belongs to both r_i and c_i or c_i only.

An excellent advantage of TPMs is that λ can adjust its parameters (A, B, π) to maximize $P(O|\lambda)$ using an iterative procedure. It implies that the model can be trained more powerfully by compiled trajectories of an object as time goes on. Hence, it can reflect the trends of the object's motions precisely.

5   Experiments

The experiments in this study were designed for comparison of discovery accuracy and effectiveness of data management between a space-partitioning method and our scheme. In order to evaluate them, we implemented a popular method of spatiotemporal data mining using observable Markov models (OMM), which was based on the space-partitioning scheme. We also did our approach, trajectory pattern models (TPM), using hidden Markov models. Both methods were implemented in the C++ language on a Windows XP operating system and run on a dual processor Intel Pentium4 3.0 Ghz system with a memory of 512MB.

Due to the lack of accumulated real datasets, we generated three synthetic datasets having different number of sub-trajectories and timestamps (Bike-small had 50 sub-trajectories and 50 timestamps, 200 and 200 for Bike-medium, and 500 and 500 for Bike-large). For the generation, we developed a data generator that produced similar trajectories to a given trajectory. We used a real trajectory as an input of the generator, which was measured by a GPS mounted bicycle over an eight hour period in a small town of Australia. The extent of each dataset was normalized to [0,10000] in both x and y axis.

In the first set of experiments, we studied how the space granularity affected discovery accuracies of OMM and TPM. To measure them, we computed the distances between each actual point of the input trajectory for the dataset generation and the center point of a cell for OMM, which the actual point belonged to. We did the same process for obtaining TPM's errors.

Figure 4 demonstrates the changes of discovery accuracies along the number of partitioned cells. As expected, the errors of OMM were significant when the number of cells were small since one cell had a too wide coverage to describe

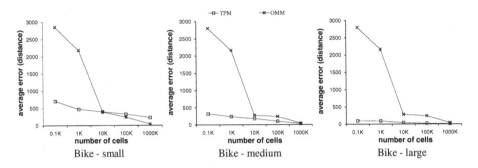

Fig. 4. Discovery accuracy changes along the number of partitioned cells

details of the object's movements. On the contrary, our approach was slightly affected by the space granularity. An interesting observation was that TPM's errors decreased as the dataset size grew. The reason was because TPM had more trajectories in bigger datasets that can obtain stronger and more precise trajectory patterns.

Another aspect about space-partitioning techniques to be considered is that a large number of cells needs more complicated management schemes so as to handle the partition effectively. For example, the data distribution is skew, many cells are not related to objects' movements but they are still needed to be managed. Therefore, though a large number of cells has lower errors, it involves a bigger number of partition cells, which causes overheads of systems.

We also explored size changes of discovered trajectory patterns along different size of datasets. As shown in Figure 5, our method showed relatively small pattern size compared to OMM. In fact, TPM's size was computed by the number of transition items and probabilities for each matrix of (A, B, π), however, the system did not need to store all the transition probability values. Recall the example of TPM in Section 4.2. Due to the time order of frequent regions, the matrix of transition probabilities are always filled with zero values for more than half of the matrix (left-bottom part), which are not physically stored in the system.

On the contrary, discovered pattern sizes of OMM were very large especially when it had higher orders (i.e., $A \rightarrow B$: 1st order, $ABC \rightarrow D$: 3rd order, and $ABCDE \rightarrow F$: 5th order). Some applications need to handle higher order transitions, thus, the size of higher order OMM is also important. Our approach can handle various length of sequences (i.e., higher order chains) without any extra storage consumption. This means our method can be applicable to many applications that have storage limits.

6   Conclusion

In this paper, we addressed space-partitioning problems which were the spatial granularity problem and the answer loss problem. In fact, many spatiotemporal

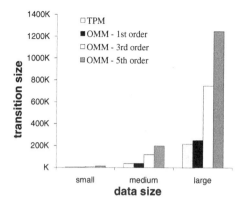

Fig. 5. Changes of trajectory pattern size along dataset size

mining methods are based on the space-partitioning scheme, thus, their pattern discoveries were not effective. In order to overcome the problems, we first revealed frequent regions where an object frequently visited by applying a data-centric approach. We then described the relationships between the partitioned cells and the frequent regions by using the *trajectory pattern models* based on hidden Markov process. Therefore, applications based on our scheme can still use space-partitions but have more precise discovery results. As shown in the experiments, our approach outperformed previous studies, especially when data space had a large size.

Acknowledgements. National ICT Australia is funded by the Australian Government's Backing Australia's Ability initiative, in part through the Australian Research Council.

References

1. Ishikawa, Y., Tsukamoto, Y., Kitagawa, H.: Extracting mobility statistics from indexed spatio-temporal datasets. In: STDBM, pp. 9–16 (2004)
2. Peng, W.C., Chen, M.S.: Developing data allocation schemes by incremental mining of user moving patterns in a mobile computing system. TKDE 15(1), 70–85 (2003)
3. Verhein, F., Chawla, S.: Mining spatio-temporal association rules, sources, sinks, stationary regions and thoroughfares in object mobility databases. In: DASFAA (2006)
4. Tao, Y., Kollios, G., Considine, J., Li, F., Papadias, D.: Spatio-temporal aggregation using sketches. In: ICDE, p. 214 (2004)
5. Jensen, C.S., Lin, D., Ooi, B.C., Zhang, R.: Effective density queries on continuously moving objects. In: ICDE, p. 71 (2006)
6. Mamoulis, N., Cao, H., Kollios, G., Hadjieleftheriou, M., Tao, Y., Cheung, D.W.: Mining, indexing, and querying historical spatiotemporal data. In: SIGKDD, pp. 236–245 (2004)
7. Ishikawa, Y., Machida, Y., Kitagawa, H.: A dynamic mobility histogram construction method based on markov chains. In: SSDBM, pp. 359–368 (2006)

8. Song, M., Ryu, J., Lee, S., Hwang, C.S.: Considering mobility patterns in moving objects database. In: ICPP, p. 597 (2003)
9. Yang, M.H., Chen, L.W., Sheu, J.P., Tseng, Y.C.: A traveling salesman mobility model and its location tracking in pcs networks. In: ICDCS, pp. 517–523 (2001)
10. Bhattacharya, A., Das, S.K.: Lezi-update: an information-theoretic approach to track mobile users in pcs networks. In: MobiCom, pp. 1–12 (1999)
11. Tsoukatos, I., Gunopulos, D.: Efficient mining of spatiotemporal patterns. In: SSTD, pp. 425–442 (2001)
12. Rabiner, L.: A tutorial on hidden markov models and selected applications in speech recognition. Proceedings of the IEEE 77, 257–286 (1989)
13. Ester, M., Kriegel, H.P., Sander, J., Xu, X.: A density-based algorithm for discovering clusters in large spatial databases with noise. In: SIGKDD, pp. 226–231 (1996)

Author Index

Lecture Notes in Computer Science

Sublibrary 3: Information Systems and Application, incl. Internet/Web and HCI

For information about Vols. 1– 4286
please contact your bookseller or Springer

Vol. 4558: M.J. Smith, G. Salvendy (Eds.), Human Interface and the Management of Information, Part II. XXIII, 1162 pages. 2007.

Vol. 4557: M.J. Smith, G. Salvendy (Eds.), Human Interface and the Management of Information, Part I. XXII, 1030 pages. 2007.

Vol. 4541: T. Okadome, T. Yamazaki, M. Makhtari (Eds.), Pervasive Computing for Quality of Life Enhancement. IX, 248 pages. 2007.

Vol. 4537: K.C.-C. Chang, W. Wang, L. Chen, C.A. Ellis, C.-H. Hsu, A.C. Tsoi, H. Wang (Eds.), Advances in Web and Network Technologies, and Information Management. XXIII, 707 pages. 2007.

Vol. 4531: J. Indulska, K. Raymond (Eds.), Distributed Applications and Interoperable Systems. XI, 337 pages. 2007.

Vol. 4526: M. Malek, M. Reitenspieß, A. van Moorsel (Eds.), Service Availability. X, 155 pages. 2007.

Vol. 4524: M. Marchiori, J.Z. Pan, C.d.S. Marie (Eds.), Web Reasoning and Rule Systems. XI, 382 pages. 2007.

Vol. 4519: E. Franconi, M. Kifer, W. May (Eds.), The Semantic Web: Research and Applications. XVIII, 830 pages. 2007.

Vol. 4518: N. Fuhr, M. Lalmas, A. Trotman (Eds.), Comparative Evaluation of XML Information Retrieval Systems. XII, 554 pages. 2007.

Vol. 4508: M.-Y. Kao, X.-Y. Li (Eds.), Algorithmic Aspects in Information and Management. VIII, 428 pages. 2007.

Vol. 4506: D. Zeng, I. Gotham, K. Komatsu, C. Lynch, M. Thurmond, D. Madigan, B. Lober, J. Kvach, H. Chen (Eds.), Intelligence and Security Informatics: Biosurveillance. XI, 234 pages. 2007.

Vol. 4505: G. Dong, X. Lin, W. Wang, Y. Yang, J.X. Yu (Eds.), Advances in Data and Web Management. XXII, 896 pages. 2007.

Vol. 4504: J. Huang, R. Kowalczyk, Z. Maamar, D. Martin, I. Müller, S. Stoutenburg, K.P. Sycara (Eds.), Service-Oriented Computing: Agents, Semantics, and Engineering. X, 175 pages. 2007.

Vol. 4500: N.A. Streitz, A.D. Kameas, I. Mavrommati (Eds.), The Disappearing Computer. XVIII, 304 pages. 2007.

Vol. 4495: J. Krogstie, A. Opdahl, G. Sindre (Eds.), Advanced Information Systems Engineering. XVI, 606 pages. 2007.

Vol. 4480: A. LaMarca, M. Langheinrich, K.N. Truong (Eds.), Pervasive Computing. XIII, 369 pages. 2007.

Vol. 4471: P. Cesar, K. Chorianopoulos, J.F. Jensen (Eds.), Interactive TV: A Shared Experience. XIII, 236 pages. 2007.

Vol. 4469: K.-c. Hui, Z. Pan, R.C.-k. Chung, C.C.L. Wang, X. Jin, S. Göbel, E.C.-L. Li (Eds.), Technologies for E-Learning and Digital Entertainment. XVIII, 974 pages. 2007.

Vol. 4443: R. Kotagiri, P. Radha Krishna, M. Mohania, E. Nantajeewarawat (Eds.), Advances in Databases: Concepts, Systems and Applications. XXI, 1126 pages. 2007.

Vol. 4439: W. Abramowicz (Ed.), Business Information Systems. XV, 654 pages. 2007.

Vol. 4430: C.C. Yang, D. Zeng, M. Chau, K. Chang, Q. Yang, X. Cheng, J. Wang, F.-Y. Wang, H. Chen (Eds.), Intelligence and Security Informatics. XII, 330 pages. 2007.

Vol. 4425: G. Amati, C. Carpineto, G. Romano (Eds.), Advances in Information Retrieval. XIX, 759 pages. 2007.

Vol. 4412: F. Stajano, H.J. Kim, J.-S. Chae, S.-D. Kim (Eds.), Ubiquitous Convergence Technology. XI, 302 pages. 2007.

Vol. 4402: W. Shen, J.-Z. Luo, Z. Lin, J.-P.A. Barthès, Q. Hao (Eds.), Computer Supported Cooperative Work in Design III. XV, 763 pages. 2007.

Vol. 4398: S. Marchand-Maillet, E. Bruno, A. Nürnberger, M. Detyniecki (Eds.), Adaptive Multimedia Retrieval: User, Context, and Feedback. XI, 269 pages. 2007.

Vol. 4397: C. Stephanidis, M. Pieper (Eds.), Universal Access in Ambient Intelligence Environments. XV, 467 pages. 2007.

Vol. 4380: S. Spaccapietra, P. Atzeni, F. Fages, M.-S. Hacid, M. Kifer, J. Mylopoulos, B. Pernici, P. Shvaiko, J. Trujillo, I. Zaihrayeu (Eds.), Journal on Data Semantics VIII. XV, 219 pages. 2007.

Vol. 4365: C.J. Bussler, M. Castellanos, U. Dayal, S. Navathe (Eds.), Business Intelligence for the Real-Time Enterprises. IX, 157 pages. 2007.

Vol. 4353: T. Schwentick, D. Suciu (Eds.), Database Theory – ICDT 2007. XI, 419 pages. 2006.

Vol. 4352: T.-J. Cham, J. Cai, C. Dorai, D. Rajan, T.-S. Chua, L.-T. Chia (Eds.), Advances in Multimedia Modeling, Part II. XVIII, 743 pages. 2006.

Vol. 4351: T.-J. Cham, J. Cai, C. Dorai, D. Rajan, T.-S. Chua, L.-T. Chia (Eds.), Advances in Multimedia Modeling, Part I. XIX, 797 pages. 2006.

Vol. 4328: D. Penkler, M. Reitenspiess, F. Tam (Eds.), Service Availability. X, 289 pages. 2006.

Vol. 4321: P. Brusilovsky, A. Kobsa, W. Nejdl (Eds.), The Adaptive Web. XII, 763 pages. 2007.

Vol. 4317: S.K. Madria, K.T. Claypool, R. Kannan, P. Uppuluri, M.M. Gore (Eds.), Distributed Computing and Internet Technology. XIX, 466 pages. 2006.

Vol. 4312: S. Sugimoto, J. Hunter, A. Rauber, A. Morishima (Eds.), Digital Libraries: Achievements, Challenges and Opportunities. XVIII, 571 pages. 2006.

Vol. 4306: Y. Avrithis, Y. Kompatsiaris, S. Staab, N.E. O'Connor (Eds.), Semantic Multimedia. XII, 241 pages. 2006.

Vol. 4302: J. Domingo-Ferrer, L. Franconi (Eds.), Privacy in Statistical Databases. XI, 383 pages. 2006.

Vol. 4299: S. Renals, S. Bengio, J.G. Fiscus (Eds.), Machine Learning for Multimodal Interaction. XII, 470 pages. 2006.

Vol. 4295: J.D. Carswell, T. Tezuka (Eds.), Web and Wireless Geographical Information Systems. XI, 269 pages. 2006.